Progress in Mathematics

Volume 169

European Congress of Mathematics

Budapest, July 22–26, 1996

Volume II

A. Balog

G.O.H. Katona

A. Recski

D. Sza'sz

Editors

Birkhäuser Verlag

Basel · Boston · Berlin

Editors:

A. Balog
Mathematical Institute
Hungarian Academy of Sciences
Reáltanoda str. 13–15
H-1053 Budapest Hungary

G.O.H. Katona
Mathematical Institute
Hungarian Academy of Sciences
Reáltanoda str. 13–15
H-1053 Budapest Hungary

A. Recski
Mathematical Institute
Technical University of Budapest
H-1521 Budapest Hungary

D. Sza'sz
Mathematical Institute
Hungarian Academy of Sciences
Reáltanoda str. 13–15
H-1053 Budapest Hungary

1991 Mathematics Subject Classification 00B25

A CIP catalogue record for this book is available from the Library of Congress, Washington D.C., USA

Deutsche Bibliothek Cataloging-in-Publication Data

European Congress of Mathematics <2, 1996, Budapest>:
European Congress of Mathematics: Budapest, July 22 - 26, 1996 / A. Balog... ed. - Basel ; Boston ; Berlin : Birkhäuser.
ISBN 3-7643-5496-8 (Basel...)
ISBN 0-8176-5496-8 (Boston)

Vol. 1. - (1998
(Progress in mathematics ; Vol. 168)
ISBN 3-7643-5497-6 (Basel...)
ISBN 0-8176-5497-6 (Boston)

Vol. 2. - (1998
(Progress in mathematics ; Vol.169)
ISBN 3-7643-5498-4 (Basel...)
ISBN 0-8176-5498-4 (Boston)

Printed on acid-free paper produced of chlorine-free pulp. TCF ∞
Printed in Germany

ISBN 3-7643-5497-6 (Vol. I/PM 168)
ISBN 3-7643-5498-4 (Vol. II/PM 169)
ISBN 3-7643-5496-8 (Set)

ISBN 0-8176-5497-6 (Vol. I/PM 168)
ISBN 0-8176-5498-4 (Vol. II/PM 169)
ISBN 0-8176-5496-8 (Set)

9 8 7 6 5 4 3 2 1

Table of Contents of Volume I

Table of Contents of Volume II

Contributions (continued)

Round Tables

Progress in Mathematics, Vol. 169, © 1998 Birkhäuser Verlag Basel/Switzerland

Geometric Set Systems

JIŘÍ MATOUŠEK*

Department of Applied Mathematics, Charles University
Malostranské nám. 25, 118 00 Praha 1, Czech Republic
e-mail: matousek@kam.mff.cuni.cz

ABSTRACT. Let X be a finite point set in the plane. We consider the set system on X whose sets are all intersections of X with a halfplane. Similarly one can investigate set systems defined on point sets in higher-dimensional spaces by other classes of simple geometric figures (simplices, balls, ellipsoids, etc.). It turns out that simple combinatorial properties of such set systems (most notably the *Vapnik-Chervonenkis dimension* and related concepts of *shatter functions*) play an important role in several areas of mathematics and theoretical computer science. Here we concentrate on applications in discrepancy theory, in combinatorial geometry, in derandomization of geometric algorithms, and in geometric range searching. We believe that the tools described might be useful in other areas of mathematics too.

1. Introduction

For a set system $\mathcal{S} \subseteq 2^X$ on an arbitrary ground set X and for $A \subseteq X$, we write $\mathcal{S}|_A = \{S \cap A;\ S \in \mathcal{S}\}$ for the *set system induced by* $\mathcal{S}$ *on* A (or the *trace* of $\mathcal{S}$ on A). Let $\mathcal{H}$ denote the system of all closed halfplanes in the plane, and let $\mathcal{T}$ be the system of all triangles in the plane. For a finite set $A \subset \mathbb{R}^2$, $\mathcal{H}|_A$ is thus the the system of all subsets of A that can be "cut off" by a halfplane. We will be interested in combinatorial properties of set systems of this type. They are far from being understood. For instance, if we ask for the maximum possible number of sets of size exactly k in $\mathcal{H}|_A$ for an n-point set A, we get the notoriously difficult *k-set problem* of combinatorial geometry ([PSS92], [ABFK92], [DE94], [Dey97], [AAH+97]).

On the other hand, many interesting results can be derived from quite simple combinatorial properties of such set systems; one such important property is the so-called Vapnik-Chervonenkis dimension (or VC-dimension for short).

The *VC-dimension* is defined for any set system $\mathcal{S} \subseteq 2^X$ on an arbitrary set X. It is the supremum of the sizes of all shattered subsets $A \subseteq X$; here A is called *shattered* if $\mathcal{S}|_A = 2^A$, i.e. for any $B \subseteq A$ there exists a set $S \in \mathcal{S}$ such that $B = A \cap S$.

*Part of this survey was written while the author was visiting ETH Zürich, whose support is gratefully acknowledged. Also supported by Czech Republic Grant GAČR 0194/1996 and by Charles University grants No. 193,194/1996.

For example, it is not difficult to check that the VC-dimension of the set system $\mathcal{H}$ is 3 (no 4-point set can be shattered). Determining the VC-dimension of $\mathcal{T}$ exactly requires some work, but using simple tools presented below it is easily seen that this dimension is bounded by a constant. Similarly set systems defined by other simple geometric figures in a Euclidean space have typically a bounded VC-dimension (a precise formulation will be given later).

On the other hand, the system of all convex sets in the plane, say, has an infinite VC-dimension.

The notion now commonly called VC-dimension was introduced by Vapnik and Chervonenkis [VC71][1]. Numerous applications and extensions of the VC-dimension concept have been developed in statistics (in the theory of so-called empirical processes; some relevant references are [Vap82], [Dud84], [GZ84], [Dud85], [AT89], [Pol90]), in learning theory (where VC-dimension is one of the main concepts; e.g., [BEHW89], [AB92], [Hau92], [KW93], [ABCBH93], [DHS94]), but also for example in program testing [RV96].

In the combinatorics of hypergraphs, set systems of VC-dimension d can be viewed as a class of hypergraphs with a certain forbidden subhypergraph (the complete hypergraph on $d+1$ points), which puts this topic into a broader context of extremal hypergraph theory (see for instance [Fra83], [WF94], [DSW94]). Here we do not consider these areas.

This survey is mainly focused on the directions of the author's own work; we review some general results of a combinatorial nature about set systems of bounded VC-dimension, and present applications in geometric discrepancy theory, combinatorial geometry, and computational geometry. We also mention more geometric notions and results (which have no good analogue for arbitrary set systems of bounded VC-dimension), namely *cuttings* (sec. 4.) and *simplicial partitions* (sec. 5.).

2. Set systems of bounded VC-dimension

Shatter functions. These are parameters of a set system related to VC-dimension but often giving more information and easier to work with.

The *primal shatter function* of a set system $(X, \mathcal{S})$ is a function, denoted by $\pi_{\mathcal{S}}$, whose value at m ($m = 0, 1, 2, \ldots, |X|$) is defined by

$$\pi_{\mathcal{S}}(m) = \max_{A \subseteq X,\ |A|=m} \left| \mathcal{S}|_A \right| .$$

In words, $\pi_{\mathcal{S}}(m)$ is the maximum possible number of distinct intersections of the sets of $\mathcal{S}$ with an m-point subset of X. Yet another way to understand the primal shatter function is via the incidence matrix, M, of the set system $\mathcal{S}$ (with rows indexed by sets of $\mathcal{S}$ and columns indexed by points of X): $\pi_{\mathcal{S}}(m)$ is the maximum number of distinct rows appearing in any m-column submatrix of M. An alternative definition of VC-dimension of $\mathcal{S}$ is then $\sup\{d;\ \pi_{\mathcal{S}}(d) = 2^d\}$.

[1] Under different names, this also appears in other papers ([Sau72],[She72]) but the work [VC71] was probably the most influential for the subsequent developments.

The following simple, but basic, result bounds the primal shatter function in terms of VC-dimension:

LEMMA 2.1 *(Vapnik and Chervonenkis [VC71]; Sauer [Sau72]; Shelah [She72]) For any set system $\mathcal{S}$ of VC-dimension at most d, we have $\pi_{\mathcal{S}}(m) \leq \Phi_d(m)$, where $\Phi_d(m) = \binom{m}{0} + \binom{m}{1} + \cdots + \binom{m}{d}$, and this bound is tight in the worst case.*

Hence the behavior of the primal shatter function is fairly restricted: either it is 2^m for all m (the case of infinite VC-dimension) or it is bounded by a fixed polynomial (the case of finite VC-dimension). Set systems of bounded VC-dimension can thus be characterized as ones with a "hereditarily polynomial" number of sets.

An easy, but important, consequence of Lemma 2.1 is that if we take several set systems $\mathcal{S}_1, \mathcal{S}_2, \ldots, \mathcal{S}_k$ on X, each of bounded VC-dimension, and form a new set system $\mathcal{S}$ from the $\mathcal{S}_i$'s by a fixed set-theoretic formula (such as $\mathcal{S} = \{(S_1 \cup S_2) \setminus S_3;\ S_1 \in \mathcal{S}_1, S_2 \in \mathcal{S}_2, S_3 \in \mathcal{S}_3\}$, then such an $\mathcal{S}$ has a bounded VC-dimension again (because its primal shatter function is polynomially bounded).

It may often happen (e.g., in many geometric situations) that the primal shatter function is actually considerably smaller than the bound implied by Lemma 2.1. For instance, the set system $\mathcal{H}$ of all halfplanes has VC-dimension 3 but $\pi_{\mathcal{H}}$ is only quadratic. More generally, the set system $\mathcal{H}_d$ of all halfspaces in $\mathbb{R}^d$ has primal shatter function of the order m^d (for a fixed d). A yet more general result is the following:

THEOREM 2.2 *Let $f(x_1, x_2, \ldots, x_d, a_1, a_2, \ldots, a_p)$ be a fixed real polynomial in $d+p$ variables, and for $a = (a_1, \ldots, a_p) \in \mathbb{R}^p$, let $S_f(a)$ be the set of all points $x = (x_1, \ldots, x_d) \in \mathbb{R}^d$ such that $f(x_1, \ldots, x_d, a_1, \ldots, a_p) \geq 0$. Finally let $\mathcal{S}_f = \{S_f(a);\ a \in \mathbb{R}^p\}$. Then $\pi_{\mathcal{S}_f}(m) = O(m^p)$, where the constant of proportionality depends on p, d, and the degree of f.*

Theorem 2.2 is a consequence of results in real algebraic geometry on the number of sign patterns of a system of polynomials ([Ole51], [Mil64], [Tho65]; for recent more precise bounds see also [BPR96]). We may thus say that the primal shatter function mainly depends on the number of "degrees of freedom" of the surface delimiting the sets.

Together with the above remark on set systems defined by a fixed set-theoretic formula from other set systems, Theorem 2.2 implies that any set system in $\mathbb{R}^d$ each set of which is defined by at most k polynomial inequalities of maximum degree D has VC-dimension bounded in terms of d, k, D. This subsumes many (but not all) interesting examples with bounded VC-dimension. On the other hand, precise estimates for the VC-dimension may be difficult in such cases.

Next, we define the dual shatter function. This is just the primal shatter function of the set system "dual" to $\mathcal{S}$, whose incidence matrix arises by transposing the incidence matrix of $\mathcal{S}$ and deleting multiple rows. Explicitly, the *dual shatter function* of a set system $(X, \mathcal{S})$ is a function, denoted by $\pi^*_{\mathcal{S}}$, whose value at m is the maximum number of equivalence classes on X defined by an m-element subfamily $\mathcal{A} \subseteq \mathcal{S}$, where two points $x, y \in X$ are *equivalent with respect to* $\mathcal{A}$ if x belongs to the same sets of $\mathcal{A}$ as y does.

As observed by Assouad [Ass83], the set system dual to a set system of VC-dimension d has VC-dimension at most $2^{d+1}-1$. (This is not difficult to see if we adopt the incidence matrix point of view: From a $2^d \times d$ zero-one matrix containing all possible row patterns, we can select a $\lfloor \log_2 d \rfloor \times d$ submatrix containing all possible column patterns.) Consequently, the dual shatter function is polynomially bounded iff the primal one is.

Geometrically, the dual shatter function depends mainly on the space dimension. More precisely, *if f and $\mathcal{S}_f$ have the same meaning as in Theorem 2.2 then $\pi^*_{\mathcal{S}_f}(m) = O(m^d)$, where the constant of proportionality depends on p, d, and the degree of f.* This is actually quite intuitive: given m sets from $\mathcal{S}_f$, their bounding surfaces partition $\mathbb{R}^d$ into $O(m^d)$ cells, and these cells correspond to the equivalence classes from the definition of the dual shatter function.

More examples. Let X be a compact and simply connected set in the plane. For a point $x \in X$, define $V(x) \subseteq X$ as the set of all points y visible from x within X, i.e. such that the segment xy is completely contained in X. The set system $\{V(x); x \in X\}$ is a nontrivial example of a geometrically defined set system of bounded VC-dimension [KM95a], of a different type from the geometric examples considered above. The current best estimates for the maximum possible VC-dimension of such a set system are 6 from below and 23 from above [Val95].

If we allow a bounded number, h, of holes in X, the VC-dimension is still bounded by a function of h.

For other interesting and nontrivial bounds on VC-dimension see Karpinski and Macintyre [KM95b].

The examples mentioned so far were all of a geometric nature. A very non-geometric example (in some sense) of a set system of VC-dimension 2 is the system of lines of a finite projective plane; some of its properties in this context have been investigated by Alon et al. [AHW87].

Pseudohyperplanes and oriented matroids. We have just remarked that a finite projective plane provides a "non-geometric" example of a set system of bounded VC-dimension. Can some geometrically defined set systems be combinatorially distinguished from other set systems with the same VC-dimension (and shatter functions, say)? Partial positive results are known for the case of set systems defined by cells in hyperplane arrangements (which are closely related, via geometric duality, to set systems defined by halfspaces on finite point sets in $\mathbb{R}^d$).

Let H be a collection of n hyperplanes in $\mathbb{R}^d$ in general position. These hyperplanes cut $\mathbb{R}^d$ into a collection C of d-dimensional open cells. Suppose that for every hyperplane $h \in H$, one of the halfspaces bounded by h is distinguished as positive. Then each cell $c \in C$ defines a subset of H, namely the set of hyperplanes for which c lies in their positive halfspace. All such sets together define a set system $\mathcal{C}$ on H which has VC-dimension d. The number of sets in $\mathcal{C}$ is $\Phi_d(n)$, i.e. it attains the bound in Lemma 2.1. As is well known in the theory of oriented matroids (see e.g., [BVS+92]), there is no good way of combinatorially distinguishing arrangements of hyperplanes from arrangements of *pseudohyperplanes* (these are possibly curved surfaces such that the arrangement of any $d + 2$ of them is topologically equivalent to an arrangement of $d+2$ hyperplanes, but globally an equivalent arrangement need not be realizable by hyperplanes). Also, a cleaner picture is obtained by passing from an "affine" to a "projective" case; in our situation this means considering an arrangement of $(d-1)$-dimensional (pseudo)spheres on the d-dimensional sphere. For set systems corresponding to simple pseudospherical arrangements (i.e. with the pseudospheres being in general position), Gärtner and Welzl [GW94] found the following nice characterization: a set system $(X, \mathcal{S})$ comes from a simple arrangement of pseudospheres on S^d iff it has VC-dimension d, it is closed on complements ($S \in \mathcal{S} \Rightarrow X \setminus S \in \mathcal{S}$), and $|\mathcal{S}| = 2\Phi_{d-1}(|X|-1)$ (which is the maximum number of sets a set system of VC-dimension d closed on complements may have).

[GW94] also contains a number of other characterizations and connections to the oriented matroids theory.

Sampling properties. If X is equipped with a probability measure μ and $\mathcal{S}$ is a system of μ-measurable sets on X, we are interested in finite samples $A \subseteq X$ which approximate μ on the sets of $\mathcal{S}$ in a suitable sense. The weakest such notion is an ε-net ([HW87]; not to be confused with the synonymous notion for metric spaces), which is a set intersecting all "μ-big" sets of $\mathcal{S}$: A set $N \subseteq X$ (not necessarily one of the sets of $\mathcal{S}$) is called an *ε-net for* $(X, \mathcal{S}, \mu)$ ($\varepsilon \in [0,1]$ is a real number) if any set $S \in \mathcal{S}$ with $\mu(S) > \varepsilon$ intersects N. The most important case is that of X finite and μ uniform; then N is required to intersect each set $S \in \mathcal{S}$ with $|S| > \varepsilon|X|$ (in the sequel, if we speak about ε-nets etc. for a set system on a finite set, we always mean the uniform measure unless stated otherwise).

A straightforward probabilistic argument shows that if $\mathcal{S}$ is finite, then a subset of $(1/\varepsilon)\ln|\mathcal{S}|$ points randomly chosen from X (according to μ) is an ε-net with a positive probability. A key result of Haussler and Welzl [HW87] says that if $\mathcal{S}$ has a bounded VC-dimension d, then ε-nets exist of size depending only on d and ε. To make the formulas look simpler, let us write $1/r$ instead of ε.

THEOREM 2.3 (EPSILON-NET THEOREM) *(Haussler and Welzl [HW87], Komlós et al. [KPW92]) Let $N(d,r)$ be the minimum number such that any set system of VC-dimension d admits an $(1/r)$-net of size at most $N(d,r)$. For any $d \geq 2$ there exists an $r_0 > 0$ such that for all $r > r_0$, $(d-2)r\ln r \leq N(d,r) \leq dr\ln r$.*

(The case $d = 1$ is special; set systems of VC-dimension 1 can be completely described and have $(1/r)$-nets of size $O(r)$.) Both the upper and lower bounds on $N(d,r)$ are proved by a probabilistic argument; for the upper bound, it turns out that a random sample of $dr\ln r$ points of X is a $(1/r)$-net with high probability. (A proof of a slightly weaker upper bound is sketched in Section 3.) Since the lower bound uses a random set system, a challenging, and probably very difficult, open problem is whether some improvement in the ε-net size is possible for set systems defined geometrically. For instance, does the set system $\mathcal{T}|_A$ defined by triangles on a finite set A in the plane admit $(1/r)$-nets of size $O(r)$?

A stronger requirement on a sample than being an ε-net is to be an ε-approximation. A subset $A \subseteq X$ is an *ε-approximation* for $(X, \mathcal{S}, \mu)$ provided that

$$\left|\frac{|A \cap S|}{|A|} - \mu(S)\right| \le \varepsilon$$

for every set $S \in \mathcal{S}$. Clearly, an ε-approximation is also an ε-net, but not conversely.

The ε-approximation was the notion of a "good" sample considered by Vapnik and Chervonenkis [VC71]. They showed that for a set system of VC-dimension bounded by a constant d, a random sample A of size $Cr^2 \log r$ with a suitable constant $C = C(d)$ is a $(1/r)$-approximation for $(X, \mathcal{S}, \mu)$ with high probability. The size of the $(1/r)$-approximation obtained in this way is thus roughly quadratic compared to a $(1/r)$-net. Although a random sample of fewer than $const.r^2 \log r$ points typically is not a $(1/r)$-approximation, $(1/r)$-approximations of a somewhat smaller size can be constructed by other methods (based on a connection of ε-approximations with discrepancy) [MWW93]: If the primal shatter function is bounded by Cm^d for some fixed C, d $(d > 1)$, then there exist $(1/r)$-approximations of size $O(r^{2-2/(d+1)})$ ([MWW93],[Mat95b]). Similarly if $\pi^*_{\mathcal{S}}(m) \le Cm^d$ for all m, $(1/r)$-approximations of size $O(r^{2-2/(d+1)}(\log r)^{1-1/(d+1)})$ exist.

Ball packing. Consider a set system $(X, \mathcal{S})$ of VC-dimension d on an n-point set X. Define a metric on $\mathcal{S}$: The distance of two sets $S_1, S_2 \in \mathcal{S}$ is the cardinality of their symmetric difference $S_1 \triangle S_2$. An interesting and quite useful fact is that as far as the packing of balls of some given radius is concerned, this metric space behaves in a way similar to the d-dimensional Euclidean space. Precisely:

LEMMA 2.4 (PACKING LEMMA) *(Haussler [Hau95]) Let $(X, \mathcal{S})$ have VC-dimension d, and let $k \ge 1$ be an integer. Let $\mathcal{P} \subseteq \mathcal{S}$ be a set such that any two distinct sets of $\mathcal{P}$ have distance at least k. Then $|\mathcal{P}| \le (cn/(k+d))^d$ for some constant c.*

This means that for a fixed d, at most $O((n/\rho)^d)$ disjoint balls of radius ρ can be packed into $\mathcal{S}$ (which is an analogy to packing Euclidean balls of radius ρ into a d-cube of side n, say). In fact, it is not difficult to verify that Haussler's proof also implies that if the primal shatter function $\pi_{\mathcal{S}}(m)$ is at most Cm^d for all m and for some constants $C, d > 1$ then the $O((n/\rho)^d)$ bound for the number of disjoint ρ-balls remains valid.

It is instructive to prove a weaker bound, namely $O((n/\rho)^d \log^d(n/\rho))$, using ε-nets. So let d be a constant, and let $\mathcal{P} \subseteq \mathcal{S}$ satisfy $|S_1 \triangle S_2| > 2\rho$ for all $S_1 \ne S_2 \in \mathcal{P}$. Consider the set system $\mathcal{D} = \{S_1 \triangle S_2;\, S_1, S_2 \in \mathcal{S}\}$. Set $r = n/2\rho$, and fix a $(1/r)$-net N of size $O(r \log r)$ for $\mathcal{D}$ ($\mathcal{D}$ has a bounded VC-dimension since its primal shatter function is polynomially bounded). Whenever the symmetric difference of any two sets $S_1, S_2 \in \mathcal{S}$ has more than $n/r = 2\rho$ elements then it contains a point of N. In particular, we get $S_1 \cap N \ne S_2 \cap N$ for any two distinct sets $S_1, S_2 \in \mathcal{P}$. Therefore the set system induced by $\mathcal{S}$ on N has at least $|\mathcal{P}|$ elements, and so we get $|\mathcal{P}| \le \pi_{\mathcal{S}}(|N|) = O((n/\rho)^d \log^d(n/\rho))$ as claimed.

Haussler's proof of Lemma 2.4 was simplified by Chazelle [unpublished note].

Spanning trees with low crossing number. Let $(X, \mathcal{S})$ be a set system, and let G be a (simple, undirected) graph with vertex set X. We say that a set $S \in \mathcal{S}$ *crosses* an edge $\{u,v\}$ of G if $|S \cap \{u,v\}| = 1$. The *crossing number of G with respect to the set S* is the number of edges of G crossed by S, and the *crossing number of G* is the maximum of the crossing numbers of G with respect to all sets of $\mathcal{S}$. The following theorem is essentially due to Welzl [Wel88]. (That paper has a somewhat weaker bound. The bound presented arises by plugging Haussler's lemma 2.4 into Welzl's original proof; see also [CW89] for arguments proving particular geometric cases without Haussler's lemma.)

THEOREM 2.5 *Let $\mathcal{S}$ be a set system on an n-point set X, with $\pi^*_{\mathcal{S}}(m) \leq Cm^d$ for all m, where $C, d > 1$ are constants. Then there exists a spanning tree T with vertex set X whose crossing number is at most $C_1 n^{1-1/d}$, where $C_1 = C_1(C,d)$ is another constant. (In fact, T can be chosen as a path.)*

For a proof, we need

LEMMA 2.6 (SHORT EDGE LEMMA) *Let $\mathcal{S}$ be a set system as in Theorem 2.5. Then for any set (or multiset[2]) $Q \subseteq \mathcal{S}$, points $x, y \in X$ exist such that the edge $\{x,y\}$ is crossed by at most $C_2\,|Q|/n^{1/d}$ sets of Q, C_2 a suitable constant.*

Lemma 2.6 follows by applying the Packing lemma 2.4 on the set system dual to (X, Q).

We sketch the proof of Theorem 2.5. Imagine that we tried to construct a spanning tree T on X as required in Theorem 2.5 by the greedy (Kruskal's) algorithm, i.e. by selecting edges into T one by one and always choosing an edge of the smallest weight connecting distinct components of the already constructed part of T, where the weight of an edge is the number of sets of $\mathcal{S}$ crossing it. Having selected i edges, we have $n-i$ components in the graph constructed so far. Let X_i be a set containing one point from each component; applying Lemma 2.6 with $X = X_i$ and $Q = \mathcal{S}|_{X_i}$ we get that there exists an edge of weight $O(|\mathcal{S}|/(n-i)^{1/d})$ connecting two distinct components. A simple calculation shows that this algorithm produces a spanning tree which is good on the average, i.e. an average set of $\mathcal{S}$ crosses the right number of edges; however, it might happen that a few exceptional sets of $\mathcal{S}$ could cross many more edges.

To circumvent this, the edge selection strategy is improved to penalize sets of $\mathcal{S}$ which already cross many of the edges selected so far (this "reweighing strategy" is useful in several proofs in different areas [Lit88], [Cla93]). Specifically, start with all sets of $\mathcal{S}$ having weight 1, and having selected a new edge e_i in the ith step, double the current weight of all sets of $\mathcal{S}$ crossing e_i.

In each step, we select the lightest edge connecting two distinct components, where the weight of an edge e is the sum of current weights of sets crossing e.

We need to bound the crossing number, κ, of the resulting spanning tree, T. Let $w_i(S)$ be the weight of a set $S \in \mathcal{S}$ after i edges have been selected, and put $W_i = \sum_{S \in \mathcal{S}} w_i(S)$. We have $W_{n-1} \geq \max_{S \in \mathcal{S}} w_{n-1}(S) = 2^\kappa$. On the other hand, from Lemma 2.6 (with Q containing $w_i(S)$ copies of each $S \in \mathcal{S}$) one can derive

[2]Multiset means that Q may contain several copies of the same set $S \in \mathcal{S}$. The cardinality of Q is counted with these multiplicities.

$W_{i+1} \le W_i + O(W_i/(n-i)^{1/d})$. Calculation then gives $\kappa = O(\log |\mathcal{S}| + n^{1-1/d})$. Since the dual shatter function of $\mathcal{S}$ is polynomially bounded, $\mathcal{S}$ has a constant-bounded VC-dimension and hence $|\mathcal{S}|$ is polynomial in n. This concludes a proof sketch for Theorem 2.5.

3. Bounds for geometric discrepancy

The notion of *uniformly distributed sequences* and *uniformly distributed sets* is important in many branches of mathematics (measure theory, ergodic theory, diophantine approximation theory, statistics, numerical integration). Together with the theory of uniform distribution, questions about *irregularities of distribution* have been studied, for instance: how uniformly can n points be placed in the unit cube, with respect to a given collection of simple geometric shapes (such as axis-parallel boxes, balls, convex sets etc.). Discrepancy theory is a rich discipline today, with lots of difficult open problems, and here we can only touch upon it (more comprehensive sources are [BC87], [BS95]; discrepancy, from a combinatorial point of view, is treated in some of the chapters of [Spe87], [AS93], [PA95]).

Discrepancy theory, or the theory of irregularities of distribution, starts with the following Van Aardenne-Ehrenfest theorem (proving a conjecture of Van der Corput; see [BC87] for references):

For each infinite sequence of real numbers in the interval $[0,1]$ and for any $k > 0$ there exists an initial segment $(x_1, \ldots, x_n)$ of the sequence considered and a subinterval $(\alpha, \beta) \subseteq [0,1]$ such that the number of elements of $\{x_1, x_2, \ldots, x_n\}$ belonging to (α, β) deviates from $n(\beta - \alpha)$ (i.e. the expected number for a total uniformity) by at least k. This result has been improved several times. Schmidt proved that the right order of magnitude of k is $const. \log n$. The one-dimensional formulation for initial segments of sequences is equivalent to a two-dimensional formulation with a discrepancy of an n-point set with respect to axis-parallel rectangles.

Discrepancy with respect to Lebesgue measure. A general definition of discrepancy of a point set in the d-dimensional unit cube $U = [0,1]^d$ can be formulated as follows (although numerous variations have been considered in the literature): Let $\mathcal{S}$ be a system of (measurable) subsets of $\mathbb{R}^d$, and A an n-point set in U. The *discrepancy of A with respect to Lebesgue measure on $\mathcal{S}$* is

$$D(A, \mathcal{S}) = \max_{S \in \mathcal{S}} \left| n \cdot \lambda^d(S \cap U) - |A \cap S| \right|,$$

where λ^d stands for the d-dimensional Lebesgue measure. Further we let $D(n, \mathcal{S}) = \min_{A \subset U;\, |A|=n} D(A, \mathcal{S})$. By comparing the definition of $D(A, \mathcal{S})$ and the definition of an ε-approximation given earlier, we see that $D(A, \mathcal{S}) \le \Delta$ holds iff A is a (Δ/n)-approximation for the set system $\mathcal{S}|_U$ with the measure μ being λ^d restricted to U.

The above mentioned result of Schmidt is then equivalent to a statement about discrepancy of axis-parallel rectangles: $D(n, \mathcal{R}_2) \ge c \log n$, where $c > 0$ is a constant and $\mathcal{R}_d$ denotes the system of all axis-parallel boxes in $\mathbb{R}^d$ (an apparently very difficult open problem is to determine the asymptotics of $D(n, \mathcal{R}_d)$ for $d \ge 3$).

In subsequent works (for instance, of Schmidt, Roth, Halász, Beck, Alexander) the discrepancy with respect to Lebesgue measure has been studied in higher dimensions and for other classes of shapes. Surprisingly, while the discrepancy for axis-parallel rectangles is of the order $\log n$, the discrepancy for circular discs or for arbitrarily rotated rectangles is of the order (approximately) $n^{1/4}$.

Combinatorial discrepancy. A purely combinatorial notion of discrepancy for set systems has also been investigated in the literature. Let $(X, \mathcal{S})$ be a finite set system and $\chi : X \to \{-1, +1\}$ a mapping; in this context it is called a *coloring* of X. For a set $A \subseteq X$, we write $\chi(A) = \sum_{x \in A} \chi(x)$. The *discrepancy* of χ for $\mathcal{S}$ is defined by $\mathrm{disc}(\mathcal{S}, \chi) = \max_{S \in \mathcal{S}} |\chi(S)|$, and the *discrepancy* of $\mathcal{S}$ is $\mathrm{disc}(\mathcal{S}) = \min\{\mathrm{disc}(\mathcal{S}, \chi);\ \chi : X \to \{-1, +1\}\}$. Intuitively, we want a coloring such that each set has nearly equal numbers of pluses and minuses.

If A is a set of n points in $\mathbb{R}^d$ and $\mathcal{S}$ a system of subsets of $\mathbb{R}^d$, we may consider the discrepancy of the set system induced on A by $\mathcal{S}$. Such discrepancy is sometimes called the *red-blue discrepancy* or *combinatorial discrepancy* of A, in order to distinguish it from the "continuous" discrepancy with respect to Lebesgue measure.

These two types of geometric discrepancy are closely related; roughly speaking, lower bounds for the continuous discrepancy yield lower bounds for the red-blue discrepancy and upper bounds for the red-blue discrepancy give upper bound for the continuous discrepancy. One possible quantitative formulation due to Beck [Bec84] is the following:

PROPOSITION 3.1 *Let $\mathcal{S}$ be a system of measurable sets in $\mathbb{R}^d$ and let k, n be natural numbers. Moreover, assume that there is an $S_0 \in \mathcal{S}$ containing the whole unit cube U. Then $D(n, \mathcal{S}) \leq \frac{1}{2^k} D(2^k n, \mathcal{S}) + \sum_{i=1}^{k} \frac{\mathrm{disc}(2^i n, \mathcal{S})}{2^{i-1}}$.*

The idea of the proof is to take a set A_0 with $2^k n$ points and with $D(A_0, \mathcal{S})$ as small as possible. Then we fix a ± 1 coloring χ_0 of A_0 with minimum discrepancy, we take the smaller of the color classes $\chi_0^{-1}(1)$, $\chi_0^{-1}(-1)$ and we add a small number of points to it so that we obtain a set A_1 of precisely $|A_0|/2$ points. This "halving step" is repeated k times, and calculation gives the claimed inequality. Beck used Proposition 3.1 (with $k \approx \log n$) to derive $\mathrm{disc}(n, \mathcal{R}_2) \geq c' \log n$ from Schmidt's result, mentioned above, that $D(n, \mathcal{R}_2) \geq c \log n$ (determining the red-blue discrepancy of axis-parallel rectangles is known as Tusnády's problem and has not yet been satisfactorily solved).

Discrepancy from shatter functions. Lower bounds for geometric discrepancy are proved, with few exceptions, in the continuous setting, and they employ analytic methods like orthogonal functions, Fourier transform, etc. (see [BC87], [Ale90]). Upper bounds seem to go in two main directions: for axis-parallel boxes, various explicit constructions based on number theory, finite fields, and so on are known; they are studied in great detail also from a practical point of view, since such sets with a low discrepancy turned out to be very important for higher-dimensional numerical integration and computer simulations ([Nie92]). On the other hand, upper bounds for sets such as disks or rectangles with rotation allowed have been obtained via semi-random combinatorial constructions. A large number of known upper bounds in this direction are subsumed by the following theorem bounding discrepancy in terms of shatter functions.

THEOREM 3.2 *Let $\mathcal{S}$ be a set system on an n-point set X, and let C and $d > 1$ be constants.*

(i) [Mat95b] *If the primal shatter function satisfies $\pi_{\mathcal{S}}(m) \le Cm^d$ for all $m = 1, 2, \ldots, |X|$, then*

$$\operatorname{disc}(\mathcal{S}) = O\left(n^{\frac{1}{2}-\frac{1}{2d}}\right).$$

(ii) [MWW93] *If the dual shatter function satisfies $\pi^*_{\mathcal{S}}(m) \le Cm^d$ for all $m = 1, 2, \ldots, |\mathcal{S}|$, then*

$$\operatorname{disc}(\mathcal{S}) = O\left(n^{\frac{1}{2}-\frac{1}{2d}}\sqrt{\log n}\right).$$

As a consequence of (i), the red-blue discrepancy of an n-point set in $\mathbb{R}^d$ with respect to halfspaces is $O(n^{1/2-1/2d})$, which asymptotically matches a lower bound due to Alexander [Ale90]; hence (i) is asymptotically sharp. On the other hand, for the set system of all circular discs in the plane, say, the primal shatter function is cubic, while the dual shatter function is quadratic, and hence a better upper bound, of $O(n^{1/4}\sqrt{\log n})$, is obtained from (ii). It is not known whether this upper bound for the discrepancy of disks can be improved, but the bound of Theorem 3.2(ii) in general cannot be improved (a proof will be sketched below).

Dual shatter function bound. To understand the connection of shatter functions to discrepancy, it is perhaps best to begin with the following simple probabilistic lemma, showing that if we have polynomially many sets on an n-point set, then the discrepancy is at most about $\sqrt{n}$:

LEMMA 3.3 (RANDOM COLORING LEMMA) *Let $\mathcal{S}$ be a set system on an n-point set X, such that all sets of $\mathcal{S}$ have size at most k. For a random coloring $\chi : X \to \{+1, -1\}$, $\operatorname{disc}(\mathcal{S}, \chi) \le \sqrt{2k \ln(4|\mathcal{S}|)}$ holds with probability at least $\frac{1}{2}$.*

Indeed, for any set $S \in \mathcal{S}$ we have $\Pr(|\chi(S)| > \lambda\sqrt{k}\,) < 2e^{-\lambda^2/2}$ ($\lambda > 0$ a parameter) by the simplest Chernoff-type tail estimate, and setting $\lambda = \sqrt{2\ln(4|\mathcal{S}|)}$ does the job. There exist set systems with k sets of size k and with discrepancy $const.\sqrt{k}$, hence the bound in the Random coloring lemma cannot be much improved without further assumptions on the set system. The point of Theorem 3.2 is that a substantial improvement is possible if the shatter functions are polynomially bounded.

From Theorem 2.5 and Lemma 3.3, the "dual" discrepancy bound in Theorem 3.2(ii) follows quite simply.

Namely, given X and $\mathcal{S}$ as in Theorem 3.2(ii), choose a spanning path with crossing number $O(n^{1-1/d})$ on X. For simplicity, let $|X|$ be even. Delete every second edge of this path, which leaves us with a matching $M = \{\{u_1, v_1\}, \{u_2, v_2\}, \ldots,$

$\{u_k, v_k\}\}$ (a set of pairwise vertex-disjoint edges) covering all vertices. Define a random coloring $\chi : X \to \{+1, -1\}$, by coloring the points $u_1, \ldots, u_k$ randomly and independently, and by setting $\chi(v_i) = -\chi(u_i)$ for all i.

Look at a fixed set $S \in \mathcal{S}$, and classify the edges of M into two types: those with both points inside S or both points outside S, and those crossed by S. The edges of the former type contribute 0 to $\chi(S)$. The contributions of edges of the latter type to $\chi(S)$ are, by definition of χ, independent random variables attaining values $+1$ and -1 with equal probability. The number of these variables is $O(n^{1-1/d})$. Thus, the situation is as if we had a random coloring of $|\mathcal{S}|$ sets of size $O(n^{1-1/d})$ each, and the Random coloring lemma 3.3 implies that $\mathrm{disc}(\mathcal{S}, \chi) = O(n^{1/2-1/2d}\sqrt{\log n})$ with a positive probability.

Before turning to the "primal" discrepancy bound, let us remark that Lemma 3.3 also gives a simple proof of the existence of $(1/r)$-nets of size $O(r \log r)$ for set systems of constant-bounded VC-dimension (a rough form of Theorem 2.3). The idea is similar to the one in the proof of Proposition 3.1. For simplicity, we consider the case of a finite X with the uniform measure. Let the primal shatter function of $\mathcal{S}$ be bounded by a fixed polynomial.

Form a new system $\mathcal{S}' = \mathcal{S} \cup \{X\}$. By the Random coloring lemma 3.3, we know that any subsystem of $\mathcal{S}'$ induced by an m-point set has discrepancy $O(\sqrt{m \log m})$. Start with a ± 1 coloring of X with discrepancy $\Delta_1 \leq C\sqrt{n \log n}$, take the smaller color class and add some points to get a set X_1 of $n_1 = \lfloor n/2 \rfloor$ points, color it with discrepancy at most $\Delta_2 = C\sqrt{n_1 \log n_1}$, obtain X_2, etc. Iterate as long as it can be guaranteed that the current set, X_i, intersects all sets of $\mathcal{S}$ of size $> n/r$. Take the last such X_i for the $(1/r)$-net. Calculation shows that this last X_i has size $O(r \log r)$.

As was mentioned above, the $\sqrt{\log n}$ factor cannot in general be removed from the bound in Theorem 3.2(ii), at least for $d = 2, 3$. The probabilistic construction showing it [Mat97] starts with some set system $\mathcal{S}_0$ on an n-point set with at least $n^{1+\delta}$ sets for some fixed $\delta > 0$ such that the sets in $\mathcal{S}_0$ have size roughly $n^{1-1/d}$, and any d of them intersect in at most a bounded number, $t = t(d)$, of points. The set system $\mathcal{S}$ serving as a lower-bound example for Theorem 3.2(ii) is $\mathcal{S} = \{R(S);\ S \in \mathcal{S}_0\}$, where $R(S)$ denotes a random subset of S (the random selections are mutually independent). It is straightforward to show that such an $\mathcal{S}$ always satisfies $\pi^*_{\mathcal{S}}(m) = O(m^d)$. If we set $\Delta = cn^{1/2-1/2d}\sqrt{\log n}$ for a sufficiently small constant c, it can be calculated that for any fixed coloring χ, $\Pr[\mathrm{disc}(\mathcal{S}, \chi) \leq \Delta] < 2^{-n}$, and hence there exists a specific $\mathcal{S}$ for which none of the 2^n possible colorings gives discrepancy at most Δ.

To make the proof work we still need to construct a suitable $\mathcal{S}_0$. Unfortunately, this is equivalent to an old open problem of constructing certain asymptotically extremal bipartite graphs without a $K_{d,t}$ subgraph (the complete bipartite graph on d and t vertices), and no such construction is known for a general d. For $d = 3$, one can take $X = F \times F \times F$ (where F is a finite field) and define $\mathcal{S}_0$ as the set of graphs of $n^{1+\delta}$ random bivariate polynomials of a suitable constant degree D over F; applying Bézout's theorem, one can show that $\mathcal{S}_0$ has the required properties with high probability (for $d = 2$, a similar but simpler construction works). For $d \geq 4$, the construction of a suitable $\mathcal{S}_0$ is not known (this problem is not solved by a recent result of Kollár et al. [KRS96] since we need a bipartite graph where one class has significantly more vertices than the other class).

Primal shatter function bound. The first step towards a proof of Theorem 3.2(i) is Beck's "Partial coloring lemma" [Bec81]. By a *partial coloring* of a set X we mean any mapping $\chi : X \to \{-1, +1, 0\}$. One possible formulation is the following:

LEMMA 3.4 (PARTIAL COLORING LEMMA) *Let $\mathcal{F}$, $\mathcal{M}$ be set systems*[3] *on an n-point set X, $|\mathcal{M}| > 1$, such that $|M| \leq k$ for every $M \in \mathcal{M}$ and*

$$\prod_{F\in\mathcal{F}} (|F| + 1) \leq 2^{(n-1)/5}. \tag{1}$$

Then there exists a partial coloring $\chi : X \to \{-1, 0, +1\}$, such that the value of χ is nonzero for at least $n/10$ elements of X, $\chi(F) = 0$ for every $F \in \mathcal{F}$ and $|\chi(M)| \leq \sqrt{2k \ln(4|\mathcal{M}|)}$ for every $M \in \mathcal{M}$.

Intuitively, the situation is as follows. We have the "few" sets of $\mathcal{F}$, for which we insist that the discrepancy of χ is 0. Each such $F \in \mathcal{F}$ thus puts one condition on χ. It seems plausible that if we do not impose too many conditions, then a χ randomly selected among the ones satisfying these conditions will still be "random enough" to behave as a true random coloring on the sets of $\mathcal{M}$. In the lemma, we claim something weaker, however: Instead of a "true" coloring $\chi : X \to \{+1, -1\}$ we obtain a partial coloring χ, which is only guaranteed to be nonzero at a constant fraction of points. If we want to get a full coloring, the lemma has to be applied iteratively: Color 10% of points, restrict the set system on the remaining points, color 10% of them, etc.; the primal shatter function condition is well-suited for such an iterative method since it is inherited by subsets.

To prove the Lemma, let $\mathcal{C}_1$ be the set of all colorings χ with $\mathrm{disc}(\mathcal{M}, \chi) \leq \sqrt{2k \ln(4|\mathcal{M}|)}$. By the Random coloring lemma 3.3, we have $|\mathcal{C}_1| \geq 2^{n-1}$. Define a mapping $v : \mathcal{C}_1 \to \mathbb{Z}^{|\mathcal{F}|}$ by setting $v(\chi) = (\chi(F);\ F \in \mathcal{F})$. Counting shows that there exists a value $v_0 \in \mathbb{Z}^{|\mathcal{F}|}$ with $\mathcal{C}_2 = v^{-1}(v_0)$ big, and hence $\mathcal{C}_2$ contains two colorings χ_1, χ_2 differing in at least 10% of components. The desired partial coloring is then defined as $\chi = (\chi_1 - \chi_2)/2$.

Let us remark that this proof (a combination of probabilistic method with the pigeonhole principle) is nonconstructive, and it is a challenging open problem to find an algorithm (randomized or deterministic) for computing a partial coloring χ as in the Lemma in polynomial time.

To apply Lemma 3.4 for bounding the discrepancy of a set system $\mathcal{S}$ with a bounded primal shatter function, we need to find suitable set system $\mathcal{F}$ with not too many sets, such that for any $S \in \mathcal{S}$ there exists an $F = F_S \in \mathcal{F}$ with $|S \triangle F_S|$ small, smaller than some parameter k. Then we set $\mathcal{M} = \{S \setminus F_S;\ S \in \mathcal{S}\} \cup \{F_S \setminus S;\ S \in \mathcal{S}\}$ and apply Lemma 3.4; for the resulting partial coloring we have, for each $S \in \mathcal{S}$, $|\chi(S)| \leq |\chi(F_S)| + |\chi(S \setminus F_S)| + |\chi(F_S \setminus S)| = O(\sqrt{k} \log n)$. Since we have the Packing lemma 2.4 at our disposal, we can simply choose $\mathcal{F}$ as a maximal collection

[3] $\mathcal{F}$ for "few" sets, $\mathcal{M}$ for "many" sets.

of sets in $\mathcal{S}$ such that any two have symmetric difference larger than k. Then $|\mathcal{F}| = O((n/k)^d)$, and we have to set k in such a way that (1) holds. Calculation leads to the bound $\operatorname{disc}(\mathcal{S}) = O(n^{1/2-1/2d}(\log n)^{1/2+1/2d})$.

This is nearly the correct bound; the next improvement is based on an idea of Spencer [Spe85]. Namely, the set system $\mathcal{F}$ need not be colored perfectly (with zero discrepancy); it suffices to color it with about the same discrepancy as the sets of $\mathcal{M}$, and this observation allows us to weaken the condition (1) somewhat. Technically, one incorporates this observation into the proof by estimating the entropy of a suitable vector function of a random coloring χ (see [AS93][4]). As a final trick, instead of just two auxiliary set systems, $\mathcal{F}$ and $\mathcal{M}$, we define a whole sequence of such set systems, say $\mathcal{S}_0, \mathcal{S}_1, \ldots, \mathcal{S}_q$ with $q \approx \log n$. The $\mathcal{S}_i$ has roughly 2^{id} sets of size at most $n/2^i$, and each set of $\mathcal{S}$ can be written as $S = (\cdots((((A_1 \cup A_2) \setminus B_2) \cup A_3) \setminus B_3) \cup \cdots \cup A_q) \setminus B_q$, where $A_i, B_i \in \mathcal{S}_i$, the unions are disjoint and the subtracted sets B_i are fully contained in the sets they are subtracted from. For each $\mathcal{S}_i$, we then prescribe a suitable discrepancy bound Δ_i, and show that there exists a partial coloring under which each $\mathcal{S}_i$ has discrepancy at most Δ_i; see [Mat95b] for details.

Another problem in discrepancy theory concerns the discrepancy of the set system formed by all arithmetic progressions on the set $\{1, 2, \ldots, n\}$. Roth [Rot64] proved a lower bound of *const.*$n^{1/4}$. The upper bound was improved several times (e.g, by Beck [Bec81] to $O(n^{1/4} \log^3 n)$), and [MS96] proves a tight upper bound of $O(n^{1/4})$ using a part of the technique from the proof of Theorem 3.2(ii). The first partial coloring for the set system of arithmetic progressions is constructed in nearly the same way as for a set system with a quadratic primal shatter function, but for the subsequent iterations one has to work somewhat more since here the situation is not hereditary (the structure of subsystems induced by various subsets may be more complicated).

Another application of similar methods led to improved upper bounds for the following discrepancy-type problem: For a given $\varepsilon > 0$, approximate the unit ball B in $\mathbb{R}^d$ (considering d fixed) with error at most ε by a convex polytope A of the form $A = \{x_1 + x_2 + \cdots + x_n;\, x_1 \in I_1, \ldots, x_n \in I_n\}$, where $I_1, \ldots, I_n$ are segments in $\mathbb{R}^d$ (such a polytope is called a *zonotope*), with the number n of summands as small as possible. This problem has several interpretations; one of them deals with a numerical integration of a special class of functions over the unit sphere S^d, and another can be formulated as a "tomography" problem. Generalizations have also been considered where one approximates a general zonoid (a convex body approximable by zonotopes arbitrarily precisely) by a zonotope with few summands. Lower bounds for this problem have been established by Bourgain, Lindenstrauss, and Milman [BLM89] (using harmonic analysis). Nearly tight upper bounds for the ball approximation were given by Bourgain and Lindenstrauss [BL88],[BL93], and an application of the methods described above led to tightening these bounds and generalizing them for arbitrary zonoids [Mat96b].

[4] Results similar to Spencer's [Spe85] were also obtained by Gluskin [Glu89] via Minkowski's lattice point theorem and by Giannopoulos [Gia97] using the Gaussian measure.

4. Derandomizing geometric algorithms

The topics of this and the subsequent section (geometric range searching) belong to the field of *computational geometry.* This is a branch of theoretical computer science considering the design and analysis of efficient algorithms for computing with configurations of simple geometric objects in a Euclidean space. The space dimension is usually considered a constant (many problems are studied mainly in the plane or in 3-dimensional space); this distinguishes computational geometry from geometric aspects of combinatorial optimization, for instance (where the dimension is usually comparable to the number of objects).

Computational geometry as a subject arose sometimes around 1980. In the period until, say, 1986, a number of basic algorithmic problems were solved (particularly in the plane), and it was also clearly demonstrated that for good algorithmic results one needs to understand the combinatorial properties of geometric configurations. Since that time, combinatorial and computational geometry have influenced each other substantially. The status of computational geometry in the early years is described in the monographs [PS85], [Ede87].

The algorithms described in these books are, with minor exceptions, *deterministic.* The subsequent more extensive introduction of *randomized algorithms* into computational geometry worked almost as a miraculous medicine. Methods elaborated in the period (approximately) 1986–1989 provide, relatively easily, efficient probabilistic algorithms for problems which seemed hopelessly difficult before. Even for problems with previously known optimal deterministic algorithms, the probabilistic algorithms are simpler, practically more efficient and easier to implement in most cases. A monograph picturing this new look computational geometry is [Mul94].

In connection with this development, a question arose naturally, namely to what extent are the probabilistic algorithms more powerful than the deterministic ones. In other words: can one eliminate the use of randomness from a given probabilistic algorithm, possibly preserving the asymptotic efficiency or making it only a little worse? This *derandomization* was of course also investigated, approximately in the same period, in other areas of theoretical computer science. In computational geometry, specific techniques have been obtained, which yield, for most of the randomized algorithms studied there, deterministic algorithms with the same or only slightly larger asymptotic complexity. A detailed survey on this subject is [Mat96a], and here we restrict ourselves to a few remarks.

A randomized algorithm can be viewed as a deterministic algorithm which is allowed to read a finite sequence x of random bits from some special device. Usually, we know that for an overwhelming majority of sequences $x \in \{0,1\}^n$, the algorithm works well, and the task of derandomization is to compute at least one good x deterministically. Two general techniques for this purpose go by the names "method of conditional probabilities" and "(approximate) k-wise independence". The former technique determines the bits of x one by one, depending on the value of suitable conditional expectations; it can be viewed as a "binary search" in the probability space $\{0,1\}^n$. The latter technique tries to replace the (exponentially large) probability space $\{0,1\}^n$ of random sequences by a suitable smaller probability space Ω, such that the algorithm is still guaranteed to work well on average if its random bit sequence x is chosen from Ω, but Ω is small enough so that it can be searched exhaustively. Both these general ideas have many variations and subtleties in applications (see e.g., [MR95] for references).

In computational geometry, most algorithms can be easily derandomized by the method of conditional probabilities in its basic form; the running time increases by a factor which is polynomial but usually quite large [CF90]. Methods specific to computational geometry yield much better asymptotic complexities.

Let us consider a specific algorithmic problem: P is a given set of n points in the plane, L is a given set of n lines in the plane, and we want to decide whether any of the points of P lies on any of the lines of L (this is so-called *Hopcroft's problem*; its algorithmic complexity is not known, but it is suspected to be of the order $n^{4/3}$). The known asymptotically efficient algorithms for this problem all use the following strategy ("geometric divide-and-conquer"). They divide the plane into some suitable number, m, of regions $\Delta_1, \Delta_2, \ldots, \Delta_m$ (usually the Δ_i are triangles, possibly unbounded ones). Let P_i be the subset of P lying in Δ_i and L_i the set of lines of L intersecting Δ_i. In this way, the original problem is subdivided into m subproblems (it suffices to check whether for any i, a point of P_i lies on a line of L_i), and these subproblems are then solved recursively or by another method. We will not go into any further details of the algorithm here; we concentrate on the part constructing the regions Δ_i. For the algorithm's efficiency it is crucial that the sets L_i be as small as possible. To avoid some technicalities, we let L_i be only the lines intersecting the *interior* of Δ_i (we suppose that the lines touching only the boundary of Δ_i are somehow handled separately). We introduce the following definition: We say that regions $\Delta_1, \ldots, \Delta_m$ form a $(1/r)$*-cutting for* L if they cover the whole plane and the interior of each Δ_i is intersected by at most $|L|/r$ lines of L.

A quite good construction of a $(1/r)$-cutting by random sampling is the following [HW87], [Cla87]. We fix a suitable number s and choose a random sample $S \subset L$ of s lines from L. The lines of S divide the plane into convex polygons. We further subdivide these polygons into triangles; an easy calculation shows that we get $O(s^2)$ triangles in total, and these will be our Δ_i's. Next, we consider the set system $\mathcal{L}$ on L, consisting of all sets of the form L_T, where T is some triangle and L_T is the set of lines of L intersecting the interior of T. The set system $\mathcal{L}$ has a constant-bounded VC-dimension, and hence our random sample S of s lines is, with high probability, an ε-net for $\mathcal{L}$ by Theorem 2.3, with $\varepsilon = C \log s/s$ for an appropriate constant C. This means that any open triangle intersected by no line of S is intersected by $O((n/s)\log s)$ lines of L only. In particular, since the interiors of the Δ_i's are not intersected by lines of S, we get $|L_i| = O((n/s)\log s)$ for all i.

Therefore, we get a $(1/r)$-cutting for L, with $r = \mathit{const}.s/\log s$, consisting of $O(r^2 \log^2 r)$ triangles. It turns out that this construction is nearly best possible (up to the logarithmic factor; see below).

One way of making the just described construction deterministic is to compute an ε-net for the set system $\mathcal{L}$ deterministically. An algorithm for the deterministic computation of ε-nets was given in [Mat95a]; the key point is that one actually computes with ε-approximations (which have some more pleasant algebraic properties than ε-nets) and in the end, the desired ε-net is computed from an ε-approximation. The algorithm works for an arbitrary set system $\mathcal{S}$ of bounded VC-dimension on a finite set X, provided that the set system $\mathcal{S}$ is given by a *subsystem oracle*. This means that given a subset $A \subseteq X$, we can compute the system $\mathcal{S}|_A$ (in the form of an incidence matrix) in time $O(|A|^{d+1})$, where d is a constant called the *dimension* of the subsystem oracle. The current most efficient version of the algorithm was given in [BCM93]. For a fixed oracle of dimension d, it can compute a $(1/r)$-net for $\mathcal{S}$ of size $O(r \log r)$ in time $O(|X| r^d \log^d r)$; in particular, if r is a constant, a $(1/r)$-net is found in time linear in the cardinality of X. This result allows one to make most of the divide-and-conquer algorithms in computational geometry deterministic with only an $O(n^\delta)$ loss in asymptotic efficiency, where $\delta > 0$ is an arbitrarily small constant.

It can also be used to derandomize the above probabilistic construction of a $(1/r)$-cutting for a given set of n lines. However, for the construction of $(1/r)$-cuttings in the plane and in higher dimensions, special methods have been developed to make it still more efficient. Let us recall that the above-sketched randomized construction gave a $(1/r)$-cutting consisting of $O(r^2 \log^2 r)$ triangles. It is not difficult to show that a $(1/r)$-cutting for n lines in the plane has to consist of at least $const.r^2$ regions (count the total number of polygons into which the lines of L divide the plane, and compare it with the maximum number of polygons incident to a single region of a $(1/r)$-cutting). On the other hand, $(1/r)$-cuttings of size $O(r^2)$ actually exist [CF90]. Let us formulate a d-dimensional version of this result:

THEOREM 4.1 (CUTTING LEMMA) *Given a set H of n hyperplanes in $\mathbb{R}^d$ (d considered constant) and a parameter $r > 1$, there exist a $(1/r)$-cutting of size $O(r^d)$ for H, that is, a collection $\Delta_1, \ldots, \Delta_m$ of (possibly unbounded) simplices covering $\mathbb{R}^d$, such that $m = O(r^d)$ and the interior of each Δ_i is intersected by at most n/r hyperplanes of H [CF90].*

Such a $(1/r)$-cutting can be deterministically computed in time $O(nr^{d-1})$ [Cha93], and if $r < n^\alpha$ for a suitable $\alpha = \alpha(d) > 0$ then even in $O(n \log r)$ time [Mat92].

The $(1/r)$-cuttings have also found various non-algorithmic applications; one is in discrepancy theory [Mat96b] and few others will be mentioned in Section 6.

5. Geometric range searching

Here we consider a specific class of algorithmic problems, which is quite important for the design of various geometric algorithms.

As a prototype question, consider the following *triangle range counting* problem. Let P be a given n-point set in the plane. We want an algorithm which, given a triangle T, quickly determines the number of points of P lying in T. The point set P is given in advance, and we can prepare some auxiliary information about it and store it in a suitable data structure. Then we will be repeatedly given triangles T as queries.

Assuming that the number of queries will be large, it will be advantageous to invest some computing time into building the data structure (this phase is called the *preprocessing*) if this makes the query answering faster.

This problem can of course be generalized in many directions. We may consider a higher-dimensional space $\mathbb{R}^d$ instead of the plane, and use sets from some class $\mathcal{S}$ of simple geometric shapes in $\mathbb{R}^d$ instead of triangles as queries (natural choices are axis-parallel boxes, simplices, balls, etc.). Also the type of a query might be different from point counting: we might require a list of all points of P lying in a query range, or we might only ask if the query range contains any point of P at all, or each point of P might be assigned some real weight and we can be interested in the maximum weight of a point in a given range, etc.

In general, we assume that every point $p \in P$ is assigned a weight $w(p) \in \Sigma$, where $(\Sigma, +)$ is some commutative semigroup (common to all the points). The objective of a query is to find the sum of weights of all points of P lying in a given range $S \in \mathcal{S}$, that is, $\sum_{p \in S \cap P} w(p)$. For example, for counting queries, $(\Sigma, +)$ will be the natural numbers with addition, and all weights will be equal to 1. For queries on maximum weight, the appropriate semigroup will be the real numbers with the operation of taking a maximum of two numbers.

It turns out that for many classes $\mathcal{S}$ of ranges (e.g., all the examples named above), one can achieve a query time $O(\log n)$, where n is the number of points in P. In most cases, however, this fast query answering requires a huge amount of memory and preprocessing time. The problem is thus to achieve a reasonable compromise between storage and preprocessing on the one hand and query time on the other hand.

Mathematically most interesting and most studied was the case when we insist that the data structure is stored in memory space $O(n)$ or close to $O(n)$, and we ask what can be saved in the query time compared to the trivial $O(n)$ solution (store the points without any preprocessing and, given a query range, examine each point in turn).

Unfortunately, from a practical point of view, it seems that except for several simple low-dimensional situations ($d = 1, 2$, say), the results are disappointing — the possible savings in query time with a linear space are meager. Theoretically, though, the problem led to nice mathematical developments, mainly in attempts to prove lower bounds.

For definiteness, let us consider triangle range searching with $O(n)$ memory space, where the points of P have weights from some semigroup S, and the query is to find the sum of weights of points in a given triangle. The most interesting aspects of known results already appear in this simple situation. Moreover, various more complicated-looking range searching problems can often be reduced to searching with simplices by a suitable transformation. A more comprehensive information can be found, e.g., in [Mat95c], [Mul94], [PA95].

Lower bounds. As shown by Chazelle [Cha89], an algorithm for answering queries with triangles and working for arbitrary semigroup weights must essentially operate as follows: In the preprocessing phase, weights are pre-computed for a suitable collection $\mathcal{C} = \{C_1, C_2, \ldots, C_m\}$ of subsets of the set P; here m is a lower bound on the memory space used by the algorithm. When answering a query with a triangle T, the set $P \cap T$ is expressed as a disjoint union of some number, k, of sets of the system $\mathcal{C}$, and the total weight of points lying in T is computed by summing the weights of these k sets of $\mathcal{C}$. The maximum value of k (over all query triangles) is a lower bound for the worst-case time complexity of a query. (This is admittedly inexact, since a precise statement is somewhat technical.) The sets in $\mathcal{C}$ are called the *canonical sets* (for the particular algorithm and set P).

Chazelle [Cha89] shows (among other things) that if we allow only $O(n)$ canonical subsets in $\mathcal{C}$, then the number of sets in the decomposition, k, must be at least $\Omega(\sqrt{n})$ for some query triangle. The proof requires a construction of a suitable point set P and a suitable finite set $\mathcal{T}_0$ of triangles, such that the bipartite incidence graph between P and $\mathcal{T}_0$ has sufficiently many edges, but it contains no large complete bipartite subgraphs (which means that no large canonical set $C \in \mathcal{C}$ can contribute to the decomposition of many triangles from $\mathcal{T}_0$). In Chazelle's proof, P is chosen as a uniformly distributed set in the unit square, and $\mathcal{T}_0$ is chosen as a random set of strips of width approximately $n^{-1/2}$ intersecting the unit square.

For dimension $d > 2$, the expected generalization would be that with $O(n)$ canonical sets, we need at least about $n^{1-1/d}$ sets in the decomposition. Chazelle's proof only gives an $n^{1-1/d}/\log n$ lower bound, and the possibility of improving the lower bound using the same method has an interesting relation to a generalization of a famous problem of combinatorial geometry, the so-called Heilbronn problem. The Heilbronn problem itself can be formulated as follows: *For a set $P \subset \mathbb{R}^2$, let $a(P)$ denote the area of a smallest triangle with vertices in P. What is the asymptotic behavior of the function $a(n) = \sup\{a(P);\ P \subset [0,1]^2, |P| = n\}$?*

Despite considerable attention paid to this nice problem (see [KSP82], [Rot76] for a survey and references), a complete solution still seems to be remote. The following is a generalization related to Chazelle's lower bound proof method: *For a set $P \subset \mathbb{R}^d$, denote by $a_d(P,j)$ the smallest volume of the convex hull of a j-tuple of points of P. What is the asymptotic behavior of the function $a_d(n,j) = \sup\{a_d(P,j);\ P \subset [0,1]^d, |P| = n\}$?* Chazelle has shown that for a suitably chosen set P in the unit cube the volume of the convex hull of each j-tuple is at least proportional to j/n for any $j \geq c \log n$, with a sufficiently large constant c. In other words,

$$a_d(n,j) = \Omega(j/n) \tag{2}$$

for $j \geq c \log n$. This is essentially a result about uniform distribution of the set P, saying that no j-tuple is too clustered (in the sense of volume). If (2) could also be proved for smaller j, an improvement of the lower bound for decomposing sets in triangles would follow immediately. Such an improvement may not be easy. Known results for the Heilbronn problem imply that (2) is false for $d = 2$, $j = 3$ (which may contradict intuition somewhat). However, it is not known that (2) could not hold for larger but constant j.

An algorithm for triangle range searching with $O(n)$ space and $O(\sqrt{n})$, which matches Chazelle's lower bound, was given in [Mat93], and no asymptotically better algorithm is known even for various special cases of the problem. On the other hand, the situation is not as satisfactory as it might seem, since the lower bound was shown under the condition that subtraction of point weights is not allowed in the algorithm. This is a somewhat unnatural restriction (say for the case of counting the points in a query range), although it is not known how to actually use weight subtraction to get a faster algorithm. Hence it would be desirable to also have lower bounds for stronger models of computation. Some progress was made by Chazelle again [Cha94]; he shows that n queries together require at least $\Omega(n \log n)$ operations even if subtractions are allowed. His proof uses eigenvalue estimates for a suitable matrix based on known lower bounds for the combinatorial discrepancy of rectangles. Other papers considering lower bounds for geometric range searching and leading to very interesting geometric problems are [BCP93] and [Eri96].

Upper bounds. Let us begin with a simple but nontrivial one-dimensional range-searching example, namely with intervals on the real line.

Let $P \subset \mathbb{R}$ be an n-point set, and let $\mathcal{I}$ be the family of all intervals in $\mathbb{R}$. For the purpose of canonical decompositions of sets of the form $P \cap I$, $I \in \mathcal{I}$, we might as well assume that $P = \{0, 1, \ldots, n-1\}$. Then a suitable system of canonical sets (subintervals of P, in this case) is $\{\{q2^k, q2^k + 1, \ldots, (q+1)2^k - 1\}; k = 0, 1, \ldots, \lfloor \log_2 n \rfloor, q = 0, 1, \lfloor n/2^k \rfloor\}$. This set system appears very often in mathematics and in computer science under various names and disguises (for instance, it corresponds to the nodes of a complete binary tree with leaves $0, 1, \ldots, n-1$). It has $O(n)$ sets, and it is easy to see that any interval on P can be partitioned into $O(\log n)$ canonical sets.

Let us now consider range searching with triangles in the plane. Spanning trees with low crossing number were originally invented exactly for this purpose [Wel88]. Since the dual shatter function for triangles in the plane is quadratic, Theorem 2.5 guarantees the existence of a spanning path with crossing number $O(\sqrt{n})$ on our given n-point set P. Geometrically, the crossing number condition means that the boundary of a query triangle T cuts this path into $O(\sqrt{n})$ pieces, so that each piece lies either completely inside T or completely outside T. Hence the set $P \cap T$ can be expressed as a disjoint union of $O(\sqrt{n})$ intervals along the path, and this reduces the triangle range searching problem into $O(\sqrt{n})$ one-dimensional queries with intervals on the path. These one-dimensional queries can be handled in a manner sketched above; hence we obtain a decomposition of $P \cap T$ into $O(\sqrt{n} \log n)$ canonical sets. To make this method into an actual algorithm, we have to specify how to find the edges of the spanning path crossed by the query triangle. In the plane, this algorithmic problem can be solved satisfactorily, as well as an analogous problem for tetrahedra in dimension 3, but in dimension 4 and higher it is not known how to use spanning trees with low crossing numbers efficiently in an algorithm (although they still provide good decompositions into canonical sets).

Let us sketch one more approach to triangle range searching; we demonstrate it on the simpler case of halfplane searching (presenting the geometric ideas only and omitting all algorithmic details). The approach was suggested by Willard [Wil82]; it was the first known algorithm for triangle range searching with a near-linear space and a sublinear query time. A geometric construction of the following type is required: given an n-point set $P \subset \mathbb{R}^2$, partition the plane into some number, a, of regions, in such a way that each region contains roughly n/a points of P, and no line intersects more than b of the regions, for some $b < a$. The simplest such construction is to take two lines such that each of the four "quadrants" R_1, R_2, R_3, R_4 determined by them contains about $n/4$ points of P (the first line can be chosen as any line slicing P into two equal-size parts, and the second line is then found using the so-called *ham-sandwich theorem*). Here we have $a = 4$, $b = 3$. Let $P_i = P \cap R_i$ be the points in the ith region. We declare the four sets $P_1, \ldots, P_4$ as canonical sets, and we apply the same construction recursively for each P_i, until one-point sets are reached. This procedure produces $O(n)$ canonical sets in total.

We claim that for any halfplane H, the set $H \cap P$ can be expressed as a disjoint union of $O(n^\alpha)$ canonical sets thus defined, where $\alpha = \log_4 3 \approx 0.774$. Indeed, since the boundary of H crosses only 3 of the 4 regions R_i, one of the sets P_i lies either completely outside H (then we may ignore it) or completely within H, and then we use it as one of the canonical sets in the decomposition. If $f(n)$ denotes the maximum number of canonical sets in a decomposition used by this method for an n-point set P, we get the recurrence $f(n) \leq 1 + 3f(n/4)$ (ignoring some rounding to integers), and this leads to the claimed bound. In general, having a construction with a regions, at most b of them intersected by a line, we get a bound of $O(n^\alpha)$ with $\alpha = \log_a b$. To get a good construction of this type by ad hoc methods was not easy, and the exponent has been improved many times (we remark that the paper [HW87] which introduced the key notion of ε-nets was one of the attempts in this direction!). A near-optimal and relatively simple method of this type was given in [Mat92]. We formulate the main geometric result for an arbitrary dimension d. First, a definition: a *simplicial partition* of a finite set $P \subset \mathbb{R}^d$ is a collection $\Pi = \{(P_1, \Delta_1), \ldots, (P_m, \Delta_m)\}$, where the P_i are disjoint subsets of P (called the *classes*) forming a partition of P, and each Δ_i is a simplex containing the set P_i.

THEOREM 5.1 *Let P be an n-point set in $\mathbb{R}^d$ ($d > 1$ considered as a constant), and let r be a parameter, $1 < r \ll n$. Then there exists a simplicial partition Π of P, whose classes have sizes between n/r and $2n/r$ (hence there are $O(r)$ classes), and such that any hyperplane crosses at most $O(r^{1-1/d})$ simplices of Π (a hyperplane h crosses a simplex Δ if Δ intersects both the open halfspaces bounded by h).*

In particular, for $d = 2$, we get a partition into about r subsets such that any line crosses at most $O(\sqrt{r})$ of their respective triangles. In the above notation, we thus have $a \approx r$, $b \approx const.\sqrt{r}$, so that $\log_a b \to \frac{1}{2}$ for $r \to \infty$, and we get close to the optimal $O(\sqrt{n})$ query time.

Theorem 5.1 is a geometric generalization of Theorem 2.5 on spanning trees with low crossing numbers. Indeed, Theorem 2.5 also provides a partition of the given set into pairs such that no set crosses too many pairs; in Theorem 5.1 we have chunks of about n/r points instead of pairs. Also the proof is quite similar to the proof of Theorem 2.5; while in the spanning tree construction, we select "short edges" one by one, in the construction of a simplicial partition one selects "small chunks" (ones crossed by few hyperplanes from a suitable "test set"). The existence of these small chunks at every step, i.e. an analogue of Lemma 2.6, is proved using $(1/r)$-cuttings (defined in the preceding section). As was shown by Alon et al. [AHW87], an analogue of Theorem 5.1 does not hold for general set systems of bounded VC-dimension; at least for r bounded by a constant no nontrivial "partition scheme" exists in general.

6. Combinatorial geometry applications

A large part of combinatorial geometry deals with properties of geometrically defined set systems. Here we select just few applications of the notions and tools discussed above. We refer to the book [PA95] for more material.

Pach [Pac91] applies spanning trees with a low crossing number to give a simple proof of a result of Aronov et al. [AEG$^+$91]: for any set of n points in the plane, one can choose $const.\sqrt{n}$ disjoint pairs of these points such that any two of the segments determined by these pairs intersect. It is a nice open problem whether the bound of $const.\sqrt{n}$ can be replaced by $const.n$.

Geometric applications of the results of [DSW94] can be found in Pach [Pac97].

Haussler and Welzl [HW87] introduced, besides ε-nets, also *weak ε-nets*. The following particular geometric case of this notion found surprising applications: let $A \subset \mathbb{R}^d$ be a finite point set; a set $N \subset \mathbb{R}^d$ is a *weak ε-net for A with respect to convex sets* if $N \cap C \neq \emptyset$ for any convex set C with $|C \cap A| > \varepsilon |A|$. This looks similar to the definition of ε-net, but the difference is that here N may consist of any points of $\mathbb{R}^d$ and is not restricted to points of A (as would be the case for an ε-net). While ε-nets of a bounded size for convex sets do not exist in general, weak ε-nets do exist. The current best bounds for the size of a weak $(1/r)$-net for convex sets are $O(r^2)$ in the plane [ABFK92] and $O\left(r^d (\log r)^{b_d}\right)$ for a suitable constant b_d in $\mathbb{R}^d$, $d > 2$ [CEG$^+$95] (see also [BC95] for related results). The truth is probably much smaller, maybe around $r^{\lfloor d/2 \rfloor}$. Weak ε-nets for convex sets have some connections with work on the k-set problem mentioned in the introduction [ABFK92], and they have been applied by Alon and Kleitman [AK92] as one of three key ingredients in a beautiful solution to the Hadwiger-Debrunner (p, q)-problem.

The $(1/r)$-cuttings (defined in Section 4.) and a number of their generalizations have been used in problems of bounding the number of incidences, complexity of cells in arrangements, etc. A prototype problem of this kind is the following question: consider a set P of n points and a set L of m lines in the plane. What is the maximum possible number of their *incidences*, i.e. pairs (p, ℓ) such that $p \in P$, $\ell \in L$ and $p \in \ell$?

A tight upper bound, $O(n^{2/3}m^{2/3} + m + n)$, was proved by Szemerédi and Trotter [ST83]. A considerably simpler proof and various generalizations have been given by Clarkson et al. [CEG$^+$90] using methods related to $(1/r)$-cuttings; other similar results may be found in [EGS90], [CEGS89], [AEGS92]. The approach via $(1/r)$-cuttings is by now one of the standard methods in proving combinatorial complexity bounds for geometric configurations. For numerous other methods and more information about the extensive realm of combinatorial complexity of arrangements in a fixed dimension we refer to a recent monograph [AS95]. Recently, a yet simpler proof of the Szemerédi-Trotter theorem and of some related results was given by Székely [Szé] by a different approach (see also [PS] for generalizations), but so far the approach via $(1/r)$-cuttings has its advantages (it generalizes to higher-dimensional questions, and it usually provides efficient algorithms).

Let us sketch a proof of the Szemerédi-Trotter bound using $(1/r)$-cuttings. Denote the number of incidences for specific sets L of lines and P of points by $I(L,P)$, and let $I(n,m)$ be the maximum of $I(L,P)$ over all choices of an n-element L and an m-element P. First consider the bipartite graph on vertex set $L \cup P$ with edges corresponding to incidences. No two lines share two points and hence this graph has no $K_{2,2}$ subgraph. A well-known extremal graph theory result then implies

$$I(n,m) = O(\min(n\sqrt{m} + m, m\sqrt{n} + n)). \tag{3}$$

This bound is already asymptotically tight if $m \geq n^2$ or $n \geq m^2$, i.e. if the number of lines and points is highly "unbalanced". Let us treat the "balanced" case; for simplicity let $m = n$ and let n-element sets L and P be given. With foresight, we set $r = n^{1/3}$ and by Theorem 4.1 we find a $(1/r)$-cutting $\Delta_1, \ldots, \Delta_t$ for L with $t = O(r^2) = O(n^{2/3})$. Let P_i denote the points of P lying in Δ_i except for its vertices, and let L_i be the lines of L intersecting the interior of Δ_i. We have

$$\begin{array}{rcll} I(L,P) & \leq & \sum_{i=1}^{t} I(L_i, P_i) & (A) \\ & & +I(n, 3t) & (B) \\ & & +I(3t, n)\,. & (B') \end{array}$$

The term (B) covers the incidences of points lying at vertices of some Δ_i. The term (B') covers all cases when a line of L contains an edge of a triangle Δ_i but does not intersect its interior. By (3), terms (B) and (B') are both $O(t\sqrt{n} + n) = O(n^{7/6})$.

Concerning the main term (A), we use $|L_i| \leq n/r = n^{2/3}$ and $\sum_{i=1}^{t} |P_i| \leq 2n$ (since each point of P goes into at most two P_i). An average P_i has about $n^{1/3}$ points while L_i has about $n^{2/3}$ lines, so the subproblems are unbalanced enough for (3) to be tight. More precisely, applying (3) for each $I(L_i, P_i)$ we get $\sum_{i=1}^{t} |I(L_i, P_i)| \leq \sum_{i=1}^{t} O(|P_i| n^{1/3} + n^{2/3}) = O(n^{4/3})$, which proves the $n = m$ case of the Szemerédi-Trotter theorem.

References

[AAH⁺97] A. Andrzejak, B. Aronov, S. Har-Peled, R. Seidel, and E. Welzl. Results on k-sets and j-facets via continuous motions. Manuscript, 1997.

[AB92] M. Anthony and N. Biggs. *Computational Learning Theory.* Cambridge University Press, Cambridge, 1992.

[ABCBH93] N. Alon, S. Ben-David, N. Cesa-Bianchi, and D. Haussler. Scale-sensitive dimension, uniform convergence and learnability. In *Proc. 34th IEEE Symposium on Foundations of Computer Science*, pages 292–300, 1993.

[ABFK92] N. Alon, I. Bárány, Z. Füredi, and D. Kleitman. Point selections and weak ε-nets for convex hulls. *Combin., Probab. Comput.*, 1(3):189–200, 1992.

[AEG⁺91] B. Aronov, P. Erdős, W. Goddard, D. J. Kleitman, M. Klugerman, J. Pach, and L. J. Schulman. Crossing families. In *Proc. 7th Annu. ACM Sympos. Comput. Geom.*, pages 351–356, 1991.

[AEGS92] B. Aronov, H. Edelsbrunner, L. J. Guibas, and M. Sharir. The number of edges of many faces in a line segment arrangement. *Combinatorica*, 12(3):261–274, 1992.

[AHW87] N. Alon, D. Haussler, and E. Welzl. Partitioning and geometric embedding of range spaces of finite Vapnik-Chervonenkis dimension. In *Proc. 3rd Annu. ACM Sympos. Comput. Geom.*, pages 331–340, 1987.

[AK92] N. Alon and D. Kleitman. Piercing convex sets and the hadwiger debrunner (p,q)-problem. *Adv. Math.*, 96(1):103–112, 1992.

[Ale90] R. Alexander. Geometric methods in the theory of uniform distribution. *Combinatorica*, 10(2):115–136, 1990.

[AS93] N. Alon and J. Spencer. *The probabilistic method.* J. Wiley & Sons, 1993.

[AS95] P. K. Agarwal and M. Sharir. *Davenport-Schinzel Sequences and Their Geometric Applications.* Cambridge University Press, Cambridge, UK, 1995.

[Ass83] P. Assouad. Densite et dimension. *Ann. Inst. Fourier (Grenoble)*, 33:233–282, 1983.

[AT89] K. S. Alexander and M. Talagrand. The law of the iterated logarithm for empirical processes on Vapnik-Chervonenkis classes. *J. Multivariate Anal.*, 30:155–166, 1989.

[BC87] J. Beck and W. Chen. *Irregularities of distribution.* Cambridge University Press, 1987.

[BC95] P. G. Bradford and V. Capoyleas. Weak ϵ-nets for points on a hypersphere. Report MPI-I-95-1-029, Max-Planck-Institut Inform., Saarbrücken, Germany, 1995.

[BCM93] H. Brönnimann, B. Chazelle, and J. Matoušek. Product range spaces, sensitive sampling, and derandomization. In *Proc. 34th Annu. IEEE Sympos. Found. Comput. Sci.*, pages 400–409, 1993.

[BCP93] H. Brönnimann, B. Chazelle, and J. Pach. How hard is halfspace range searching. *Discrete Comput. Geom.*, 10:143–155, 1993.

[Bec81] J. Beck. Roth's estimate on the discrepancy of integer sequences is nearly sharp. *Combinatorica*, 1(4):319–325, 1981.

[Bec84] J. Beck. Some upper bounds in the theory of irregularities of distribution. *Acta Arith.*, 43:115–130, 1984.

[BEHW89] A. Blumer, A. Ehrenfeucht, D. Haussler, and M. Warmuth. Classifying learnable geometric concepts with the Vapnik-Chervonenkis dimension. *J. ACM*, 36:929–965, 1989.

[BL88] J. Bourgain and J. Lindenstrauss. Distribution of points on the sphere and approximation by zonotopes. *Isr. J. Math.*, 64:25–31, 1988.

[BL93] J. Bourgain and J. Lindenstrauss. Approximating the ball by a Minkowski sum of segments with equal length. *Discr. & Comput. Geom.*, 9:131–144, 1993.

[BLM89] J. Bourgain, J. Lindenstrauss, and V. Milman. Approximation of zonoids by zonotopes. *Acta Math.*, 162:73–141, 1989.

[BPR96] S. Basu, R. Pollack, and M.-F. Roy. On the number of cells defined by a family of polynomials on a variety. *Mathematika* 43,1:120–126, 1996.

[BS95] J. Beck and V. Sós. Discrepancy theory. In *Handbook of Combinatorics*, pages 1405–1446. North-Holland, Amsterdam, 1995.

[BVS+92] A. Björner, M. Las Vergnas, B. Sturmfels, N. White, and G. M. Ziegler. *Oriented matroids (Encyclopedia of Mathematics).* Cambridge University Press, 1992.

[CEG+90] K. Clarkson, H. Edelsbrunner, L. Guibas, M. Sharir, and E. Welzl. Combinatorial complexity bounds for arrangements of curves and spheres. *Discrete Comput. Geom.*, 5:99–160, 1990.

[CEG+95] B. Chazelle, H. Edelsbrunner, M. Grigni, L. Guibas, M. Sharir, and E. Welzl. Improved bounds on weak ϵ-nets for convex sets. *Discrete Comput. Geom.*, 13:1–15, 1995.

[CEGS89] B. Chazelle, H. Edelsbrunner, L. J. Guibas, and M. Sharir. Lines in space: combinatorics, algorithms, and applications. In *Proc. 21st Annu. ACM Sympos. Theory Comput.*, pages 382–393, 1989.

[CF90] B. Chazelle and J. Friedman. A deterministic view of random sampling and its use in geometry. *Combinatorica*, 10(3):229–249, 1990.

[Cha89] B. Chazelle. Lower bounds on the complexity of polytope range searching. *J. Amer. Math. Soc.*, 2:637–666, 1989.

[Cha93] B. Chazelle. Cutting hyperplanes for divide-and-conquer. *Discrete Comput. Geom.*, 9(2):145–158, 1993.

[Cha94] B. Chazelle. A spectral approach to lower bounds. In *Proc. 35th Ann. IEEE Symp. on Foundations of Computer Science*, pages 674–682, 1994.

[Cla87] K. L. Clarkson. New applications of random sampling in computational geometry. *Discrete Comput. Geom.*, 2:195–222, 1987.

[Cla93] K. L. Clarkson. Algorithms for polytope covering and approximation. In *Proc. 3rd Workshop Algorithms Data Struct.*, volume 709 of *Lecture Notes in Computer Science*, pages 246–252, 1993.

[CW89] B. Chazelle and E. Welzl. Quasi-optimal range searching in spaces of finite VC-dimension. *Discrete Comput. Geom.*, 4:467–489, 1989.

[DE94] T. K. Dey and H. Edelsbrunner. Counting triangle crossings and halving planes. *Discrete Comput. Geom.*, 12:281–289, 1994.

[Dey97] T. K. Dey. Improved bounds for planar k-sets and related problems. *Proc. 38th IEEE Sympos. Foundat. Comput. Sci.*, 1997, pages 156–161, 1997.

[DHS94] M. Kearns D. Haussler and R. Schapire. Bounds on the sample complexity of Bayesian learning using information theory and the VC dimension. *Machine Learning*, 14:84–114, 1994.

[DSW94] G.-L. Ding, P. Seymour, and P. Winkler. Bounding the vertex cover number of a hypergraph. *Combinatorica*, 14:23–34, 1994.

[Dud84] R. M. Dudley. *A course on empirical measures.* Lecture Notes in Math. 1097. Springer, 1984.

[Dud85] R. M. Dudley. The structure of some Vapnik-Chervonenkis classes. In *Proc. of Berkeley Conference in honor of Jerzy Neyman and Jack Kiefer, Volume II, (LeCam and Olshen, Eds.)*, pages 495–507, 1985.

[Ede87] H. Edelsbrunner. *Algorithms in Combinatorial Geometry*, volume 10 of *EATCS Monographs on Theoretical Computer Science.* Springer-Verlag, Heidelberg, West Germany, 1987.

[EGS90] H. Edelsbrunner, L. Guibas, and M. Sharir. The complexity of many cells in arrangements of planes and related problems. *Discrete Comput. Geom.*, 5:197–216, 1990.

[Eri96] J. Erickson. New lower bounds for convex hull problems in odd dimensions. In *Proc. 12th Annu. ACM Sympos. Comput. Geom.*, pages 1–9, 1996.

[Fra83] P. Frankl. On the trace of finite sets. *J. Comb. Theory, Ser. A*, 34:41–45, 1983.

[Gia97] A. Giannopoulos. On some vector balancing problems. *Studia Math.* 122,3:225–234, 1997.

[Glu89] E. D. Gluskin. Extremal properties of orthogonal parallelpipeds and their applications to the geometry of Banach spaces. *Math. USSR Sbornik*, 64(1):85–96, 1989.

[GW94] B. Gärtner and E. Welzl. Vapnik-Chervonenkis dimension and (pseudo-) hyperplane arrangements. *Discr. Comput. Geom.*, 12:399–432, 1994.

[GZ84] E. Giné and J. Zinn. Some limit theorems for empirical processes. *Annals of Probability*, 12:929–989, 1984.

[Hau92] D. Haussler. Decision theoretic generalizations of the PAC model for neural net and other learning applications. *Information and Computation*, 100(1):78–150, 1992.

[Hau95] D. Haussler. Sphere packing numbers for subsets of the Boolean n-cube with bounded Vapnik-Chervonenkis dimension. *Journal of Combinatorial Theory Ser. A*, 69:217–232, 1995.

[HW87] D. Haussler and E. Welzl. Epsilon-nets and simplex range queries. *Discrete Comput. Geom.*, 2:127–151, 1987.

[KM95a] G. Kalai and J. Matoušek. Guarding galleries where every point sees a large area. KAM Series (tech. Report) 95–293, Charles University, 1995. To appear in Israel J. Math.

[KM95b] M. Karpinski and A. Macintyre. Polynomial bounds for VC dimension of sigmoidal neural networks. In *Proceedings of the 27th Annual ACM Symposium on Theory of Computing*, pages 200–208, 1995.

[KPW92] J. Komlós, J. Pach, and G. Woeginger. Almost tight bounds for ϵ-nets. *Discrete Comput. Geom.*, 7:163–173, 1992.

[KRS96] J. Kollár, L. Rónyai, and T. Szabó. Norm-graphs and bipartite Turán numbers. *Combinatorica*, 16,3:399–406, 1996.

[KSP82] J. Komlós, E. Szemerédi, and J. Pintz. A lower bound for Heilbronn's problem. *J. London Math. Soc.*, 25(2):13–24, 1982.

[KW93] M. Karpinski and T. Werther. VC dimension and uniform learnability of sparse polynomials and rational functions. *SIAM Journal on Computing*, 22(6):1276–1285, 1993.

[Lit88] N. Littlestone. Learning quickly when irrelevant attributes abound: A new linear-threshold algorithm. *Machine Learning*, 2:285–318, 1988.

[Mat92] J. Matoušek. Efficient partition trees. *Discrete Comput. Geom.*, 8:315–334, 1992.

[Mat93] J. Matoušek. Range searching with efficient hierarchical cuttings. *Discrete Comput. Geom.*, 10(2):157–182, 1993.

[Mat97] J. Matoušek. On discrepancy bounds via dual shatter function. *Mathematika*, 44:42–49, 1997.

[Mat95a] J. Matoušek. Approximations and optimal geometric divide-and-conquer. *J. of Computer and System Sciences*, 50:203–208, 1995.

[Mat95b] J. Matoušek. Tight upper bounds for the discrepancy of halfspaces. *Discr. & Comput. Geom.*, 13:593–601, 1995.

[Mat95c] J. Matoušek. Geometric range searching. *ACM Comput. Surveys*, 26:421–461, 1995.

[Mat96a] J. Matoušek. Derandomization in computational geometry. *J. Algorithms*, 20:545–580, 1996.

[Mat96b] J. Matoušek. Improved upper bounds for approximation by zonotopes. *Acta Math.* 177:55–73, 1996.

[Mil64] J. Milnor. On the Betti numbers of real algebraic varieties. *Proc. Amer. Math. Soc.*, 15:275–280, 1964.

[MR95] R. Motwani and P. Raghavan. *Randomized Algorithms*. Cambridge University Press, 1995.

[MS96] J. Matoušek and J. Spencer. Discrepancy in arithmetic progressions. *J. Amer. Math. Soc.*, 9:195–204, 1996.

[Mul94] K. Mulmuley. *Computational Geometry: An Introduction Through Randomized Algorithms.* Prentice Hall, Englewood Cliffs, NJ, 1994.

[MWW93] J. Matoušek, E. Welzl, and L. Wernisch. Discrepancy and ε-approximations for bounded VC-dimension. *Combinatorica*, 13:455–466, 1993.

[Nie92] H. Niederreiter. *Random number generation and quasi-Monte Carlo methods.* CBMS-NSF Regional Conference Series in Applied Mathematics. 63. Philadelphia, PA: SIAM, 1992.

[Ole51] O.A. Oleinik. Estimates of the Betti numbers of real algebraic hypersurfaces (in Russian). *Mat. Sbornik (N. S.)*, 28(70):635–640, 1951.

[PA95] J. Pach and P. K. Agarwal. *Combinatorial Geometry.* John Wiley & Sons, New York, NY, 1995.

[Pac91] J. Pach. Geometric graphs. In J.E. Goodman, R. Pollack, and W. Steiger, editors, *Computational Geometry: papers from the DIMACS special year.* Amer. Math. Soc., 1991.

[Pac97] J. Pach. A remark on transversal numbers. In: The Mathematics of Paul Erdős, Vol. II, pages 310–317, Springer-Verlag, Berlin etc., 1997.

[PS] J. Pach and M. Sharin. On the number of incidences between points and curves. *Combinatorics, Probability, and Computing*, to appear.

[Pol90] D. Pollard. *Empirical processes: Theory and applications.* NSF-CBMS Regional Conference Series in Probability and Statistics vol. 2. Institute of Math. Stat. and Am. Stat. Assoc., 1990.

[PS85] F. P. Preparata and M. I. Shamos. *Computational Geometry: An Introduction.* Springer-Verlag, New York, NY, 1985.

[PSS92] J. Pach, W. Steiger, and E. Szemerédi. An upper bound on the number of planar k-sets. *Discrete Comput. Geom.*, 7:109–123, 1992.

[Rot64] K. F. Roth. Remark concerning integer sequences. *Acta Arithmetica*, 9:257–260, 1964.

[Rot76] K. Roth. Developments in the triangle problem. *Adv. Math.*, 22:364–385, 1976.

[RV96] K. Romanik and J. Vitter. Using Vapnik-Chervonenkis dimension to analyze the testing complexity of program segments. *Information and Computation* 128,2:87–108, 1996.

[Sau72] N. Sauer. On the density of families of sets. *Journal of Combinatorial Theory Ser. A*, 13:145–147, 1972.

[She72] S. Shelah. A combinatorial problem, stability and order for models and theories in infinitary languages. *Pacific J. Math.*, 41:247–261, 1972.

[Spe85] J. Spencer. Six standard deviations suffice. *Trans. Amer. Math. Soc.*, 289:679–706, 1985.

[Spe87] J. Spencer. *Ten lectures on the probabilistic method.* CBMS-NSF. SIAM, 1987.

[ST83] E. Szemerédi and W. T. Trotter, Jr. Extremal problems in discrete geometry. *Combinatorica*, 3:381–392, 1983.

[Szé] L. Székely. Crossing numbers and hard Erdős problems in discrete geometry. *Combinatorics, Probability, and Computing.*

[Tho65] R. Thom. Sur l'homologie des variétés algébriques reélles. In *Differential and Combinatorial Topology (S.S. Cairns ed.).* Princeton Univ. Press, 1965.

[Val95] P. Valtr. Guarding galleries where no point sees a small area. KAM Series (tech. Report) 95–305, Charles University, 1995.

[Vap82] V. N. Vapnik. *Estimation of dependences based on empirical data.* Springer-Verlag, Berlin, 1982.

[VC71] V. N. Vapnik and A. Ya. Chervonenkis. On the uniform convergence of relative frequencies of events to their probabilities. *Theory Probab. Appl.*, 16:264–280, 1971.

[Wel88] E. Welzl. Partition trees for triangle counting and other range searching problems. In *Proc. 4th Annu. ACM Sympos. Comput. Geom.*, pages 23–33, 1988.

[WF94] M. Watanabe and P. Frankl. Some best possible bounds concerning the traces of finite sets. *Graphs Comb.*, 10:283–292, 1994.

[Wil82] D. E. Willard. Polygon retrieval. *SIAM J. Comput.*, 11:149–165, 1982.

Progress in Mathematics, Vol. 169, © 1998 Birkhäuser Verlag Basel/Switzerland

Recent Developments in Symplectic Topology

DUSA MCDUFF*

State University of New York at Stony Brook
e-mail: dusa@math.sunysb.edu

1. Introduction

Symplectic topology explores the topological and geometric implications of the existence of a skew-symmetric form. As such, it is very relevant to different parts of mathematics, including dynamical systems, differential equations, representation theory and gauge theory. Thus the symplectic theme is itself a unifying idea in mathematics.

Until recently, very little was known about symplectic topology since there were no obvious invariants to get one's hands on. However, the past fifteen years have seen remarkable developments. Two early achievements were Conley-Zehnder's use of variational methods to solve the Arnold conjecture for the torus (in 1983) and Gromov's introduction of elliptic methods (in 1985) to prove the non-squeezing theorem. In a series of papers starting in 1988, Floer showed how to use elliptic methods to study the kind of variational problem that arises in the context of the Arnold conjecture. The concept of Floer homology that he introduced turns out to be crucial, not only in symplectic topology but also in gauge theory for manifolds with boundary.

In this talk we briefly describe the different mathematical ideas which have contributed to these developments, and then discuss some very recent applications.

1.1. Basic notions

Recall that a **symplectic structure** on an even-dimensional smooth manifold (M, ω) is a **closed, nondegenerate** 2-form ω. This definition has two parts. Closedness is a geometric property (we will see its first implications in §1.3.), while nondegeneracy is algebraic. It means that at every point $x \in M$ the induced map $T_xM \to T_x^*M$ from tangent to cotangent space given by

$$X \mapsto \iota_X\omega = \omega(X, \cdot)$$

is an isomorphism. The basic example is the linear form

$$\omega_0 = dp_1 \wedge dq_1 + \ldots + dp_n \wedge dq_n$$

on Euclidean space $\mathbb{R}^{2n}$ with coordinates $(p_1, q_1, \ldots, p_n, q_n)$. Note that under the above isomorphism the vector field $\frac{\partial}{\partial p_i}$ corresponds to the 1-form dq_i, and so the

*Partially supported by NSF grant DMS 9401443.

identification of tangent space with cotangent space differs from the usual one by a quarter turn.

THEOREM 1.1 (DARBOUX) *Every symplectic form on M is locally diffeomorphic to the standard form ω_0 on $\mathbb{R}^{2n}$.*

This result is characteristic of symplectic topology. It is well known that at a point (ie in the tangent space) there is only one symplectic form up to isomorphism. Darboux's theorem says that this uniqueness remains true near the point. Similarly, one can easily show that the symplectic structure of a neighborhood of a symplectic (or Lagrangian) submanifold Q is entirely determined by the restriction of the form ω to $TM|_Q$. (For proofs of this and other basic results, the reader can consult [1, 30], for example.) Many of the results of modern symplectic topology (including the Arnold conjecture) can be understood as saying that local properties in fact hold globally. This is in marked contrast to Riemannian geometry where there are many local invariants derived from curvature. Nevertheless, it was Gromov's insight that there are important connections between symplectic and Riemannian geometry via almost complex structures.

1.2. Almost complex structures Recall that an **almost complex structure** J is an automorphism of the tangent bundle (over the identity) such that $J^2 = -\mathrm{Id}$. J is said to be **compatible** with ω if the form

$$g(x, y) := \omega(x, Jy)$$

is symmetric and positive definite. It is not hard to prove that the space $\mathcal{J}_\omega$ of ω-compatible almost complex structures on (M, ω) is nonempty and contractible. (This is equivalent to the statement that the homogeneous space $\mathrm{Sp}(2n, \mathbb{R})/\mathrm{U}(n, \mathbb{C})$ is contractible, where $\mathrm{Sp}(2n, \mathbb{R})$ is the linear symplectic group, and $\mathrm{U}(n, \mathbb{C})$ denotes the unitary group.)

As an example consider a Kähler manifold such as complex projective space $\mathbb{C}P^n$, with the triple (ω, J, g) where ω is a Kähler form, J is the almost complex structure induced by the underlying complex structure on the manifold and g is the corresponding Kähler metric. Here J is integrable, but for a general symplectic manifold it will not be. As Gromov realised, for many purposes this does not matter.

1.3. Hamiltonian flows Another important fact about symplectic geometry is that symplectic manifolds have many symmetries: more precisely, the group $\mathrm{Symp}(M, \omega)$ of symplectomorphisms[1] is always infinite dimensional. To see this, recall that if

$$H : M \to \mathbb{R}$$

is a function on M then H generates a 1-parameter subgroup of $\mathrm{Symp}(M)$ (called the **Hamiltonian flow** of H) as follows. Starting with the 1-form dH construct the corresponding vector field X_H given by:

$$\iota_{X_H}\omega = \omega(X_H, \cdot) = dH.$$

[1]Recall that a symplectomorphism is a diffeomorphism ϕ such that $\phi^*(\omega) = \omega$.

The flow $\phi_t^H, t \in \mathbb{R}$ of this vector field preserves ω because the Lie derivative $\mathcal{L}_{X_H}$ satisfies:

$$\mathcal{L}_{X_H}\omega = d(\iota_{X_H}\omega) + \iota_{X_H}d\omega = ddH = 0.$$

(Note that here it is essential that ω be closed.[2]) It is also easy to see that the function H, which is often called the **energy function** or **Hamiltonian**, is preserved by the flow, i.e.

$$H \circ \phi_t^H = H.$$

This corresponds to the classical principle of **conservation of energy**. In fact, symplectic geometry was first studied because the time flows of classical mechanical systems are the Hamiltonian flows of functions H on standard Euclidean space $(\mathbb{R}^{2n}, \omega_0)$. This is one reason why symplectic structures are ubiquitous: many of the equations of interest today have a physical origin which is reflected in the fact that their solutions preserve some symplectic structure.

In the above construction it is also possible to allow the Hamiltonian H to depend on t. The corresponding flow $\phi_t = \phi_t^H$ then satisfies the equation

$$\frac{d}{dt}\phi_t(x) = X_t(\phi_t(x)),$$

where the tangent vector field X_t is given by

$$\iota_{X_t}\omega = dH_t.$$

Such flows are also called Hamiltonian flows. Just as in the physical situation, their trajectories can be described in terms of variational principles, such as the principle of least action.

1.4. The Flux homomorphism We define

$$\mathrm{Ham}(M) = \mathrm{Ham}(M, \omega)$$

to be the set of all symplectomorphisms ϕ that are the time 1-map of some Hamiltonian flow (generated by a time dependent Hamiltonian). It is not hard to see that the elements of $\mathrm{Ham}(M, \omega)$ form a group. In fact, $\mathrm{Ham}(M, \omega)$ is a normal subgroup of $\mathrm{Symp}_0(M, \omega)$, the identity component of the group of symplectomorphisms: viz.

$$\mathrm{Ham}(M, \omega) \subset \mathrm{Symp}_0(M, \omega).$$

When M is simply connected, these two groups coincide, but this is not true in general. For example, the translation $(x, y) \mapsto (x + t, y)$ of $\mathbb{R}^2$ is generated by the function $H(x, y) = y$, but when considered as a map of the 2-torus it is not Hamiltonian. Rather, it is generated by the closed 1-form dy, which is of course no longer exact.

[2]Here, as elsewhere, we will assume for simplicity that M is compact. Most results can be extended to the noncompact case by considering functions, symplectomorphisms etc of compact support. Note also that we always suppose that M together with all objects defined on it are C^∞ smooth.

To explain the difference in general we must pass to the universal cover $\widetilde{\mathrm{Symp}}_0(M)$ of $\mathrm{Symp}_0(M)$. The elements of this space consists of paths $\{\phi_t\}_{t\in[0,1]}$ starting at the identity in $\mathrm{Symp}_0(M)$ which are well-defined modulo homotopies fixing the endpoints. The Flux homomorphism

$$\mathrm{Flux} : \widetilde{\mathrm{Symp}}_0(M) \to H^1(M;\mathbb{R})$$

is defined as follows. Given $\gamma \in H_1(M)$,

$$\mathrm{Flux}(\{\phi_t\})(\gamma) = \int_{\phi_{[0,1]}(\gamma)} \omega,$$

where $\phi_{[0,1]}(\gamma) \in H_2(M)$ is represented by the chain

$$[0,1] \times S^1 \to M : \quad (s,t) \mapsto \phi_s(\gamma(t)).$$

It is not hard to check that the path $\{\phi_t\}$ is homotopic with fixed endpoints to a path in $\mathrm{Ham}(M)$ if and only if

$$\mathrm{Flux}(\{\phi_t\}) = 0.$$

In other words there is an exact sequence

$$\widetilde{\mathrm{Ham}}(M) \to \widetilde{\mathrm{Symp}}_0(M) \stackrel{\mathrm{Flux}}{\to} H^1(M;\mathbb{R}),$$

where $\widetilde{\mathrm{Ham}}(M)$ is the universal cover of $\mathrm{Ham}(M)$.[3] This descends to an exact sequence

$$\mathrm{Ham}(M) \to \mathrm{Symp}_0(M) \to H^1(M;\mathbb{R})/\Gamma$$

where $\Gamma = \mathrm{Flux}(\pi_1(\mathrm{Symp}_0(M)))$. It follows that $\mathrm{Ham}(M)$ is C^1-closed in $\mathrm{Symp}_0(M)$ if and only if Γ is a discrete subgroup (lattice) in $H^1(M;\mathbb{R})$.

One of the first major achievements of symplectic topology was to show that the group $\mathrm{Symp}(M)$ is C^0-closed in the group of all diffeomorphisms. This means that symplectic diffeomorphisms may be distinguished from ordinary diffeomorphisms by their topological properties, and is really the foundation stone of symplectic topology. By analogy, one can make the following conjecture.

CONJECTURE 1.1 (FLUX CONJECTURE) *The group Γ is always discrete.*

Notice that if $[\omega]$ is an integral cohomology class, the homomorphism Flux takes integral values, and the discreteness of Γ is obvious. However, it is not at all clear whether Γ is always discrete. One could of course go further and conjecture:

CONJECTURE 1.2 (C^0 FLUX CONJECTURE) *The group $\mathrm{Ham}(M,\omega)$ is C^0-closed in $\mathrm{Symp}_0(M,\omega)$.*

At present this is known in special cases (such as for the standard torus, where it is an easy consequence of Arnold's conjecture), but not in general. If this were true it would mean that the elements of the Hamiltonian subgroup could be distinguished by their topological properties, and one could try to understand what their special topological features are.

[3]This was first proved by Calabi in [5]: see also Banyaga [2].

2. Arnold's conjecture

Arnold's celebrated conjectures, which were made in the mid 60s, have been one of the main motivating problems for much of the recent development of symplectic topology. For simplicity, we will concern ourselves only with the nondegenerate case, and will state the conjecture in its simplest form. (Recall that a fixed point x of a diffeomorphism ϕ is called nondegenerate if the derivative $d\phi_x$ does not have 1 as an eigenvalue.)

CONJECTURE 2.1 (ARNOLD) *If $\phi \in \mathrm{Ham}(M)$ has only nondegenerate fixed points, then the number $\#\mathrm{Fix}(\phi)$ of its fixed points should satisfy the following estimate:*

$$\#\mathrm{Fix}(\phi) \geq \sum_i \dim H^i(M;\mathbb{R}).$$

As motivation for this conjecture, consider the following points.

- In dimension 2, there are classical results due to Poincaré, Birkhoff and others which describe situations in which symplectic (ie area-preserving) maps have more fixed points than do ordinary diffeomorphisms.
- If ϕ is generated by a time-independent Hamiltonian H, then clearly the critical points of H give rise to zeros of the tangent vector field X_H and hence to fixed points of ϕ. Therefore Arnold's estimate follows from the Morse inequalities for H.
- If ϕ is the time 1-map of a Hamiltonian flow ϕ_t that remains C^1-close to the identity, then it is not hard to prove Arnold's conjecture.

Thus we can think of this conjecture as claiming that a phenomenon (existence of "many" fixed points) that holds locally near the identity element extends globally, to all elements of $\mathrm{Ham}(M)$. This conjecture was proved in the 2-dimensional case by Eliashberg in the 1970s by very geometric methods. More germane to our story is the first proof in higher dimensions [6]:

THEOREM 2.2 (CONLEY-ZEHNDER, 1983) *The Arnold conjecture holds for the standard torus $(\mathbb{T}^{2n} = \mathbb{R}^{2n}/\mathbb{Z}^{2n}, \omega_0)$.*

Their proof reduces the problem to a variational one, which they then solve using methods (such as Fourier series) which are special to Euclidean space. Here is a sketch of their idea. A more detailed survey (with more references) of many of the ideas presented here may be found in [28].

2.1. The loop space and action functional Let $\mathcal{L}$ denote the space of all contractible loops $x : S^1 \to M$ on M, and suppose that ϕ is the time 1-map of the flow generated by $H_t, t \in \mathbb{R}$. It is not hard to arrange for H_t to be 1-periodic: i.e. $H(t+1) = H(t)$. Then, it is easy to check that the fixed points of $\phi = \phi_1$ correspond precisely to the periodic orbits of the flow ϕ_t, at least when $\pi_1(M) = 0$. (In the general case, this method only takes into account those fixed points x for which the loop $\phi_t(x)$ is contractible.)

Define the **action functional**

$$a_H : \mathcal{L} \to M$$

by:

$$a_H(x) = \int_{D^2} \hat{x}^*(\omega) - \int_{S^1} H_t(x(t))dt,$$

where $\hat{x} : D^2 \to M$ extends $x : \partial D^2 \to M$.[4] Then it is not hard to check that the critical points of a_H on $\mathcal{L}$ are in $1-1$ correspondence with the periodic orbits of H_t and hence to $\mathrm{Fix}(\phi)$. Hence one only has to find a way to count the critical points of a_H. This is not a trivial matter since all (nondegenerate) fixed points have infinite index and coindex, so that usual Morse theoretic methods do not apply. However, as mentioned above, special methods apply when M (or its universal cover) is Euclidean space. These methods exploit the linear structure of $\mathbb{R}^{2n}$ and generalize to cotangent bundles. They were developed further by many authors, for example Viterbo, Ekeland-Hofer, Hofer-Zehnder. For detailed references, see Hofer-Zehnder [15].

3. Gromov's nonsqueezing theorem

The next key event was the publication of Gromov's 1985 paper [12] which introduced elliptic methods into symplectic geometry. Gromov proved many wonderful theorems in this paper. Here we content ourselves with one of them. It concerns the question of what can happen to the standard Euclidean ball $B^{2n}(r)$ under a local symplectomorphism of $\mathbb{R}^{2n}$. By the cylinder $B^2(R) \times \mathbb{R}^{2n-2}$ we mean the subset of $(\mathbb{R}^{2n}, \omega_0)$ defined by $p_1^2 + q_1^2 \le R^2$.

THEOREM 3.1 (GROMOV, 1985) *There is a symplectic embedding of the standard ball $B^{2n}(r)$ of radius r into the cylinder $B^2(R) \times \mathbb{R}^{2n-2}$ only if $r \le R$.*

Note that this result distinguishes symplectic from volume preserving diffeomorphisms since, when $n \ge 2$ it is clearly possible to find a volume preserving embedding of $B^{2n}(r)$ into an arbitrarily thin cylinder.

The nonsqueezing theorem actually holds for all symplectic manifolds (M, ω). In other words:

THEOREM 3.2 (LALONDE-MCDUFF [18]) *If (M, ω) is any symplectic manifold of dimension $2n$, there is a symplectic embedding of the standard ball $B^{2n+2}(r)$ of radius r into the cylinder $(B^2(R) \times M, dp \wedge dq \oplus \omega)$ only if $r \le R$.*

Observe that here (in contrast to the rest of this article) we make no assumption on (M, ω). In fact, if M is closed or if ω is tame at infinity, this result follows by adapting the argument of Gromov that we explain briefly below. All that is needed is to extend the theory of J-holomorphic curves to arbitrary symplectic manifolds, which has now been done, thanks to the recent work of Fukaya-Ono [10] and Li-Tian [22]. However, to extend the result to completely arbitrary M one needs a rather delicate embedding argument.

[4] Note that this is independent of the choice of extension $\hat{x}$ if and only if $\omega = 0$ on the image of the Hurewicz homomorphism $\pi_2 \to H_2$, which is the case for $M = \mathbb{T}^{2n}$.

3.1. Applications of the nonsqueezing theorem This deceptively simple result is in fact strong enough to permit the characterization of symplectomorphisms. To do this, it is convenient to use the Ekeland-Hofer idea of a **capacity** which was introduced in [7]. Roughly speaking, this is a symplectically invariant way of measuring the size of a subset of a symplectic manifold. There are many possible capacities. Here we will use the Gromov area $c(U)$. For $U \subset (M^{2n}, \omega)$,

$$c(U) = \sup\{\pi r^2 : B^{2n}(r) \text{ embeds symplectically in } U\}.$$

It is not hard to see that any linear map L of $\mathbb{R}^{2n}$ that preserves capacity of ellipsoids must either be symplectic or anti-symplectic (ie $L^*(\omega_0) = \pm\omega_0$.) Further, by rescaling, it is easy to see that if ψ is a diffeomorphism of $\mathbb{R}^{2n}$ that preserves capacity, i.e. if

$$c(\psi(U)) = c(U), \qquad \text{for all } U,$$

then the derivative $d\psi_x$ of ψ at any point x preserves capacity. Hence we find:

PROPOSITION 3.1 (EKELAND-HOFER) *Every orientation preserving diffeomorphism of (M, ω) that preserves capacity is symplectic.*

Proof: If n is odd, this follows immediately by combining the above remarks with Darboux's theorem. If n is even, consider $\mathrm{id} \times \phi$ on $\mathbb{R}^2 \times \mathbb{R}^{2n}$. □

Because capacity is a C^0-invariant, we also find:

PROPOSITION 3.2 (ELIASHBERG [8], EKELAND-HOFER [7])
The group $\mathrm{Symp}(M)$ *is C^0-closed in* $\mathrm{Diff}(M)$.

This result is known as **symplectic rigidity**. It is quite unexpected since the defining equations for a symplectomorphism (namely that $\phi^*\omega = \omega$) involve the first derivatives of ϕ. It means that there are interesting topological invariants of symplectic structures and spaces, and so is a crucial foundation for symplectic topology.

It is also very interesting to look for other objects whose size can be measured in a symplectically invariant way. One of the most fruitful ideas has been that of the **Hofer norm** of a Hamiltonian symplectomorphism [13][5] which is defined to be

$$\|\phi\| = \inf \int_0^1 \left(\max_{x \in M} H_t(x) - \min_{x \in M} H_t(x) \right) dt,$$

where the infimum is taken over all (compactly supported) Hamiltonians H_t with time-1 map ϕ. One can think of $\|\phi\|$ as the minimal energy needed to generate ϕ. It is related to the capacity defined above by the **energy-capacity inequality**:

PROPOSITION 3.3 (HOFER) *If ϕ is a compactly-supported symplectomorphism of $\mathbb{R}^{2n}$ such that $\phi(U) \cap U = \emptyset$ for some $U \subset \mathbb{R}^{2n}$, then*

$$\|\phi\| \geq c(U).$$

[5] A related norm defined using generating functions was proposed by Viterbo in [41].

Note that this result is new even in the case $n = 1$. It also extends (in a slightly modified form) to any manifold: see Lalonde-McDuff [18]. Using this inequality, it is not hard to show that the function

$$\rho(\phi, \psi) = \|\psi^{-1}\phi\|$$

is a biinvariant metric on the group $\mathrm{Ham}(M)$. Its geometric properties (particularly its geodesics) have been explored in Bialy-Polterovich [3] and Lalonde-McDuff [19]. One would like to use this norm to investigate the dynamics of ϕ. Recently there has been some success in this area for special classes of maps ϕ: see Bialy-Polterovich [4] in connection with KAM-theory and Siburg [37] in connection with Aubry-Mather theory.

This norm is still rather little understood. In particular, it is not known whether $\mathrm{Ham}(S^2)$ has bounded Hofer diameter. (Note that one can use the energy-capacity inequality to show that $\mathrm{Ham}(\Sigma)$ has infinite diameter for any other oriented closed surface: see [19].) Added in proof: Polterovich [33] has recently shown that $\mathrm{Ham}(S^2)$ has unbounded diameter.

3.2. ***J*-holomorphic curves** There are now many different proofs of the non-squeezing theorem. In fact, it is such a central result that it can be proved by any of the new methods in symplectic topology, whether they be variational (Conley-Zehnder, Viterbo [40], Ekeland-Hofer, Hofer-Zehnder [15]), use generating functions (Viterbo [41]), or use elliptic analysis (Gromov). We now sketch Gromov's approach using J-holomorphic curves and elliptic geometry. More details can be found in [28, 29]. Given an ω-compatible almost complex structure J on (M, ω) Gromov considered the associated parametrized J-holomorphic curves:

$$u : (\Sigma, j) \to (M, J) \quad du \circ j = J \circ du.$$

Here (Σ, j) is a fixed Riemann surface, often a sphere or disc. Since the defining equation for these curves is a generalized Cauchy-Riemann equation, it is elliptic. Moreover, the fact that J is compatible with the closed form ω implies that, when Σ is closed, the energy (or $W^{1,2}$-norm) $E(u)$ of u is bounded by topological data:

$$E(u) = \langle \omega, A \rangle$$

where $A \in H_2(M, \mathbb{Z})$ is the homology class $u_*[\Sigma]$ of $\mathrm{Im}\, u$. It follows that, for generic J, the space of such curves in class A is a finite-dimensional manifold. Moreover, Gromov proved an important compactness theorem which allows one to understand the compactification of these moduli spaces, and to show that in many cases J-holomorphic curves persist under deformation of J.[6]

In the case at hand, what Gromov proved is the following. Let

$$\psi : B^{2n}(r) \to B^2(R) \times \mathbb{R}^{2n-2}(R)$$

[6]Note that Gromov developed a deformation theory. Results on the *existence* of nontrivial families of J-holomorphic curves have taken longer, and are sofar only known in 4-dimensions, where one uses Taubes-Seiberg-Witten theory [39]. The existence of these curves has many interesting consequences for the structure of symplectic 4-manifolds: see Liu [24], for example.

be a symplectic embedding, and pick an ω_0-tame J on the cylinder which is standard near the boundary and which coincides with the pushforward $\psi_*(J_0)$ on the image of the ball. Because, when J is the product almost complex structure, there is a J-holomorphic disc of the form

$$z \mapsto (z, x_0)$$

through every point (z_0, x_0) of the cylinder, Gromov showed that there is such a disc through every point for an arbitrary J. Moreover, one can arrange that these discs all have symplectic area πR^2. (Strictly speaking, one should compactify the cylinder to $S^2 \times \mathbb{R}^{2n-2}$, and consider A-curves where A is the class of $[S^2 \times \mathrm{pt}]$ so that $\omega(A) = \pi R^2$.) In particular, there is a J-holomorphic disc C through the point $\psi(0)$ for our special J. Let S be the pullback via ψ of this disc C. Then, by choice of J, S is a properly embedded curve in the ball which goes through the origin 0 and is holomorphic in the usual sense. Hence it is a minimal surface with respect to the usual metric. By the monotonicity theorem for minimal surfaces, the area of S has to be at least πr^2. But, because S is holomorphic, this area is just the integral of ω_0 over S. Hence:

$$\pi r^2 \leq \int_S \omega_0 = \int_{\psi(S)} \omega < \int_C \omega = \pi R^2.$$

4. Floer theory and the Arnold conjecture

Floer theory combines the elliptic with the variational approach. Here we will just give a very brief sketch of the main ideas. For simplicity, we will assume that ω vanishes on $\pi_2(M)$ so that the action functional is well-defined. In general, one has to pass to a suitable covering of $\mathcal{L}$ and then the algebra becomes more involved: see Hofer-Salamon [14].

As we pointed out in § 2.1., the action functional a_H is defined in such as way that its critical points on the loop space $\mathcal{L}$ correspond bijectively to periodic orbits of the flow ϕ_t of H_t and hence to (a subset of) the fixed point set of the time 1 map ϕ. Floer's idea is to try to estimate the number of critical points of a_H using Morse theory. To do this, one should put a Riemannian metric on $\mathcal{L}$, consider the associated gradient vector field of a_H and then build a complex out of its flow lines. If one can show that the homology of this complex is independent of the choice of metric, then estimations of its size give lower bounds for the number of critical points of a_H.

The problem, of course, is that the loop space $\mathcal{L}$ is infinite dimensional so that the gradient flow may not exist globally on $\mathcal{L}$. Floer showed that, nevertheless, the elliptic nature of the gradient flow equation for a_H allows the construction of a suitable complex. Here the choice of metric on $\mathcal{L}$ is crucial, since it heavily influences the analytic properties of the gradient flow equation. The appropiate metric here is the most natural choice in the context of Gromov theory. One chooses an ω-compatible J on M. As explained in §1.2., this gives rise to a Riemannian metric g_J on M and hence to an inner product on $\mathcal{L}$ defined by:

$$\langle \alpha, \beta \rangle = \int_0^1 g_J(\alpha(t), \beta(t))dt, \quad \alpha, \beta \in \mathcal{L}.$$

(To a first approximation, one can think of $\mathcal{L}$ as made up of C^∞-paths, but to carry out the details of the proofs one has to complete it in a suitable Sobolev norm.)

Paths in $\mathcal{L}$ may be thought of as maps

$$u : \mathbb{R} \times S^1 \to M : \quad (\tau, t) \mapsto u(\tau, t),$$

and it is not hard to check that the gradient flow equation for a_H is precisely

$$\frac{\partial u}{\partial \tau} + J(u)\frac{\partial u}{\partial t} = -\mathrm{grad}H(u, t).$$

Note that the left-hand side of this equation is given by the action of the Cauchy-Riemann operator on u. Hence this can be thought of as a perturbed Cauchy-Riemann equation. Although arbitrary trajectories of this flow might behave quite wildly, the important fact is that trajectories of **bounded energy** behave well. Here one defines the energy $E(u)$ to be the square of the L^2-norm of the first derivative of u, that is:

$$E(u) = \int_{\mathbb{R}\times S^1} \left(\|\frac{\partial u}{\partial \tau}\|^2 + \|\frac{\partial u}{\partial t}\|^2 \right) d\tau dt.$$

To be more precise, Floer showed that for generic pairs J, H:

- There are finitely many critical points of a_H (corresponding to periodic orbits of the flow of H_t);
- Each critical point $x \in \mathrm{Crit}(a_H)$ has infinite index and coindex. However one can define the **relative** index $\mu(x, y) \in \mathbb{Z}$ of any pair.
- Each gradient flow line u of finite energy converges at its two ends to critical points $x_{\pm\infty}$, that is
$$\lim_{\tau\to\pm\infty} u(\tau, t) = x_{\pm\infty}.$$
We can think of such a trajectory as flowing from $x_{-\infty}$ to x_∞, and will denote the set of all such trajectories by $\mathcal{M}(x_{-\infty}, x_\infty)$.
- For each pair of critical points x, y of relative index $\mu(x, y) \geq 0$ the space $\mathcal{M}(x, y)$ is a smooth manifold of dimension $\mu(x, y)$ with a finite number of connected components.

In this situation one can define the **Floer complex** as follows. The underlying chain group $CF^*(\phi)$ is the free abelian group with generators $[x] : x \in \mathrm{Crit}(a_H)$. The boundary operator ∂ is given by

$$\partial([x]) = \sum_{[y]:\mu(x,y)=1} n(x, y)[y],$$

where $n(x, y)$ is the number of components of $\mathcal{M}(x, y)$ counted with appropriate sign. Note first that one can ignore the sign at the expense of working over $\mathbb{Z}/2\mathbb{Z}$ instead of over $\mathbb{Z}$. Second, observe that the elements of $\mathcal{M}(x, y)$ are parametrized trajectories, and that the group $\mathbb{R}$ acts as a natural reparametrization group via $u(\tau, t) \mapsto u(\tau + \sigma, t)$. Hence when $\dim \mathcal{M}(x, y) = 1$ each component of $\mathcal{M}(x, y)$ consists of a single trajectory. Hence $n(x, y)$ can be described as the number of isolated trajectories going from x to y. The important point is that $\partial^2 = 0$, i.e. ∂ is a boundary operator. The proof involves a detailed study of the compactification of the trajectory spaces $\mathcal{M}(x, y)$, and goes significantly beyond Gromov's compactness theorem since one needs to show how to create trajectories by gluing two of them end to end. Once this is established, one can define the **Floer cohomology** $HF^*(\phi)$ to be the homology of this chain complex:

$$HF^*(\phi) = H(CF^*(\phi), \partial) = \frac{\ker \partial}{\mathrm{Im} \partial}.$$

As the notation suggests, one can show that this homology is independent of the choice of J and of H. In fact, it is even independent of ϕ (provided that this is in $\mathrm{Ham}(M)$.[7]) Floer showed in 1989 that in the situation considered here, that is when $\omega = 0$ on $\pi_2(M)$,

$$HF^*(\phi) = H^*(M, \mathbb{Z}).$$

The Arnold conjecture follows almost immediately from this, since one has

$$\#\mathrm{Fix}(\phi) \geq \#\mathrm{Crit}(a_H) = \mathrm{rank}\, CF^*(\phi) \geq \mathrm{rank}\, HF^*(\phi) = \sum_i \dim H^i(M; \mathbb{R}).$$

For more details of this argument see Floer [9] or [28].

The situation gets significantly more complicated when ω does not vanish on $\pi_2(M)$ since in this case there can be J-holomorphic spheres and one must define Floer homology to have coefficients in a suitable Novikov ring: see Hofer-Salamon [14]. Over the years many authors have made significant contributions to this problem, especially Hofer-Salamon and Ono [31]. The last step involves understanding the possible contributions of multiply-covered curves which create orbifold singularities in the compactified moduli space. Inspired by ideas of Kontsevich [16], this (very technical) problem has now been solved completely by Fukaya-Ono [10] and Liu-Tian [25]. (See also Ruan [34], Siebert [38], Li-Tian[22].) Thus we have

THEOREM 4.1 (FUKAYA-ONO, LIU-TIAN) *The Arnold conjecture for nondegenerate $\phi \in \mathrm{Ham}(M)$ holds for all compact symplectic manifolds (M, ω).*

The degenerate case, plus versions in which one estimates the number of fixed points by means of the minimal number of critical points of a function on M, is still not known in all cases.

There are several other very interesting situations in which one can define Floer homology, most notably in the setting of gauge theory for 4-dimensional manifolds with boundary. The most recent development in this connection is Floer-Seiberg-Witten theory (see Kronheimer-Mrowka [17], Marcolli [26], Wang [42] and also Lisca-Matic [23]), which is closely connected with the contact geometry of 3-manifolds.

[7] One can in fact define $HF^*(\phi)$ for more general manifolds M and for any symplectomorphism ϕ. Le-Ono [21], for example, treats the case when ϕ is isotopic to the identity. Observe also that if one applies the same procedure using a pair (X, f) instead of $(\mathcal{L}, a_H)$, where f is a Morse function on the closed manifold X, then the associated Floer homology is precisely the homology of X. A detailed proof of this may be found in Schwarz [35].

5. Floer homology and the Flux group

Let us look at the Flux homomorphism in a slightly different way. Given a loop $\phi_t \in Maps(M, M), t \in S^1$, define the map

$$\partial_\phi : H_i(M) \to H_{i+1}(M)$$

to be the map on homology which is induced by the map $c \mapsto \phi_{[0,1]}(c)$ on cycles, where, as above, if c is given by $f : Z \to M$, the cycle $\phi_{[0,1]}(c)$ is given by

$$S^1 \times Z \to M : \quad (t, z) \mapsto \phi_t(f(z)).$$

Further, define $\partial_\phi^* : H^i(M) \to H^{i-1}(M)$ to be the dual of ∂_ϕ. Then it is easy to check that ∂_ϕ^* is a derivation with respect to the cup product. Moreover, our definitions immediately imply that

$$\mathrm{Flux}(\{\phi_t\}) = \partial_\phi^*([\omega]).$$

Recall that a symplectic manifold (M, ω) is said to be monotone if the homomorphism $\pi_2(M) \to \mathbb{R}$ induced by integrating the form ω is a positive multiple of that induced by the first Chern class c_1 of (TM, J).

PROPOSITION 5.1 (LALONDE-MCDUFF-POLTEROVICH) *If (M, ω) is monotone, then the loop $\{\phi_t\}$ is isotopic to a loop in* $\mathrm{Ham}(M)$ *if and only if $\partial_\phi : H_*(M) \to H_*(M)$ is identically zero.*

This is a strong rigidity result for Hamiltonian loops. A priori, to say that a loop is Hamiltonian only means that the derivation ∂_ϕ^* vanishes on the one class $[\omega]$. The proposition says that if it vanishes on this one class, it must vanish on them all. In particular, this means that the elements of $\pi_1(\mathrm{Ham}(M))$ are invariant under perturbation of ω. More precisely, given a loop $\{\phi_t\}$ in $\mathrm{Symp}(M, \omega)$ and a symplectic form ω' sufficiently close to ω, it is very easy to perturb $\{\phi_t\}$ (by isotopy) to a loop $\{\phi_t'\}$ in $\mathrm{Symp}(M, \omega')$. The above result implies that $\{\phi_t\}$ is ω-Hamiltonian, if and only if $\{\phi_t'\}$ is ω'-Hamiltonian. Since Γ is discrete whenever $[\omega] \in H^2(M, \mathbf{Q})$ we find:

COROLLARY 5.2 *When (M, ω) is monotone, the rank of the Flux group*

$$\Gamma = \mathrm{Flux}(\pi_1(\mathrm{Symp}_0(M, \omega))$$

is lower semi-continuous with respect to $[\omega]$ and is always $\leq \dim H^1(M, \mathbf{Q})$.

REMARK 5.3

(i) The argument sketched below which proves Proposition 5.1 almost surely works without the assumption of monotonicity. However the details have not yet been fully checked.

(ii) Seidel pointed out that if (M, ω) is Kähler and $\{\phi_t\}$ is holomorphic so that P_ϕ is also Kähler, then Proposition 5.1 holds by a result of Deligne without the assumption of monotonicity: see [11, Ch 3.5].

(iii) It is an elementary fact that the identity component of the closure of Γ in $H^1(M,\mathbb{R})$ is contained in the kernel of the map $\wedge[\omega]^{n-1} : H^1 \to H^{2n-1}$. Apparently much deeper is the fact that Γ is discrete whenever the restriction of $[\omega]$ to the image of the Hurewicz homomorphism $\pi_2(M) \to H_2(M,\mathbb{Z})$ takes values in a discrete subgroup of $\mathbb{R}$. The proof (given in [20]) follows easily from the weak Arnold conjecture, which states that for every Hamiltonian isotopy ϕ_t there is at least one point $x \in M$ that is fixed by ϕ_1 and is such that the loop $\{\phi_t(x)\}_{t\in[0,1]}$ is contractible.

The proof of Proposition 5.1 uses recent ideas of Seidel and Piunikhin-Salamon-Schwarz connecting Floer homology to quantum cohomology. The idea, first of all, is to look at the associated fibration

$$P_\phi = M \times D^+ \cup_\phi M \times D^- \quad \to S^2,$$

which is obtained by thinking of the loop $\{\phi_t\}$ as a clutching map. In other words, the point $(x, e^{-2\pi it})$ on $M \times \partial D^-$ is identified with $(\phi_t(x), e^{2\pi it})$ on $M \times \partial D^+$. It is not hard to see that:

Lemma 5.4 *The derivation ∂_ϕ is the connecting homomorphism in the Wang sequence*

$$\cdots \to H_{i+2}(P_\phi) \stackrel{\cap M}{\to} H_i(M) \stackrel{\partial_\phi}{\to} H_{i+1}(M) \to H_{i+1}(P_\phi) \to \cdots$$

Corollary 5.5 *There is a symplectic form Ω on P_ϕ that restricts to ω on each fiber if and only if $\{\phi_t\}$ is in the kernel of the Flux homomorphism.*

Following Seidel, we now choose an Ω-compatible J on P_ϕ such that the projection $\pi : P_\phi \to S^2$ is J-holomorphic, and then look at the moduli space $\mathcal{M}$ of all J-holomorphic sections of P_ϕ. If we pick a point z_0 in S^2 and identify the fiber over z_0 with M, the evaluation map

$$ev_P : \mathcal{M} \times S^2 \to P_\phi, \quad (u, z) \mapsto u(z)$$

composes with the intersection map $\cap M$ to give rise to a map

$$ev_M : \mathcal{M} \to M, \quad u \mapsto u(z_0).$$

Therefore, in order to show that $\partial_\phi \equiv 0$, it suffices to prove:

Lemma 5.6 *When M is monotone, the evaluation map $ev_M : \mathcal{M} \to M$ induces a surjection on homology.*

This crucial result follows almost immediately from Seidel's work in [36]. The proof constructs elements of $\mathcal{M}$ by using the gluing techniques for J-holomorphic discs which are described in Piunikhin-Salamon-Schwarz [32]. The essential point here is that, although there is no sensible gluing theory for J-holomorphic discs with free boundary, one can construct such a theory if one perturbs the discs near their boundaries so that they become Floer trajectories converging to a periodic solution of a generic Hamiltonian.

References

[1] V.I. Arnold and A. Givental, Symplectic topology. In *Dynamical systems*, Vol. IV. (1990), Springer-Verlag, Berlin.

[2] A. Banyaga, Sur la structure du groupe des difféomorphisms qui preservent une forme symplectique, *Comm. Math. Helv.* **53** (1978), 174–227.

[3] M. Bialy and L. Polterovich, Geodesics of Hofer's metric on the group of Hamiltonian diffeomorphisms, *Duke Math. J.* **76** (1994), 273–292.

[4] M. Bialy and L. Polterovich, Invariant tori and Symplectic Topology, *Amer. Math. Soc. Transl.*, 2 **171** (1996), 23–33.

[5] E. Calabi, On the group of automorphisms of a symplectic manifold, *Problems in Analysis* ed Gunning, Princeton Univ. Press (1970), 1–26.

[6] C. Conley and E. Zehnder, The Birkhoff-Lewis fixed point theorem and a conjecture of V.I. Arnold, *Inventiones Mathematicae*, **73**, (1983) 33–49.

[7] I. Ekeland and H. Hofer, Symplectic topology and Hamiltonian dynamics. *Mathematische Zeitschrift*, **200** (1989), 355–78.

[8] Y. Eliashberg, A theorem on the structure of wave fronts and its applications in symplectic topology. *Functional Analysis and Applications*, **21**, (1987) 65–72.

[9] A. Floer, Symplectic fixed points and holomorphic spheres. *Communications in Mathematical Physics*, **120** (1989), 575–611.

[10] K. Fukaya and K. Ono, Arnold conjecture and Gromov–Witten invariants, preprint (1996).

[11] P. Griffiths and J. Harris, *Principles of Algebraic Geometry*, Wiley (1978).

[12] M. Gromov, Pseudo holomorphic curves in symplectic manifolds, *Inventiones Mathematicae*, **82** (1985), 307–47.

[13] H. Hofer, On the topological properties of symplectic maps. *Proceedings of the Royal Society of Edinburgh*, **115** (1990), 25–38.

[14] H. Hofer and D.A. Salamon, Floer homology and Novikov rings. In *Floer Memorial volume*, Birkhäuser, Basel, (1996).

[15] H. Hofer and E. Zehnder, *Symplectic capacities and Hamiltonian Dynamics*, Birkhäuser, Basel, (1994).

[16] M. Kontsevich, Enumeration of Rational Curves by Torus Action, in *Moduli spaces of surfaces* ed Dijkgraaf, Faber and v.d.Geer (1995), Birkhäuser, 335–368.

[17] P. Kronheimer and T. Mrowka, Monopoles and contact structures, preprint (1996).

[18] F. Lalonde and D. McDuff, The geometry of symplectic energy, *Annals of Mathematics*, **141** (1995), 349–371.

[19] F. Lalonde and D. McDuff, Hofer's L^∞ geometry: geodesics and stability, I, II. *Invent. Math.* **122** (1995), 1–33, 35–69.

[20] F. Lalonde, D. McDuff and L. Polterovich, Topological rigidity of Hamiltonian loops and quantum homology, preprint (1997).

[21] H.Y. Le and K. Ono, Symplectic fixed points, Calabi invariant and Novikov homology, *Topology* **34** (1995), 155–176.

[22] J. Li and G. Tian, Virtual moduli cycles and Gromov–Witten invariants of general symplectic manifolds, preprint (1996).

[23] P. Lisca and G. Matic, Tight contact structures and the Seiberg–Witten invariants, *Invent. Math.* **129** (1997), 509–525.

[24] A. Liu, Some new applications of general wall crossing formula, Math. Res. Letters **3** (1996), 569–585.

[25] G. Liu and G. Tian, Floer homology and Arnold conjecture, preprint (1996).

[26] M. Marcolli, Seiberg–Witten–Floer homology and Heegaard splittings, *Intern. Journ. of Math.* **7** (1996), 671–696.

[27] D. McDuff, Symplectic diffeomorphisms and the flux homomorphism, *Inventiones Mathematicae*, **77**, (1984), 353–66.

[28] D. McDuff, Elliptic methods in symplectic geometry, *Bulletin of the American Mathematical Society*, **23** (1990), 311–58.

[29] D. McDuff and D.A. Salamon, *J-holomorphic curves and quantum cohomology.* University Lecture Series, American Mathematical Society, Providence, RI, (1994).

[30] D. McDuff and D.A. Salamon, *Introduction to Symplectic Topology*, OUP, Oxford, (1995).

[31] K. Ono, On the Arnold conjecture for weakly monotone symplectic manifolds, *Invent. Math.* **119** (1995), 519–537.

[32] S. Piunikhin, D. Salamon and M. Schwarz, Symplectic Floer–Donaldson theory and quantum cohomology, Proceedings of the 1994 Newton Institute Conference, CUP, Cambridge (1996).

[33] L. Polterovich, Hofer's Diameter and Lagrangian intersections, preprint (1998).

[34] Y. Ruan, Virtual neighborhood and monopole equations, preprint (1996), alg-geom/9611021.

[35] M. Schwarz, *Morse Homology*, Birkhäuser, Basel (1995).

[36] P. Seidel, π_1 of symplectic automorphism groups and invertibles in quantum cohomology rings, *Geom. and Funct. Anal.* **7** (1997), 1046–1095.

[37] K.F. Siburg, Bi-invariant metrics for symplectic twist mappings on $T^*(T^{2n})$ and an application in Aubry–Mather theory, preprint (1996).

[38] B. Siebert, Gromov–Witten invariants for general symplectic manifolds, preprint (1996).

[39] C.H. Taubes, The Seiberg–Witten and the Gromov invariants, *Math. Research Letters* **2** (1995), 221–238.

[40] C. Viterbo, A proof of the Weinstein conjecture in $\mathbb{R}^{2n}$. *Annales de l' Institut Henri Poincaré – Analyse nonlinéaire*, **4** (1987), 337–57.

[41] C. Viterbo, Symplectic topology as the geometry of generating functions. *Mathematische Annalen*, **292** (1992), 685–710.

[42] B.-L. Wang, Seiberg–Witten–Floer theory for homology 3-spheres, preprint (1996).

Progress in Mathematics, Vol. 169, © 1998 Birkhäuser Verlag Basel/Switzerland

K-Theory and Algebraic Groups

A.S. Merkurjev

Dept. of Mathematics
University of California at Los Angeles
405 Hilgard ave.
Los Angeles, California, 90095-1555, USA

The present paper is devoted to the description of certain interactions between the algebraic K-theory and the theory of algebraic groups. It consists of two parts. In the first part we consider an application of the algebraic K-theory to the rationality problem for algebraic groups. In the second part, conversely, we show how the theory of algebraic groups allows to compute K-groups of algebraic varieties with relatively large automorphism group.

In the first section we recall basic definitions and results on rationality and stable rationality of algebraic group varieties. In the next section we define the group of R-equivalence classes, an important birational invariant of an algebraic group. This notion was introduced by Manin in [18] and studied for linear algebraic groups in [10] by Colliot-Thélène and Sansuc. In particular, the group of R-equivalence classes for algebraic tori was computed in this paper. We review this computation and also the stable birational classification of algebraic tori in section 3.

By a classical result of Weil [49], an adjoint simple group of classical series is the connected component of the automorphism group of some algebra with involution (except for some "non-classical" groups of the type D_4). In section 4 we give a computation of the group of R-equivalence classes for such algebraic groups. The result is given in terms of certain invariants of the corresponding algebra with involution.

In the following sections we consequently consider the classical types A, B, C and D of simply connected and adjoint absolutely simple groups. In some cases we compute the group of R-equivalence classes and prove the (stable) rationality of group varieties. For types A and D we give sufficient conditions for corresponding groups to be not stably rational. The algebraic K-theory comes out, for example, as the group $\mathrm{SK}_1(A)$, being the group of R-equivalence classes of $\mathbf{SL}_1(A)$, where A is a central simple algebra. The proof of Theorem 5.4 is based on Rost's computation of the group SK_1 for a biquaternoin algebra, which, in its turn, uses in particular Swan's computation of the higher algebraic K-theory of a projective quadric hypersurface. In the proof of Theorems 5.14 and 8.9 the general index reduction formula is used (see [26]), which is based on the computation of the algebraic K-theory of certain projective homogeneous varieties given by Panin in [30].

By Corollary 1.8, algebraic groups of rank at most 2 are rational. In section 9, the complete stably rational classification of all absolutely simple groups of types $A_3 = D_3$ (rank 3) is given.

In the second part of the paper we consider the problem of the computation of the algebraic K-groups for algebraic varieties on which a "large" algebraic groups acts. The K-groups $K'_n(X)$ of the category of coherent sheaves on an algebraic variety X are known at least for the following classes of varieties:

1. Homogeneous projective varieties [30].
2. Varieties of simply connected semisimple algebraic groups and principal homogeneous spaces over these groups [17], [29].
3. Toric varieties and toric models [27].

In all these cases there is a reductive algebraic group acting non-trivially on a variety. The basic objects of the second part of the present paper are reductive algebraic group G, defined over an arbitrary field F, and a G-scheme X, i.e. a scheme X over F, on which G acts. The problem we study here is to compare the equivariant K-groups $K'_n(G;X)$ of the category of G-modules on X and the ordinary ones $K'_n(X)$ of the category of coherent sheaves on X via the *restriction homomorphism*

$$\text{res} \; : K'_n(G;X) \to K'_n(X).$$

It turns out that sometimes the group $K'_n(G;X)$ is easy to compute. For example, if $X = G/H$ is a homogeneous variety with the natural action of G, where $H \subset G$ is a subgroup, then the category of G-modules on X is equivalent to the category of finite dimensional representations of the group H. Since in the latter category any object has finite length, the computation of the K-groups of this category reduces to the computation of the K-groups of the endomorphism rings of simple objects. For example, $K'_0(G;G/H) = R(H)$ is the representation ring of H, which is a free abelian group generated by the classes of irreducible representations of H. Therefore, the knowledge of the kernel and cokernel of the restriction homomorphism sheds some light on the structure of $K'_n(X)$.

In the second part of the paper we review the main results of [24]. In section 10 we give certain basic definitions and results in the theory of algebraic groups. In particular, we give an algebraic definition of the fundamental group of an arbitrary algebraic group, which is not based on the choice of a maximal torus. In section 11 we introduce equivariant algebraic K-theory developed by Thomason in [42]. In the following section for a split reductive group G with $\pi_1(G)$ torsion free and any G-scheme X a spectral sequence with the E^2-term related to the equivariant K-groups of X, which converges to the ordinary K-groups of X, is given (Theorem 12.1). In particular, the restriction homomorphism induces an isomorphism

$$\mathbf{Z} \otimes_{R(G)} K'_0(G;X) \to K'_0(X).$$

For example,

$$K_0(G/H) \simeq \mathbf{Z} \otimes_{R(G)} R(H)$$

for any subgroup $H \subset G$.

In section 13 we consider the case of reductive groups of inner type. We describe the kernel of the restriction homomorphism in terms of Tits algebras of the group G. At the end of the section we compute the group $K_0(G)$ for any adjoint semisimple group G of inner type.

A simple argument (see the beginning of section 14) shows that the surjectivity of the restriction homomorphism for $n = 0$ and all G-schemes X implies that the Picard group $\mathrm{Pic}(G \otimes_F E)$ is trivial for any finite field extension E/F (we call groups G with this property *factorial*). It turns out that the converse statement holds for quasi-projective varieties (Theorem 14.1): If G is factorial and X is quasi-projective, then the restriction homomorphism is surjective for $n = 0$. In particular, for any subgroup $H \subset G$, the group $K_0(G/H)$ is generated by the classes of vector bundles $(G \times V)/H$ over G/H for all linear representations $H \to \mathbf{GL}(V)$ (Corollary 14.2). Note that for a projective homogeneous variety X under a simply connected semisimple group G the surjectivity of of the restriction homomorphism has been proven by I. Panin in [31, Th. 2].

The last section is devoted to some applications.

We use the following notation. For a central simple algebra A over F by $\mathrm{Nrd} : A^\times \to F^\times$ we denote the reduced norm homomorphism. The degree $\deg(A)$ is the square root of the dimension of A and index $\mathrm{ind}(A)$ is the square root of the dimension of the division algebra similar to A. For a field extension E/F the algebra $A \otimes_F E$ over E is denoted by A_E.

$\mathrm{Br}(F)$ is the Brauer group of a field F.

We use the word *scheme* for a separated scheme of finite type over a field. If X is a scheme over F and E/F is a field extension, then the scheme $X \otimes_F E$ we simply denote by X_E and the set of E-points $\mathrm{Mor}_{\mathrm{Spec}\, F}(\mathrm{Spec}\, E, X)$ – by $X(E)$. For a morphism of schemes $\alpha : X \to Y$ over F, the map of sets of points $X(E) \to Y(E)$ is denoted by α_E.

If E/F is a finite separable field extension and Y is an algebraic variety defined over a field E, then $R_{E/F}(Y)$ is the variety over F, obtained from Y by the restriction of scalars.

$\mathbf{A}_F^n$ is the affine space over F.

An *algebraic group* is a smooth affine group scheme of finite type over a field. The one-dimensional split torus $\mathrm{Spec}\, F[t, t^{-1}]$ is denoted by $\mathbf{G}_{m,F}$ and the additive group $\mathrm{Spec}\, F[t]$ by $\mathbf{G}_{a,F}$.

For an algebraic group G over F by $H^1(F, G)$ we denote the pointed set $H^1\Big(\mathrm{Gal}(F_{\mathrm{sep}}/F), G(F_{\mathrm{sep}})\Big)$ ([39]).

1. Rational and stably rational algebraic groups

Let F be an arbitrary field. An irreducible algebraic variety X, defined over F, is called *rational* if X is birationally isomorphic to an affine space, or equivalently, the function field $F(X)$ is a purely transcendental extension of F. An irreducible algebraic variety X, defined over F, is called *stably rational* if the variety $X \times_F \mathbf{A}_F^n$ is rational for some n.

If G is a connected algebraic group, defined over a field F, we simply say that G is (stably) rational if the variety of G is (stably) rational. It is still unknown if there exists a stably rational group which is not rational.

The following well-known statement is often useful.

LEMMA 1.1. *Let $1 \to H \to G \xrightarrow{\varphi} T \to 1$ be an exact sequence of algebraic groups defined over F. Assume that for any field extension E/F the map $H^1(E,H) \to H^1(E,G)$ has trivial kernel. Then the sequence splits rationally, i.e. there exists a rational morphism $\psi : T \to G$, such that $\varphi \circ \psi = \mathrm{id}_T$. In particular, G is birationally isomorphic to the product $H \times T$.*

Proof. Let $E = F(T)$ be the function field of T. It follows from the exactness of the sequence

$$G(E) \to T(E) \to H^1(E,H) \to H^1(E,G)$$

that the generic point in $T(E)$ belongs to the image of $G(E) \to T(E)$, hence the sequence splits rationally. □

A connected solvable group G over F is called *split* if there is a series of subgroups $G = G_0 \supset G_1 \supset \cdots \supset G_n = 1$ such that G_i/G_{i+1} is isomorphic either to $\mathbf{G}_{a,F}$ or to $\mathbf{G}_{m,F}$ for $i = 0, 1, \ldots, n-1$ (see [4, Def. 15.1]). Since $H^1(F, \mathbf{G}_{a,F}) = 1 = H^1(F, \mathbf{G}_{m,F})$, Lemma 1.1 gives

COROLLARY 1.2. ([4, Cor. 15.8]) *A connected solvable split group is rational.* □

Example 1.3 A connected unipotent group over a perfect field splits ([4, Cor. 15.5(ii)]) and hence is rational.

An algebraic torus T over F is called *quasi-trivial* if it is isomorphic to $\mathbf{GL}_1(C)$ for some étale F-algebra C. Since T is an open subset in the affine space of C, it is a rational variety.

A reductive group G over F is called *split* (resp. *quasi-split*) if it has a maximal torus which splits over F (resp. a Borel subgroup defined over F) (see [4, Th. 18.7]).

PROPOSITION 1.4. ([4]) *A split reductive group is rational.*

Proof. Let B be a Borel subgroup in G defined over F, B^- be the opposite Borel subgroup, U (resp. U^-) the unipotent redical of B (resp. of B^-), T a maximal torus in B defined over F. Then the product morphism

$$U \times T \times U^- \to G$$

is a birational isomorphism and U, U^- and T are split solvable groups. □

Remark 1.5 The statement of the Proposition 1.4 is still valid for a quasi-split semisimple simply connected or adjoint group, since in this case T is a quasi-trivial torus.

If F is a perfect field, then the unipotent radical U of G is defined over F. Since U splits, by induction $H^1(E,U) = 1$ for any field extension E/F and, by

Lemma 1.1, the sequence $1 \to U \to G \to G/U \to 1$ splits rationally, hence G is birationally isomorphic to the product $U \times (G/U)$, where G/U is a reductive group. If G/U splits (e.g. if F is algebraically closed), then it is rational by Proposition 1.4. Since U and G/U are split over an algebraically closed field, we get

PROPOSITION 1.6. ([9]) *Any connected algebraic group over an algebraically closed field is rational.* □

Denote by $\mathcal{T}$ the variety of maximal tori in G. It is proved in [9, Prop.3] (in characteristic zero), [12, Exp.XIV, Th. 6.2] and [5, Th. 7.9] that $\mathcal{T}$ is a rational variety and if G is reductive, then the function field $F(G)$ is isomorphic to the function field of the generic maximal torus T in G defined over $F(\mathcal{T})$. Hence we get

PROPOSITION 1.7. *If all maximal tori in a reductive group G are rational, then G is rational.* □

The *rank* of a reductive group G is the dimension of a maximal torus in G. Since all algebraic tori of dimension at most 2 are rational ([46, Th.4.74]), we get the following

COROLLARY 1.8. *Any reductive group of rank at most 2 is rational.* □

The first example of a non-rational algebraic group (3-dimensional torus) was constructed by Chevalley in [9]. In [37] Rosenlicht has given an example of a 1-dimensional unipotent group G over an infinite field with finite group of rational points. Clearly, G is not rational.

Serre has found an example of a semisimple group which does not satisfy the weak approximation property (and hence is not rational, see [39]). Note that the weak approximation holds for simply connected and adjoint groups over global fields (see [35]). The rationality problem over local fields was studied in [8].

An example of a simply connected semisimple non-rational group was given by Platonov in [32]. The first example of adjoint semisimple non-rational group was constructed in [23].

2. R-equivalence for algebraic groups

In this section we define an important birational invariant of an algebraic group: the group of R-equivalence classes. The notion of R-equivalence was introduced by Manin in [18] and studied for algebraic groups by Colliot-Thélène and Sansuc in [10].

Let F be a field. Denote by $\mathcal{O}$ the semilocal subring in the field of rational functions $F(t)$ consisting of all functions defined at points $t = 0$ and $t = 1$. Evaluation at these points gives two F-algebra homomorphisms $e_i : \mathcal{O} \to F$, $i = 0, 1$.

Let X be an algebraic variety defined over a field F. For any element $x(t) \in X(\mathcal{O})$ and $i = 0, 1$ denote by $x(i)$ the *value of $x(t)$ at i*, i.e. the image of $x(t)$ under the map
$$X(e_i) : X(\mathcal{O}) \to X(F).$$

Let G be an algebraic group defined over F. An element $g \in G(F)$ is called *R-trivial* if there is $g(t) \in G(\mathcal{O})$ such that $g(0) = 1$ and $g(1) = g$. The set of all R-trivial elements in $G(F)$ we denote by $RG(F)$.

LEMMA 2.1. *$RG(F)$ is a normal subgroup in $G(F)$.*

Proof. Let $g_{1,2} \in RG(F)$. Choose $g_{1,2}(t) \in G(\mathcal{O})$ such that $g_{1,2}(0) = 1$, $g_{1,2}(1) = g_{1,2}$ and consider $g(t) = g_1(t) \cdot g_2(t)^{-1} \in G(\mathcal{O})$. Then $g(0) = 1$ and $g(1) = g_1 g_2^{-1}$, hence $g_1 g_2^{-1} \in RG(F)$, i.e. $RG(F)$ is a subgroup in $G(F)$. Let $g \in G(F)$ and $g' \in RG(F)$. Choose $g'(t) \in G(\mathcal{O})$ such that $g'(0) = 1$ and $g'(1) = g'$ and consider $g''(t) = g \cdot g'(t) \cdot g^{-1} \in G(\mathcal{O})$. Then $g''(0) = 1$ and $g''(1) = gg'g^{-1}$, hence $gg'g^{-1} \in RG(F)$ and $RG(F)$ is a normal subgroup in $G(F)$. □

We simply denote the factor-group $G(F)/RG(F)$ by $G(F)/R$ and call it the *group of R-equivalence classes.* An algebraic group G, defined over F, is called *R-trivial* if $G(E)/R = 1$ for any field extension E/F.

The relation between R-triviality and stable rationality of an algebraic group is given by the following

PROPOSITION 2.2. ([10]) *If a connected algebraic group G, defined over a field F, is stably rational, then the group G is R-trivial.*

Proof. If G is stably rational over F, then it is stably rational over an arbitrary field extension of F. Hence, it is sufficient to show that $G(F)/R$=1. Assume first that G is rational. The case of a finite field F is considered in [10, Corollary 6]. Over an infinite field there is an open subset $U \subset G$ with $U(F) \neq \emptyset$ isomorphic to an open subset V in some affine space $\mathbf{A}^n_F$. Let $x \in G(F)$ be any rational point. Consider two translations U_1 and U_2 of U containing points x and 1 respectively. Choose any rational point y in the intersection $U_1 \cap U_2$. Translating the intersections with V of two appropriate straight lines from $\mathbf{A}^n_F$ to U_1 and U_2 we get elements $g(t), h(t) \in G(\mathcal{O})$ such that $g(0) = h(0) = y$, $g(1) = x$ and $h(1) = 1$. Then for $f(t) = g(t)h(t)^{-1}$ we have: $f(0) = 1$ and $f(1) = x$, hence $x \in RG(F)$.

If G is stably rational then $G \times_F \mathbf{A}^n_F$ is rational for some $n \in \mathbf{N}$. Since $\mathbf{A}^n_F$ is a rational algebraic group, it follows from the first part of the proof that $G(F)/R = G(F)/R \times \mathbf{A}^n_F(F)/R = (G \times_F \mathbf{A}^n_F)(F)/R = 1$. □

Let $\alpha : G \to H$ be an algebraic group homomorphism over a field F. It induces homomorphisms

$$\alpha_F : G(F) \to H(F) \quad \text{and} \quad \alpha_{\mathcal{O}} : G(\mathcal{O}) \to H(\mathcal{O}).$$

Let $g \in RG(F)$, so that there exists $g(t) \in G(\mathcal{O})$ such that $g(0) = 1$ and $g(1) = g$. Denote by $h(t)$ the element $\alpha_{\mathcal{O}}\big(g(t)\big) \in H(\mathcal{O})$. Clearly, $h(0) = \alpha_F\big(g(0)\big) = \alpha_F(1) = 1$ and $h(1) = \alpha_F\big(g(1)\big) = \alpha_F(g)$, hence, $\alpha_F(g) \in RH(F)$. In other words, f_F takes $RG(F)$ to $RH(F)$ and therefore induces a group homomorphism $G(F)/R \to H(F)/R$. Thus, the correspondence $G \mapsto G(F)/R$ induces a functor from the category of algebraic groups over F to the category of (abstract) groups.

3. Stably birational classification of algebraic tori

Let L/F be a finite Galois field extension with the Galois group Π, T an algebraic torus, defined over a field F and split by L. Denote by T^* the Π-module of characters

$$\mathrm{Hom}_L(T_L, \mathbf{G}_{m,L}).$$

It is a free abelian group of finite rank.

A Π-module M, being a free abelian group of finite rank, we call a Π-*lattice*. A Π-lattice M is a *permutation module* if there there is a **Z**-base of M which is invariant under Π. A torus T is quasi-trivial iff T^* is a permutation Π-module. Two Π-modules M and N are called *stably isomorphic*, if $M \oplus P_1 \simeq N \oplus P_2$ for some permutation Π-modules $P_{1,2}$. The monoid (with respect to the direct sum) of stable isomorphism classes of Π-lattices we denote by $C(\Pi)$.

Let M be a Π-lattice. An exact sequence of Π-lattices

$$0 \to M \to P \to N \to 0$$

is called a *flasque resolution of* M if P is a permutation Π-module and $H^{-1}(\Pi', N) = 0$ for any subgroup $\Pi' \subset \Pi$. A flasque resolution always exists and the stable isomorphism class of the module N in $C(\Pi)$ is uniquely determined by M (see [10]). We denote it by $p(M)$.

THEOREM 3.1. (Colliot-Thélène – Sansuc [10], Voskresenskiǐ [46]) *Let T_1 and T_2 be two algebraic tori split by L/F. They are stably birationally isomorphic iff $p(T_1^*) = p(T_2^*) \in C(\Pi)$. A torus T, split by L/F, is stably rational iff $p(T^*) = 0 \in C(\Pi)$.* □

The group of R-equivalence classes of a torus T can be also computed by using a flasque resolution

$$0 \to T^* \to P \to N \to 0$$

of the Π-module of characters T^*.

PROPOSITION 3.2. (Colliot-Thélène – Sansuc [10]) *The group of R-equivalence classes $T(F)/R$ is naturally isomorphic to $H^1\big(\Pi, S(L)\big)$, where S is a torus, such that $S^* \simeq N$.* □

The notion of R-triviality for algebraic tori is very close to the stable rationality. It is interesting whether a similar statement holds for reductive groups.

PROPOSITION 3.3. (Colliot-Thélène – Sansuc [10]) *Let T be an algebraic tori split by a finite Galois extension L/F, $\Pi = \mathrm{Gal}(L/F)$. Then the following conditions are equivalent*

(1) *T is R-trivial;*
(2) *There is a torus S, such that $T \times S$ is a rational torus;*
(3) *$p(T^*)$ is an invertible element in $C(\Pi)$.* □

4. Semisimple classical groups

Let F be any field of characteristic different from 2, Z be either a field F or a quadratic etale extension of F (not necessarily a field). Consider an Azumaya algebra A over Z with involution σ such that F coincides with the subfield of σ-invariant elements in Z. Of course, A is a central simple Z-algebra if Z is a field. By Wedderburn theorem, $A \simeq \mathrm{End}_D(V)$ for some skewfield D and a vector space V over D. There is an ε-hermitian form h on V over D with respect to some involution on D, such that σ is *adjoint to* h, i.e.

$$h\big(a(v), u\big) = h\big(v, \sigma(a)(u)\big)$$

for all $a \in A$ and $u, v \in V$ [14]. The involution σ is called *isotropic* (resp. *hyperbolic*), if the hermitian form h is isotropic.

An element $a \in A^\times$ is called a *similitude* if $\sigma(a)a \in F^\times$. Denote by $\mathrm{Sim}(A, \sigma)$ the group of all similitudes. For any $a \in \mathrm{Sim}(A, \sigma)$ the element $\mu(a) = \sigma(a)a$ in $F^\times$ is called the *similarity factor* of the similitude a. Moreover, we have a group homomorphism

$$\mu : \mathrm{Sim}(A, \sigma) \to F^\times.$$

The image of μ we denote by $G(A, \sigma)$.

An element $a \in A^\times$ is called an *isometry* if $\sigma(a)a = 1$. The group of all isometries $\mathrm{Iso}(A, \sigma)$ coincides with the kernel of μ.

We consider the groups $\mathrm{Sim}(A, \sigma)$ and $\mathrm{Iso}(A, \sigma)$ as the groups of F-points of the corresponding algebraic groups $\mathbf{Sim}(A, \sigma)$ and $\mathbf{Iso}(A, \sigma)$. The latter algebraic group is the kernel of the algebraic group homomorphism

$$\mu : \mathbf{Sim}(A, \sigma) \to \mathbf{G}_{m,F}.$$

Since $Z^\times$ is in the center of $\mathrm{Sim}(A, \sigma)$, it follows that the torus $R_{Z/F}(\mathbf{G}_{m,Z})$ is a central subgroup in $\mathbf{Sim}(A, \sigma)$. The group of *projective similitudes* is the factor-group $\mathbf{Sim}(A, \sigma)/R_{Z/F}(\mathbf{G}_{m,Z})$ which we denote by $\mathbf{PSim}(A, \sigma)$. By Hilbert's theorem 90, the group of F-points of $\mathbf{PSim}(A, \sigma)$ equals

$$\mathbf{PSim}(A, \sigma)(F) = \mathrm{Sim}(A, \sigma)/Z^\times.$$

For any $a \in A^\times$ the inner automorphism $\mathrm{Int}(a)$ of A commutes with σ iff $a \in \mathrm{Sim}(A, \sigma)$. By Skolem-Noether theorem, the correspondence $a \mapsto \mathrm{Int}(a)$ induces an isomorphism of the group $\mathrm{Sim}(A, \sigma)/Z^\times$ and the group $\mathrm{Aut}_Z(A, \sigma)$ of Z-automorphisms of A commuting with σ.

Denote by $\mathbf{Sim}_+(A, \sigma)$, $\mathbf{PSim}_+(A, \sigma)$, $\mathbf{Aut}_+(A, \sigma)$ and $\mathbf{Iso}_+(A, \sigma)$ the connected components of the identity in the corresponding algebraic groups. We have the canonical isomorphism

$$\mathbf{PSim}_+(A, \sigma) \simeq \mathbf{Aut}_+(A, \sigma).$$

A relation between classical semisimple groups and algebras with involutions is given by the following classical

THEOREM 4.1. (Weil, [49]) *Any adjoint absolutely simple classical algebraic group, defined over F, is isomorphic to $\mathbf{PSim}_+(A,\sigma)$ for a suitable algebra with involution (A,σ).* □

Denote the subgroup $\mu_F\big(\mathbf{Sim}_+(A,\sigma)(F)\big) \subset F^\times$ by $G_+(A,\sigma)$. In the case, when A splits, σ is adjoint to a symmetric bilinear form (quadratic form) q, $G_+(A,\sigma) = G(A,\sigma) = G(q)$ is the group of similarity factors of q.

Denote by $\mathrm{Hyp}(A,\sigma)$ the subgroup in $F^\times$ generated by the norms in all finite extensions E/F, such that σ_E is a hyperbolic involution.

The following Theorem computes the group $\mathbf{PSim}_+(A,\sigma)$ modulo R-equivalence.

THEOREM 4.2. ([23]) *There is a natural isomorphism*

$$\mathbf{PSim}_+(A,\sigma)(F)/R \simeq G_+(A,\sigma)/\Big(NZ^\times \cdot \mathrm{Hyp}(A,\sigma)\Big).\square$$

The group of isometries behaves much better.

LEMMA 4.3. ([49]) *The group $\mathbf{Iso}_+(A,\sigma)$ is rational.*

Proof. The Cayley transformation $a \mapsto (1-a)/(1+a)$ establishes a birational isomorphism between $G = \mathbf{Iso}_+(A,\sigma)$ and the affine space of all skew-symmetric elements in A with respect to σ (the Lie algebra of G). □

Remark 4.4 In the case A splits, $A = \mathrm{End}(V)$, and σ is adjoint to a quadratic form q, the group $\mathbf{Iso}_+(A,\sigma)$ is isomorphic to the special orthogonal group $\mathbf{O}_+(q)$, so that the latter group is rational.

At the end of this section we prove the following useful technical result.

PROPOSITION 4.5. *Let $1 \to H_i \to G_i \xrightarrow{f^i} T \to 1$, $i = 1,2$, be two exact sequences of connected algebraic groups. Assume that for any field extension E/F the images of f^1_E and f^2_E coincide in $T(E)$. Then the groups $H_1 \times G_2$ and $H_2 \times G_1$ are birationally isomorphic.*

Proof. Let $G = G_1 \times_T G_2$, $E = F(G_1)$ and $x \in G_1(E)$ be the generic point. By assumption, $f^1_E(x)$ belongs to the image of f^2_E. By the universal property of the fibred product, x belongs to the image of $G(E) \to G_1(E)$, i.e. the exact sequence $1 \to H_2 \to G \to G_1 \to 1$ splits rationally and G is birationally isomorphic to $H_2 \times G_1$. Analogously, G is birationally isomorphic to $H_1 \times G_2$. □

COROLLARY 4.6. *Let G be a connected stably rational algebraic group over a field F and $\alpha : G \to \mathbf{G}_{m,F}$ be a homomorphism defined over F. Assume that*
(1) The kernel of α is a connected stably rational algebraic group defined over F.
(2) For any field extension E/F the image of α_E in $E^\times$ equals $G_+(A_E,\sigma_E)$.
Then the variety of the group $\mathbf{PSim}_+(A,\sigma)$ is stably rational.

Proof. We can apply Proposition 4.5 to the following exact sequences

$$1 \to \ker(\alpha) \to G \to \mathbf{G}_{m,F} \to 1, \quad 1 \to \mathbf{Iso}_+(A,\sigma) \to \mathbf{Sim}_+(A,\sigma) \xrightarrow{\mu} \mathbf{G}_{m,F} \to 1$$

and conclude that $\mathbf{Sim}_+(A,\sigma)$ is stably rational by Lemma 4.3. By Lemma 1.1 and Hilbert's theorem 90, the exact sequence

$$1 \to R_{Z/F}(\mathbf{G}_{m,Z}) \to \mathbf{Sim}_+(A,\sigma) \to \mathbf{PSim}_+(A,\sigma) \to 1$$

splits rationally, hence the claim. □

5. Type A_{n-1}

An arbitrary semisimple simply connected (resp. adjoint) algebraic group over a field F is isomorphic to a direct product of several groups of the type $G_1 = R_{E/F}(G)$, where E/F is a finite separable field extension and G is absolutely simple simply connected (resp. adjoint) adjoint algebraic group over E ([43, 38, 3.1.2]). Since R-equivalence commutes with direct products and $G_1(F)/R = G(E)/R$, the computation of the group of R-equivalence classes for a semisimple simply connected (resp. adjoint) algebraic group reduces to the case of absolutely simple algebraic groups of the same type.

Below we compute the group of R-equivalence classes for all classical absolutely simple simply connected and adjoint groups of types A, B, C and D. We follow the classification of adjoint groups given in [49], [44] and [15]. In this section we consider groups of type A over a field F of arbitrary characteristic.

5.1 Simply connected groups

An arbitrary absolutely simple simply connected algebraic group of the type A_{n-1} is isomorphic to the *special unitary group* $\mathbf{SU}(A,\sigma)$, where A is an Azumaya algebra of degree n over an etale quadratic extension Z of F and σ is an involution of the second kind trivial on F. The group of F-points of $\mathbf{SU}(A,\sigma)$ consists of all $a \in A$, such that $\sigma(a)a = 1$ and $\mathrm{Nrd}(a) = 1$.

Consider first the case when Z splits, $Z = F \times F$. Then A is isomorphic to $B \times B^{op}$ with the switch involution, where B is a central simple algebra of degree n over F. It is easy to see that the map $b \mapsto \big(b, (b^{-1})^{op}\big)$ gives rise to an isomorphism between $\mathbf{SL}_1(B)$ and $\mathbf{SU}(A,\sigma)$. The group of R-equivalence classes is computed in the following Proposition. Here comes the algebraic K-theory.

PROPOSITION 5.1. ([46]) $SL_1(B)/R \simeq SK_1(B) = SL_1(B)/[B^\times, B^\times]$. □

Example 5.2 If $\mathrm{ind}(B)$ is square-free, then $SK_1(B) = 1$ ([48]). Hence the group $\mathbf{SL}_1(B)$ is R-trivial in this case. It is not known whether these groups are rational.

By Wedderburn theorem $B \simeq \mathrm{M}_n(D)$ for some skewfield D over F. Then the group $\mathbf{SL}_1(B)$ is isomorphic to $\mathbf{SL}_n(D)$. Let U (resp. U^-) be the subgroup of upper (resp. lower) strictly triangular matrices, T be the maximal torus of diagonal matrices. Since $\mathbf{SL}_1(B)$ is birationally isomorphic to $U^- \times T \times U$ and $T \simeq \mathbf{SL}_1(D) \times \mathbf{GL}_{n-1}(D)$, it follows that $\mathbf{SL}_1(B)$ is birationally isomorphic to $\mathbf{SL}_1(D) \times \mathbf{A}^m$ for some m. Hence, by Corollary 1.8, we get

PROPOSITION 5.3. *If* $\mathrm{ind}(B) \le 3$, *then* $\mathbf{SL}_1(B)$ *is a rational group.* □

Suslin has conjectured ([41]) that if $\mathrm{ind}(B)$ is not square-free, then the group $\mathbf{SL}_1(B)$ is not R-trivial and hence is not stably rational. The following statement confirmes this conjecture in a special case.

THEOREM 5.4. ([19]) *If* $\operatorname{ind}(B)$ *is divisible by* 4, *then the group* $\mathbf{SL}_1(B)$ *is not* R*-trivial and hence is not stably rational.* □

Remark 5.5 The proof of the theorem is based on Rost's computation of the group SK_1 for a biquaternoin algebra, which, in its turn, uses in particular Swan's computation of the higher algebraic K-theory of a projective quadric hypersurface.

Remark 5.6 Examples of central simple algebras B of arbitrary index divisible by 4 and hence of non-rational groups $\mathbf{SL}_1(B)$ exist over any number field. Note that by [48], the group of R-equivalence classes $\mathrm{SL}_1(B)/R$ is trivial in this case.

Now consider the case when Z is a quadratic field extension of F. Denote by Σ' the group of all $a \in A^\times$, such that $\operatorname{Nrd}(a) \in F^\times$. Any invertible σ-symmetric element $a \in A^\times$ belongs to Σ'. Denote by Σ the subgroup in Σ', generated by invertible σ-symmetric elements in A, so that $\Sigma \subset \Sigma'$. The group Σ does not depend on the choice of involution of the second kind on A.

Let $a \in \mathrm{SU}(A,\sigma)$. Choose an element $z \in Z$ with the (Z/F)-trace 1 (one can take $z = 1/2$ if $\operatorname{char}(F) \neq 2$), such that the element $za + \sigma(z)$ is invertible. Since for any a in $\mathbf{SU}(A,\sigma)$ one has $\sigma(a)a = 1$ and $\operatorname{Nrd}(a) = 1$, it follows that

$$\sigma\big(za + \sigma(z)\big) \cdot a = za + \sigma(z)$$

and hence $za + \sigma(z) \in \Sigma'$.

Consider the map

$$\varphi : \mathrm{SU}(A,\sigma) \to \Sigma'/\Sigma,$$

defined by the formula $\varphi(a) = \big(za + \sigma(z)\big) \cdot \Sigma$.

THEOREM 5.7. ([7, Th. 5.4]) *The map* φ *is a well-defined homomorphism, which induces an isomorphism* $SU(A,\sigma)/R \xrightarrow{\sim} \Sigma'/\Sigma$. □

By [50], the group Σ'/Σ is trivial if $\operatorname{ind}(A)$ is square-free. Thus, we obtain

COROLLARY 5.8. *If* $\operatorname{ind}(A)$ *is square-free, then the group* $SU(A,\sigma)$ *is* R*-trivial.* □

Let D be a skewfield over Z Brauer equivalent to A. One can prove the following

THEOREM 5.9. ([7, Th. 5.4]) *There is an involution* τ *of the second kind on* D, *such that* $\mathbf{SU}(A,\sigma)$ *is birationally isomorphic to* $\mathbf{SU}(D,\tau) \times \mathbf{A}_F^m$ *for some* m. □

Corollary 1.8 then gives

COROLLARY 5.10. ([7, Cor. 5.5]) *Let* A *be a central simple* Z*-algebra with an involution* σ *of the second kind trivial on* F. *If* $\operatorname{ind}(A) \leq 3$, *then the variety of the group* $\mathbf{SU}(A,\sigma)$ *is rational.* □

Remark 5.11 Since $\mathbf{SU}(A,\sigma) \otimes_F Z \simeq \mathbf{SL}_1(A)$, it follows from Theorem 5.4 that the group $\mathbf{SU}(A,\sigma)$ is not rational if $\operatorname{ind}(A)$ is divisible by 4.

Let $T = \ker\big(N_{Z/F} : R_{Z/F}(\mathbf{G}_{m,Z}) \to \mathbf{G}_{m,F}\big)$ be the norm 1 torus in the quadratic extension Z/F. The group $\mathbf{SU}(A,\sigma)$ is the kernel of the reduced norm homomorphism Nrd : $\mathbf{U}(A,\sigma) \to T$, where $\mathbf{U}(A,\sigma) = \mathbf{Iso}(A,\sigma)$ is the *unitary group*. By [22, Prop. 6.1], for any field extension E/F, the image of this homomorphism on the groups of E-points depends only on the Brauer class of A. Applying Lemma 4.3 and Proposition 4.5 to exact sequences

$$1 \to \mathbf{SU}(A,\sigma) \to \mathbf{U}(A,\sigma) \to T \to 1, \quad 1 \to \mathbf{SU}(A',\sigma') \to \mathbf{U}(A',\sigma') \to T \to 1,$$

we get the following

COROLLARY 5.12. ([7, Cor. 5.7]) *Let (A,σ) and (A',σ') be central simple Z-algebras with involutions of the second kind. If A and A' are Brauer equivalent algebras, then the groups $\mathbf{SU}(A,\sigma)$ and $\mathbf{SU}(A',\sigma')$ are stably birationally isomorphic.* □

Remark 5.13 Corollary shows that the stable birational type of the group $\mathbf{SU}(A,\sigma)$ depends only on the Brauer class of A and does not depend on the involution σ.

5.2 Adjoint groups

An arbitrary absolutely simple adjoint algebraic group of the type A_{n-1} is isomorphic to the connected group $\mathbf{PSim}(A,\sigma)$, where A is an Azumaya algebra of degree n over an etale quadratic extension Z of F and σ is an involution of the second kind trivial on F.

Consider first the case when Z splits, i.e A is isomorphic to $B \times B^{op}$ with the switch involution, where B is a central simple algebra of degree n over F. It is easy to see that the map $b \mapsto \big(b,(b^{-1})^{op}\big)$ gives rise to an isomorphism of $\mathbf{PGL}_1(B)$ and $\mathbf{PSim}(A,\sigma)$. The group $\mathbf{PGL}_1(B)$ embeds as an open subset in the projective space $\mathbf{P}(B)$, and hence is rational.

Consider now the case when Z is a field. The group $\mathbf{PSim}(A,\sigma)$ is called the *projective unitary group* and is denoted by $\mathbf{PGU}(A,\sigma)$. The involution σ_Z is hyperbolic since $Z \otimes_F Z$ splits, hence $NZ^\times \subset \operatorname{Hyp}(A,\sigma)$. It follows from Theorem 4.2 that

$$\mathbf{PGU}(A,\sigma)(F)/R = G(A,\sigma)/\operatorname{Hyp}(A,\sigma).$$

Consider the following particular cases:

Case 1: $n = 2$.
In this case A is a quaternion algebra over Z. Denote by $(a \mapsto \bar{a})$ the canonical involution on A and consider the quaternion F-subalgebra

$$Q = \{a \in A \text{ such that } \sigma a = \bar{a}\}$$

in A. Then $\mathbf{PGU}(A,\sigma) \simeq \mathbf{PGL}_1(Q)$ is a rational algebraic group (see [15]).

Case 2: n is odd.
In this case $G(A,\sigma) = \operatorname{Hyp}(A,\sigma) = NZ^\times$ ([23]), in particular

$$\mathbf{PGU}(A,\sigma)(F)/R = 1.$$

This reflects the fact that the algebraic group $\mathbf{PGU}(A,\sigma)$ is rational if n is odd ([47, Cor. of Th. 8]). This statement is proved by showing that all maximal tori in $\mathbf{PGU}(A,\sigma)$ are rational (see Proposition 1.7).

Case 3: n is even.
Denote by $D(A,\sigma)$ the *discriminant algebra* of (A,σ) (see [15], [25]).

THEOREM 5.14. *Assume that*

(1) $\deg(A)$ *is divisible by* 4 ;
(2) $\operatorname{ind}(A)$ *is even;*
(3) *The algebra* $D(A,\sigma)$ *is not split.*

Then the group $\mathbf{PGU}(A,\sigma)$ *is not* R*-trivial and hence is not stably rational.*

Proof. We will show that there is a field extension E/F such that

$$\mathrm{PGU}(A_E,\sigma_E)/R \neq 1.$$

Let $X = \mathrm{SB}(A,2)$ be the generalized Severi-Brauer variety of right ideals in A of dimension $2\cdot\deg(A)$ over Z ([3]) and $X' = R_{Z/F}(X)$. The kernel of $\mathrm{Br}(Z) \to \mathrm{Br}\big(Z(X)\big)$ is denerated by the class of $A\otimes_Z A$ ([3]). Since the corestriction of A in the extension Z/F is trivial ([11]), by [21, Cor. 2.12], the natural homomorphism $\mathrm{Br}(F) \to \mathrm{Br}\big(F(X')\big)$ is injective, hence $D(A,\sigma)$ is not split over $F(X')$. Since $X'\otimes_F Z \simeq \mathrm{SB}(A,2)\times\mathrm{SB}(A^{op},2)$, by the index reduction formula (see [3], [26]), the index of $A\times_F F(X') = A\otimes_Z Z(X')$ equals 2. Replacing the field F by $F(X')$, we may assume that $\operatorname{ind}(A)=2$, so that $A\simeq \mathrm{End}_Q(V)$ for some quaternion algebra Q and a Q-vector space V of even dimension $2k$. The involution σ is adjoint to some hermitian form h on V (with respect to some involution of the second kind on Q).

Let Y be the variety over F of isotropic subspaces in V of dimension $(k-1)$ over Q. By [21, Cor. 2.8], the natural homomorphism $\mathrm{Br}(F)\to\mathrm{Br}\big(F(Y)\big)$ is injective, hence $D(A,\sigma)$ is not split over $F(Y)$. Since Y has a rational point over Z, it follows that $A\otimes_F F(Y) = A\otimes_Z Z(Y)$ is not split. Hence, replacing F by $F(Y)$, we may assume that $V = V'\perp V''$, where V'' is a hyperbolic space of dimension $2(k-1)$ and $\dim(V')=2$.

Consider the algebra $A' = \mathrm{End}_Q(V')$ of degree 4 and the involution σ' of the second kind on A' adjoint to the hermition form $h|_{V'}$ on V'. By Theorem 4.2, the groups of R-equivalence classes for $\mathbf{PGU}(A,\sigma)$ and $\mathbf{PGU}(A',\sigma')$ coincide over any field extension of F. The discriminant algebra $D(A',\sigma')$, being similar to $D(A,\sigma)$, is not trivial, hence by Theorem 9.3 in Section 9 (see also Remark 9.5), the group $\mathbf{PGU}(A',\sigma')$, and therefore $\mathbf{PGU}(A,\sigma)$, is not R-trivial. □

Remark 5.15 Theorem gives examples of non-rational adjoint groups of types A_{4k-1} for all k over any number field.

6. Type B_n

We assume that characteristic of the base field F is different from 2.

6.1 Simply connected groups

An arbitrary absolutely simple simply connected algebraic group of the type B_n is isomorphic to the group $\mathbf{Spin}(q) = \mathbf{Spin}(V,q)$, where (V,q) is a quadratic form (uniquely determined up to a scalar) of dimension $2n+1\geq 3$. In this section we collect some results for arbitrary forms (not necessarily of odd dimension).

PROPOSITION 6.1. ([33]) *If q is an isotropic form, then the group* $\mathbf{Spin}(q)$ *is rational.*

Proof. Since q is isotropic, there are two vectors $u, v \in V$, such that $v^2 = u^2 = 0$ and $uv + vu = 1$ in the Clifford algebra. The image of the homomorphism

$$\mathbf{G}_{m,F} \to \mathbf{Spin}(q), \quad a \mapsto a^{-1}(uv + a^2vu)$$

we denote by T and the factor-variety $\mathbf{Spin}(q)/T$ by X. Since $T \simeq \mathbf{G}_{m,F}$, by Hilbert's theorem 90 and Lemma 1.1, the exact sequence

$$1 \to T \to \mathbf{Spin}(q) \to X \to 1$$

splits rationally, hence $\mathbf{Spin}(q)$ is birationally isomorphic to $T \times X$. The kernel $K = \{\pm 1\}$ of the natural homomorphism $\mathbf{Spin}(q) \to \mathbf{O}_+(q)$ belongs to T and the square endomorphism of T induces an isomorphism $T/K \simeq T$. Hence, $\mathbf{Spin}(q)$ is birationally isomorphic to $T/K \times X$, and therefore, to the rational group $\mathbf{Spin}(q)/K \simeq \mathbf{O}_+(q)$ (see Lemma 4.3). □

For a quadratic form q denote by $D(q)$ the set of non-zero values of q.

LEMMA 6.2. *Let (V, q) be a quadratic form over F of dimenrsion at least 2, (V', q') subform of codimension 1, L/F quadratic extension, such that $q_L = q \otimes_F L$ is isotropic. Then $N_{L/F}(L^\times) \subset D(q) \cdot D(q')$.*

Proof. We may clearly assume that q is anisotropic. Consider first the case $\dim(V) = 2$, i.e. $q' = \langle b \rangle$ and $q = \langle b, -ab \rangle$ for some $a, b \in F^\times$. Then $L = F(\sqrt{a})$ and the result immediately follows.

Consider now the general case. Since q_L is isotropic, there is a 2-dimensional subspace $U \subset V$, such that $q|_U$ is isotropic over L ([16, Lemma VII.3.1]). If $U \subset V'$, then q' is isotropic over L and by Knebusch's norm principle ([16, Th. VII.5.1]),

$$N_{L/F}(L^\times) \subset D(q') \cdot D(q') \subset D(q) \cdot D(q').$$

If U is not contained in V', then by the 2-dimensional case, applied to the subform $q'|_{U \cap V'}$ in $q|_U$, we have

$$N_{L/F}(L^\times) \subset D(q|_U) \cdot D(q'|_{U \cap V'}) \subset D(q) \cdot D(q'). \qquad \square$$

A *Pfister form* is a tensor product of binary forms representing 1. A *Pfister neighbor* is a form similar to a subform of a Pfister form f of dimension bigger than one half of $\dim(f)$.

LEMMA 6.3. *Let q be a Pfister neighbor of a Pfister form f, q' subform in q of codimension 1. Then $D(f) = D(q) \cdot D(q')$.*

Proof. We may assume that f is anisotropic. Let $x \in D(f)$. Then there exists a quadratic extension L/F, such that f_L is isotropic and $x \in N_{L/F}(L^\times)$. Since q is a Pfister neighbor, q_L is also isotropic and by Lemma 6.2, $x \in D(q) \cdot D(q')$, i.e. $D(f) \subset D(q) \cdot D(q')$. The inverse inclusion follows from the fact that $D(f)$ is a group ([16, Cor. X.1.7]). □

THEOREM 6.4. (Chernousov–Merkurjev–Rost) *Let q be a Pfister neighbor and p quadratic form of dimension at most 2. Then the group* $\mathbf{Spin}(q \perp p)$ *is rational.*

Proof. In the space V of the form $q \perp p$ choose subspaces $V'' \subset V' \subset V$ of consecutive codimensions 1 so that V'' is contained in the space of q. An element x in a certain open subset in $\mathbf{Spin}(q \perp p)$ can be written in the form $x = y \cdot v \cdot (v')^{-1}$, where y belongs to the Clifford group of V'' and $v' \in V'$, $v \in V$, so that the class $[y]$ in $\mathbf{O}_+(V'')$ is uniquely determined. We get a rational dominant morphism

$$\mathbf{Spin}(q \perp p) \to \mathbf{O}_+(V''), \quad x \mapsto [y].$$

The generic fiber of this morphism is the projective quadric in the space $V \oplus V'$, given by the quadratic form $g = (-q_1) \perp \mathrm{Sn}(y) \cdot q_2$, where $q_2 = q \perp p$ and $q_1 = q_2|_{V'}$. Since q is a Pfister neighbor of some Pfister form f, $\mathrm{Sn}(y) \in D(f)$. The form q_1 contains the Pfister neighbor q and q_1 is of codimension 1 in q_2, hence, by Lemma 6.3, g is isotropic, i.e. the generic fiber is a rational quadric. It remains to notice that the group $\mathbf{O}_+(V'')$ is rational by Lemma 4.3. □

Example 6.5 Any form q of dimension at most 5 satisfies conditions of Theorem 6.4, hence the group $\mathbf{Spin}(q)$ is rational in this case. Of course, this result also follows from Corollary 1.8, since the rank of the group is at most 2. Note that if $\dim(q) = 6$ and q is anisotropic over the discriminant quadratic extension, then $\mathbf{Spin}(q)$ is never stably rational by Corollary 9.2.

Example 6.6 The canonical quadratic form $q = \langle 1, 1, \ldots, 1 \rangle$ is a Pfister neighbor, hence $\mathbf{Spin}(q)$ is rational. This result was obtained in [34].

6.2 Adjoint groups

An arbitrary absolutely simple adjoint algebraic group of the type B_n is isomorphic to a connected group $\mathbf{O}_+(q)$, where q is a quadratic form (uniquely determined up to a scalar) of dimension $2n + 1$. This group is known to be rational by Lemma 4.3.

7. Type C_n

7.1 Simply connected groups

An arbirtrary absolutely simple simply connected algebraic group of the type C_n is isomorphic to the *symplectic group* $\mathbf{Sp}(A, \sigma) = \mathbf{Iso}(A, \sigma)$ where A is a central simple algebra of degree $2n$ over F with a symplectic involution σ. This group is rational by Lemma 4.3.

7.2 Adjoint groups

An arbirtrary absolutely simple adjoint algebraic group of the type C_n is isomorphic to the *projective symplectic group* $\mathbf{PGSp}(A, \sigma)$, where A is a central simple algebra of degree $2n$ over F with a symplectic involution σ.

Consider the following particular cases:

Case 1: $n = 1$.
A quaternion algebra A has the unique (standard) symplectic involution σ, hence any automorphism of A preserves σ and therefore $\mathbf{PGSp}(A,\sigma) = \mathbf{PGL}_1(A)$ is a rational group.

Case 2: $n = 2$.
The space of skew-symmetric elements of trivial reduced trace

$$V = \{a \in A \text{ such that } \sigma a = -a \text{ and } \mathrm{Trd}(a) = 0\}$$

carries the 5-dimensional quadratic form $q(a) = a^2 \in F$. The map

$$\mathrm{Sim}(A,\sigma) \to \mathrm{O}_+(q), \quad a \mapsto \mathrm{Int}(a)|_V$$

induces an isomorphism of simple adjoint algebraic groups $\mathbf{PGSp}(A,\sigma) \simeq \mathbf{O}_+(q)$ of the types $C_2 = B_2$ ([20, Prop. 5.4], [15]). In particular, the group $\mathbf{PGSp}(A,\sigma)$ is rational. Of course, the rationality of the groups in the first two cases follows also from Corollary 1.8.

Case 3: n is odd.

LEMMA 7.1. ([23, Lemma 3]) *If n is odd, then $G(A,\sigma) = \mathrm{Hyp}(A,\sigma) = \mathrm{Nrd}(A^\times)$.* □

PROPOSITION 7.2. ([23, Prop. 4]) *Any absolutely simple adjoint group of the type C_n with odd n is stably rational and hence is R-trivial.*

Proof. We apply Corollary 4.6 to the rational algebraic group $G = \mathbf{GL}(Q)$ and to the reduced norm homomorphism $\mathrm{Nrd} = \alpha : G \to \mathbf{G}_{m,F}$. The kernel of α is the rational algebraic group $\mathbf{SL}_1(Q)$ (affine quadric with a rational point) and, by Lemma 7.1, $G(A_E,\sigma_E) = \mathrm{Nrd}(A_E^\times) = \mathrm{im}\,\alpha_E$ for any field extension E/F. □

8. Type D_n

We assume that $\mathrm{char}(F) \neq 2$.

8.1 Simply connected groups

An arbitrary absolutely simple simply connected algebraic group of the type D_n (except for some non-classical groups of the type D_4) is isomorphic to the group $\mathbf{Spin}(A,\sigma)$, where A is a central simple algebra of degree $2n$ over F with an orthogonal involution σ. The group $\mathbf{Spin}(A,\sigma)$ is the kernel of the *spinor norm* homomorphism

$$\mathrm{Sn} : \mathbf{\Gamma}(A,\sigma) \to \mathbf{G}_{m,F},$$

where $\mathbf{\Gamma}(A,\sigma)$ is the *Clifford group* (see [44], [15]).

We will use the following

LEMMA 8.1. ([22, Prop. 6.2]) *The image of* Sn_F *on the groups of* F*-points is generated by* $F^{\times 2}$ *and the norms in all finite field extensions over which* A *splits and* σ *is isotropic.* $\square$

Consider the subgroup H in the product $\mathbf{GL}_1(A) \times \mathbf{G}_{m,F}$ consisting of pairs (a, x), such that $\mathrm{Nrd}(a) = x^2$. The group H is the kernel of the homomorphism

$$f : \mathbf{GL}_1(A) \times \mathbf{G}_{m,F} \to \mathbf{G}_{m,F}, \quad (a, x) \mapsto \mathrm{Nrd}(a) \cdot x^{-2}.$$

By Lemma 8.1, for any field extension E/F, the image of Sn_E is contained in the image of f_E. Hence, by [7], there is a natural homomorphism

$$\theta : \mathbf{Spin}(A, \sigma)/R \to H(F)/R.$$

In order to compute the group $RH(F)$ consider any symplectic involution τ on A and the subgroup $\Sigma \subset A^\times$ generated by the set S of τ-symmetric invertible elements. The group Σ does not depend on the choice of τ. There is the *pfaffian* map $\mathrm{pf} : S \to F^\times$, such that $\mathrm{pf}(s)^2 = \mathrm{Nrd}(s)$ for any $s \in S$ (see [15]). Denote by Δ the subgroup in $H(F)$ generated by pairs $\big(s, \mathrm{pf}(s)\big)$ for all $s \in S$. We also consider the commutant $[A^\times, A^\times]$ as a subgroup in $H(F)$, identifying an element a in the commutant with the pair $(a, 1)$.

LEMMA 8.2. $RH(F) = \Delta \cdot [A^\times, A^\times]$.

Proof. Let $a, b \in A^\times$. Consider $a(t) = 1 + (a-1)t \in (A \otimes_F \mathcal{O})^\times$ and $c(t) = [a(t), b]$. Clearly, $c(0) = [a(0), b] = [1, b] = 1$ and $c(1) = [a(1), b] = [a, b]$, i.e. the commutator $[a, b]$ is R-trivial and $[A^\times, A^\times] \subset RH(F)$. Similarly, if s is a τ-symmetric element in $A^\times$, then for the pair $h(t) = \big(s(t), \mathrm{pf}(s(t))\big)$ in $H(\mathcal{O})$, where $s(t) = 1 + (s-1)t$, we have $h(0) = (1, 1)$ and $h(1) = \big(s, \mathrm{pf}(s)\big)$, hence $\Delta \subset RH(F)$.

In order to prove the desired equality, consider the subgroup $\Sigma' \subset A^\times$, consisting of all a, such that $\mathrm{Nrd}(a) \in F^{\times 2}$. There is the exact sequence

$$\{\pm 1\} \xrightarrow{\alpha} H(F)/\Big(\Delta \cdot [A^\times, A^\times]\Big) \xrightarrow{\beta} \Sigma'/\Sigma \to 1,$$

where $\alpha(-1) = (1, -1)$ and $\beta(a, x) = a \cdot \Sigma$. Since the latter group in the sequence does not change under purely transcendental extensions (see [28, Th. 3.2]), so does the middle one. As in [28] and [7] we conclude that $\Delta \cdot [A^\times, A^\times] = RH(F)$. $\square$

The homomorphism θ is not an isomorphism in general. For example, if A splits, then $H(F) = \Delta \cdot [A^\times, A^\times]$, whereas the group $\mathbf{Spin}(A, \sigma)$ is not necessarily R-trivial in this case (see Corollary 9.2). Nevertheless, there is a case when θ is an isomorphism.

PROPOSITION 8.3. *If* σ *is an isotropic involution, then the map* $\theta : \mathbf{Spin}(A, \sigma)/R \to H(F)/\Big(\Delta \cdot [A^\times, A^\times]\Big)$ *is an isomorphism.*

Proof. By Lemma 8.1, for any field extension E/F the images of Sn_E and f_E coincide. Hence, θ is an isomorphism (see [7]). $\square$

Remark 8.4 Proposition 8.3 is a generalization of [28, Th. 1.6].

8.2 Adjoint groups

An arbitrary absolutely simple adjoint algebraic group of the type D_n (except for some non-classical groups of the type D_4) is isomorphic to the *projective orthogonal group* $\mathbf{PGO}_+(A,\sigma) = \mathbf{PSim}_+(A,\sigma)$, where A is a central simple algebra of degree $2n$ over F with an orthogonal involution σ. The group of F-points of $\mathbf{PGO}_+(A,\sigma)$ equals $\mathrm{Sim}_+(A,\sigma)/F^\times$, where $\mathrm{Sim}_+(A,\sigma)$ consists of *proper similitudes*, i.e similitudes $a \in \mathrm{Sim}(A,\sigma)$, such that $\mathrm{Nrd}(a) = \mu(a)^n$.

Denote by $C = C(A,\sigma)$ the Clifford algebra of algebra with involution (A,σ) with the center L/F, being an etale quadratic extension over F, and by $\mathrm{disc}(\sigma)$ the discriminant of σ (see [15]).

Case 1: $n = 2$.

PROPOSITION 8.5. ([21, Prop. 1.15], [23, Prop. 5], [15])
(1) *Any adjoint group of the type $D_2 = A_2$ is isomorphic to $\mathbf{PGO}_+(A,\sigma) \simeq R_{L/F}\big(\mathbf{PGL}_1(C)\big)$ and hence is rational.*
(2) $\mathrm{Hyp}(A,\sigma) = N_{L/F}\,\mathrm{Nrd}\,C$ *and* $G_+(A,\sigma) = F^{\times 2} \cdot N_{L/F}\,\mathrm{Nrd}\,C$. □

Case 2: $n = 3$.

PROPOSITION 8.6. ([23, Prop. 6], [15]) *Let $\mathrm{disc}(\sigma)$ be trivial. Then $C(A,\sigma) \simeq C^+ \times C^-$ and*
(1) *There is an isomorphism of algebraic groups $\mathbf{PGO}_+(A,\sigma) \simeq \mathbf{PGL}_1(C^+)$ of the types $D_3 = A_3$. In particular, the former group is rational.*
(2) $\mathrm{Hyp}(A,\sigma) = \mathrm{Nrd}(C^+)$ *and* $G_+(A,\sigma) = F^{\times 2} \cdot \mathrm{Nrd}(C^+)$. □

Case 3: The algebra A splits, $A = \mathrm{End}_F(V)$.
In this case the involution σ is adjoint to some non-degenerate quadratic form q on the space V. The group $\mathbf{PGO}_+(A,\sigma)$ is equal to the *projective special orthogonal group* $\mathbf{PGO}_+(q)$ of the quadratic form q. The following statement gives examples of stably rational groups $\mathbf{PGO}_+(q)$.

PROPOSITION 8.7. ([23, Prop. 7]) *If $q = f \otimes_F g$ is the tensor product of a Pfister form f and a form g of odd dimension over F, then the group $\mathbf{PGO}_+(q)$ is stably rational.* □

Example 8.8 If q is the standard form $\langle 1, 1, \dots, 1\rangle$, then the group $\mathbf{PGO}_+(q)$ is stably rational (see also [6]).

Examples of non-rational groups are given by the following

THEOREM 8.9. ([23, Th. 2]) *Let A be a central simple algebra of even degree over a field F with an orthogonal involution σ. If $\mathrm{disc}(\sigma)$ is not trivial and $\mathrm{ind}\, C(A,\sigma) \geq 4$, then the group $\mathbf{PGO}_+(A,\sigma)$ is not R-trivial and hence is not stably rational.* □

Remark 8.10 One of the main ingredients of the proof is the index reduction formula (see [25], [26]) which is based on the computation of the algebraic K-theory of certain projective homogeneous varieties given by Panin in [30].

Remark 8.11 Examples of groups of type D_n with odd n, satisfying conditions of the Theorem, exist over any number field (see [23, Sec. 4]).

9. Stably birational classification of absolutely simple groups of type $A_3 = D_3$

By Corollary 1.8, all semisimple groups of rank at most 2 are rational. In this section we give complete birational classification of semisimple groups (of rank 3) of type $A_3 = D_3$.

Let A be a central simple algebra of degree 6 over a field F with orthogonal involution σ, $C = C(A, \sigma)$ the Clifford algebra of (A, σ). It is a central simple algebra of degree 4 over the discriminant quadratic extension L/F with the canonical involution τ of the second kind ([15]).

Consider first the simply connected case. An absolutely simple group of the type $D_3 = A_3$ is $\mathbf{Spin}(A, \sigma) \simeq \mathbf{SU}(C, \tau)$ (see [15]). From Theorem 5.4, Proposition 5.3, Corollary 5.10 and Remark 5.11 one deduces:

THEOREM 9.1. I. *If* $\operatorname{disc}(\sigma)$ *is trivial, i.e.* $C \simeq C^+ \times C^-$, *where* $C^\pm$ *are central simple algebras of degree 4 over* F, *then*

(1) *If* C^+ *is not a skewfield, then* $\mathbf{Spin}(A, \sigma)$ *is a rational group.*
(2) *If* C^+ *is a skewfield, then* $\mathbf{Spin}(A, \sigma)$ *is not* R*-trivial and hence is not stably rational.*

II. *If* $\operatorname{disc}(\sigma)$ *is not trivial, then*

(1) *If* C *is not a skewfield, then* $\mathbf{Spin}(A, \sigma)$ *is a rational group.*
(2) *If* C *is a skewfield, then* $\mathbf{Spin}(A, \sigma)$ *is not* R*-trivial and hence is not stably rational.* □

COROLLARY 9.2. *Let* q *be a non-degenerate quadratic form of dimension 6 over* F. *Then* $\mathbf{Spin}(q)$ *is not stably rational iff* q *is anisotropic over the discriminant extension* $F(\sqrt{\operatorname{disc}(\sigma)})$ *of* F.

Proof. If $\operatorname{disc}(q)$ is trivial, then $C = C_0(q) = C^+ \times C^-$ and q is anisotropic iff C^+ is a skewfield by [2].

Assume that $\operatorname{disc}(q)$ is not trivial. Let L/F be the discriminant quadratic extension of q. The result follows again from the fact that q_L is anisotropic iff the algebra $C = C^+(q_L)$ is a skewfield. □

Now consider the adjoint case. An absolutely simple group of the type $D_3 = A_3$ is $\mathbf{PGO}_+(A, \sigma) = \mathbf{PGU}(C, \tau)$.

THEOREM 9.3. ([23, Th. 3]) I. *If* $\operatorname{disc}(\sigma)$ *is trivial, then the group* $\mathbf{PGO}_+(A, \sigma)$ *is rational.*

II. *If* $\operatorname{disc}(\sigma)$ *is not trivial, then*

(1) *If* C *splits, or if* $\operatorname{ind} C = 2$ *and* A *splits, then the group* $\mathbf{PGO}_+(A, \sigma)$ *is stably rational.*
(2) *If* $\operatorname{ind} C = 4$, *or if* $\operatorname{ind} C = 2$ *and* A *is not split, then the group* $\mathbf{PSim}_+(A, \sigma)$ *is not* R*-trivial and hence is not stably rational.* □

COROLLARY 9.4. *Let* q *be a non-degenerate quadratic form of dimension 6 over* F. *Then* $\mathbf{PGO}_+(q)$ *is not stably rational iff the discriminant of* q *is not trivial and* q *is anisotropic over the discriminant extension* $F(\sqrt{\operatorname{disc}(\sigma)})$ *of* F. □

Remark 9.5 The algebra A can be recovered from (C, τ) as the discriminant algebra $D(C, \tau)$ (see [15]).

Finally, let G be an absolutely simple group of type D_3, which is neither simply connected nor adjoint. Then $G = \mathbf{Iso}_+(A, \sigma)$ is a rational group by Lemma 4.3.

10. Algebraic groups

In this section we collect definitions and preliminary results needed in the second part of the paper. We give an algebraic definition of the fundamental group $\pi_1(G)$ for an algebraic group G.

10.1 Fundamental group

Let G be an algebraic group defined over an algebraically closed field F. A *loop in* G is a group homomorphism (co-character) $p : \mathbf{G}_{m,F} \to G$. Two loops p and p' are called *conjugate* if there is $g \in G(F)$ such that $p'(t) = g \cdot p(t) \cdot g^{-1}$ for all t in $\mathbf{G}_{m,F}$. A loop p is called *contractible* if p extends to a group homomorphism $\mathbf{SL}_2(F) \to G$ (we identify $\mathbf{G}_{m,F}$ with the maximal torus of diagonal matrices in $\mathbf{SL}_2(F)$ by $t \mapsto \operatorname{diag}(t, t^{-1})$). If p and p' are two loops with commuting images, then the product pp' given by $(pp')(t) = p(t)p'(t)$ is also a loop.

The *fundamental group* $\pi_1(G)$ of the group G is the group defined by generators and relations as follows. Generators are loops in G. For a loop p we denote by $[p]$ the corresponding generator in $\pi_1(G)$. The relations are the following:

1. $[p] = [p']$ if p and p' are conjugate;
2. $[p] = 1$ if p is contractible;
3. $[pp'] = [p] \cdot [p']$ if the images of p and p' commute.

Example 10.1 Let G be a commutative algebraic group. Since conjugate loops are equal and contractible loops are trivial ($\mathbf{SL}_2(F)$ coincides with its commutant), the group $\pi_1(G)$ equals the group $G_* = \operatorname{Hom}(\mathbf{G}_{m,F}, G)$ of co-characters.

Now let G be an algebraic group over an arbitrary field F. The fundamental group $\pi_1(G)$ of G is, by definition, the group $\pi_1(G_{F_{\mathrm{alg}}})$, where F_{alg} is an algebraic closure of F.

A group homomorphism $G \to G'$ clearly induces a homomorphism of fundamental groups $\pi_1(G) \to \pi_1(G')$.

Proposition 10.2. *Let G be a reductive group over F. Then*

(1) *The fundamental group $\pi_1(G)$ of a reductive group G is an abelian finitely generated group.*
(2) *$\pi_1(G)$ is finite iff G is a semisimple group.*
(3) *$\pi_1(G)$ is trivial iff G is a simply connected semisimple group.*
(4) *If G' is the commutant of G, then $\pi_1(G') = \pi_1(G)_{tors}$.*
(5) *$\pi_1(G)$ is torsion free iff G' is a simply connected group.* □

Remark 10.3 If G is an algebraic group over the field of complex numbers, then $\pi_1(G)$ coincides with the ordinary fundamental group of the corresponding Lie group [1, Th.5.47]. The algebraic definition of the fundamental group, we present here, is not based on the choice of a maximal torus. That makes evident the functorial properties of the fundamental group.

10.2 Factorial groups

An algebraic group G over a field F is called *factorial* if for any finite field extension E/F the Picard group $\mathrm{Pic}(G_E)$ is trivial.

Example 10.4 Let T be an algebraic torus over F, L/F be a finite Galois splitting field, $\Pi = \mathrm{Gal}(L/F)$. The torus T is factorial iff T is *coflasque*, i.e. $H^1(\Pi', T^*) = 0$ for any subgroup $\Pi' \subset \Pi$ (see [10]). A quasi-trivial torus is factorial.

Example 10.5 Let G be a semisimple algebraic group over F, $\pi : \widetilde{G} \to G$ be the universal covering, $C = \ker(\pi)$. By [38, Lemme 6.9], there is a natural isomorphism $\mathrm{Pic}(G) \simeq C^*$. Hence G is factorial iff G is simply connected.

The following Proposition reduces the case of a reductive group to these two examples.

PROPOSITION 10.6. *Let G be a reductive group over a field F, G' be the commutant of G. Then G is factorial iff G' and the torus G/G' are factorial, i.e. G' is simply connected and the torus G/G' is coflasque. The group $\pi_1(G)$ of a factorial reductive group G is torsion free.* □

10.3 Reductive groups of inner type

Let G and H be algebraic groups over F. The group G is called a *twisted form of* H if G_{sep} and H_{sep} are isomorphic over F_{sep}. Let G be a reductive group over F, G^d be its split twisted form. Denote by $\overline{G}^d$ the corresponding split adjoint group $G^d/Center(G^d)$. The group G is of *inner type* if the class of the 1-cocycles in $H^1(F, \mathrm{Aut}(G^d_{\mathrm{sep}}))$, corresponding to G, belongs to the image of the map

$$H^1(F, \overline{G}^d(F_{\mathrm{sep}})) \to H^1(F, \mathrm{Aut}(G^d_{\mathrm{sep}}))$$

induced by the conjugation homomorphism

$$\mathrm{Int} \ : \overline{G}^d(F_{\mathrm{sep}}) \to \mathrm{Aut}(G^d_{\mathrm{sep}}).$$

A quasi-split reductive group of inner type is split.

PROPOSITION 10.7. *Let G be a reductive group, G' be the commutant of G. Then G is of inner type iff G' is of inner type and the torus G/G' splits.* □

COROLLARY 10.8. *Let G be a reductive group of inner type. Then G is factorial iff $\pi_1(G)$ is torsion free.* □

Example 10.9 The group $\mathbf{GL}_1(A)$ for a central simple F-algebra A is a factorial reductive group over F of inner type.

10.4 Category of representations

Let G be an algebraic group defined over a field F. Denote by $\mathrm{Rep}(G)$ the category of finite dimensional linear representations of G. It is an abelian category in which any object has finite length. Hence by [36, Th.4, Cor.1]

$$K_n\Big(\mathrm{Rep}(G)\Big) = \coprod_{V \text{ irreducible}} K_n\Big(\mathrm{End}_G(V)^{op}\Big).$$

In particular, $K_0(\mathrm{Rep}(G))$ is a free abelian group freely generated by the classes of irreducible representations. We denote this group by $R(G)$. It has a ring structure with respect to the tensor product.

There is the *rank homomorphism* of rings $\mathrm{rk} : R(G) \to \mathbf{Z}$, taking the class of a G-module V to $\dim V$. We denote its kernel by $J(G)$.

A homomorphism $f : H \to G$ of algebraic groups over F induces a ring homomorphism $f^* : R(G) \to R(H)$. Thus, we can consider $R(H)$ as a $R(G)$-module. If f is the embedding of a subgroup H to G, the homomorphism f is called the *restriction map*.

Example 10.10 For a split solvable group B over F, $R(B) = \mathbf{Z}[B^*]$ is a Laurent polynomial ring over $\mathbf{Z}$.

Example 10.11 Let G be a split reductive group over F, $T \subset G$ a split maximal torus. The Weyl group W acts naturally on $R(T)$ and the restriction homomorphism induces an isomorphism (see [45])

$$R(G) \tilde{\to} R(T)^W = \mathbf{Z}[T^*]^W.$$

Example 10.12 Let G be a split simply connected semisimple group over F. Denote by $\rho_1, \ldots, \rho_r$ the classes of the fundamental representations in $R(G)$ (with the highest weights being the fundamental ones, [13, Ch.XI]). Then $R(G)$ is the polynomial ring $\mathbf{Z}[\rho_1, \ldots, \rho_r]$ (see [1, Th.6.41], [40, Lemma 3.1], [45]).

Example 10.13 Let ρ_i be the class in $R\big(\mathbf{GL}_n(F)\big)$ of the ith-exterior power representation of $\mathbf{GL}_n(F)$. Then

$$R\Big(\mathbf{GL}_n(F)\Big) = \mathbf{Z}[\rho_1, \ldots, \rho_{n-1}, \rho_n^{\pm 1}].$$

11. Equivariant K-theory

The equivariant K-theory was developed by Thomason in [42].

Let G be a group scheme over a field F. A scheme X over F is called a *G-scheme* if an action morphism $\theta : G \times X \to X$ of the group scheme G on X over F is given.

A *G-module M over X* is a coherent $\mathcal{O}_X$-module M together with the isomorphism of $\mathcal{O}_{G\times X}$-modules

$$\rho : \theta^*(M) \tilde{\to} p_2^*(M),$$

(where $p_2 : G \times X \to X$ is the projection), satisfying the cocycle condition

$$p_{23}^*(\rho) \circ (\mathrm{id}_G \times \theta)^*(\rho) = (m \times \mathrm{id}_X)^*(\rho),$$

where $m : G \times G \to G$ is the multiplication (see [42]).

The abelian category of G-modules over a G-scheme X we denote by $\mathcal{M}(G; X)$. We set also

$$K'_n(G; X) \overset{def}{=} K_n(\mathcal{M}(G; X)).$$

If G is a trivial group scheme, then $K'_n(G; X) = K'_n(X)$ are the ordinary K-groups of the category of coherent sheaves on X.

Example 11.1 Let E be a G-vector bundle over a G-scheme X. Then the sheaf of sections of E is endowed with a natural structure of a G-module. Thus, we can associate to E an object of $\mathcal{M}(G;X)$.

Example 11.2 Let $\mu : G \to \mathbf{GL}(V)$ be a finite dimensional representation of an algebraic group G over a field F. One can view V as a G-module over $\operatorname{Spec} F$. Clearly, we obtain an equivalence of categories $\operatorname{Rep}(G)$ and $\mathcal{M}(G; \operatorname{Spec} F)$. Hence there is a natural isomorphism

$$R(G) \xrightarrow{\sim} K_0'(G; \operatorname{Spec} F).$$

Therefore, for any G-scheme X over F the inverse image map with respect to the structure morphism $X \to \operatorname{Spec} F$ is a homomorphism $R(G) \to K_0'(G;X)$, making the latter group a module over $R(G)$.

Example 11.3 Let H be a subgroup in G. For any representation $\rho : H \to \mathbf{GL}(V)$ one can associate the G-vector bundle $P_\rho = (G \times V)/H$ over $Y = G/H$, where the H-action on $G \times V$ is given by $h \cdot (g, v) = (gh^{-1}, \rho(h)(v))$. We get an equivalence of categories $\operatorname{Rep}(G)$ and $\mathcal{M}(G;Y)$. Hence, the assignment $\rho \mapsto [P_\rho]$ induces an isomorphism

$$R(H) \xrightarrow{\sim} K_0'(G;Y).$$

12. Split reductive groups

In this section we consider the case of a split reductive group. The main result is the following

THEOREM 12.1. *Let G be a split reductive group defined over F with $\pi_1(G)$ torsion free and X be a G-scheme. Then there exists a spectral sequence*

$$E^2_{p,q} = \operatorname{Tor}_p^{R(G)}\Big(\mathbf{Z}, K_q'(G;X)\Big) \Rightarrow K'_{p+q}(X). \square$$

The idea of the proof is as follows. We construct the spectral sequence in three steps. First of all, we compare $K_n'(G;X)$ with $K_n'(B;X)$, where $B \subset G$ is a Borel subgroup, using the technique developed by Panin in [30]. We prove that the natural homomorphism

$$R(B) \otimes_{R(G)} K_n'(G;X) \to K_n'(B;X)$$

is an isomorphism. In the second step we prove that the restriction homomorphism $K_n'(B;X) \to K_n'(T;X)$, where $T \subset B$ is a maximal torus, is an isomorphism, using a version of the homotopy invariance theorem proved by Thomason [42]. In the last step we compare $K_n'(T;X)$ and $K_n'(X)$ by using the spectral sequence associated to a family of closed subschemes, constructed by Levine in [17]. This sequence generalizes the localization exact sequence in the equivariant algebraic K-theory.

Corollary 12.2.

(1) *The natural homomorphism*

$$\mathbf{Z} \otimes_{R(G)} K_0'(G;X) \to K_0'(X)$$

is an isomorphism, i.e. $K_0'(X) \simeq K_0'(G;X)/J(G) \cdot K_0'(G;X)$.

(2) *There is an exact sequence*

$$K_2'(X) \to \operatorname{Tor}_2^{R(G)}(\mathbf{Z}, K_0'(G;X)) \to \mathbf{Z} \otimes_{R(G)} K_1'(G;X) \to$$

$$K_1'(X) \to \operatorname{Tor}_1^{R(G)}(\mathbf{Z}, K_0'(G;X)) \to 0.$$

□

Corollary 12.3. *Let* $H \subset G$ *be a closed subgroup of a split reductive group* G *with* $\pi_1(G)$ *torsion free. Then*

$$K_0(G/H) \simeq \mathbf{Z} \otimes_{R(G)} R(H) \simeq R(H)/J(G) \cdot R(H).$$

In particular, the group $K_0(G/H)$ *is generated by the classes of* $P_\rho = (G \times V)/H$ *for all representations* $\rho : H \to \mathbf{GL}(V)$. □

Example 12.4 Let $H \subset \mathbf{GL}_n(F)$ be an embedding. Then for the "classifying variety" $X = \mathbf{GL}_n(F)/H$ of H one has

$$K_0(X) \simeq \mathbf{Z} \otimes_{R\left(\mathbf{GL}_n(F)\right)} R(H).$$

Remark 12.5 If $X = G$, one has $E_{p,q}^2 = \operatorname{Tor}_p^{R(G)}(\mathbf{Z}, K_q(F))$. This formula together with the fact that the spectral sequence degenerates gives the computation of $K_n(G)$ (see [17]).

Assume now that X is a smooth projective G-scheme. One can show that

$$E_{p,q}^2 = \operatorname{Tor}_p^{R(G)}(\mathbf{Z}, K_n'(G;X)) = \begin{cases} K_n'(X), & \text{if } p = 0; \\ 0, & \text{if } p > 0, \end{cases}$$

in other words, the spectral sequence in Theorem 12.1 degenerates, i.e. $E_{p,q}^2 = 0$ if $p > 0$. Hence, we get

Theorem 12.6. *Let* G *be a split reductive group with* $\pi_1(G)$ *torsion free,* X *be a smooth projective* G*-scheme. Then the natural homomorphism*

$$\mathbf{Z} \otimes_{R(G)} K_n'(G;X) \to K_n'(X)$$

is an isomorphism. Thus, the homomorphism $\operatorname{res} : K_n'(G;X) \to K_n'(X)$ *is a surjection with the kernel* $J(G) \cdot K_n'(G;X)$. □

13. Semisimple groups of inner type

Let G be a simply connected semisimple group over F and $\rho : G \to \mathbf{GL}_1(A)$, be an algebraic group homomorphism, where A is a central simple F-algebra. Since A_{sep} is a matrix algebra over F_{sep}, ρ_{sep} is a representation of the group G_{sep}. We call A a *Tits algebra of the group* G, if ρ_{sep} is an irreducible representation over F_{sep}. A Tits algebra A is uniquely determined, up to an isomorphism, by the representation ρ_{sep} (see [45]).

Denote by C the center of G. It is a group scheme of multiplicative type. The restriction of the homomorphism ρ to C is given by the multiplication by some character $\chi \in C^* = \mathrm{Hom}(C, \mathbf{G}_{m,F})$. Moreover, any character in C^* can be obtained in such a way. The class of the corresponding Tits algebra A in the Brauer group $\mathrm{Br}(F)$ does not depend on the choice of ρ, but only on the restriction of ρ to the center C. Thus, we get the *Tits homomorphism*

$$\beta : C^* \to \mathrm{Br}(F).$$

A Tits algebra A is called a *fundamental Tits algebra* if ρ_{sep} is a fundamental representation. If G is a group of inner type, then the fundamental Tits algebra exists for any fundamental representation of G_{sep}.

Example 13.1 The fundamental Tits algebras of $\mathbf{SL}_1(A)$, where A is a central simple F-algebra of degree n, are the λ-powers $\lambda^i A$ of A, $i = 1, 2, \dots, n-1$ (see [15]).

Let $\rho : G \to \mathbf{GL}_1(A)$ be a group homomorphism, where A is a central simple F-algebra. We denote by ${}^{\rho}A$ the algebra A together with the G-action given by the formula

$$ga = \rho(g) \cdot a \cdot \rho(g)^{-1},$$

where $a \in A$ and g is in G. We simply write A for the algebra A together with the trivial G-action.

Let X be a G-scheme. We consider the abelian category $\mathcal{M}(G; X, A)$ of G-A-modules and morphisms of $A \otimes_F \mathcal{O}_X$- and G-modules and set

$$K'_n(G; X, A) = K_n\Big(\mathcal{M}(G; X, A)\Big).$$

The exact functor (being an equivalence of categories)

$$\mathcal{M}(G; X, A) \to \mathcal{M}(G; X, {}^{\rho}A),$$

taking $M \in \mathcal{M}(G; X, A)$ to the same $A \otimes \mathcal{O}_X$-module M with the G-action given by

$$g \bullet m = \rho(g) \cdot gm,$$

induces an isomorphism

$$K'_n(G; X, A) \xrightarrow{\sim} K'_n(G; X, {}^{\rho}A).$$

The image of an element $u \in K'_n(G; X, A)$ in $K'_n(G; X, {}^{\rho}A)$ we denote by ${}^{\rho}u$.

Denote by $\langle u\rangle$ the image of an element $u \in K'_n(G;X,{}^{\rho}A)$ under the homomorphism

$$\mathrm{for} : K'_n(G;X,{}^{\rho}A) \to K'_n(G;X)$$

induced by the forgetful functor

$$\mathcal{M}(G;X,{}^{\rho}A) \to \mathcal{M}(G;X).$$

THEOREM 13.2. *Let G be a simply connected semisimple group of inner type, X be a G-scheme. Then the restriction homomorphism $K'_0(G;X) \to K'_0(X)$ is surjective, and its kernel is generated by the elements $\langle {}^{\rho}u\rangle - \langle u\rangle$ for all fundamental Tits representations $\rho : G \to \mathbf{GL}_1(A)$ and all $u \in K'_0(G;X,A)$.* □

As an application we compute the Grothendieck group of an adjoint group of inner type.

THEOREM 13.3. *Let G be an adjoint semisimple group of inner type over F, $\pi : \widetilde{G} \to G$ the universal covering, $C = \ker(\pi)$. Then the natural homomorphism*

$$R(C)/J(\widetilde{G}) \cdot R(C) = \mathbf{Z} \otimes_{R(\widetilde{G})} R(C) \to K_0(G)$$

is an isomorphism. □

Consider the isomorphism in the Theorem in detail. First of all, note that $R(C) \simeq \mathbf{Z}[C^*]$. For any character $\chi \in C^*$ consider the line bundle $P_\chi = (\widetilde{G} \times \mathbf{A}^1_F)/C$ on G, where the C-action is given by $c \cdot (\tilde{g}, x) = (\tilde{g}c^{-1}, \chi(c)x)$ and denote by n_χ the greatest common divisor of dimensions of all representations of $\widetilde{G}_{\mathrm{sep}}$ over F_{sep} with the restriction on C given by the multiplication by χ. These numbers depend only on the Dynkin diagram of G and were computed in [26].

COROLLARY 13.4. *The assignment $\chi \mapsto [P_\chi]$ induces an isomorphism*

$$\mathbf{Z}[C^*]/A \xrightarrow{\sim} K_0(G),$$

where A is the subgroup in $\mathbf{Z}[C^*]$ generated by

$$n_{\chi\psi^{-1}} \cdot \operatorname{ind} \beta(\chi\psi^{-1}) \cdot (\chi - \psi)$$

for all $\chi, \psi \in C^*$ (here $\beta : C^* \to \mathrm{Br}(F)$ is the Tits map). In particular, the group $K_0(G)$ is generated by the classes of linear vector bundles P_χ, $\chi \in C^*$. □

Example 13.5 Let (V,q) be a quadratic form over F of dimension $2n+1$, $G = \mathbf{O}_+(V,q)$ be the special orthogonal group. Then $\widetilde{G} = \mathbf{Spin}(V,q)$ is the universal covering group, $C = \boldsymbol{\mu}_2$, $C^* = \{0, \chi\}$ and $n_\chi = 2^n$ (see [26]). The fundamental Tits algebra corresponding to the character χ is the even Clifford algebra $C_0(V,q)$. Hence

$$K_0\Big(\mathbf{O}_+(V,q)\Big) = \mathbf{Z} \cdot 1 \oplus \Big(\mathbf{Z}/2^n \operatorname{ind} C_0(V,q)\mathbf{Z}\Big) \cdot ([P_\chi] - 1).$$

14. Arbitrary reductive groups

Let G be an arbitrary reductive group over a field F. We would like to find a condition on G under which the restriction homomorphism

$$K_0'(G;X) \to K_0'(X)$$

is surjective for all G-schemes X. Assume that taking $X = G_E$ for some finite field extension E/F, we get a surjection, i.e. the homomorphism

$$\mathbf{Z} = K_0(E) = K_0'(G;G_E) \to K_0'(G_E)$$

is surjective. Therefore, $K_0'(G_E) = \mathbf{Z} \cdot 1$. Hence, the first term of the topological filtration $K_0'(G_E)^{(1)}$ of $K_0'(G_E)$ (see [36, §7.5]), being the kernel of the rank homomorphism $K_0'(G_E) \to \mathbf{Z}$, is trivial, in particular,

$$\mathrm{Pic}(G_E) \simeq K_0'(G_E)^{(1/2)} = 0,$$

i.e. G is a factorial group.

Surprisingly, the converse is true.

THEOREM 14.1. *Let G be a reductive group defined over a field F. Then the following condition are equivalent:*
(1) G is factorial.
(2) For any quasi-projective G-scheme X the restriction homomorphism

$$K_0'(G;X) \to K_0'(X)$$

is surjective. □

COROLLARY 14.2. *Let H be a subgroup in G. Then $K_0(G/H)$ is generated by the classes of vector bundles $(G \times V)/H$ over G/H for all linear representations $H \to \mathbf{GL}(V)$.* □

Remark 14.3 One can drop the assumption for X to be quasi-projective in the case when the factor group G/G' modulo the commutant is a quasi-trivial torus. For example, a semisimple group satisfies this property.

At the end of this section we consider the smooth projective case:

THEOREM 14.4. *In the conditions of Theorem 14.1 let X be a smooth projective G-scheme over F. Then the restriction homomorphism*

$$K_n'(G;X) \to K_n'(X)$$

is split surjective. □

15. Applications

In this section we consider some applications of the mentioned results.

THEOREM 15.1. *Let G be a reductive group over F, H be a connected subgroup in G, $X = G/H$. Then the group $K_0(X)$ is finitely generated.* □

Corollary 15.2. *Let G be a reductive group over F. Then the kernel $K_0(G)^{(1)}$ of the rank homomorphism $K_0(G) \to \mathbf{Z}$ is a finite group. In particular, $K_0(G)$ is a finitely generated group of $\mathbf{Q}$-rank 1.* □

Remark 15.3 The group $K_0(G)$ for an algebraic torus G was computed in terms of generators and relations in [27].

Let G be an algebraic group and X be a G-scheme over F. For any $g \in G(F)$ denote the automorphism $x \mapsto g \cdot x$ of X by λ_g. The group $G(F)$ acts naturally on $K'_n(X)$ by the inverse images λ_g^*.

Proposition 15.4. *Let G be a reductive group. Then*

(1) *For any quasi-projective G-scheme X the group $G(F)$ acts trivially on $K'_0(X)$.*
(2) *For any smooth projective G-scheme X the group $G(F)$ acts trivially on $K'_n(X)$ for any $n \geq 0$.* □

Corollary 15.5. *The classes of rational points from the same $G(F)$-orbit in $X(F)$ coincide in $K'_0(X)$. In particular, the classes of rational points of G in $K_0(G)$ are pairwise equal.* □

References

[1] J.F. Adams. *Lectures on Lie Groups,* Benjamin, New York, 1969.

[2] A.A. Albert. Tensor products of quaternion algebras. *Proc. Amer. Math. Soc.* **35** (1972), 65–66.

[3] A. Blanchet. Function fields of generalized Brauer-Severy varieties. *Comm. Algebra.* **19** (1991), 97–118.

[4] A. Borel. Linear Algebraic Groups, Second Enlarged Edition. *Graduate Texts in Mathematics.* Vol. 126, Springer, Berlin, 1991.

[5] A. Borel, T.A. Springer. Rationality properties of linear algebraic groups II. *Tôhoku Math. J.* **20** (1968), 443–497.

[6] V.I. Chernousov. The group of multipliers of the canonical quadratic form and stable rationality of the variety *PSO* (in Russian). *Matematicheskie zametki.* **55** (1994), 114–119.

[7] V.I. Chernousov, A.S. Merkurjev. *R*-equivalence and special unitary groups. Preprint, Universität Bielefeld, SFB343, 96-083 (1996).

[8] V.I. Chernousov, V.P. Platonov. The rationality problem for group varieties. *C.R. Acad. Sci. Paris.* **22** (1996), 245–250.

[9] C. Chevalley. On algebraic group varieties. *J. Math. Soc. Jap.* **6** (1954), 303–324.

[10] J.-L. Colliot-Thélène, J.-J. Sansuc. La *R*-équivalence sur les tores. *Ann. Scient. Éc. Norm. Sup., 4e série.* **10**, (1977), 175–230.

[11] P. Deligne, D. Sullivan. Division algebras and the Hausdorff–Banach–Tarski paradox. *L'Enseignement Math.* **29** (1983), 145–150.

[12] M. Demazure, A. Grothendieck. *Schémas et Groupes III.* (SGA 3, Tome III). Lect. Notes in Math. **153**, Springer, Heidelberg, 1970.

[13] J.E. Humphreys. *Linear Algebraic Groups.* Springer, New York, 1975.

[14] M.-A. Knus. *Quadratic and Hermitian Forms over Rings.* Grundlehren Math. Wiss. 294, Springer-Verlag, Berlin, 1991.

[15] M.-A. Knus, A.S. Merkurjev, M. Rost, J.-P. Tignol. *Book of involutions,* in preparation.

[16] T.-Y. Lam. *The Algebraic theory of Quadratic Forms.* Benjamin, Reading, Massachusetts, 1973.

[17] M. Levine. *The algebraic K-theory of the classical groups and some twisted forms,* Duke Math. J., **20**, (1993), 405–443.

[18] Yu.I. Manin. *Cubic forms.* North-Holland, Amsterdam, 1974.

[19] A.S. Merkurjev. Generic element in SK_1 for simple algebras. *K-theory.* **7** (1993), 1–3.

[20] A.S. Merkurjev. Involutions and algebraic groups. Preprint. Recherches de mathématique, No. 36, Université Catholique de Louvain, Louvain-la-Neuve, Belgium, 1993.

[21] A.S. Merkurjev, J.-P. Tignol. The multipliers of similitudes and the Brauer group homogeneous varieties. *J. reine angew. Math.* **461** (1995), 13–47.

[22] A.S. Merkurjev. Norm principle for algebraic groups, *St. Petersburg Math.J.* **7** (1996), 243–264.

[23] A.S. Merkurjev. R-equivalence and rationality problem for semisimple adjoint classical algebraic groups, *Pub. Math. I.H.E.S.* vol. 84 (1996), 189–213.

[24] A.S. Merkurjev. Comparison of equivariant and ordinary K-theory of algebraic varieties, to appear in St. Petersburg Math. J.

[25] A.S. Merkurjev, I.A. Panin, A. Wadsworth. A list of index reduction formulas. Preprint 94–079, SFB preprint series, University of Bielefeld, Bielefeld, Germany, 1994.

[26] A.S. Merkurjev, I.A. Panin, A. Wadsworth. Index reduction formulas for twisted flag varieties, I, *K-theory* vol. 10 (1996), 517–596.

[27] A.S. Merkurjev, I.A. Panin. *K-theory of algebraic tori and toric varieties, K-theory* vol. 12 (1997), 101–143.

[28] A.P. Monastyrnyĭ, V.I. Yanchevskiĭ. Whitehead groups of spinor groups. *Math. USSR Izvestiya* **36** (1991), 61–100.

[29] I.A. Panin. *A splitting principle,* Preprint, 94–036, Bielefeld University, 1994.

[30] I.A.Panin, *On the algebraic K-theory of twisted flag varieties*, K-Theory, **8**, (1994), 541–585.

[31] I.A. Panin. K-theory of twisted flags and a general index reduction formula. *Mathematisches Forschungsinstitut, Oberwolfach, Tagungsbericht Quadratische Formen 18.06.95 – 24.06.95* **24** (1995), 11 – 12.

[32] V.P. Platonov. Algebraic groups and reduced K-theory. *Proc. Inter. Congr. Math., V.1. Helsinki,* 1978, 311–317.

[33] V.P. Platonov. On the problem of rationality of spinor varieties. *Soviet Math. Dokl.* **20** (1979), 1027–1031.

[34] V.P. Platonov, V.I. Chernousov. On the rationality of canonical spinor varieties. *Soviet Math. Dokl.* **21** (1980), 830–834.

[35] V.P. Platonov, A. Rapinchuk. *Algebraic Groups and Number Theory.* Academic Press, 1994.

[36] D. Quillen. *Higher algebraic K-theory I,* Lect. Notes Math., Springer-Verlag, Berlin **341**, (1972), 85–147.

[37] M. Rosenlicht. Some rationality questions on algebraic groups. *Annali di Mat.(IV)* **43** (1957), 25–50.

[38] J.-J. Sansuc. *Group de Brauer et arithmétique des groupes algébrique linéaires sur un corps de nombres,* J. Reine Angew. Math., **327**, (1981), 12–80.

[39] J.-P. Serre. *Cohomologie Galoisienne*, Lect. Notes Math., **5**, Cinquième édition, révisée et complétée, (1994), Springer-Verlag, Berlin.

[40] R. Steinberg. *On a theorem of Pittie,* Topology, **14**, (1975), 173–177.

[41] A.A. Suslin. SK_1 of division algebras and Galois cohomology. *Advances in Soviet Math.* **4** (1991), 75–100.

[42] R. Thomason. *Algebraic K-theory of group scheme action,* in: Algebr. topol. and algebr. K-theory: Proc. conf. Princeton, oct.24–28, 1983, Princeton, N.J., (1987), 539–563.

[43] J. Tits. Classification of algebraic semisimple groups. *in: Algebraic Groups and discontinuous Subgroups, (eds. A.Borel and G.D.Mostow), Proc. Symp. Pure Math., Vol. 9, 1966*, 33–62

[44] J. Tits. Formes quadratiques, groupes orthogonaux et algèbres de Clifford. *Invent. Math.* **5** (1968), 19–41.

[45] J. Tits. *Représentations linéaires irréductibles d'un groupe réductif sur un corps quelconque,* J. Reine Angew. Math., **247**, (1971), 196–220.

[46] V.E. Voskresenskiĭ. *Algebraic tori,* Nauka, Moscow, 1977 (Russian).

[47] V.E. Voskresenskiĭ, A.A. Klyachko. Toroidal Fano varieties and root systems. *Math. USSR Izvestiya.* **24** (1985), 221–244.

[48] S. Wang. On the commutator group of a simple algebra. *Amer. J. Math.* **72** (1950), 323–334.

[49] A. Weil. Algebras with involutions and the classical groups. *J. Indian Math. Soc.* **24** (1960), 589–623.

[50] V.I. Yanchevskiĭ. Simple algebras with involution and unitary groups. *Math. Sbornik* **22** (1974), 372–385.

Progress in Mathematics, Vol. 169,

Surprising Geometric Phenomena in High-Dimensional Convexity Theory

V. Milman

Sackler Faculty of Exact Sciences
Tel Aviv University
Tel Aviv 69978, Israel

Abstract. We shall describe asymptotic phenomena for high-dimensional objects, which violate intuition based on traditional studies of objects of low dimension. We will illustrate these phenomena with examples which involve convex bodies and homogeneous structures (both continuous and discrete) for which other parameters play the role of dimension. Our aim will be to convey some of the new intuition about high-dimensional structures, that underlies such counterintuitive results. We will also discuss the notion of "isomorphic" geometry as the study of common geometric features of families of, say, convex bodies, and analyze their unexpected asymptotic behavior as the dimension increases to infinity. The underlying methods use different mathematical tools and are useful in a variety of apparently unrelated mathematical areas.

1. Introduction

In this lecture we discuss results which stand between Geometry and Functional Analysis. The theory was built during the last two decades. It considers geometric problems via a Functional Analysis point of view. Consequently, the "isometric" problems and point of view typical for geometry are substituted by "isomorphic" ones. This became possible with the *asymptotic* approach (with respect to dimension increasing to infinity) to the study of high-dimensional spaces. The goal of this lecture is to demonstrate, with some examples and results, a new intuition which corresponds to high dimensional spaces and present some discoveries and highlights of this theory. We avoid the technical part of the theory and will emphasize the geometric insight. In fact, we restrict this presentation to one facet of the theory: isomorphic (asymptotic) geometry and related phenomena. We do not discuss (for example) the very important type-cotype theory with the deep probabilistic connections behind it. To fill these and other omissions I recommend the following books: [MSch86], [P86a], [P86b], [P89a], [TJ88] and surveys [M88a], [M92], [L92], [LM93].

Consider a finite dimensional normed space $X = (\mathbb{R}^n, \|\cdot\|)$. Such a space is defined by its unit ball $K_X = \{x \in \mathbb{R}^n, \|x\| \leq 1\}$. Inversely, if K is a convex

*) Partially supported by a BSF Grant.

centrally-symmetric compact body in $\mathbb{R}^n$, then $X_K = (\mathbb{R}^n, \|\cdot\|_K)$ is the normed space with the unit ball K. Let $|x|$ be the canonical euclidean norm in $\mathbb{R}^n$, (x,y) the standard inner product, and denote $\mathcal{D}$ the standard euclidean ball, i.e. $\mathcal{D} = K_{(\mathbb{R}^n,|\cdot|)}$.

A few preliminary examples. We start with three observations:

a) ([GrM84], [MP89]) Fix $0 < \delta < \frac{1}{2}$ (say, $\delta = \frac{1}{4}$). Define the *floating body* K_δ of K as the intersection of the half spaces $\{x \in \mathbb{R}^n \mid (x,\theta) \le m_\delta(\theta)\}$, $\theta \in \mathbb{R}^n$, where $m_\delta(\theta)$ is defined by

$$\operatorname{Vol}\big\{x \in K \bigm| (x,\theta) > m_\delta(\theta)\big\} = \delta \operatorname{Vol} K\ .$$

Then, *there is a number $C(\delta)$, independent of dimension n or $K \subset \mathbb{R}^n$, such that for any symmetric convex body K the floating body K_δ is uniformly, up to a factor $C(\delta)$, isomorphic to an ellipsoid; this means that there is an ellipsoid $\mathcal{E}$, s.t.*

$$\mathcal{E} \subset K_\delta \subset C(\delta)\mathcal{E}\ .$$

[Note that the initial body K could be very far from any ellipsoid; it could be, say, a cross-polytope (= the unit ball of ℓ_1^n) or a cube, but the "regularization" described above, which is obtained by cutting a fixed portion of volume in any direction, brings us to a $C(\delta)$-neighborhood of an ellipsoid.]

Moreover, this ellipsoid is homothetic to the Legendre ellipsoid of inertia $\mathcal{L}(K)$ of K.

Let us formally introduce a multiplicative geometric distance $d(\cdot,\cdot)$ between convex bodies K and T: $d(K,T) = \inf\{b/a \mid K \subset bT \text{ and } aT \subset K\}$. The Banach-Mazur distance between two normed spaces X_K and X_T is $d(X_K, X_T) = \inf\{d(K,uT) \mid u \in GL_n\}$.

b) *Centroid body* (of Petty [Pet61]) (or "zonoid of inertia" by [MP89]). Consider a body $Z(K) = \frac{1}{|K|}\int_K [0,x]dx$ which is a "Minkowski" integral, meaning that the finite sums approximating the integral are Minkowski sums of sets: intervals $[0,x_i]$ with suitable weights. Geometrically, $\partial Z(K)$ is the collection of centroid points (centers of gravity) of "halves" of our body K obtained by cutting K by hyperplanes passing through the origin.

Again, as in a), there is a universal constant $C > 0$, such that $\forall\, n$, $\forall\, K \subset \mathbb{R}^n$, $d(Z(K), \mathcal{L}(K)) \le C$ (and $C \le \sqrt{8}$).

c) *Lattice tiling*. Let the inner part $\overset{\circ}{K} \neq \emptyset$. Denote $K_i = K + x_i$ for $x_i \in \mathbb{R}^n$. We call $\{K_i\}$ a tiling of $\mathbb{R}^n$ (and say that K produces a tiling by shifts) if (i) $\bigcup\limits_i K_i = \mathbb{R}^n$ and (ii) $\overset{\circ}{K}_i \cap \overset{\circ}{K}_j = \emptyset$ for $i \neq j$. A lattice tiling is a tiling such that the set $\{x_i\}$ is a lattice, i.e. $\{x_i\} = A\mathbb{Z}^n$, where A is an invertible linear map and $\mathbb{Z}^n \subset \mathbb{R}^n$ is the set of all integer vectors of $\mathbb{R}^n$.

Trivially, an affine image of the cube $[-1,1]^n$ produces a lattice tiling and the euclidean ball does not. However, do uniformly isomorphic versions K_n of the euclidean balls $\mathcal{D}_n$ exist which produce a tiling? Surprisingly, the answer is "Yes". More precisely, *for any integer n there is a convex symmetric body $K_n \subset \mathbb{R}^n$, $\mathcal{D}_n \subset K_n \subset 3 \cdot \mathcal{D}_n$, such that K_n produces a lattice tiling of* $\mathbb{R}^n$. (This observation of Alon-Milman follows immediately from some known results of [R50], [Bu72], [Bou87], [Ban90], on lattice covering-packing; see [M92] for details and references.)

Isomorphic Geometry. The three examples above lead us to a notion of "isomorphic ellipsoid": a family of convex bodies $\{K_\alpha\}$ of infinitely increasing dimension represents an "isomorphic ellipsoid" if there is a constant C and a family of ellipsoids $\{\mathcal{E}_\alpha\}$ such that $\mathcal{E}_\alpha \subset K_\alpha \subset C\mathcal{E}_\alpha$ for every K_α in the family, i.e. $d(K_\alpha, \mathcal{E}_\alpha) \le C$. So, in example a) we can state that for a fixed δ, $0 < \delta < \frac{1}{2}$, the family of δ-floating bodies $\{K_\delta \mid \forall n\ \forall K \subset \mathbb{R}^n\}$ is an isomorphic ellipsoid. (We may add: $C(\delta)$-isomorphic ellipsoid).

So, we naturally derive *isomorphic* geometric objects and *isomorphic* geometric results. Of course, an "isomorphic" geometric object is, in fact, a family of objects in different spaces of increasing dimension and by "isomorphic" geometric properties of such an "isomorphic" object we mean a common property of this family. Classical Geometry (in a fixed dimension) is an isometric theory (or, at most, "almost" isometric, ε-isometric for small $\varepsilon > 0$). "Isomorphic" study means asymptotic behavior determined by some parameter (most often it is dimension $n \to \infty$) and exact control of dependence of the constants involved on this parameter (say, dimension in most examples in this lecture).

The appearance of such an isomorphic geometric object is a new feature of asymptotic high-dimensional theory. Geometry and Analysis meet here in a non-trivial way. We will meet later *geometric inequalities in isomorphic form*. The study of special examples of such inequalities, "isomorphic isoperimetric inequalities" led to the discovery of the "Concentration Phenomenon" – one of the most powerful tools of asymptotic theory, responsible for many counter-intuitive results in the theory.

2. Another example – theorem; correction of intuition

Our next example is already a non-trivial theorem originally proved in 1988 (see [M91]).

THEOREM 2.1. *There is a universal constant C so that for any n and any centrally-symmetric, compact, convex body $K \subset R^n$, there are two linear operators $u_1, u_2 \in SL_n$ so that if $P = K \cap u_1 K$ and $Q = \mathrm{conv}(P \cup u_2 P)$, then the body Q is within distance C of an ellipsoid: that is, there is an ellipsoid $\mathcal{E}$ for which $\mathcal{E} \subset Q \subset C\mathcal{E}$.*

In the language of Functional Analysis, the same fact can be reformulated in the following form:

THEOREM 2.2. *For every finite dimensional normed space $X = (\mathbb{R}^n, \|\cdot\|)$ there are three linear operators $T_1, T_2, T_3 \subset GL_n$, such that the following relation holds:*

Consider $p(x) = \|T_1 x\| + \|T_2 x\|$ and an inf-convolution

$$q(x) = p(x) * p(T_3 x) \quad [\text{by definition } p(x) * p(ux) \equiv \inf_{y+z=x} \{p(y) + p(uz)\}] \ ;$$

then $q(x)$ is C-isomorphic to the standard euclidean norm in $\mathbb{R}^n$:

$$|x| \le q(x) \le C|x| \ .$$

Note that C does not depend on the dimension n or on the initial norm $\|\cdot\|$. In fact, an even stronger statement is correct:

THEOREM 2.3. *For any four centrally-symmetric convex bodies $K_i \subset \mathbb{R}^n$, $i = 1,2,3,4$, $|K_i| = |\mathcal{D}|$, there are $\{u_i\}_1^4 \subset \mathrm{SL}_n$ such that if $P_1 = u_1K_1 \cap u_2K_2$, $P_2 = u_3K_3 \cap u_4K_4$ and $Q = \mathrm{Conv}\, P_1 \cup P_2$, then $\frac{1}{C}\mathcal{D} \subset Q \subset C\mathcal{D}$ for some universal constant C independent of the dimension n and $\{K_i\}_1^4$.*

There are many remarkable ellipsoids associated with a given body K. They recover different traces of hidden symmetries which exist in any high-dimensional convex body. If K has a "large" group of symmetries then all these ellipsoids coincide. Well-known, for example, are the maximal volume ellipsoid inscribed in K (introduced by Löwner and studied by F. John), the minimal volume ellipsoid circumscribed around K or the Legendre and Binet ellipsoids of inertia. There are many other important constructions of ellipsoids associated with K and an ellipsoid which Theorem 2.1 provides is one of them. We call it the M-ellipsoid and know how to identify it a priori through some geometric inequality. Changing coordinates we may consider this ellipsoid $\mathcal{E}$ to be the standard euclidean ball $\mathcal{D} \subset \mathbb{R}^n$; let $w \in GL_n$ be such that $w\mathcal{E} = \mathcal{D}$. We call the family of *position* of K the family $\{vK\}_{v \in GL_n}$. So the affine image $wK = \widehat{K}$ is a *position* of K. Applying Theorem 2.1 to K in such a position, i.e. to $\widehat{K}$, we may additionally claim that $(u_1; u_2)$ are orthogonal pairs of operators and "most" (meaning with high probability with respect to the Haar probability measure of orthogonal groups) pairs $(u_1; u_2) \in O(n) \times O(n)$ will regularize $\widehat{K}$ to an isomorphic euclidean ball $\mathcal{D}$.

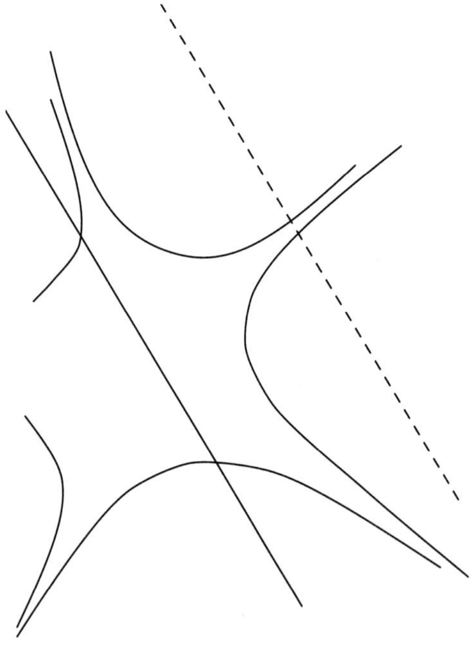

Fig. 1

Clearly, every position of K produces the unit ball of isometrically the same normed space as X_K. It is an interesting feature of an (asymptotic) high dimensional theory of convex sets which we will meet throughout the lecture that we are, in fact, forced to consider the family of all positions of a given K (that is all affine images uK, $u \in \mathrm{GL}_n$) even when we are aiming at some volume inequalities or other properties of an individual K. (See K. Ball [B91] for the use of the family of positions in isoperimetric type problems.)

Let us try to develop an intuition of high-dimensional spaces. To start with, we should understand how to draw "high-dimensional" pictures, how high-dimensional convex bodies "look".

The first non-intuitive point is that 2 or 3 dimensional "pictures" of a high-dimensional convex body should have a "hyperbolic" form! The reason is that the volume of parallel hypersections decays exponentially after passing the median level (see Fig.1).

This is a *fact* observed in [GrM84]. Yes, K is a convex set but the rate of volume decay has a crucial influence on the geometry and we should find a way to visualize it in our pictures.

Now, with such a picture in mind, let us try, intuitively, to understand the theorem.

We use the following notation: $|K|$ defines the volume of K and $K^\circ = \{x \in \mathbb{R}^n \mid (x, y) \leq 1 \text{ for every } y \in K\}$ is the polar body of K.

Let $|K| = |\mathcal{D}|$; for some $R > 1$ almost no $(n-1)$-dimensional volume belongs to the intersection $R \cdot \partial\mathcal{D} \cap K$. Then find a turn u_1 s.t. (see Fig. 2)

$$(R\partial D \cap K) \cap (R\partial D \cap u_1 K) = \emptyset \ .$$

So $P = K \cap u_1 K \subset R \cdot D$.

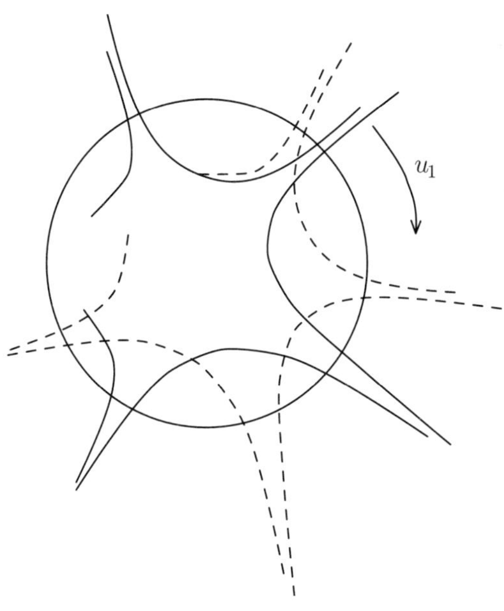

Fig. 2

But K *is convex* (finally, we recall this fact), so consider $P^\circ \supset \frac{1}{R}D$. Do the same operation with the body P°: $Q^\circ = P^\circ \cap u_2 P^\circ \subset RD$ and it looks as though we succeeded:

$$\frac{1}{R} D \subset Q \subset R \cdot D .$$

However: When does exponential decay start? What is the value of R? For many classical ellipsoids (say, maximal vol. ellipsoids), R may be as large as $\sim \sqrt{n/\log n}$. For some other ellipsoids $R \sim \log n$. It was necessary to construct a special ellipsoid (better to say, to prove its existence) for which $R \sim$ Const. However, this is not yet enough. The result above is a "metric one": we want $(K \cap u_1 K \cap \partial(R \cdot D))$ to be empty, not just to have small volume; so more corrections are necessary, but the idea is correct. Just that instead of estimating decay of measure, we should consider decay of *coverings*.

3. Isomorphic geometric inequalities

Let us return to the ellipsoid which appeared in Theorem 2.1. It was originally constructed in connection with some geometric inequality (but, again, inequality in isomorphic form).

THEOREM 3.1. *There exists a universal constant C, such that for any convex symmetric body K in $\mathbb{R}^n$, there is an ellipsoid $\mathcal{M}_K$, $|\mathcal{M}_K| = |K|$, such that for any (not only convex) set T*

$$\text{(i)} \qquad \frac{1}{C} |\mathcal{M}_K + T|^{1/n} \le |K + T|^{1/n} \le C |\mathcal{M}_K + T|^{1/n}$$

where $K+T = \{x+y \mid x \in K, y \in T\}$ is the Minkowski addition of sets. (We ignore measurability problems; they are not crucial here). And for any convex symmetric set T

$$\text{(ii)} \qquad \frac{1}{C} |\mathcal{M}_K \cap T|^{1/n} \le |K \cap T|^{1/n} \le C |\mathcal{M}_K \cap T|^{1/n} .$$

Define *volume radius* of K by $\operatorname{vrad} K = (|K|/|D|)^{1/n}$. Then Theorem 3.1 states that substituting K by some special ellipsoid $\mathcal{M}_K$ in (i) and (ii) essentially preserves volume radii.

Of course, it immediately implies reversion of the Brunn-Minkowski inequality (which was the initial result in this direction). Recall that the Brunn-Minkowski inequality states that for any two compact sets (say, convex) K and T in $\mathbb{R}^n$,

$$|K + T|^{1/n} \ge |K|^{1/n} + |T|^{1/n} .$$

The reverse Brunn-Minkowski inequality is the following statement:

THEOREM 3.2 [M86]. *$\exists C$ such that for any integer n and any two convex bodies K and T there is a relative position $\widehat{K} = uK$, $u \in SL_n$, of say, K such that*

$$|\widehat{K} + T|^{1/n} \le C\big(|K|^{1/n} + |T|^{1/n}\big) .$$

(Also the same inequality holds for the polars $\widehat{K}^\circ, T^\circ$.)

This is a very easy consequence of Theorem 3.1(i) (but it precedes it).

The original proof of Theorem 3.2 was followed by three other proofs: two by G. Pisier ([P89a], [P89b]) and another one of [M88b]. All four proofs being different in their technical presentations added a number of facets to the understanding of the phenomenon behind Theorems 3.1 and 3.2. One such addition is the following *covering property*.

Let $N(K,T) = \min\{N \mid \exists \{x_i\}_1^N \subset \mathbb{R}^n, K \subset \bigcup_1^N (x_i + T)\}$.

An equivalent statement for Theorem 3.2 (or Theorem 3.1) is the following fact: *There is a number C such that for any $n \in \mathbb{N}$, any K (convex symmetric) there is an ellipsoid $\mathcal{M}_K$, $|K| = |\mathcal{M}_K|$, such that*

$$N(K, \mathcal{M}_K) \le e^{Cn} \quad \text{and} \quad N(\mathcal{M}_K, K) \le e^{Cn} .$$

This is, of course, the best possible estimate (up to a value of the universal constant C). Note that, for any ellipsoid $\mathcal{E}$, $N(2\mathcal{E}, \mathcal{E}) \sim C_0^n$ (for some $C_0 > 1$) but K may be very far from any ellipsoid.

Of course, the ellipsoid described above is not unique and there are many with different additional properties.

Such additional important information is a decay of the functions $N(K, t\mathcal{M}_K)$ and $N(\mathcal{M}_K, tK)$ when t increases. Pisier [P89a] showed that, for any $p < 2$ there is an M-ellipsoid $\mathcal{E}$ such that $N(\mathcal{E}, tK) + N(K, t\mathcal{E}) \le e^{C_p n/t^p}$ for some constant C_P depending on $p < 2$ only.

Remarkably, in fact, ellipsoids do not play a special role in these facts. We may take *any* convex symmetric body P and the family of its positions $\mathcal{P} = \{uP\}_{u \in GL_n}$. Then the same facts are true for $\mathcal{P}$ instead of the family of ellipsoids. Say:

$\exists C$ such that for any convex body K, there is a position $P_K \in \mathcal{P}$ such that $|K| = |P_K|$ and $N(P_K, K) + N(K, P_K) \le e^{Cn}$.

Also P_K can be put in inequalities (i) and (ii) of Theorem 3.1 instead of $\mathcal{M}_K$. So, we have some (high-dimensional) *Principle*:

The affine family of **any** *fixed convex symmetric body (say P) is rich enough to substitute any other body (K above) in an essential part of volume computations.*

This reflects a new intuition about high dimension: instead of the expected increasing of essentially different bodies in $\mathbb{R}^n$ with increasing $n \to \infty$ (diversity of possibilities) we observe their *decreasing* to essentially one (any) body with its affine class.

In many of the results we have described till now, convexity is not a crucial assumption (but *high* dimension is). One may substitute convexity by another, very weak condition, and Theorems 2.1, 2.2 and 3.1, 3.2 still hold. Define a convolution of two bodies $K \square T = \bigcup_{\substack{x \in K \\ y \in T}} [x, y]$, were $[x, y]$ is the interval joining points x and y.

Let K be a *centrally-symmetric quasi-convex* star body, i.e.

(i) $tK \subset K$, $0 \le t \le 1$ (star-body condition); and $K = -K$

(ii) $K \square K \subset C \cdot K$

(We say that K is C-quasi-convex. Example: $K_{\ell_p^n}$, the unit ball of ℓ_p^n space for $0 < p < 1$, is a $C(p)$-quasi-convex for some constant $C(p)$ independent of dimension n.) Clearly, $C = 1$ iff K is convex.

Then Theorems 3.1 and 3.2 are true for quasi-convex bodies with constant C in these theorems depending on the quasi-convexity constant *only* ([BBP95]). Using this fact, Theorems 2.1 and 2.2 are also extended to a quasi-convex setting ([M95]):

Let $f \in C(S^{n-1})$, $f > 0$ and even. Consider its homogeneous extension to $\mathbb{R}^n$ by $\widehat{f}(x) = |x| f\left(\frac{x}{|x|}\right)$, for $x \neq 0$. Let $K = \{x \in \mathbb{R}^n \mid \widehat{f}(x) \leq 1\}$ be a C-quasi-convex body. Apply Theorem 2.2 to the function $\widehat{f}(x) \equiv \|x\|$ (it is not a norm now but a *quasi-norm*). Then the function $q(x)$ we obtained is C_1-isomorphic to the standard euclidean norm in $\mathbb{R}^n$ where C_1 depends *only* on the C-quasi-convexity constant of K.

As we will see below, these facts reflect the probabilistic nature of high dimension. And I mean by this more than just the fact that we are using probabilistic techniques in many steps of the proofs. In the next section we will discuss the so-called *concentration phenomenon* but here I would like to mention one very recent result which gives a probabilistic description for the norm of an operator

$$A : \ell_2^n \to X .$$

Of course, the norm $\|A\| = \sup\{\|Ax\| \mid |x| \leq 1\}$ is not a probabilistic notion. However, up to a universal factor, i.e. in isomorphic form, it is (!) (see [MSch96]): there is a universal constant C such that

$$\frac{1}{C}\sqrt{n/k}M(A) \leq \|A\| \leq C\sqrt{n/k}M(A)$$

where $M(A) = \int_{S^{n-1}} \|Ax\| d\mu(x)$, $\mu(x)$ is the rotation invariant probability measure on the sphere $S^{n-1} = \{x \mid |x| = 1\}$, and $k = k(A)$ has also a probability meaning:

$$k = k(A) = \max \Big[k \in \mathbb{N} \mid Prob\{E \in G_{n,k} \mid \|Ax\| \leq 2M(A)|x| \quad \text{for} \quad x \in E\} \geq 1 - \frac{k}{n+k} \Big] .$$

Here $G_{n,k}$ is the Grassmann manifold of k-dimensional subspaces of $\mathbb{R}^n$ and the probability measure on $G_{n,k}$ is defined by the euclidean structure ℓ_2^n on $\mathbb{R}^n$.

4. Isoperimetric inequalities in isomorphic form; concentration phenomenon

One of the major tools in the development of asymptotic high dimensional theory are isomorphic forms of isoperimetric inequalities. Let us introduce a general framework for such problems and then let us describe a few examples to feel what we mean by *isomorphic* form of inequalities.

Let (X, ρ, μ) be a metric space (ρ is a metric) with a probability measure μ. For $A \subset X$, let $A_\varepsilon = \{x \in X \mid \rho(x, A) \leq \varepsilon\}$. Define

$$\alpha(X; \varepsilon) = \sup\left\{\mu(X \backslash A_\varepsilon) \,\middle|\, \forall A\ ,\ \mu(A) \geq \tfrac{1}{2}\right\} .$$

#6020 NOVA BROW SLIP 5/21/02 Y7

38132102675409

ORDER: APP200204110298 004

3764354984 QTY: 1

EUROPEAN CONGRESS OF MATHEMATICS

RECEIVED: V#2

ORDERED: VOL# 02

EP199.00

FBC PO# X4410-32 2032228

PREFERS CLOTHBACK
MUST BAND ALL BOXES-NOTE TO PACKER
DO NOT SHIP PAPERBACK SEND TO BIND
USE B FOR BINDERY BILLING
NO LARGE PRINT TITLES W/O CUS APPR
CALL IF PRICE IS MORE THAN $50 HIGH
GREEN INVOICE/LABEL FORMS

So, the value $\alpha(\varepsilon)$ corresponds to a solution of the "ε-isoperimetric problem" for sets of $\frac{1}{2}$-measure.

Only in very few cases, can solutions be described explicitly and $\alpha(X,\varepsilon)$ calculated exactly. The case of a euclidean sphere S^n – solved by P. Lévy [L51] in 1919 is such a lucky case and the most classical one after $\mathbb{R}^n$. The calculation gives an estimate $\alpha(S^{n+1},\varepsilon) \le \sqrt{\frac{\pi}{8}}e^{-\varepsilon^2 n/2}$. We consider here $S^n \subset \mathbb{R}^{n+1}$ equipped with geodesic distance ρ_r and rotation invariant probability measure μ.

This means that for (any) $A_n \subset S^{n+1}$, $\mu(A_n) = \frac{1}{2}$, and $\varepsilon > 0$, $\mu((A_n)_\varepsilon)$ tends to 1 when $n \to \infty$ (and exponentially quickly). This is not a geometric point of view. Classical geometry fixes a space (meaning "n") and considers the dependence on $\varepsilon > 0$ and then it does not observe the phenomenon! However, we do not consider one set "A" but a family of sets $\{A_n \subset S^{n+1}, \mu(A_n) \ge \frac{1}{2}\}$ and a statement which is an isomorphic geometric inequality is a uniform bound for ε-extensions of sets of this family.

At the same time in many (most) interesting cases, we know how to estimate $\alpha(X;\varepsilon)$ without any idea of the structure of the extremal sets. The above estimate for S^n happens to be typical and implies the so called "concentration phenomenon". Typical estimates for many natural families $\{X_n\}$ have a similar form:

$$\alpha(X_n,\varepsilon) \le c_1 e^{-c_2\varepsilon^2 n} . \qquad (*)$$

However, the exact structure of extremal sets or exact value of $\alpha(X;\varepsilon)$ are not known even for such "simple" cases as the torus $X = \mathbb{T}^n = \prod_i^n S_i^1$ equipped with Haar probability measure and distance $d(x,y) = \frac{1}{\sqrt{n}}\sqrt{\sum_1^n \rho(x_i,y_i)^2}$ where ρ is the geodesic distance on S^1 and $x = (x_i)_{i=1}^n$, $y = (y_i)_{i=1}^n$, or the Stiefel manifold $W_{n,2}$ – the unit tangent bundle of S^{n-1}. (Both these examples satisfy $(*)$.)

Another important geometric example (where no extremal sets are known, as well as the exact value of $\alpha(\varepsilon)$) ([GrM83]):

$$\alpha(\mathrm{SO}_n,\varepsilon) \le \sqrt{\frac{\pi}{8}}e^{-\varepsilon^2 n/8} .$$

Such geometric estimates, asymptotic by increasing "dimension" n, form one of the major technical tools of Asymptotic Theory, and the isomorphic form of the inequalities is enough for all applications [and also the only one typically available]. So the isomorphic (=asymptotic) view on isoperimetric problems has freed us from the necessity of solving (exactly) the isoperimetric problems.

[It is an extremely well-developed direction which uses different methods and with many people involved; besides those mentioned above there are Alon, Amir, Bobkov, Borell, Ledoux, Marton, Maurey, Schechtman, Sudakov, Talagrand, Tsirelson and others. (See the books [MSch86], [LT91] and surveys [M88a] and [T95] for details.)]

Let us outline why a bound of the form $(*)$ is so crucial. Consider a 1-Lip function $f(x)$ defined on (X,ρ,μ), i.e. $|f(x) - f(y)| \le \rho(x,y)$.

Denote by L_f the median of $f(x)$, i.e. $\mu\{x \in X \mid f(x) \ge L_f\} \ge \frac{1}{2}$ and $\mu\{x \in X \mid f(x) \le L_f\} \ge \frac{1}{2}$. Then

$$\mu\{x \in X \mid |f(x) - L_f| < \varepsilon\} \ge 1 - 2\alpha(X,\varepsilon) .$$

So, if the value of $\alpha(X,\varepsilon)$ is very small, then values of the Lipschitz function "concentrate" by a measure around one value. This is the case when $X = S^n$ and dimension n is large, as well as for $\mathbb{T}^n$ or SO_n for large n. It is, in fact, a general property of high-dimensional metric probability spaces which we call the "concentration phenomenon". Many different techniques were developed to treat different examples of metric probability spaces and to prove concentration phenomenona for them. Let us describe one discrete example considered by B. Maurey [Mau79].

Let Π_n be the group of permutations of $[1,\dots,n]$ equipped with the counting probability measure $\mu(A)=\frac{\#A}{n!}$ (for any $A \subset \Pi_n$) and Hamming distance $\rho(s,t)=\frac{\#\{i\in[1,\dots,n] | s(i)\neq t(i)\}}{n}$ for any two permutations $s,t \subset \Pi_n$. Then $\alpha(\Pi_n,\varepsilon) \leq c_1 e^{-c_2\varepsilon^2 n}$ for absolute constants c_1 and $c_2 > 0$. Again, we have an inequality of "isoperimetric type" but in a new, isomorphic version, which is enough for what we call the "concentration phenomenon".

This phenomenon is responsible for many "unexpected", "strange" properties of high-dimensional spaces. Behind intuitively "expected" properties of high-dimensional spaces stands the behavior of ε-entropy which increases exponentially with increase in dimension. However, this exponential increase is compensated by the exponential effect of the concentration phenomenon. As a result, we often observe only linear behavior where a priori intuition expects an exponential. Some Examples:

Approximation by Minkowski sums. Let $A+B = \{x+y \mid x \in A, y \in B\}$ be the Minkowski sum of two sets A and B in $\mathbb{R}^n$. Let $I_i = [-x_i, x_i] \subset \mathbb{R}^n$ be intervals of length, say, 1. Consider $T = \sum_{i=1}^N I_i$. We want to approximate a Euclidean ball by such sums, that is, for a given $\varepsilon > 0$ we would like to have $d(T,D) \leq 1+\varepsilon$. Obviously, if $N = n$ then $d(T,D) \geq \sqrt{n}$ and, by an entropy consideration, it looks as if we need at least an exponential by n number of intervals to achieve a good approximation to D. However, an easy geometric interpretation of an old result from [FLM77] (see also [Go85], [Sch88]) shows that there exist intervals $I_i \subset \mathbb{R}^n$, $i = 1,\dots,N_0$, for $N_0 \leq c\frac{n}{\varepsilon^2}$ (c is a numerical constant) such that $d\big(\sum_1^{N_0} I_i, D\big) \leq 1+\varepsilon$.

This direction was later treated intensively as part of a program of estimating the number of intervals needed for approximating zonoids by a sum of intervals (in [Sch87], [BLM89a] and [BLM88]). It is shown there that the above situation is essentially preserved when we substitute intervals by other convex bodies or if we consider approximation by sums of other convex bodies instead of D. For example [BLM88], [Schm91].

THEOREM 4.1. *Let a convex compact body $K \subset \mathbb{R}^n$ be given. There exist orthogonal operators $A_i \in SO_n$, $i = 1,\dots,N_0$ for $N_0 < c\frac{n}{\varepsilon^2}$ such that $T = \frac{1}{N_0}\sum_1^{N_0} A_i K$ satisfies $d(T,D) \leq 1+\varepsilon$ (as usual, c is a numerical constant).*

Classical symmetrizations. Surprisingly very few symmetrizations are necessary to transform any convex body K to ε-neighborhood of a euclidean ball. Just $\frac{1}{2}n\log n + c(\varepsilon)n$ Minkowski symmetrizations will do the job [BLM88].

The situation with Steiner symmetrization is less clear: $\exists\, c > 0$ s.t. starting with $\forall\, K \subset \mathbb{R}^n$, $|K| = |\mathcal{D}|$, we derive T, $c\mathcal{D} \leq T \leq (1+\varepsilon)\mathcal{D}$, after properly chosen $c(\varepsilon)n \log n$ Steiner symmetrizations (note it is again an isomorphic result) [BLM89b]. However, there is resistance for deriving a $(1+\varepsilon)$-approximation by a small number of symmetrizations. Some partial results for higher dimensional symmetrizations are due to Tsolomitis [Ts96].

Another application of the concentration estimate $(*)$ is the following: again, let $f(x) \in C(X)$ be a Lipschitz function $|f(x_1) - f(x_2)| \le \rho(x_1, x_2)$. Then $\mathbb{E}|f| \equiv \|f\|_{L_1(X)}) \sim L_{|f|}$ (median of $|f|$) if $\alpha(X, \varepsilon)$ is very small. (So as we have already observed, f is "almost constant" with high probability.) But also

$$\|f\|_{L_p(X)} \sim L_{|f|} \sim \|f\|_{L_1(X)} ,$$

and we see that L_p-norms are equivalent for Lip. functions on X with small $\alpha(X;\varepsilon)$.

The Brunn-Minkowski inequality implies similar properties for any convex body K ([Bor75], [GrM84], [MP89]):

$$\forall n,\ \forall K \subset \mathbb{R}^n\ ,\ |K| = 1 \quad \text{and} \quad 0 < p, q < \infty$$

a) $\|f\|_{L_p(K)} \equiv \left(\int_K |f(x)|^p dx\right)^{1/p} \le c_{p,q} \|f\|_{L_q(K)}$ for $\forall$ linear function $f(x)$

b) For any $\|\cdot\|$ and $0 < p, q < \infty : \|(\|x\|)\|_{L_p(K)} \le c_{p,q} \|(\|x\|)\|_{L_q(K)}$

One side of such inequalities are just Hölder inequalities; so another side is "reverse" Hölder inequalities which we also call Khinchine type inequalities because $K = \left[-\frac{1}{2}, \frac{1}{2}\right]^n$ in a) gives Khinchine inequalities and in b) gives Kahane's inequalities.

These cases have a probability interpretation (they correspond to independent random variables uniformly distributed on $\left[-\frac{1}{2}, \frac{1}{2}\right]$). However, from our point of view, the role of the "independence" condition, as well as other standard probability type conditions such as, say, "martingale", is to create a "high-dimensional" convex body; but *any* n-dimensional convex body will serve the same purpose. The reason behind applications of these conditions is the high dimension created and not, for example, independence.

So, probability type ideas and approach are combined here with high-dimensional convexity theory. In fact, the starting point for P. Lévy's study of isoperimetric inequality for S^n (and its consequence – the concentration phenomenon for the sphere S^n) was a geometric interpretation by E. Borel ($\sim$1914) of the law of large numbers.

Let $C^n = [-1, 1]^n$ be a cube in $\mathbb{R}^n$ with the standard euclidean distance "dist". Then $\operatorname{diam} C^n = 2\sqrt{n}$. Consider a linear functional f, $f(x) = \sum_1^n x_i$ (i.e. $\operatorname{Ker} f = (1, \dots, 1)^\perp$). Then $(1/2^n)\operatorname{Vol}\{x \in C^n : \operatorname{dist}(x, \operatorname{Ker} f) \ge \varepsilon\sqrt{n}\} = P\{\left|\frac{1}{n}\sum_1^n \xi_i\right| > \varepsilon \mid (\xi_i)|_{i=1}^n$ are uniformly distributed on $[-1, 1]$ independent random variables$\}] \le c \exp\left(-\frac{\varepsilon^2 n}{2}\right)$. Therefore, "most" of the volume of C^n is concentrated near a "small slice" (relative to the diameter). However, the concentration phenomenon for a cube C^n implies a similar estimate for any 1-Lip. function on C^n and not only for a particular linear functional. So, in a sense, the concentration phenomenon is a "non-linear" version of deviation-inequality type results in Probability.

Returning to a) and b) equivalences above, Bourgain [Bou91] extended a) to any polynomials (with a constant $c_{p,q}(d)$ depending also on the degree d of the polynomial). An interesting case of a) and b) is $q \to 0$. It was shown in [MP89] that $c_{p,q}$ are independent of q for linear functionals and recently Latala [La96] showed that the constants $c_{p,q}$ are independent of q also in b).

The last example of family of V spaces X_n with a concentration property I would like to demonstrate is also discrete. It is taken from [AlM85] and is useful in Computer Science (see also [Al86]).

Let $G = (V, E)$ be a connected graph on $|V| = n$ vertices. We equip V with the counting probability measure $\mu(A \subset V) = |A|/|V|$ and with the path metric: $\rho(x, y) = \{$the smallest number of edges in a path which joins x and $y\}$.

Let $L^2(V)$ denote the space of real-valued functions on V with the usual scalar product (f, g) and the usual norm $\|f\| = \sqrt{(f, f)}$ induced by it. Consider the quadratic form

$$(Qf, f) = \sum_{e \in E} \left(f(e^+) - f(e^-)\right)^2$$

defined on $f \in L^2(V)$ where e^+ and e^- denote the vertices joined by the edge e. Let $0 = \lambda_0 < \lambda_1 = \lambda_1(G) \leq \lambda_2 \cdots \leq \lambda_{n-1}$ be the eigenvalues of Q. Clearly, Q plays a role of Laplacian for the graph $G = (V, E)$. Let $d = \max\{d(v) \mid v \in V\}$ be the maximum degree of a vertex of the graph G. Then the concentration function of (V, ρ, μ) is estimated by

$$\alpha(V; \varepsilon) \leq \tfrac{1}{2} \exp\left\{ -\varepsilon \sqrt{\lambda_1/2d} \log 2 \right\} .$$

(Note that ε is an integer in this case and ranges from 1 to the diameter of V.)

The most important examples of the use of this estimate are the so-called Cayley graphs. Let V be a finite group and $S \subset V$ some set of generators. Assume $S = S^{-1}$ and the identity $e \notin S$. We join v and u from V by an edge if $u = s^{-1}v$ for some $s \in S$. Then a path distance in such a graph, $G = (V; E)$, is the word distance in V induced by S. The degree d of G is equal to $|S|$. Consider $L_2(V)$ and let π be the left regular representation of $V : \pi(t)f(v) = f(t^{-1}v)$. So if $\{e_t\}_{t \in V}$ is the natural basis of L_2 then $\pi(t)e_v = e_{tv}$ and the Laplacian $Q = |A|Id - \sum_{s \in S} \pi(s)$. This is a self-adjoint operator and $\lambda_1(Q) = \{|S|$ – the second largest positive eigenvalue of $A(G) = \sum_{s \in S} \pi(s)\} \geq |S| - \|A|_{L_2^0}\|$ where $L_2^0 \oplus \{\text{const}.\} = L_2(V)$.

To use this construction successfully we have to find natural families of groups V_i, $|V_i| = n_i \to \infty$, and generators $S_i \subset V_i$, $|S_i| = d_i \leq$ Const, such that $\|A(G_i)|_{L_2^0}\| \leq |S_i| - \varepsilon$ for some positive fixed ε. This brings us to the T-property of Kazhdan [K67] well-known in Representation Theory, which gives us a number of most interesting examples, say $V_m = \mathrm{SL}_k(\mathbb{Z}/m\mathbb{Z})$, $k \geq 2$ fixed and $m \to \infty$.

There is a (right) choice of generators $S_m \subset V_m$, $|S_m| = 4$, such that $\alpha(V_m; t) \leq \frac{1}{2}\exp(-c_k t)$. Note that the bound does not depend on p. It is a very strong concentration property and implies families of expanders – special bipartite graphs which play a central role in the construction of fast algorithms (see, e.g. [Al86], [AlSp92]). I also recommend the books [S90] and [Lu94] which show how to apply delicate results from Number Theory to choose the best possible sets of generators S_m. Many other fascinating related results can be found in these books.

5. Duality and connected isomorphic inequalities

Let K be a convex compact body in $\mathbb{R}^n$ and $K = -K$. Then the *polar* body (or the *dual* body) of K is $K^\circ = \{x \in \mathbb{R}^n \mid (x, y) \leq 1 \text{ for } \forall y \in K\}$. Obviously $X_K^* = X_{K^\circ}$. It is well-known how important duality is in the study of convexity and also in Functional Analysis. The notion has its roots, in fact, in mechanics. Legendre and Binet ellipsoids of inertia of K are, up to a factor, the dual of each other (see [J37] or [MP89]). Clearly, a larger body has a smaller polar. It is possible to give (and in many different forms) a precise quantitative form to this relation.

5.1. For example, the product of the volume radii of K and K° remains between two universal constants: there is a universal constant $c > 0$ such that $\forall n$ and $\forall K \subset \mathbb{R}^n$ as above

$$c \leq \left(\frac{|K| \cdot |K^\circ|}{|\mathcal{D}|^2}\right)^{1/n} \leq 1 \, .$$

(The right side is the classical Blaschke-Santaló inequality, and the left side is [BM85].)

These inequalities imply, perhaps more surprisingly, a similar relation for covering numbers ([KM87]): there are universal constants C and $c > 0$ such that $\forall\, n$ and for every two convex symmetric bodies K and $T \subset \mathbb{R}^n$

$$c \leq \left(\frac{N(K,T)}{N(T^\circ, K^\circ)}\right)^{1/n} \leq C \, .$$

5.2. The next two results are an interpretation of facts from Local Theory in the language of classical convexity. However, I do not know of any other proofs than the very delicate and non-trivial analysis of the corresponding normed spaces by purely Functional Analysis methods.

5.2a. [FLM77] Let P be a centrally symmetric polytope, $v(P) = \#$ of its vertices and $f(P) = \#$ of its faces. Then $\log v(P) \cdot \log f(P) \geq cn$ for some universal constant $c > 0$ (and for some polytopes it is exact). (Note the importance of the symmetry condition $P = -P$; otherwise, for simplex S,

$$v(S) = n + 1 \quad \text{and} \quad f(S) = n + 1 \, ;$$

so exponents are a result of symmetry!)

5.2b. Let $w(K)$ be the mean width of K (again $K = -K$). (It is called $b(K)$ in Convex Geometry; we use the notation $M^*(K)$ in Local Theory; let $\|x\|^*$ be the norm with K° being the unit norm; then $\frac{1}{2}w(K) = M^*(K) = \int_{S^{n-1}} \|x\|^* d\mu(x)$.)

Then $\exists$ a position $uK = \widehat{K}$, $u \in \mathrm{SL}_n$, of K, s.t.

$$w(\widehat{K}) \cdot w(\widehat{K}^\circ) \leq C \log n$$

(or, more strongly, $\leq c \log d_K$ where $d_K = d(X_K, \ell_2^n) = \min\{d(K, \mathcal{E}) \mid \mathcal{E} \text{ is an ellipsoid}\}$).

[This is a combination of results of Lewis [Lew79]/Figiel-Tomczak [FT79]/ Pisier [P82].] Note that it is a different type of isomorphic geometric inequality (not a volume-type) and one of the most important tools in the proofs of all isomorphic volume type inequalities we discussed before. It uses solutions of some variational problems, non-trivial duality arguments and, most important, the estimate of Pisier for Rademacher projections which is a non-trivial application of very concrete Harmonic analysis methods in abstract Functional Analysis.

5.3. There is an even deeper connection between sections of K and K°.

5.3a. Fix $\kappa > 0$. Let $k + m = [(1-\kappa)n]$. $\forall K$, for "most" [=high probability] sections E of dimension k and most sections F of K° of dimension m,

$$w(K \cap E) \cdot w(K^\circ \cap F) \leq \frac{8}{\kappa} .$$

Also, in fact, for some universal C

$$\int_{G_{n,k}} w(K \cap E) d\mu(E) \cdot \int_{G_{n,m}} w(K^\circ \cap F) d\mu(F) \leq \frac{C}{\kappa} .$$

This is surprising: polarity is a *global* construction, so why does it connect so precisely with local structure, i.e. with the structure of sections?

5.3b. In fact, even the maximal width $r(K \cap E) = \operatorname{diam}(K \cap E)$ of a central section E of K of dimension, say, $n/2$, can be estimated in average by the mean-width: for some universal constant C,

$$\int_{G_{n,n/2}} \operatorname{diam}(K \cap E) d_\mu(E) \leq C w(K) .$$

In Local Theory, we call this "the low M^*-estimate" (see [M85], [M90], [PT86], [Go88], or any of the books mentioned earlier). In its analytic form it is used as a bridge between results of a Functional Analytic nature and Geometric type Global results.

5.4. Another type of geometric (isomorphic) inequality connecting a body K with its polar K° is the following:

Let $r_\ell = \max\{r \mid r\mathcal{D} \subset \frac{1}{\ell}\sum_{i=1}^{\ell} u_i K \ , \ u_i \in O(n)\}$.

Then $\exists c > 0$ such that, $\forall n$, $\forall K \subset \mathbb{R}^n$, convex symmetric compact body

$$r_2(K) \cdot r_3(K^\circ) \geq c .$$

In fact, in this case we know that $c \geq 0.0388$ (see [M91]).

Applying this fact to, say, the unit cubes $C^n = [-1,1]^n$, we immediately see that $\exists u \in O(n)$ s.t.

$$c\sqrt{n}\mathcal{D} \subset \frac{C^n + uC^n}{2} \subset \sqrt{n}\mathcal{D}$$

for an absolute constant $c > 0$. This particular very interesting case was derived long ago by Kashin [Kas77].

6. Local and global asymptotic theories

At the heart of the global results presented in this lecture stand methods of Functional Analysis. By global properties we refer to properties of the original body in question, while the local properties pertain to the structure of lower dimensional sections and projections of the body, i.e. to the linear structure of a normed space in the spirit of functional analysis.

From the beginning of the 70's the needs of Geometric Functional Analysis led to a deep investigation of the linear structure of finite dimensional normed spaces. Dvoretzky's theorem was the initial result in this direction. But it had to develop a long way before this structure was understood well enough that it could be used for the study of the global properties of a space.

The culmination of this study was an understanding of the fact that subspaces (and quotient spaces) of proportional dimension behave very predictably. This was the bridge between the problems of Functional Analysis and the Global Asymptotic properties of convex sets. The following theorem was one of the first links between local properties and global structure [M85]:

THEOREM 6.1. *Fix* $0 < \lambda < 1$. *Every finite dimensional normed space* $X = (\mathbb{R}^n, \|\cdot\|)$ *contains a subspace* sX *and a quotient* $q(sX) = F$ *of a subspace* sX *such that*

$$\text{i) } k = \dim(F) \geq \lambda n \quad \text{ii) } d(F, \ell_2^k) \leq C(\lambda) \sim \frac{1}{1-\lambda} \log \frac{1}{1-\lambda} .$$

We know today how to rewrite almost every fact of Local Theory in a Global form. For example, the above theorem is (for $\lambda = \frac{1}{2}$) a "Local form" of the "Global" Theorems 2.1 and 2.2. (λ close to 1 has no global analogue.) Also Theorem 4.1 may be considered a global form of Dvoretzky's Theorem. However, only recently in one case did we succeed in putting this analogy in an exact quantitative form, and I end this lecture with this example.

Let $|\cdot|$ define the standard euclidean norm in $\mathbb{R}^n$, X be $\mathbb{R}^n$ with norm $\|\cdot\|$ and $M = \int_{S^{n-1}} \|x\| d\mu(x)$. Let $k = k(X; |\cdot|) \leq n$ be the largest integer such that

$$\mu_{G_{n,k}}\left(\left\{E \; ; \; \frac{M}{2}|x| \leq \|x\| \leq 2M|x| \text{ for all } x \in E\right\}\right) > 1 - \frac{k}{n+k} .$$

Let t be the smallest integer such that there are orthogonal transformations $u_1, \dots, u_t \in O(n)$ with

$$\frac{M}{2}|x| \leq \frac{1}{t}\sum_{i=1}^{t} \|u_i x\| \leq 2M|x| , \quad \text{for all } x \in \mathbb{R}^n .$$

So $k = k(X, |\cdot|)$ is a quantitative expression for a Local property of a space X equipped with the given euclidean structure; at the same time, t gives us a quantitative form for the corresponding global property of the space X.

THEOREM 6.2 [MSch96]. *For some universal constant* C

$$C^{-1} n \leq kt \leq Cn .$$

Concluding remarks

There are a number of reasons for and origins of orderal behavior. One may mention "repetition" which creates order, as statistics demonstrates. What I have attempted to show in this lecture is that very high dimensions, or more generally, high parametric families are another source of order. Of course, time limitations have restricted me to very specific objects and structures.

Let me note in conclusion a few remarks about the techniques used in this theory, which stand behind the scenes of my lecture. In addition to the concentration phenomenon which absorbed a huge diversity of methods, the probabilistic nature of a high-dimensional space implies that probability notions and parameters should play an important, perhaps central, role. Indeed, notions of the type and cotype of a space which were developed in the 70s (through works of Maurey, Pisier, Kwapien, Krivine and others) give a "right" classification of normed spaces (convex bodies) by their probabilistic behavior. Also Harmonic Analysis techniques enter the abstract theory of Convexity through these notions (Pisier; Bourgain). Let me also note the use of delicate martingale theory, p-stable probability laws (Pisier, Johnson-Schechtman) and the construction of "random normed spaces" (Gluskin; Szarek) with a very interesting behavior, not observed in concrete examples.

I would like to thank Miriam Hercberg for helping me to edit this text and artfully weaving computer graphics into TEX. I also thank Noga Alon and Leonid Polterovich for their useful remarks on the first draft of this talk.

References

Books and Surveys

[AlSp92] N. Alon, J.H. Spencer, The Probabilistic Method, Wiley Interscience, 1992.

[LT91] M. Ledoux, M. Talagrand, Probability in Banach Spaces, Ergebnisse der Mathematik und ihrer Grenzgebiete 23, Springer Verlag, 1992.

[L51] P. Levy, Problèmes concrets d'analyse fonctionnelle, Gauthier Villars, Paris 1951.

[L92] J. Lindenstrauss, Almost spherical sections, their existence and their applications, Jber. Deutsch. Math.-Vereinig. Jubiläumstagung 1990, Teubner, Stuttgart, 39–61 (1992).

[LM93] J. Lindenstrauss, V. Milman, The local theory of normed spaces and its applications to convexity, In "Handbook of Convex Geometry", Vol. B, 1149–1220 (P.M. Gruber and J.M. Wills, eds), North Holland (1993).

[Lu94] A. Lubotzky, Discrete Groups, Expanding Graphs and Invariant Measures, Progress in Mathematics 125, Birkhäuser Verlag, 1994.

[M88a] V.D. Milman, The heritage of P. Levy in geometrical functional analysis, Asterisque 157/158, 273–301 (1988).

[M92] V. Milman, Dvoretzky Theorem – 30 years later, Geometric and Functional Analysis (GAFA) 2:4, 455–479 (1992).

[MP89] V.D. Milman, A. Pajor, Isotropic position and inertia ellipsoids and zonoids of the unit ball of a normed n-dimensional space, GAFA-Seminar 87–88, Springer Lecture Notes in Math. 1376, 64–104, (1989).

[MSch86] V.D. Milman, G. Schechtman, Asymptotic theory of finite dimensional normed spaces, Springer-Verlag, Lecture Notes in Mathematics 1200, 156pp. (1986).

[P86a] G. Pisier, Factorization of linear operators and geometry of Banach spaces, CBMS Regional Conference Series no. 60, AMS, 1–154 (1986). (Second printing with corrections, 1987).

[P86b] G. Pisier, Probabilistic methods in the geometry of Banach spaces, CIME, Varenna, 1985, Springer Lecture Notes in Mathematics 1206, 167–241 (1986).

[P89a] G. Pisier, The Volume of Convex Bodies and Banach Space Geometry, Cambridge Tracts in Math. 94 (1989).

[S90] P. Sarnak, Some Applications of Modular Forms, Cambridge Tracts in Mathematics 99, Cambridge University Press, 1990.

[T95] M. Talagrand, Concentration of measure and isoperimetric inequalities in product spaces, IHES, Publications Mathématiques 81, 205pp. (1995).

[TJ88] N. Tomczak-Jaegermann, Banach-Mazur Distances and Finite Dimensional Operator Ideals, Pitman, 1988.

Articles

[Al86] N. Alon, Eigenvalues and expanders, Combinatorica 6 (1986), 83–96.

[AlM85] N. Alon, V.D. Milman, λ_1, isoperimetric inequalities for graphs and superconcentrators, J. Combinatorial Theory, Ser. B 38, N1, 73–88 (1985).

[AmM80] D. Amir, V.D. Milman, Unconditional and symmetric sets in n-dimensional normed spaces, Israel J. of Math., Vol. 37, 1–2, 3–20 (1980).

[B91] K. Ball, Volume ratios and a reverse isoperimetric inequality, J. London Math. Soc. 44 (1991).

[Ban90] W. Banaszczyk, On the lattice packing-covering ratio of finite dimensional normed spaces. Colloq. Math. 59:1, 31–33 (1990).

[BBP95] J. Bastero, J. Bernués, A. Peña, An extension of Milman's reverse Brunn-Minkowski inequality, GAFA 5:3, 572–581 (1995).

[Bob96] S.G. Bobkov, An isoperimetric inequality on the discrete cube, and an elementary proof of the isoperimentric inequality in Gauss space, preprint.

[BobH96] S.G. Bobkov, C. Houdré, Isoperimetric constants for product probability measures, preprint.

[Bor75] C. Borell, The Brunn-Minkowski inequality in Gauss spaces, Inventiones Math. 30, 207–216 (1975).

[Bou87] J. Bourgain, On lattice packing of convex symmetric sets in $\mathbb{R}^n$, Springer Lecture Notes in Math. 1267, 5–12 (1987).

[Bou89] J. Bourgain, Bounded orthogonal systems and the $\Lambda(p)$-set problem. Acta 162 (1989), 227–245.

[Bou91] J. Bourgain, On the distribution of polynomials on high dimensional convex sets, Israel Seminar (GAFA) 1989–90 (J. Lindenstrauss, V.D. Milman, eds.) Springer Lecture Notes in Math. 1469, 127–137 (1991).

[BLM88] J. Bourgain, J. Lindenstrauss, V. Milman, Minkowski sums and symmetrizations, GAFA-Seminar Notes '86–87, Springer Lecture Notes in Math. 1317, 44–66 (1988).

[BLM89a] J. Bourgain, J. Lindenstrauss, V. Milman, Approximation of zonoids by zonotopes, Acta Math. 162, 73–141 (1989).

[BLM89] J. Bourgain, J. Lindenstrauss, V. Milman, Estimates related to Steiner symmetrizations, Springer Lecture Notes in Math., v. 1376, 264–273 (1989).

[BM85] J. Bourgain, V.D. Milman, Sections euclidiennes et volume des corps symétriques convexes dans R^n, C.R Acad. Sc. Paris, t.300 Ser. 1, No. 13, 435–438 (1985) (see also, J. Bourgain, V.D. Milman, New volume ratio properties of convex symmetric bodies in R^n, Invent. Math. 88, 319–340 (1987)).

[Bu72] G.Y. Butler, Simultaneous packing and covering in euclidean space, Pro. London Math. Soc. 3:25, 721–735 (1971).

[FLM77] T. Figiel, J. Lindenstrauss, V.D. Milman, The dimension of almost spherical sections of convex bodies, Acta Math. 139:1–2, 53–94 (1977).

[FT79] T. Figiel, N. Tomczak-Jaegermann, Projections onto Hilbertian subspaces of Banach spaces, Israel J. Math. 33, 155–171 (1979).

[Gl81] E.D. Gluskin, The diameter of the Minkowski compactum is roughly equal to n, Funct. Anal. Appl. 15, 72–73 (1981).

[Go85] Y. Gordon, Some inequalities for Gaussian processes and applications, Israel J. Math. 50, 265–289 (1985).

[Go88] Y. Gordon, On Milman's inequality and random subspaces which escape through a mesh in $\mathbb{R}^n$, GAFA, Israel Funct. Analysis Seminar (86/87), Springer Lecture Notes 1317, 84–106 (1988).

[Gr83] M. Gromov, Filling Riemannian manifolds, J.D.G. 18, 1–147 (1983).

[Gr88] M. Gromov, Dimension, non-linear spectra and width, GAFA, Israel Funct. Analysis Seminar (86/87), Springer Lecture Notes 1317, 132–184 (1988).

[GrM83] M. Gromov, V.D. Milman, A topological application of the isoperimetrical inequality, American J. of Math. 105:4, 843–854 (1983).

[GrM84] M. Gromov, V.D. Milman, Brunn Theorem and a concentration of volume of convex bodies, GAFA Seminar Notes, Tel Aviv University, Israel 1983–1984, Exp. V, 12pp.

[IST76] I.A. Ibragimov, V.N. Sudakov, B.S. Tsirelson, Norms of Gaussian sample functions, Lecture Notes Math. 550, 20–41 (1976).

[J37] F. John, Polar correspondence with respect to convex region, Duke Math. J. 3:2, 355–369 (1937) (University of Kentucky).

[JS82] W.B. Johnson, G. Schechtman, Embedding ℓ_p^m into ℓ_1^n, Acta Math 149, 71–85 (1982).

[Kas77] B. Kašin, Sections of some finite dimensional sets and classes of smooth functions, Izv. Acad. Nauk SSR 41, 334–351 (1977) (in Russian).

[K67] D.A. Kazhdan, Connection of the dual space of a group with the structure of its closed subgroups, Functional Anal. Appl. 1, 63–65 (1967).

[KM87] H. König, V.D. Milman, On the covering numbers of convex bodies, GAFA-Seminar Notes, Springer Lecture Notes in Math. 1267, 82–95 (1987).

[La96] R. Latala, On the equivalence between geometric and arithmetic means for log-concave measures, preprint.

[Le92] M. Ledoux, A heat semigroup approach to concentration on the sphere and on a compact Riemannian manifold, GAFA 2, 221–224 (1992).

[Lew79] D. Lewis, Ellipsoids defined by Banach ideal norms, Mathematika 26, 18–29 (1979).

[Ma96] K. Marton, A measure concentration inequality for contracting Markov chains, GAFA 6:3, 556–571 (1996).

[Mau79] B. Maurey, Construction de suites symétriques, C.R.A.S., Paris 288, 679–681 (1979).

[M85] V.D. Milman, Almost Euclidean quotient spaces of subspaces of a finite dimensional normed space, Proceedings AMS 94:3, 445–449 (1985).

[M86] V.D. Milman, An inverse form of the Brunn-Minkowski inequality with applications to local theory of normed spaces, C.R. Acad. Sc. Paris 302, Ser. 1, No. 1, 25–28 (1986).

[M88b] V.D. Milman, Isomorphic symmetrization and geometric inequalities, GAFA-Seminar Notes '86–87, Springer Lecture Notes in Math. 1317, 107–131 (1988).

[M90] V. Milman, A note on a low M^*-estimate, In "Geometry of Banach Spaces" (eds. P.F.X. Müller and W. Schachermayer), London Mathematical Society Lecture Notes Series 158, 219–229 (1990).

[M91] V. Milman, Some applications of duality relations, GAFA Seminar 1989–90, Springer Lecture Notes in Math. 1469, 13–40 (1991).

[M95] V. Milman, Isomorphic Euclidean regularization of quasi-norms in $\mathbb{R}^n$, C.R. Acad. Sci. Paris, 321, 879–884 (1995).

[MSch96] V.D. Milman, G. Schechtman, Global vs. local asymptotic theories of finite dimensional normed spaces, preprint 1996.

[PT86] A. Pajor, N. Tomczak-Jaegermann, Subspaces of small codimension of finite dimensional Banach spaces, Pro. Amer. Math. Soc. 97, 637–642 (1986).

[Pet61] C.M. Petty, Centroid surfaces, Pacific J. Math. 11, 1535–1547 (1961).

[P82] G. Pisier, Holomorphic semi-groups and the geometry of Banach spaces, Annals of Math. 115, 375–392 (1982).

[P89b] G. Pisier, A new approach to several results of V. Milman, Journal für die Reine und Angew. Math. 393, 115–131 (1989).

[R50] C.A. Rogers, A note on coverings and packing, J. London Math. Soc. 25, 327–331 (1950).

[Sch81] G. Schechtman, Levy type inequality for a class of metric spaces, in "Martingale Theory in Harmonic Analysis and Banach Spaces", 211–215, Springer-Verlag, 1981.

[Sch87] G. Schechtman, More on embedding subspaces of L_p in ℓ_r^n, Compositio Math. 61, 159–170 (1987).

[Sch88] G. Schechtman, A remark concerning the dependence on ε in Dvoretzky's Theorem, GAFA Seminar Notes 87–88, springer Lecture Notes in Math. 1376, 274–277 (1988).

[Schm91] M. Schmuckenschlager, On the dependence on ε in a theorem of J. Bourgain, J. Lindenstrauss, V. Milman, GAFA Seminar Notes 89–90, Springer Lecture Notes in Math. 1469, 166–173 (1991).

[SuTs78] V.N. Sudakov, B. Tsirelson, Extremal properties of half-spaces for spherically invariant measures, J. Soviet. Math. 9, 9–18 (1978) [translated from Zap. Nauch. Sem. L.O.M.I. 41, 14–24 (1974)].

[Sz83] S.J. Szarek, The finite dimensional basis problem, with an appendix on nets of Grassman manifold, Acta Math. 151, 153–179 (1983).

[Ts96] A. Tsolomitis, Quantitative Steiner/Schwartz-type symmetrizations, Geometriae Dedicata 60:2, 187–206 (1996).

Progress in Mathematics, Vol. 169, © 1998 Birkhäuser Verlag Basel/Switzerland

Microstructures, Phase Transitions and Geometry

STEFAN MÜLLER

Max Planck Institute for Mathematics in the Sciences
Inselstr. 22-26, 04103 Leipzig, Germany

1. What are microstructures?

Microstructures are structures on a scale between the atomic scale and the macroscopic scale on which we usually make observations. They are abundant in natural and man-made materials, and often the microstructure optimizes a material's properties (maximum strength at given weight, minimal energy, maximum or minimum permeability, ...). Some materials can change their internal microstructure and hence their properties in response to external influences. They are sometimes referred to as 'smart materials' and are of great technological interest.

The attempt to mathematically describe and analyze the formation, interaction and the macroscopic effects of microstructures yields new, easily stated but deep mathematical problems. Their resolution is in its very beginning and involves the interaction of a variety of branches of mathematics including the calculus of variations, differential geometry, geometric measure theory, dynamical systems and nonlinear partial differential equations. In the following I try to describe some of the basic questions and how different branches of mathematics are involved in their resolution.

To simplify the exposition I focus on models of microstructures that arise in elastic crystals due to a solid-solid phase transformation. A large part of this article, in particular Sections 2-5, mainly reviews work of others. I draw heavily on the beautiful survey of Šverák [Sv 94] as well as the detailed expositions of Ball and James [BJ 87], [BJ 92]. Currently no exposition of the field in book form is available but various projects are in preparation [BJ 97], [Mu 97a],[Mu 97b], [PZ 97], [Ta 97]. Dacorogna's book [Da 89] is a good reference for the direct method and notions of convexity.

2. Microstructure in elastic crystals

Inspired by the fundamental works of Ericksen, Ball and James [BJ 87] (see also [CK 88], [Fo 89]) proposed to study crystal microstructure by analyzing minimizers and minimizing sequences for the elastic energy. They use a continuum model based on nonlinear elasticity (a theory based on linearized elasticity was proposed earlier by Khachaturyan, Roitburd and Shatalov; see [Kh 67], [Kh 83], [KSh 69], [Ro 69], [Ro 78]; a comparison of the two approaches appears in [BJ 92], Section 9, [Bh 93] and [Ko 89]). The idea to use weak convergence to relate microscopic and macroscopic quantities in continuum mechanics and to study the implications of a combination of pointwise and differential constraints goes back to the pioneering work of Tartar [Ta 79], [Ta 95], [Ta 97].

The elastic energy I needed to deform the crystal from its reference configuration (identified with a bounded domain $\Omega \subset \mathbf{R}^3$) by a map $u : \Omega \to \mathbf{R}^3$ is given by

$$I(u) = \int_\Omega W(Du)dx, \tag{2.1}$$

where $Du = (\frac{\partial u^i}{\partial x^j})$ is the deformation gradient. The Cauchy-Born rule asserts that the stored-energy density $W(F)$ is the energy (per unit reference volume) needed to perform an affine deformation $x \mapsto Fx$ of the crystal lattice. The stored energy is invariant under rigid transformations of the ambient space (frame indifference) and under changes of the independent variables that correspond to lattice invariant rotations (crystal symmetry)

$$W(QF) = W(F) \quad \forall Q \in SO(3), \tag{2.2}$$
$$W(FR) = W(F) \quad \forall R \in \mathcal{P} \subset SO(3). \tag{2.3}$$

The discrete point group $\mathcal{P}$ reflects the symmetry properties of the lattice that persist in the continuum approach (see [Er 80], [Er 89], [Pi 84] for the relevance of $\mathcal{P}$ and [Er 84], [Za 92] for a discussion of the Cauchy-Born rule).

It is convenient to normalize W such that $\min W = 0$. Then the set

$$K = \{F : W(F) = 0\}$$

contains exactly the zero-energy affine deformations of the lattice. By (2.2) and (2.3) this set consists of one or several disjoint copies of $SO(3)$

$$K = SO(3)U_1 \cup \ldots \cup SO(3)U_m, \quad U_i = U_i^T > 0. \tag{2.4}$$

In general the energy W and hence the set K depend on the temperature Θ. For materials that undergo solid-solid phase transitions K consists of one component at high temperatures and of several symmetry related (through $\mathcal{P}$) components below a critical temperature Θ_c (see [BJ 92]).
The relation between minimization of I and microstructure is discussed in Section 4 for a model problem; situations with $SO(n)$ symmetry are discussed in Section 5.

3. Basic mathematical questions

The fundamental problem is:

> Characterize minimizers and minimizing
> sequences of $I(u) = \int_\Omega W(Du)dx$
> subject to suitable boundary conditions.

This is a very difficult problem since W is typically not convex (nor rank-1 convex) for materials that undergo solid-solid phase transitions. In other words the corresponding Euler-Lagrange equations are not strongly elliptic but of mixed elliptic-hyperbolic type. Research has therefore mostly focused on (almost) zero energy states, i.e. pointwise minimizers of the integrand. Generalizing the setting slightly we consider maps

$$u : \Omega \subset \mathbf{R}^n \to \mathbf{R}^m$$

on a bounded domain Ω (with Lipschitz boundary) and a compact set $K \subset M^{m\times n}$ in the $m \times n$ matrices (e.g. the zero set of W).

Problem 1 (exact solutions): Characterize all Lipschitz maps u such that

$$Du \in K \qquad \text{a.e. in } \Omega \,. \tag{3.1}$$

Problem 2 (approximate solutions): Characterize all sequences u_j of maps whose Lipschitz constants are uniformly bounded and which satisfy

$$\operatorname{dist}(Du_j, K) \to 0 \qquad \text{a.e. in } \Omega \,. \tag{3.2}$$

Problem 3 (relaxation of K): Determine the sets K^{ex} and $K^{app} \subset M^{m\times n}$ of all matrices F such that Problem 1 and Problem 2 have a solution that satisfies in addition

$$u(x) = Fx \quad \text{on } \partial\Omega, \tag{3.3}$$

$$u_j(x) = Fx \quad \text{on } \partial\Omega, \tag{3.4}$$

respectively.

A simple covering and scaling argument shows that K^{ex} and K^{app} do not depend on Ω.

In the context of crystal microstructures the sets K^{ex} and K^{app} have an important interpretation. They represent *macroscopic* zero (or almost zero) energy deformations, while K represents *microscopic* zero energy deformations. The sets K^{ex} and K^{app} can be much larger than K; in fact it is conjectured (and proved in two dimensions) that for many sets of the form (2.4) the sets K^{ex} and K^{app} have interior points (subject to an incompressibility constraint $\det F = 1$). This would correspond to an (experimentally observed) fluid-like behavior. Some known results are reviewed in Section 5 below.

Problems 1-3 also arise in many other contexts, e.g. in the classification of isometric immersions or in the theory of nonlinear elliptic and hyperbolic partial differential equations, see Šverák's survey [Sv 94] and Gromov's treatise [Gr 86].

If minimizers do not exist there are often many different minimizing sequences. Some of those, however, only differ in rather superficial ways.

Problem 4: Develop mathematical objects that capture the 'essential features' of minimizing sequences.

One such object is the Young measure, discussed in Section 5; see also Section 8.1. Minimization of elastic energy often predicts minimizing sequences that develop finer and finer oscillations. Experimentally a limited fineness is observed.

Problem 5: Explain the length scale and fine geometry of microstructures, possibly by including other contributions to the energy, such as interfacial energy.

Some attempts in this directions are discussed in Section 7.

4. The two-gradient problem

Before reviewing a general framework for Problems 1–4 let us look at the simplest example

$$K = \{A, B\} .$$

4.1. Exact solutions

The simplest solution satisfies $Du = A$ on a halfspace H and $Du = B$ on its complement. Due to tangential continuity of u this is only possible if $B - A = a \otimes n$, where n is the normal to H. In particular one has $\mathrm{rk}(B - A) = 1$. It turns out that for an arbitrary solution of $Du \in K$, the gradient can only jump across hyperplanes with normal n.

Lemma 4.1 *([BJ 87]). Suppose that $u : \mathbf{R}^n \to \mathbf{R}^m$ is Lipschitz and*

$$Du \in \{A, B\} \text{ a.e .}$$

(i) *If $\mathrm{rk}(B - A) > 1$ then $Du \equiv A$ a.e or $Du \equiv B$ a.e*
(ii) *If $\mathrm{rk}(B - A) = 1$ then $B - A = a \otimes n$ and there exists a Lipschitz function $h : \mathbf{R} \to \mathbf{R}$ such that $h' \in \{0, 1\}$ a.e and*

$$u(x) = Ax + a\, h(x \cdot n) .$$

The proof is a simple exercise using the fact that the distributional curl of a gradient vanishes.

4.2. Approximate solutions

Lemma 4.2 *([BJ 87]). Suppose that $\mathrm{rk}(B - A) > 1$ and let $u_j : \Omega \to \mathbf{R}^n$ be a sequence of functions with uniformly bounded Lipschitz constants that satisfy*

$$\mathrm{dist}(Du_j, \{A, B\}) \to 0 \text{ a.e. in } \Omega.$$

Then (for a subsequence; not relabelled) either

$$Du_j \to A \text{ a.e. or } Du_j \to B \text{ a.e.}$$

This can be proved e.g. by using the minors relations (see Section 5).

4.3. Relaxation, nonattainment, microstructure

If $\mathrm{rk}(B - A) > 1$ then $K^{app} = K$ by Lemma 4.2.

Lemma 4.3 *([BJ 87]). Suppose that $\mathrm{rk}(B - A) = 1$. Then*
(i) $K^{ex} = \{A, B\}$.
(ii) $K^{app} = \mathrm{conv} K = \{\lambda A + (1 - \lambda) B : \lambda \in [0, 1]\}$.
Moreover for any sequence u_j that satisfies (3.2) and (3.4) one has

$$u_j \to Fx \text{ uniformly on } \Omega. \tag{4.1}$$

In particular for $\lambda \in (0, 1), F = \lambda A + (1 - \lambda) B$ there exist approximate but not exact solutions. The gradients Du_j of minimizing sequences must oscillate rapidly so that u_j can approach Fx. Nonexistence of a minimizer thus leads to fine scale oscillations and 'microstructure'.

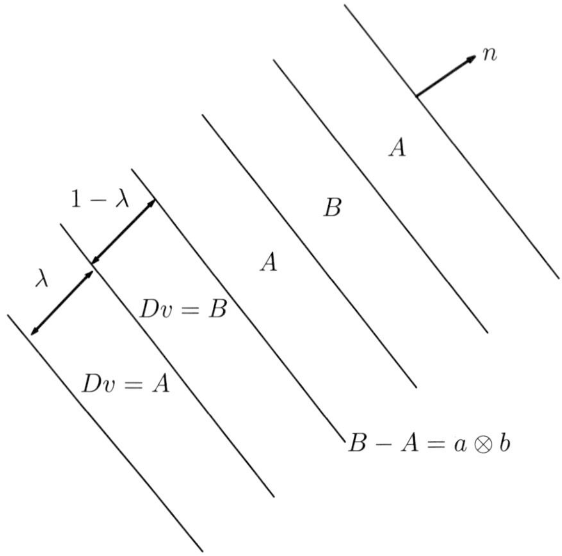

Figure 1: A simple laminate.

We sketch one possible construction of a minimizing sequence. Many variants are possible, and we return in Section 5 to the question whether all minimizing sequences are unique up to minor details.

Assume without loss of generality that $F = 0$, let h be a 1-periodic Lipschitz function with $h' = -(1-\lambda)$ on $(0, \lambda)$, $h' = \lambda$ on $(\lambda, 1)$ and let

$$v(x) = h(x \cdot n), \quad v_k(x) = k^{-1} v(kx).$$

The functions v_k are called simple laminates since the sets $\{Du = A\}$ and $\{Du = B\}$ consist of parallel stripes (see Figure 1). We have $Dv_k \in \{A, B\}$ a.e and $v_k \to 0$ uniformly. To adjust the boundary conditions consider the cut-off functions $\varphi_k(x) = \min\{k \text{dist}(x, \partial\Omega), 1\}$. Then $u_k = \varphi_k v_k$ has the desired properties.

5. Approximate solutions and Young measures

Young measures (also called 'generalized functions') were introduced by L.C. Young in the 30's to solve optimal control problems that have no classical solution but rather require an infinitesimally fast switching between states (see [Yo 37], [Yo 69]). The idea is to replace functions which take values in a (compact) set K by functions that take values in the space of probability measures on K. Thus an 'infinitesimally fine' mixture of two states is represented by a convex combination of two Dirac masses. For the problems discussed above it is crucial to characterize those Young measures that are generated by sequences of gradients. An abstract characterization is available in terms of quasiconvex functions, but already simple concrete examples lead to challenging open questions, since very few quasiconvex functions are known.

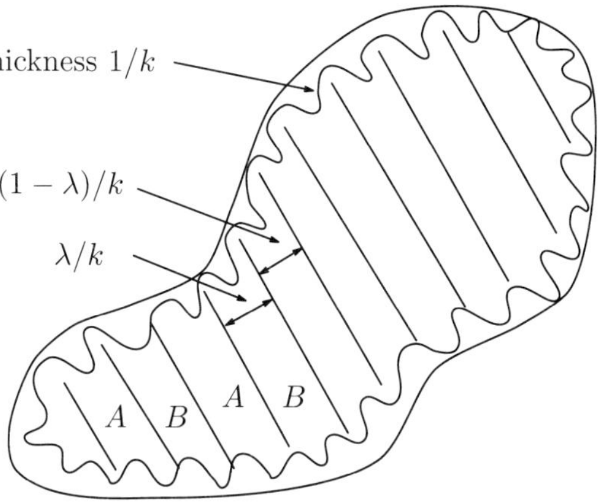

Figure 2: A minimizing sequence obtained by truncation of very fine laminates.

The presentation below follows [Ba 89]. A short selection of the rich literature on Young measures is [Ba 84], [BL 73], [Kr 94], [Va 90], [Va 94]. Tartar [Ta 79] introduced Young measures in connection with the theory of compensated compactness (developed jointly with Murat) as a fundamental tool to study existence and compactness questions for nonlinear partial differential equations (see [Ta 82], [DP 85] for applications to conservation laws). Varifolds ([Al 72]) are a generalizaton of Young measures to a geometric setting.

Let E be a measurable set in $\mathbf{R}^n, K$ a compact set in $\mathbf{R}^l$ and denote by $\mathcal{M}(K)$ and $\mathcal{P}(K)$ the space of the Radon measures and of probability measures on K. The pairing between the space $C(K)$ of continuous functions and its dual $\mathcal{M}(K)$ is denoted by $\langle \mu, f \rangle = \int_K f d\mu$. A *Young measure* $\nu : E \to \mathcal{P}(K)$ is a weak$*$ measurable map, i.e. a map such that $x \mapsto \langle \nu(x), f \rangle$ is measurable for all $f \in C(K)$.

Theorem 5.1 *(Fundamental theorem on Young measures)*
Let $w_j : E \to K$ be a sequence of measurable functions. Then there exists a subsequence w_{j_k} and a Young measure $\nu : E \to \mathcal{P}(K)$ such that for all $f \in C(K)$

$$f(w_{j_k}) \overset{*}{\rightharpoonup} f_\infty \text{ in } L^\infty(E), \qquad f_\infty(x) = \langle \nu_x, f \rangle = \int_K f d\nu_x. \tag{5.1}$$

Moreover if $K' \subset K$ is a compact subset then:

$$\operatorname{supp}\nu_x \subset K' \text{ for a.e. } x \iff \text{dist } (w_{j_k}, K') \to 0 \text{ (locally) in measure.} \tag{5.2}$$

Notation: We say that the sequence w_{j_k} generates the Young measure ν. A Young measure ν is called a $(W^{1,\infty}$-)*gradient Young measure* if E is open and there exists a sequence u_j of Lipschitz functions with uniformly bounded Lipschitz constants such that Du_j generates ν.

Remark. A similar result holds for sequences $w_j : E \to \mathbf{R}^l$ if one replaces $C(K)$ by $C_0(\mathbf{R}^l)$, the predual of $\mathcal{M}(\mathbf{R}^l)$. In this case, however, $\nu(x)$ is in general only a subprobability measure. Mild additional conditions such as $\sup_j \int_E |w_j|^s < \infty$ for some $s > 0$ ensure that $\nu(x)$ is a probability measure. Alternatively one can work in a suitable compactification of $\mathbf{R}^l$.

Proof. To each w_j we associate the elementary Young measure

$$\nu_j(x) = \delta_{w_j(x)}.$$

The sequence $\{\nu_j\}_{j\in\mathbf{N}}$ is bounded in the space $L^\infty_w(E;\mathcal{M}(K))$ of essentially bounded, weak$*$ measurable maps. Since this space is a dual of $L^1(E;C(K))$ (see [Ed 65], p. 588; [IT 69], p. 93; [Me 66], p. 244) the first assertion follows from the Banach-Alaoglu theorem and the fact that $\mathcal{P}(K)$ is weak$*$ closed in $\mathcal{M}(K)$. The second assertion is a simple consequence of the first. □

Young measures are closely related to compactness by the following fact.

Corollary 5.2 *Let w_{j_k} be as in Theorem 5.1. Then*

$$w_{j_k} \to w \text{ locally in measure} \iff \nu(x) = \delta_{w(x)} \quad \text{a.e.}$$

Young measures provide a concise description of the fine scale oscillations that may arise in minimizing sequences.

Example 5.3 As in Section 4.3 consider $A, B, F \in M^{m\times n}$ with $\mathrm{rk}(B-A) = 1$, $F = \lambda A + (1-\lambda)B, \lambda \in (0,1)$. Let u_j be a sequence with $\mathrm{dist}(Du_j, \{A,B\}) \to 0$ a.e, $u_j(x) = Fx$ on $\partial\Omega$. Then Du_j generates the (unique) Young measure $\lambda\delta_A + (1-\lambda)\delta_B$.

Proof. By (5.2), supp $\nu(x) \subset \{A,B\}$. Hence $\nu(x) = \mu(x)\delta_A + (1-\mu(x))\delta_B$. Application of (5.1) with $f = \mathrm{id}$ and Lemma 4.3 (ii) yield $\mu(x)A + (1-\mu(x))B = F$, which implies the assertion. □

Young measures can be seen as generalized solutions of the problem $Du \in K$ or $\int_\Omega W(Du) \stackrel{!}{\to} \min$ generated by minimizing sequences. A classical solution corresponds to a Young measure that is a Dirac mass at every point. Problem 2 (approximate solutions) can be rephrased as follows.

Problem 2$'$: Characterize all gradient Young measures supported in K.

To characterize gradient Young measures Morrey's notion of quasiconvexity is fundamental.

Definition 5.4 *A function $f : M^{m\times n} \to \mathbf{R}$ is quasiconvex if for every bounded domain Ω and every $F \in M^{m\times n}$*

$$\int_\Omega f(F + D\varphi) \geq \int_\Omega f(F) = |\Omega| f(F), \quad \forall \varphi \text{ Lipschitz }, \varphi_{|\partial\Omega} = 0. \tag{5.3}$$

Thus f is quasiconvex if affine functions minimize $\int_\Omega f(Du)$ subject to affine boundary values. The definition is independent of Ω, and quasiconvex functions are automatically continuous.

Morrey's crucial observation was that quasiconvexity of f is equivalent to (sequential) $W^{1,\infty}$ weak$*$ lower semicontinuity of the functional $u \mapsto \int_\Omega f(Du)dx$. Combining this fact with careful measure-theoretic and functional-analytic arguments, Kinderlehrer and Pedregal obtained the following classification.

Theorem 5.5 *([KP 91]) A Young measure $\nu : \Omega \to \mathcal{P}(M^{m\times n})$ is a ($W^{1,\infty}$-) gradient Young measure if and only if the following three conditions hold*
(i) $\operatorname{supp} \nu_x$ *is uniformly bounded (for a.e. x);*
(ii) $\langle \nu_x, \mathrm{id}\rangle$ *is the gradient of a Lipschitz function;*
(iii) $f(\langle \nu_x, \mathrm{id}\rangle) \leq \langle \nu_x, f\rangle$ *for all quasiconvex f.*
The key point is (iii) which is in nice duality with Definition 5.4. Roughly speaking, quasiconvex functions satisfy Jensen's inequality for gradients while gradient Young measures must satisfy Jensen's inequality for all quasiconvex functions. A similar result holds for Young measures generated by gradients of sequences that are bounded in the Sobolev space $W^{1,p} (p > 1)$, see [KP 94], [Kr 94], [FMP 97]; general references on Young measures for noncompact targets include [DPM 87] and [Ro 97].

As a consequence we obtain an abstract solution of Problem 2′ and Problem 3. A Young measure ν is called homogeneous if the map $x \mapsto \nu(x)$ is constant (a.e.). A blow-up argument shows that if ν is a gradient Young measure then $\nu(a)$ arises as a homogeneous gradient Young measure for a.e. $a \in \Omega$, see [KP 91].

Corollary 5.6 *Let $K \subset M^{m\times n}$ be compact.*
(i) *The set of homogeneous gradient Young measures supported on K is given by*

$$\mathcal{M}^{qc}(K) := \{\nu \in \mathcal{P}(K) : f(\langle \nu, \mathrm{id}\rangle) \leq \langle \nu, f\rangle \quad \forall f : M^{m\times n} \to \mathbf{R} \ \textit{quasiconvex}\}.$$

(ii) *The set K^{app} is the quasiconvex hull of K,*

$$\begin{aligned} K^{app} = K^{qc} &:= \{F : f(F) \leq \inf_K f, \quad \forall f : M^{m\times n} \to \mathbf{R} \ \textit{quasiconvex}\} \\ &= \{\langle \nu, \mathrm{id}\rangle : \nu \in \mathcal{M}^{qc}(K)\} \,. \end{aligned}$$

Following Šverák we say that $\mathcal{M}^{qc}(K)$ is trivial if it only contains Dirac masses. In view of Corollary 5.2 this occurs if and only if all approximating sequences are compact.

Quasiconvexity is a natural condition, but hard to verify. Therefore two algebraic conditions were introduced. A function $f : M^{m\times n} \to \mathbf{R}$ is *polyconvex* if it can be expressed as a convex function of minors (subdeterminants); it is called *rank-1-convex* if it is convex on all line segments $[A, B]$ with rank $B - A = 1$. One has the implications ([Mo 52]; [Da 89]):

$$f \text{ polyconvex} \Rightarrow f \text{ quasiconvex} \Rightarrow f \text{ rank-1 convex}.$$

For $m = 1$ or $n = 1$ all notions are equivalent to ordinary convexity. For $m, n \geq 2$ quasiconvexity does not imply polyconvexity.

Conjecture 5.7 *(Morrey, 1952). For $m, n \geq 2$ rank-1 convexity does not imply quasiconvexity.*

K	Nontrivial exact soln.	Nontrivial GYM	Method	References
$\{A, B\}$	No	No	Minors or Cauchy-Riemann eqns.	[BJ 87] [Sv 91a]
$\{A, B, C\}$	No	No	Nonconvex solns. of Monge-Ampère	[Sv 91b] [Sv 92b]
finite	?	Yes	Example with four matrices	[Ta 93], [CT 93], [BFJK 94]
countable	Yes	Yes	Example	[Sv]
$SO(2); \mathbf{R}SO(2)$	No; holomorphic fns.	No	Cauchy-Riemann eqns.	[Sv 91a]
$SO(n); \mathbf{R}SO(n)$ $n \geq 3$	No; Möbius maps	No No	Minors;degenerate elliptic eqns.	[Ki 88] [Re 68]
$\cup_{i=1}^{k} SO(2)A_i$	No	No	Minors or elliptic regularity	[Sv 93]
$SO(3)A \cup SO(3)B$	?	?	Special cases: Matos, Šverák ...	[Ma 93], [Sv 93]
scalar conservation laws	No	No	compensated compactness; div-curl lemma	[Ta 79]
$m \times 2$ elliptic systems	?	Yes, $m \geq 6$?, $m < 6$		[Sv 94]

Table 1: Nontrivial exact solutions and nontrivial gradient Young measures (GYM) of incompatible sets K (i.e. $\mathrm{rk}(A - b) \neq 1 \; \forall A, B \in K$); see [Sv 94], [BFJK 94] and the references therein for further examples.

Šverák [Sv 92a] proved the conjecture for $m \geq 3$. The case $m = 2$ is open. Kristensen [Kr 97] very recently proved that for $m \geq 3$ there is no local condition that implies quasiconvexity.

For a minor M application of Theorem 5.5 to $\pm M$ yields the minors relation

$$\langle \nu_x, M \rangle = M(\langle \nu_x, \mathrm{id} \rangle) \tag{5.4}$$

as a necessary condition for gradient Young measures. The following table illustrates some solved and unsolved problems for homogeneous gradient Young measures.

6. Exact solutions and convex integration

Approximate solutions are characterized by the quasiconvex hull K^{qc} and the set $\mathcal{M}^{qc}(K)$ of gradient Young measures. The construction of exact solutions is more delicate. In view of the negative result for the two-gradient problem (see Lemma 4.3 (i)) it was widely believed that exact solutions are rather rare. Recent results suggest that many exact solutions exist but that they have to be very complicated.

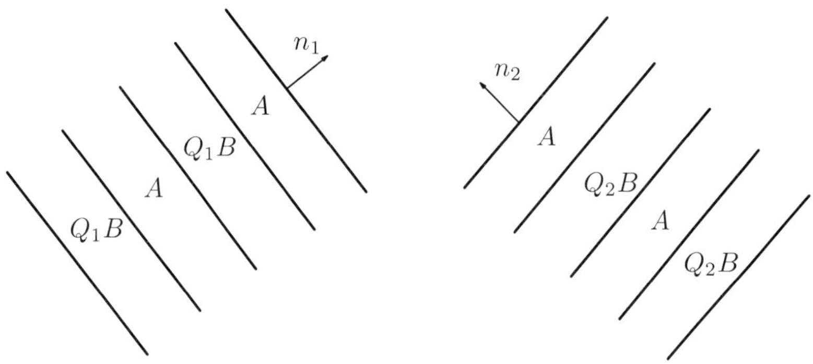

Figure 3: Two possible laminates for the two-well problem.

This is reminiscent of rigidity and flexibility results on isometric immersions and other geometric problems (see [Na 54]; [Ku 55]; [Gr 86], Section 2.4.12).

To illustrate some of the difficulties consider the two-dimensional two-well problem:

$$Du \in K \text{ a.e. in } \Omega, \quad u = Fx \text{ on } \partial\Omega, \tag{6.1}$$

$$K = SO(2)A \cup SO(2)B, \tag{6.2}$$

$$A = \text{ Id }, \quad B = \text{diag}(\lambda, \mu), \quad 0 < \lambda < 1 < \mu. \tag{6.3}$$

If we ignore boundary conditions the simplest solutions of $Du \in K$ are simple laminates, see Figure 3. A short analysis of the rank-1 connections in K shows that such laminates are perpendicular to one of the normals n_1 or n_2, determined by the two solutions of the equation

$$QA - B = a \otimes n, \quad Q \in SO(2). \tag{6.4}$$

There is, however, no obvious way to combine the two laminates (see Figure 4). It was thus believed that the problem (6.1)–(6.3) has no nontrivial solutions. This is false. The construction of nontrivial solutions is based on Gromov's method of convex integration.

6.1. Existence of solutions

First, one observes that the open version of the two-gradient problem admits a solution.

Lemma 6.1 *([MS 96]). Suppose that* $\text{rk}(B - A) = 1, F = \lambda A + (1 - \lambda)B, \lambda \in (0, 1)$. *Then, for a bounded domain* Ω *and every* $\epsilon > 0$ *there exists a piecewise linear map* u *such that*

$$u(x) = Fx \text{ on } \partial\Omega, \quad \text{dist}(Du, \{A, B\}) < \epsilon, \quad \sup |u(x) - Fx| < \epsilon.$$

Here a map $u : \Omega \to \mathbf{R}^m$ is called piecewise linear if u is Lipschitz continuous and if there exist finite or countably many disjoint open sets Ω_i whose union has full

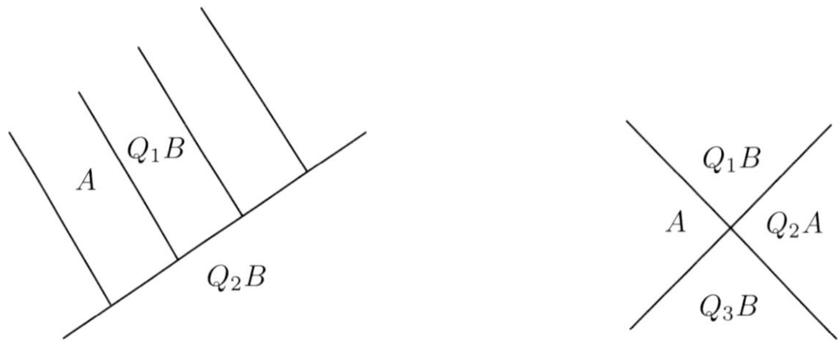

Figure 4: None of the above constructions satisfies the rank-1 condition across every interface.

measure such that $u_{|\Omega_i}$ is affine. To iterate the result of Lemma 6.1 we define the lamination convex hull K^l of a set $K \subset M^{m\times n}$ by successive addition of rank-1 segments:

$$\begin{aligned} K^{(0)} &:= K, \\ K^{(i+1)} &:= K^{(i)} \cup \{\lambda A + (1-\lambda)B, \lambda \in (0,1), A, B \in K^{(i)}, \mathrm{rk}(B-A) = 1\} \\ K^l &:= \bigcup_{i=1}^{\infty} K^{(i)}. \end{aligned}$$

Remark. In contrast with the ordinary convex hull there is no version of Carathéodory's theorem that guarantees that the union of a fixed number of the $K^{(i)}$ is sufficient. Generally, very little is known about the geometry of rank-1 convexity (but see [MP 96]).

Lemma 6.2 *Suppose that $U \subset M^{m\times n}$ is open. Let $v : \Omega \to \mathbf{R}^m$ be piecewise affine and Lipschitz continuous and suppose $Dv \in U^l$ a.e. Then there exists $u : \Omega \to \mathbf{R}^m$ such that*

$$Du \in U \text{ a.e. in } \Omega, \quad u = v \text{ on } \partial\Omega.$$

The crucial step is the passage from open to compact sets $K \subset M^{m\times n}$. Following Gromov we say that a sequence of sets U_i is an in-approximation of K if

(i) the U_i are open and contained in a fixed ball

(ii) $U_i \subset U_{i+1}^l$

(iii) $U_i \to K$ in the following sense: if $F_{i_k} \in U_{i_k}, i_k \to \infty$ and $F_{i_k} \to F$, then $F \in K$.

Theorem 6.3 *([Gr 86], p. 218; [MS 96]). Suppose that K admits an in-approximation $\{U_i\}$. Let $v \in C^1(\Omega, \mathbf{R}^m)$ with*

$$Dv \in U_1.$$

Then there exists a Lipschitz map u such that

$$Du \in K \in \Omega \text{ a.e.}, \quad u = v \text{ on } \partial\Omega.$$

The proof uses a sequence of approximations obtained by successive application of Lemma 6.2. To achieve strong convergence each approximation uses a much finer spatial scale than the previous one, similar to the construction of continuous but nowhere differentiable functions. This is one of the key ideas of convex integration.

This theorem applies to the two-well problem (6.1)–(6.3) if $\lambda\mu \neq 1$. A calculation shows that solutions exist if $F \in \operatorname{int} K^l$, and K^l is given by a simple formula (see [Sv 93]). A different construction of (many) nontrivial solutions was recently obtained by Dacorogna and Marcellini [DM 96b] using Baire's theorem. Other existence results appear in [DM 96a], [DM 97]. The case $\lambda\mu = 1$ in (6.1) – (6.3) requires (i) of the following refinement.

Theorem 6.4 *[MS 97]). Theorem 6.3 remains valid if the definition of in-approximation is modified as follows.*
 (i) *the sets U_i are relatively open in the set $\{F \in M^{n\times n} : \det F = 1\}, n > 1$.*
 (ii) *the lamination convex hull U^l is replaced by the rank-1 convex hull*

$$\begin{aligned} U^{rc} := \; & \{F \in M^{m\times n} : f(F) \leq \inf_U f \text{ for} \\ & \text{all rank}-1 \text{ convex } f : M^{m\times n} \to \mathbf{R}\}. \end{aligned}$$

Part (i) requires only a modification in the proof of Lemma 6.1. A completely open problem is how many constraints on F can be prescribed. A natural guess is that all minors can be prescribed but this is false.

Regarding part (ii) one trivially has the inclusion $K^l \subset K^{rc}$. Several authors ([AH 86], [CT 93], [Ta 93]) discovered independently an example where $K^{rl} \neq K^{rc}$. Let

$$K = \left\{ \pm\begin{pmatrix} 3 & 0 \\ 0 & 1 \end{pmatrix}, \pm\begin{pmatrix} 1 & 0 \\ 0 & 3 \end{pmatrix} \right\}.$$

One has $K^l = K$ but

$$K^{rc} \supset \left\{ \begin{pmatrix} \lambda & 0 \\ 0 & \mu \end{pmatrix} : |\lambda| \leq 1, |\mu| \leq 1 \right\}.$$

It follows from Theorem 6.4 (ii) that the relation $Du \in U$ admits nontrivial solutions for every (small) neighbourhood U of K, although U contains no rank-1 connections (i.e. no simple laminates solve the relation).
A natural question is whether one can use the quasiconvex hull U^{qc} instead of U^{rc}. One key point seems to be the resolution of the following.

Conjecture 6.5 *Let K be a compact quasiconvex set, i.e. $K^{qc} = K$ and let $\nu \in \mathcal{M}^{qc}(K)$. Then for every open set $U \supset K$ there exists a sequence $u_j : (0,1)^n \to \mathbf{R}^m$ such that Du_j generates ν and $Du_j \in U$* a.e.

The conjecture is true for compact convex sets ([Mu 97a]); this refines a result of Zhang [Zh 92] who showed that $Du_j \in B(0, R)$ for a sufficiently large ball.

The use of the rank-1 convex hull U^{rc} rather then U^l is potentially relevant for the construction of singular solutions to $m \times 2$ elliptic systems

$$\operatorname{div}\sigma(Dv) = 0, \quad v : \Omega \subset \mathbf{R}^2 \to \mathbf{R}^m.$$

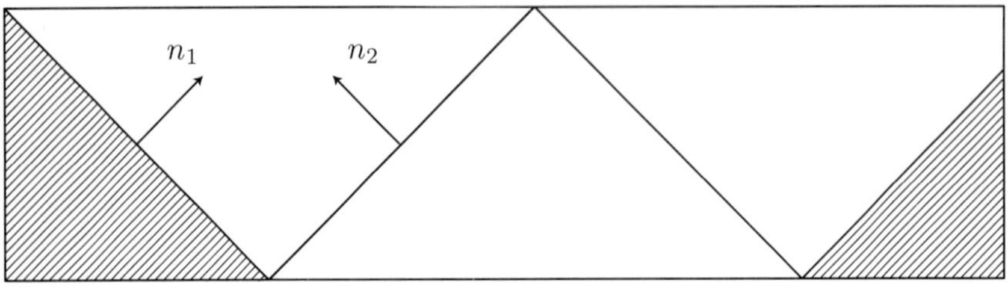

Figure 5: Structure of solutions with finite perimeter. The normals n_1, n_2 are determined by (6.4).

Such a system may be written as first order system (see [Sv 94])

$$Du \in K := \{(F,G) \in M^{2m\times 2}, \sigma(F) = G^\perp\},$$

where $G^\perp_{i1} = -G_{i2}, G^\perp_{i2} = G_{i1}$. Ellipticity implies that $K^l = K$, but nontrivial solutions may exist if one can show $K^{rc} \neq K$.

6.2. Regularity and rigidity

The construction outlined above yields very complicated solutions of the two-well problem (6.1)–(6.3). This raises the question whether the geometry of the solutions can be controlled. Consider the set

$$E = \{x \in \Omega : \ Du(x) \in SO(2)A\}$$

where Du takes values in one connected component of K (or one phase in the applications to crystals). The perimeter of a set $E \subset \Omega \subset \mathbf{R}^n$ is defined as

$$\mathrm{Per}E = \sup\{\int_E \mathrm{div}\ \varphi\ dx : \varphi \in C^1_0(\Omega, \mathbf{R}^n),\ |\varphi| \leq 1\}.$$

For smooth or polyhedral sets this agrees with the $(n-1)$ dimensional measure of ∂E.

Theorem 6.6 *[DM 95]). If u is a solution of (6.1) - (6.3) and if* Per$E < \infty$ *then u is locally a simple laminate and ∂E consists of straight line segments that can only intersect at $\partial\Omega$.*

The proof combines geometric and measure-theoretic ideas. The geometric idea is that the Gauss curvature $K(g)$ of the pull-back metric $g = (Du)^T Du$ should vanish (in a suitable sense). Since g only takes two values this should give information on E.

To implement this idea one first establishes a version of Liouville's theorem (infinitesimal rotations on an open connected set are rigid motions) for sets of finite perimeter. The analogue of connected components in this frame-work are indecomposable components (a set F of finite perimeter is indecomposable if for every $F_1 \subset F$ with Per$F$ = PerF_1 + Per$(F \setminus F_1)$ the set F_1 or $F \setminus F_1$ has zero measure). To finish the proof one decomposes Du as $e^{i\Theta} g^{1/2}$ (with the usual identification $\mathbf{C} \simeq \mathbf{R}^2$) and analyzes the jump conditions at the boundary of each component to deduce that Θ takes only two values and solves (in the distributional sense) a wave equation with characteristics n_1 and n_2. Using more subtle arguments Kirchheim [Ki 97] has recently proved a similar rigidity result for

$$K = SO(3)U_1 \cup SO(3)U_2 \cup SO(3)U_3$$

where $U_1 = \operatorname{diag}(\lambda, \mu, \mu)$ and U_2, U_3 are obtained by permuting the entries. An important additional difficulty is that $SO(3)$ is not abelian and one can thus not derive a linear equation for a suitable lift like Θ in the two-dimensional case.

7. Length scales, surface energy and singular perturbations

Minimization of elastic energy does not determine the length scale of the microstructure. If the infimum of the energy is not attained, minimizing sequences develop increasingly finer oscillations and generalized Young measure solutions correspond to an 'infinitely fine' microstructure. Also the Young measure only records the asymptotic volume fraction of each phase but not their geometric arrangement.

It is believed that addition of small surface (or higher gradient) energy contribution can remedy these deficiencies. The precise form of the surface energy term is not obvious. Popular choices for the total energy are

$$I^\epsilon(u) = \int_\Omega W(Du) + \epsilon^2 |D^2 u|^2 dx \qquad \text{or} \qquad I^\epsilon(u) = \int_\Omega W(Du)dx + \int_\Omega \epsilon |D^2 u|,$$

where $|D^2 u|$ is understood as a Radon measure. The idea is to study minimizers of these functionals in the limit $\epsilon \to 0$. Almost nothing is known about the three-dimensional model with full $SO(3)$ symmetry but results on lower dimensional models already show very interesting behaviour.

7.1. A one-dimensional model

As a caricature of the two-gradient problem consider

$$I(u) = \int_0^1 (u_x^2 - 1)^2 + u^2 dx \overset{!}{\longrightarrow} \min \tag{7.1}$$

subject to periodic boundary conditions. Clearly $I(u) > 0$ since the conditions $u = 0$ a.e. and $u_x = \pm 1$ a.e. are incompatible. On the other hand $\inf I = 0$, since a sequence of finely oscillating sawtooth functions u_j can achieve $u_{jx} \in \{\pm 1\}$, $u_j \to 0$ uniformly. For any such sequence u_{jx} generates the (unique) Young measure $\nu = \frac{1}{2}\delta_{-1} + \frac{1}{2}\delta_1$. Note that there are many 'different' sequences that generate this Young measure.

Minimizers of the singularly perturbed functional

$$I^\epsilon(u) = \int_0^1 \epsilon^2 u_{xx}^2 + (u_x^2 - 1)^2 + u^2 dx$$

yield a very special minimizing sequence for I.

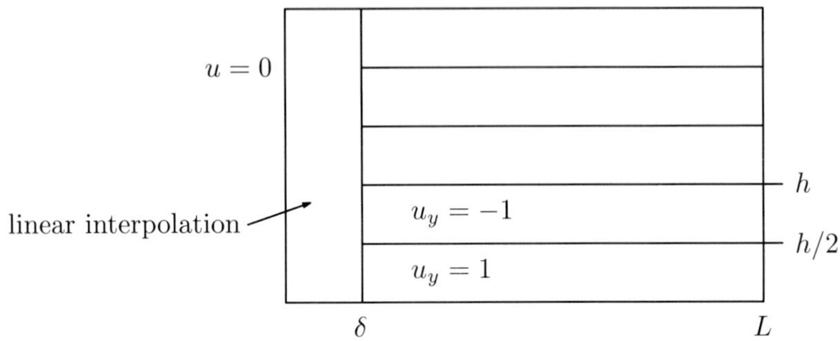

Figure 6: Construction of a minimizing sequence.

Theorem 7.1 *([Mu 93]). If $\epsilon > 0$ is sufficiently small then every minimizer of I^ϵ (subject to periodic boundary conditions) is periodic with minimal period $P^\epsilon = 4(2\epsilon)^{1/3} + \mathcal{O}(\epsilon^{2/3})$.*

7.2. Surface energy and domain branching

Consider the two-dimensional scalar model problem (see [KM 92] for the relation with three-dimensional elasticity)

$$I(u) = \int_0^1 \int_0^L u_x^2 + (u_y^2 - 1)^2 dx\, dy \stackrel{!}{\longrightarrow} \min,$$

$$u = 0 \text{ on } x = 0. \tag{7.2}$$

The integrand is minimized at $Du = (u_x, u_y) = (0, \pm 1)$. The preferred gradients are incompatible with the boundary condition. The infimum of I subject to (7.2) is zero but not attained. The gradients Du_j of any minimizing sequence generate the Young measure $\frac{1}{2}\delta_{(0,-1)} + \frac{1}{2}\delta_{(0,1)}$. One possible construction of a minimizing sequence is as follows (see Figure 6). Let s_h be a periodic sawtooth function with period h and slope ± 1 and let $u(x,y) = s_h(y)$ for $x \geq \delta$, $u(x,y) = \frac{x}{\delta} s_h(y)$ for $0 \leq x < \delta$. Then consider a limit $h \to 0, \delta \to 0$ such that h/δ remains bounded. Similar reasoning applies if we replace (7.2) by the condition that u vanishes on the whole boundary of $[0, L] \times [0, 1]$.

To understand the influence of regularizing terms on the length scale and the geometry of the fine scale structure we consider

$$I^\epsilon(u) = \int_0^1 \int_0^L u_x^2 + (u_y^2 - 1)^2 + \epsilon^2 u_{yy}^2 dx\, dy,$$

subject to (7.2). Instead of the second derivatives in y one can consider other regularizing terms, e.g. $|D^2 u|^2$. The derivatives in y are, however, the most important ones, since we expect that fine scale oscillations arise mainly in the y direction. It was widely believed that for small $\epsilon > 0$ the minimizers of I^ϵ look roughly like the

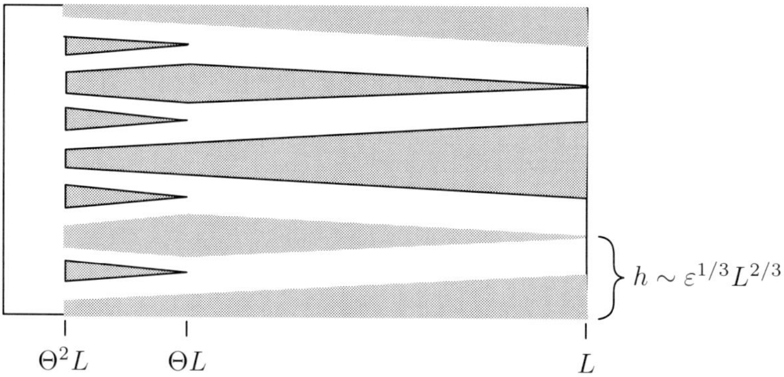

Figure 7: The self-similar construction with $1/4 < \Theta < 1/2$. Only two generations of refinement are shown.

construction $u_{h,\delta}$ depicted in Figure 6 (with the corners of the sawtooth 'rounded off' and optimal choices $\delta(\epsilon), h(\epsilon)$). This is false. Indeed a short calculation shows that $\delta(\epsilon) \sim (\epsilon L)^{1/2}$, $h(\epsilon) \sim (\epsilon L)^{1/2}$ and $I^\epsilon(u^\epsilon_{h,L}) \sim \epsilon^{1/2}L^{1/2}$. On the other hand one has

Theorem 7.2 *([Sch 94]). For $0 < \epsilon < 1$ there exists constants $c, C > 0$ such that*

$$c\epsilon^{2/3}L^{1/3} \leq \min_{u=0 \text{ at } x=0} I^\epsilon \leq C\epsilon^{2/3}L^{1/3}.$$

The upper bound is obtained by a smooth version of the self-similar construction depicted in Figure 7.

The mathematical issues become clearer if we replace I^ϵ by a sharp interface version

$$J^\epsilon(u) = \int_0^L \int_0^1 u_x^2 + \epsilon|u_{yy}|dydx \tag{7.3}$$

subject to

$$|u_y| = 1 \text{ a.e.} \tag{7.4}$$

Thus $y \mapsto u(x,y)$ is a sawtooth function and $\int_0^1 |u_{yy}|dy$ denotes twice the number of jumps of u_y. Minimization of (7.3) subject to (7.4) is in fact a purely geometric problem for the set

$$E = \{(x,y) : u_y(x,y) = 1\}.$$

The first term in J^ϵ is a nonlocal energy in terms of E, while the second is essentially the length of ∂E (more precisely its projection to the x-axis; as before we consider this to be the essential part since oscillations occur mainly in the y direction). The functional and the constraint are invariant under the scaling

$$u_\lambda(x,y) = \lambda^{-1}u(\lambda^{3/2}x, \lambda y)$$

which suggests a self-similar construction with $\Theta = \left(\frac{1}{2}\right)^{3/2}$.

Theorem 7.3 *([KM 94]). For $0 < \epsilon < 1$*

$$c\epsilon^{2/3}L^{1/3} \leq \min_{(7.2)(7.4)} J^\epsilon \leq C\epsilon^{2/3}L^{1/3} .$$

Moreover if $\bar{u}$ is a minimizer of J^ϵ subject to (7.2), (7.4) then, for all l,

$$c\epsilon^{2/3}l^{1/3} \leq \int_0^1 \int_0^l \bar{u}_x^2 + \epsilon|\bar{u}_{yy}| \leq C\epsilon^{2/3}l^{1/3} . \tag{7.5}$$

The scaling in (7.5) is exactly the scaling predicted by the self-similar construction with $\theta = (\frac{1}{2})^{3/2}$.

Conjecture 7.4 *As $\epsilon \to 0, l \to 0$ the rescaled minimizers*

$$v_\epsilon(x,y) = \epsilon^{-1/3}l^{-2/3}\bar{u}_\epsilon\,(lx, \epsilon^{1/3}l^{2/3}y)$$

converge to a self-similar function (at least for a subsequence).
The prediction of refinement of the microstructure (domain branching) towards the boundary $x = 0$ in the simple model (7.3), (7.4) inspired new experimental investigations ([Sch 93]). In closely related models for magnetization domains in ferromagnetic materials domain branching is experimentally well-established ([Li 44], [Hu 67], [Pr 76]), a rigorous mathematical analysis is just beginning to emerge. Already a quick look at some of the sophisticated constructions in [Pr 76] suggests that a lot is to be discovered.

8. Outlook

To close I briefly mention three questions in the analysis of microstructures that are widely open:

- Which mathematical object describes microstructure most efficiently?
- What is the rôle of dynamics?
- How can one efficiently compute microstructures?

8.1. Beyond Young measures

Young measures are but one way to extract 'relevant' information from a rapidly oscillating sequence. They describe the asymptotic local phase proportions in a finer and finer mixture. The Young measure contains no information about the geometry of the mixture, self correlations of the sequence or relevant length scales.

There is an intense search for new objects that record more information, and Tartar's article [Ta 95], from which the title of this subsection was borrowed, gives a recent survey. Due to space constraints I can only briefly mention two such objects: the H-measure and its variants which already found many applications and the two-scale Young measure which still has to prove its usefulness.

A typical example where knowledge of local phase proportions is insufficient arises in the theory of homogenization. A fine laminate of a good and a poor heat conductor behaves macroscopically like an anisotropic material that conducts poorly in the direction of lamination and well in the perpendicular directions.

To take the influence of the layering direction into account Tartar [Ta 90] introduced the H-measure that acts simultaneously on real space and on Fourier space. The same object was introduced independently by Gérard [Ge 91] under the name 'microlocal defect measure'. For every sequence $u_j \rightharpoonup 0$ in $L^2(\Omega)$ there exists a subsequence u_{j_k} and a Radon measure μ on $\bar{\Omega} \times S^{n-1}$ (the H-measure of $\{u_{j_k}\}$) such that for every pseudo-differential operator A of order zero with (sufficiently regular) symbol $a(x,\xi)$ one has

$$\langle A u_{j_k}, u_{j_k} \rangle_{L^2} \to \int_{\bar{\Omega} \times S^{n-1}} a \, d\mu \, .$$

For $\mathbf{R}^m$-valued sequences one similarly obtains a matrix-valued (hermitian) measure $\boldsymbol{\nu} = (\nu^{ij})_{1 \le i,j \le m}$. Applications of H-measures include small amplitude homogenization, compensated compactness with variable coefficients, compactness by averaging in kinetic equations and the propagation of energy concentrations in hyperbolic systems.

Comparing H-measures and Young measures one sees that the former predicts limits that involve pseudo-differential operators while the latter only yields limits of local expressions (cf. (5.1)). On the other hand the use of H-measures is restricted to quadratic expressions while Young measures can handle arbitrary nonlinearities. A synthesis of the two concepts is a major challenge for the future. So far even a good theory for a trilinear analogue of the H-measure is outstanding. Also the relation between H-measures and Young measures is not known (partial results were obtained by Murat and Tartar [MT 97], [Ta 95]).

As regards length scales, Gérard [Ge 90] introduced a variant of the H-measure, called semiclassical measure, that allows one to study oscillations with a typical length scale $h_j \to 0$. A similar measure was later considered by Lions and Paul [LP 93].

A different approach to the resolution of length scales is the notion of two-scale Young measures or Young measures on patterns ([AM 97]). For illustration consider a sequence of periodic functions $w_j : (0,1) \to \mathbf{R}$ that is uniformly bounded, $|w_j| \le M$. For a given sequence of scales $h_j \to 0$ one defines a new sequence

$$v_j(x,y) = w_j(x + h_j y) \, .$$

For $L > 0$ the set $K = \{g \in L^\infty(-L,L) : |g| \le M\}$ is a compact subset of $L^\infty(-L,L)$ equipped with the weak$*$ topology and v_j can be viewed as a map

$$\begin{aligned} V_j : (0,1) &\to K \\ x &\mapsto v_j(x,\cdot) \, . \end{aligned}$$

Since K is a compact metric space (a subsequence of) the sequence V_j generates a Young measure $\nu : (0,1) \to \mathcal{M}(K)$. This Young measure (called the two-scale Young measure of the original sequence w_j) is a probability measure on functions, obtained by blowing up the scale h_j. Using this idea one can describe the asymptotic behaviour as $\epsilon \to 0$ of minimizers of

$$\tilde{I}^\epsilon(u) = \int_0^1 \epsilon^2 u_{xx}^2 + W(u_x) + a(x)u^2,$$

where $0 < c \le a(x) \le C$ (cf. section 7.1). If u_{ϵ_j} is a sequence of minimizers, $\epsilon_j \to 0$, $w_j(x) = \epsilon_j^{-1/3} u_{\epsilon_j}(x), h_j = \epsilon_j^{1/3}$ and ν is the Young measure generated by V_j then $\nu(x)$ is supported on (translates of) sawtooth functions with period $4(2a(x))^{1/3}$ (cf. Theorem 7.1). Two-scale Young measures generalize the concept of two-scale convergence ([Ng 89], [Al 92]) which is very useful if the sequences involved are well-approximated by functions that are periodic (with fixed period) in the fast variable.

8.2. Dynamics

So far we have only considered time independent situations and studied minimizers or almost minimizers of the energy. A typical justification is that the time dependent problem admits a natural Liapunov function (free energy, entropy, ...) and therefore the evolution of 'generic' initial data should converge to (at least local) minimizers of the Liapunov functional.

Is such reasoning still reasonable if minimizers of the Liapunov functional do not exist and minimizing sequences have to develop microstructure? Friesecke [Fr 93] and Friesecke and McLeod [FM 96] studied a one dimensional model problem (viscoelastic bar on a foundation) whose Liapunov functional is essentially given by (7.1) (strong convergence of the kinetic energy is easy). They solved a long standing conjecture by showing that the dynamics excludes formation of finer and finer microstructure and hence energy minimization. See Ball et al. [BHJPS 91] and Pego [Pe 87] for related models and Ball [Ba 90] for a general discussion of energy minimization versus dynamics. Virtually nothing is known for the long-time dynamics of higher dimensional models, despite the interesting numerical simulations of Swart [Sw 91].

Another important issue is the evolution of microstructures. If the microstructure at each point in time is described by a Young measure (or some of the objects discussed in section 8.1) is it possible to deduce an evolution law for the Young measure? Conceptually this is similar to the closure problem in turbulence models and one might thus be pessimistic. Following Tartar's work since the early eighties (see e.g. [Ta 87]) there has nonetheless been interesting progress over the last years (see [KP 91], [FBS 94], [HR 94], [De 96], and in particular recent work of Otto [Ot 95], [Ot 97] and Theil [Th 97]). An interesting new approach to dynamics that takes into account the presence of many shallow local minima appears in [ACJ 96].

8.3. Computation

Computation of microstructures through numerical energy minimization is a very challenging task. There are presumably many local minimizers for the discrete and the continuous problem, the Euler-Lagrange equations are mixed elliptic-hyperbolic and oscillations typically develop on the scale of the discretization, which renders results very discretization dependent. For a recent review of the state of the art see Luskin [Lu 96] and the references therein. So far, most numerical methods make no use of available analytical information such as the classification of gradient Young measures, the minors relations or algorithms for rank-1 convexification (see sections 1 and 7 of [Lu 96] for the discussion of some exceptions). It seems to me that an important question is how to represent microstructures numerically in an efficient way. Ideally a good representation should both yield a high compression rate and be adapted to the numerical algorithm. This issue is closely related to the search for better analytical descriptions of microstructure (see section 8.1 above).

Acknowledgements. The writing of this article and the work reported here would not have been possible without the support and encouragement of many colleagues and friends. Particular thanks go to G. Alberti, J.M. Ball, K. Bhattacharya, G. Dolzmann, I. Fonseca, G. Friesecke, R.D. James, D. Kinderlehrer, B. Kirchheim, J. Kristensen, R.V. Kohn, V. Šverák and L. Tartar.

References

[ACJ 96] R. Abeyaratne, C. Chu and R.D. James, Kinetics of materials with wiggly energies: theory and application to the evolution of twinning microstructures in a Cu-Al-Ni shape memory alloy, *Phil. Mag.* **A 73** (1996), 457–497.

[AM 97] G. Alberti and S. Müller, in preparation.

[Al 92] G. Allaire, Homogenization and two-scale convergence, *SIAM J. Math. Anal.* **23** (1992), 1482–1518.

[Al 72] W.K. Allard, On the first variation of a varifold, *Ann. Math.* **95** (1972), 417–491.

[AH 86] R. Aumann and S. Hart, Bi-convexity and bi-martingales, *Israel J. Math.* **54** (1986), 159–180.

[Ba 84] E.J. Balder, A general approach to lower semicontinuity and lower closure in optimal control theory, *SIAM J. Control and Optimization* **22** (1984), 570–598.

[Ba 89] J.M. Ball, A version of the fundamental theorem for Young measures, in: *PDE's and Continuum Models of Phase Transitions* (M. Rascle, D. Serre, M. Slemrod, eds.), *Lecture Notes in Physics* **344**, Springer, 1989, 207–215.

[Ba 90] J.M. Ball, Dynamics and minimizing sequences, in: *Problems involving change of type* (K. Kirchgässner, ed.), *Lecture Notes in Physics* **359**, Springer, 1990, 3–16.

[BHJPS 91] J.M. Ball, P.J. Holmes, R.D. James, R.L. Pego and P.J. Swart, On the dynamics of fine structure, *J. Nonlin. Sci.* **1** (1991), 17–70.

[BJ 87] J.M. Ball and R.D. James, Fine phase mixtures as minimizers of energy, *Arch. Rat. Mech. Anal.* **100** (1987), 13–52.

[BJ 92] J.M. Ball and R.D. James, Proposed experimental tests of a theory of fine microstructure and the two-well problem, *Phil. Trans. Roy. Soc. London* **A 338** (1992), 389–450.

[BJ 97] J.M. Ball and R.D. James, book in preparation.

[BL 73] H. Berliocchi and J.M. Lasry, Intégrandes normales et mesures paramétrées en calcul des variations, *Bull. Soc. Math. France* **101** (1973), 129–184.

[Bh 93] K. Bhattacharya, Comparison of geometrically nonlinear and linear theories of martensitic transformation, *Cont. Mech. Thermodyn.* **5** (1993), 205–242.

[BFJK 94] K. Bhattacharya, N. Firoozye, R.D. James and R.V. Kohn, Restrictions on microstructure, *Proc. Roy. Soc. Edinburgh*, A **124** (1994), 843–878.

[CT 93] E. Casadio-Tarabusi, An algebraic characterization of quasi-convex functions, *Ricerche Mat.* **42** (1993), 11–24.

[CK 88] M. Chipot and D. Kinderlehrer, Equilibrium configurations of crystals, *Arch. Rat. Mech. Anal.* **103** (1988), 237–277.

[Da 89] B. Dacorogna, *Direct methods in the calculus of variations*, Springer, 1989.

[DM 96a] B. Dacorogna and P. Marcellini, Théorèmes d'existence dans les cas scalaire et vectoriel pour les équations de Hamilton-Jacobi, *C.R.A.S. Paris* **322** (1996), Serie I, 237–240.

[DM 96b] B. Dacorogna and P. Marcellini, Sur le problème de Cauchy-Dirichlet pour les systèmes d'équations non linéaires du premier ordre, *C.R.A.S. Paris* **323** (1996), Serie I, 599–602.

[DM 97] B. Dacorogna and P. Marcellini, General existence theorems for Hamilton-Jacobi equations in the scalar and vectorial cases, *Acta. Math.* **178** (1997), 1–37.

[De 96] S. Demoulini, Young measure solutions for a nonlinear parabolic equation of forward-backward type, *SIAM J. Math. Anal.* **27** (1996), 378–403.

[DP 85] R.J. DiPerna, Compensated compactness and general systems of conversation laws, *Trans. Amer. Math. Soc.* **292** (1985), 383–420.

[DPM 87] R.J. DiPerna and A.J. Majda, Oscillations and concentrations in weak solutions of the incompressible fluid equations, *Comm. Math. Phys.* **108** (1987), 667–689.

[DM 95] G. Dolzmann and S. Müller, Microstructures with finite surface energy: the two-well problem, *Arch. Rat. Mech. Anal.* **132** (1995), 101–141.

[Ed 65] R.E. Edwards, *Functional analysis*, Holt, Rinehart and Winston, 1965.

[Er 80] J.L. Ericksen, Some phase transitions in elastic crystals, *Arch. Rat. Mech. Anal.* **73** (1980), 99–124.

[Er 84] J.L. Ericksen, The Cauchy and Born hypothesis for crystals, in: *Phase transformations and material instabilities in solids* (M.E. Gurtin, ed.), Academic Press, 1984, 61–77.

[Er 89] J.L. Ericksen, Weak martensitic transformations in Bravais lattices, *Arch. Rat. Mech. Anal.* **107** (1989), 23–36.

[Fo 89] I. Fonseca, Phase transitions for elastic solid materials, *Arch. Rat. Mech. Anal.* **107** (1989), 195–223.

[FBS 94] I. Fonseca, D. Brandon and P. Swart, Dynamics and oscillatory microstructure in a model of displacive phase transformations, in: *Progress in partial differential equations, the Metz surveys* **3** (M. Chipot, ed.), Pitman, 1994, 130–144.

[FMP 97] I. Fonseca, S. Müller and P. Pedregal, Analysis of concentration and oscillation effects generated by gradients, to appear in *SIAM J. Math. Anal.*

[Fr 93] G. Friesecke, Static and dynamic problems in nonlinear mechanics, Ph.D. thesis, Heriot-Watt University, Edinburgh, 1993.

[FM 96] G. Friesecke and J.B. McLeod, Dynamics as a mechanism preventing the formation of finer and finer microstructure, *Arch. Rat. Mech. Anal.* **133** (1996), 199–247.

[Ge 90] P. Gérard, Mesures semi-classique et ondes de Bloch, in: Equations aux dérivées partielles, exposé XVI, seminaire 1990–91, Ecole Polytechnique, Palaiseau.

[Ge 91] P. Gérard, Microlocal defect measures, *Comm. PDE* **16** (1991), 1761–1794.

[Gr 86] M. Gromov, *Partial differential relations*, Springer, 1986.

[HR 94] K.-H. Hoffmann and T. Roubíček, Optimal control of a fine structure, *Appl. Math. Optim.* **30** (1994), 113–126.

[Hu 67] A. Hubert, Zur Theorie der zweiphasigen Domänenstrukturen in Supraleitern und Ferromagneten, *Phys. Status Solidi* **24** (1967), 669–682.

[IT 69] A. & C. Ionescu Tulcea, *Topics in the theory of liftings*, Springer, 1969.

[Kh 67] A. Khachaturyan, Some questions concerning the theory of phase transformations in solids, *Soviet Physics-Solid State* **8** (1967), 2163–2168.

[Kh 83] A. Khachaturyan, *Theory of structural transformations in solids*, Wiley, 1983.

[KSh 69] A. Khachaturyan and G. Shatalov, Theory of macroscopic periodicity for a phase transition in the solid state, *Soviet Physics JETP* **29** (1969), 557–561.

[Ki 88] D. Kinderlehrer, Remarks about equilibrium configurations of crystals, in *Material instabilities in continuum mechanics*, Heriot-Watt Symposium, (J.M. Ball, ed.), Oxford Univ. Press, 1988, 217–242.

[KP 91] D. Kinderlehrer and P. Pedregal, Characterization of Young mesures generated by gradients, *Arch. Rat. Mech. Anal.* **115** (1991), 329–365.

[KP 94] D. Kinderlehrer and P. Pedregal, Gradient Young measure generated by sequences in Sobolev spaces, *J. Geom. Analysis* **4** (1994), 59–90.

[Ki 97] B. Kirchheim, in preparation.

[Ko 89] R.V. Kohn, The relationship between linear and nonlinear variational models of coherent phase transitions, in: *Trans. 7th Army Conf. on applied mathematics and computing*, (F. Dressel, ed.), 1989.

[KM 92] R.V. Kohn and S. Müller, Branching of twins near an austenite/twinned-martensite interface, *Phil. Mag.* **A 66** (1992), 697–715.

[KM 94] R.V. Kohn and S. Müller, Surface energy and microstructure in coherent phase transitions, *Comm. Pure Appl. Math.* **47** (1994), 405–435.

[Kr 94] J. Kristensen, Finite functionals and Young measures generated by gradients of Sobolev functions, Ph.D. Thesis, Technical University of Denmark, Lyngby.

[Kr 97] J. Kristensen, On the non-locality of quasiconvexity, to appear in *Ann. IHP Anal. non linéaire.*

[Ku 55] N.H. Kuiper, On C^1 isometric embeddings I., *Nederl.-Akad.-Wetensch.-Proc.*, **A 58** (1955), 545–556.

[Li 44] E. Lifshitz, On the magnetic structure of iron, *J. Phys.* **8** (1944), 337–346.

[LP 93] P.-L. Lions and T. Paul, Sur les mesures de Wigner, *Rev. Mat. Iberoamericana* **9** (1993), 553–618.

[Lu 96] M. Luskin, On the computation of crystalline microstructure, *Acta Num.* **5** (1996), 191–258.

[Ma 93] J.P. Matos, Young measures: regularity results in the two well problem, in: Int. Conf. on Differential Equations, EQUADIFF 91, Barcelona 1991, (C. Perello, C. Simo, J. Sola-Morales, eds.), World Sci. Publ., 1993, 736–739.

[MP 96] J. Matoušek and P. Plechač, On functional separately convex hulls, to appear in J. Discrete Comput. Geom.

[Me 66] P.-A. Meyer, *Probability and potentials*, Blaisdell, 1966.

[Mo 52] C.B. Morrey, Quasi-convexity and the lower semicontinuity of multiple integrals, *Pacific J. Math.* **2** (1952), 25–53.

[Mo 66] C.B. Morrey, *Multiple integrals in the calculus of variations*, Springer, 1966.

[Mu 93] S. Müller, Singular perturbations as a selection criterion for periodic minimizing sequences, *Calc. Var.* **1** (1993), 169–204.

[Mu 97a] S. Müller, *Microstructure and the calculus of variations*, Nachdiplomvorlesung ETH Zürich, in preparation.

[Mu 97b] S. Müller, Variational models of microstructure, in Proc. CIME Summer School, Cetraro, 1996, in preparation.

[Mu 97c] S. Müller, A sharp version of Zhang's theorem, preprint.

[MS 96] S. Müller and V. Šverák, Attainment results for the two-well problem by convex integration, *Geometric analysis and the calculus of variations* (J. Jost, ed.), International Press, 1996, 239–251.

[MS 97] S. Müller and V. Šverák, in preparation.

[MT 97] F. Murat and L. Tartar, On the relation between Young measures and H-measures, in preparation.

[Na 54] J. Nash, C^1 isometric embeddings, *Ann. Math.* **60** (1954), 383–396.

[Ng 89] G. Nguetseng, A general convergence result for a functional related to the theory of homogenization, *SIAM J. Math. Anal.* **20** (1989), 608–623.

[Ot 95] F. Otto, Evolution of microstructure in unstable porous media flow: a relaxational approach, preprint SFB 256, Bonn, 1995.

[Ot 97] F. Otto, Dynamics of labyrinthine pattern formation in magnetic fluids: a mean-field theory, to appear in *Arch. Rat. Mech. Anal.*

[Pe 87] R.L. Pego, Phase transitions in one-dimensional nonlinear viscoelasticity: admissibility and stability, *Arch. Rat. Mech. Anal.* **97** (1987), 353–394.

[Pi 84] M. Pitteri, Reconciliation of local and global symmetries of crystals, *J. Elasticity* **14** (1984), 175–190.

[PZ 97] M. Pitteri and G. Zanzotto, *Continuum models for phase transitions and twinning in crystals*, Chapman and Hall, forthcoming.

[Pr 76] I. Privorotskii, *Thermodynamic theory of domain structures*, Wiley, 1976.

[Re 68] Yu.G. Reshetnyak, Liouville's theorem under minimal regularity assumptions, *Sib. Math. J.* **8** (1968), 631–634.

[Ro 69] A. Roitburd, The domain structure of crystals formed in the solid phase, *Soviet Physics-Solid State* **10** (1969), 2870–2876.

[Ro 78] A. Roitburd, Martensitic transformation as a typical phase transformation in solids, *Solid State Physics* **33**, Academic Press, New York, 1978, 317–390.

[Ro 97] T. Roubíček, *Relaxation in optimization theory and variational calculus*, de Gruyter, 1997.

[Sch 94] C. Schreiber, Rapport de stage de D.E.A., ENS Lyon, 1994.

[Sch 93] D. Schryvers, Microtwin sequences in thermoelastic Ni_xAl_{100-x} martensite studied by conventional and high resolution transmission electron microscopy, *Phil. Mag.* **A 68** (1993), 1017–1032.

[Sv 91a] V. Šverák, Quasiconvex functions with subquadratic growth, *Proc. Roy. Soc. London* **A 433** (1991), 723–725.

[Sv 91b] V. Šverák, On regularity for the Monge-Ampère equations, preprint, Heriot-Watt University, 1991.

[Sv 92a] V. Šverák, Rank-one convexity does not imply quasiconvexity, *Proc. Roy. Soc. Edinburgh* **120** (1992), 185–189.

[Sv 92b] V. Šverák, New examples of quasiconvex functions, *Arch. Rat. Mech. Anal.* **119** (1992), 293–300.

[Sv 93] V. Šverák, On the problem of two wells, in: *Microstructure and phase transitions*, IMA Vol. Appl. Math. **54** (D. Kinderlehrer, R.D. James, M. Luskin and J. Ericksen, eds.), Springer, 1993, 183–189.

[Sv 94] V. Šverák, Lower semincontinuity of variational integrals and compensated compactness, in: Proc. ICM 1994 (S.D. Chatterji, ed.), vol. 2 Birkhäuser, 1995, 1153–1158.

[Sv] V. Šverák, personal communication.

[Sw 91] P. Swart, The dynamical creation of microstructure in material phase transitions, Ph.D. thesis, Cornell University, 1991.

[Ta 79] L. Tartar, Compensated compactness and partial differential equations, in: *Nonlinear Analysis and Mechanics: Heriot-Watt Symposium* Vol. IV (R. Knops, ed.), Pitman, 1979, 136–212

[Ta 82] L. Tartar, The compensated compactness method applied to systems of conservations laws, in: *Systems of Nonlinear Partial Differential Equations*, (J.M. Ball, ed.), NATO ASI Series, Vol. C111, Reidel, 1983, 263–285.

[Ta 87] L. Tartar, Oscillations and asymptotic behaviour for two semilinear hyperbolic systems, in: *Dynamics of infinite dynamical systems*, NATO ASI Ser. F, Springer, 1987.

[Ta 90] L. Tartar, H-measures, a new approach for studying homogenization, oscillations and concentration effects in partial differential equations, *Proc. Roy. Soc. Edinburgh* **A 115** (1990), 193–230.

[Ta 93] L. Tartar, Some remarks on separately convex functions, in: *Microstructure and phase transitions*, IMA Vol. Math. Appl. **54**, (D. Kinderlehrer, R.D. James, M. Luskin and J.L. Ericksen, eds.), Springer, 1993, 191–204.

[Ta 95] L. Tartar, Beyond Young measures, *Meccanica* **30** (1995), 505–526.

[Ta 97] L. Tartar, *Homogenization, compensated compactness and H-measures*, CBMS-NSF conference, Santa Cruz, June 1993, lecture notes in preparation.

[Th 97] F. Theil, Young-measure solutions for a viscoelastically damped wave equation with nonmonotone stress-strain relation, *Arch. Rat. Mech. Anal.*, to appear.

[Va 90] M. Valadier, Young measures, in: *Methods of nonconvex analysis*, Lecture notes in mathematics **1446**, Springer, 1990.

[Va 94] M. Valadier, A course on Young measures, *Rend. Istit. Mat. Univ. Trieste* **26** (1994) suppl., 349–394.

[Yo 37] L.C. Young, Generalized curves and the existence of an attained absolute minimum in the calculus of variations, *Comptes Rendus de la Société des Sciences et des Lettres de Varsovie, classe III*, **30** (1937), 212–234.

[Yo 69] L.C. Young, *Lectures on the calculus of variations and optimal control theory*, Saunders, 1969 (reprinted by Chelsea 1980).

[Za 92] G. Zanzotto, On the material symmetry group of elastic crystals and the Born rule, *Arch. Rat. Mech. Anal.* **121** (1992), 1–36.

[Zh 92] K. Zhang, A construction of quasiconvex functions with linear growth at infinity, *Ann. S.N.S. Pisa* **19** (1992), 313–326.

Progress in Mathematics, Vol. 169,

Different Types of Non Uniform Hyperbolicity for Interval Maps Are Equivalent

TOMASZ NOWICKI*

Warsaw University, Institute of Mathematics
Banacha 2, PL 02-097 Warsaw

ABSTRACT. After a soft introduction to interval dynamics we describe the typical patterns of behaviour of some one dimensional dynamical systems. As an example of up-to-date results in this area we enumerate equivalent conditions of non-uniform hyperbolicity for some unimodal maps of the interval; they are all equivalent to a Collet-Eckmann condition.

By non-uniform hyperbolicity we mean that the derivatives for iterates are separated from one on some subset of the interval.

1. Patterns and exceptions

Mathematical thinking is a combination of two interacting searches: for patterns and exceptions. A mathematician considers an object, formulates some of its properties and then asks whether this properties are of a general nature. Once found, the pattern is tried on other objects and now the interst is in discovering all irregularities in the expanded domain. After some considerations such exceptions are incorporated into new patterns and the game is played again. Let us try this idea on an example.

2. Dynamical systems

Dynamical systems model time evolution. Time itself is represented by integers or reals. We assume the compatibility of the evolution i.e. given a starting point the endpoint depends only on the time difference. It is therefore possible to model time with other (semi-)groups.

The evolution takes place in so-called phase space. Depending on taste and interest the phase space may have some additional structure – a topology or a measure are the most common examples.

The evolution is a transformation from the product of the phase space and the time space back into the phase space itself, conserving the additional structure and fulfiling the compatibility condition $F(x, t + s) = F(F(x, t), s)$.

The main question in dynamical systems is the limit behaviour of the evolution.

*partially supported by the organizers of ECM_2 and Polish KBN Grant 2 P03A 02208

3. The linear model

3.1. The geometrical sequence Consider the sequence $x_{n+1} = a \cdot x_n$, with real parameter a and real starting point x_0. Reals are the phase space and the semigroup of natural numbers is the time space. The convergence is well known, so let us describe the pattern.

For $x_0 = 0$ the sequence is constant for any a. Otherwise for $|a| < 1$ the sequence converges to zero, for $|a| > 1$ the sequence diverges to infinity, for $a = 1$ it is constant and for $a = -1$ it is periodic with period 2. Is it a pattern ?

Look at the linear function $f(x) = a \cdot x$, it is the evolution of the geometrical sequence, with natural numbers representing time $F(x, n) = f^n(x)$. Compare with the evolution by some other (smooth) function g, suppose it has a fixed point $p = g(p)$. Depending on whether the derivative $|g'(p)|$ is smaller or larger than 1 the nearby points converge to p under g^n or g^{-n}. The exceptional situation in this pattern is when the derivative is 1 or -1, when the evolution of $g^n(x)$ depends on more subtle properties of g.

The situation $|g'(p)| = 1$ is also exceptional in the following sense: namely among the C^1 maps of the real line with the fixed point p the condition $|g'(p)| \neq 1$ is open and dense. This is called hyperbolicity of the fixed point and it is invariant under small perturbations.

3.2. Rotations We analyze now the exceptional case $|a| = 1$, but now let the phase space and the parameters be the complex numbers. It is enough to consider $|x| = 1$ i.e. the unit circle. We see now that $f(x) = ax$ is a rotation of the circle with angle $2\pi\alpha$, where $a = \exp(2\pi i\alpha)$. With $\alpha = p/q$ rational f^q is the identity (each point comes back to itself after p turns around the circle), otherwise each trajectory (the sequence $f^n(x)$) is dense in the circle. Is this a pattern ?

When considering a C^2 diffeomorphism g say of the circle we may distinguish two cases: g has a periodic point or for each x and each n $g^n(x) \neq x$. By a classical theorem by Denjoy, in the second case g is topologically conjugated with a rotation (i.e. there is a homeomorphism h of the circle with $f \circ h = h \circ g$) with irrational α. This α is called the rotation number and may be found as a limit of the ratio of the number of turns around the circle divided by the number of iterates and is independent of the starting point. There are some problems with diffeomorphisms which are only C^1, in this case h might be just a continuous, monotone map (constant on some intervals).

The smoothness class of the conjugacy depends on how irrational α is [29]. The expansion of α into a continuous fraction defines the order of closest returns of the trajectory to the starting point.

We remark that there are more irrational numbers than rational ones, but the irrational rotation numbers are unstable, i.e. after a small perturbation periodic points appear. On the other hand, the diffeomorphisms having a finite number of only hyperbolic periodic orbits are stable.

We remark also that the rotation leaves the length of the arc invariant.

3.3. Expansions While dealing with the complex numbers on the unit circle we may go back to the map $x \mapsto ax$, but now applied to the arguments, i.e.

$\exp(2\pi i\phi) \mapsto \exp(2\pi i a\phi)$. In order to get a continuous map of the circle we assume a to be an integer, say $a = 2$. Due to the compactness of S^1 and the expansion of the map we obtain a very complicated picture. Look at the representation of this map as $f : x \mapsto 2x \bmod 1$ of the unit interval. Its graph consists of two branches; the n-th iterate has 2^n branches covering the whole interval. This map has a dense set of periodic points.

A base 2 expansion of each number codes the trajectory under f, 0 at position n says that $f^n(x) < 0.5$ and 1 at position n otherwise. In this coding the map represents a shift in the code. By Borel's theorem on normal numbers we see that almost every trajectory is dense. This happens as the Lebesgue measure is invariant ($|f^{-1}A| = |A|$), and (contrary to the rotation) has strong stochastic properties.

It is a pattern: smooth expansion on compact manifolds implies the existence of a finite absolutely continuous invariant measure [17].

4. Nonlinear maps

4.1. The quadratic family Doubling the argument is an effect of the square function. The evolution $z \mapsto z^2$ is easily understood; the argument part was just described, and the modulus part evolves to zero or infinity depending on whether we start inside or outside the unit circle. The unit circle itself forms a border between the bassins of two attractors (0 and ∞) and is the Julia set of the map.

If we project $(x, y) = z \mapsto z^2 = (x^2 - y^2, 2xy)$ from the unit circle $x^2 + y^2 = 1$ onto the real diameter $[-1, 1]$ we obtain $x \mapsto 2x^2 - 1$. This map conserves (after some obvious changes) essential properties of the angle doubling.

In this way we considered two representations of the family $x \mapsto tx^2 - t + 1$ ($t = 1, 2$), or after linear change of variable $f_\alpha : x \mapsto 4\alpha x(1 - x)$, for both phase variable x and the parameter α from the unit interval $[0, 1]$. Such maps have one critical point 0.5, which is turning with critical value α.

Are these two behaviors unique or maybe typical for that family or for any unimodal map?

The pattern is slightly more complicated. As we have seen the behaviour of the map $4x(1 - x)$ is essentially due to the expansion, but because of critical points there is no hope for uniform expansion in this family. Similarly, because of a repelling fixed point the uniform contraction is impossible for $\alpha > 0.75$

First let us consider a set $\mathcal{A}$ of those parameters α for which f_α has a periodic point p which is attracting i.e. $f^n(p) = p$ and $|Df^n(p)| < 1$ (in fact the equality may also occur). It turns out that at most one such periodic orbit may exist; it attracts the critical trajectory. Such parameters form an open dense set in $[0, 1]$ [9], [27] (in the complex setting the analogue of $\mathcal{A}$ is called the Mandelbrot set). A similar result for general C^2 maps was proven in [16].

Next consider a set $\mathcal{T}$ of those parameters α for which f_α has a trajectory which is dense in some interval (then it is dense in a finite cycle of intervals). Such parameters form a set of positive Lebesgue measure [12]. The cited paper includes stronger result: the parameters of Collet-Eckmann maps (see below) form a set with positive Lebesgue measure.

But there is also another set $\mathcal{R}$ of parameters α such that f_α has an infinite cascade of nested invariant cycles of intervals, their intersection being a zero measure Cantor set called a solenoid. The best known example of this set is a Feigenbaum parameter at the end of a period doubling bifurcations [8]. This set of parameters is uncountable, but its Lebesgue measure is zero for a quadratic family [19].

The sets $\mathcal{A}, \mathcal{T}, \mathcal{R}$ are disjoint and contain all possibilities; they have different dynamical properties [10, 1]. We say that a property is typical (in a topological sense) when it occurs on a residual set (this is a thick set in the sense of Baire category or a countable intersection of dense open sets), and we say that it is typical (in a metrical sense) when it occurs on a set of full Lebesgue measure. For the quadratic family the statements below hold for both senses of the word 'typical'.

For $\alpha \in \mathcal{A}$ a typical trajectory is attracted by the periodic attractor. In the case of $\mathcal{T}$ a typical trajectory is dense in the (smallest) invariant cycle of intervals. In the case of $\mathcal{R}$ a typical trajectory is attracted by the solenoid.

A topological invariant, called the kneading sequence [21], fully describes to which of the above classes a unimodal function belongs. The kneading sequence is the coding of the trajectory of the critical point 0.5 by 0 if $f^n(0.5) < 0.5$ and 1 if $f^n(0.5) > 0.5$ (and a special symbol for $f^n(0.5) = 0.5$, if this happens we say that f has a superstable periodic attractor). It plays a similar role in the unimodal class to the rotation number in circle diffeomorphisms; the role of rotations is played by tent maps $x \mapsto 1 - s|x - 0.5|$.

This correspondence between combinatorics and topology is clearly due to the existence of the natural order in one dimensional manifolds.

4.2. General maps All of this pattern extends to those unimodal maps with negative Schwarzian derivative ($f'''/f' - 3(f''/f')^2/2 \leq 0$) except that for some maps in the transitive case a metrically typical trajectory may be attracted by some minimal Cantor set. Two topologically conjugated maps may have different metrical behaviour [2]. While $z \mapsto z^2$ expands the argument, negative Schwarzian maps expand the cross-ratio, or the Poincarré metric.

Roughly speaking, in the multimodal case, one can attach one of the three types of behaviors to each critical point.

In general for C^2 maps it is believed (and in many cases proved) that for large iterates (and in particular for asymptotics) the behaviour is very much the same as for S-unimodal maps. For example, there are no wandering intervals, i.e. the semi-conjugacy to a tent maps is in fact a homeomorphism [5].

However, the pattern is strongly destroyed if the critical points are flat (all the derivatives zero). The structure of maps with solenoids is even not conjectured.

5. Invariant measures

When a system has a finite invariant measure $\mu = \mu \circ f^{-1}$ then one can split the phase space into its so-called ergodic components. On each such component the time mean of an integrable function G for a μ-typical point tends to the space average

$$\frac{1}{n}\sum_{i=0}^{n-1} G\circ f^i(x) \to \int G d\mu \,.$$

There are trivial invariant measures such as the Dirac measure on the fixed point, so we are mostly interested in invariant measures with some additional properties: such as the measures which are spread by the trajectory of a typical point. This happens in particular when f has a hyperbolic periodic attractor or a solenoid, and also when a finite invariant measure absolutely continuous with respect to the Lebesgue measure (*facim* in short) exists (by Birkhoff's ergodic theorem).

The absolutely continuous invariant measures may appear in only in the case $\mathcal{T}$, but there are maps in $\mathcal{T}$ without a *facim* [13, 11]. For a quadratic family however this is an exception (i.e. parameters of maps without a *facim* in $\mathcal{T}$ form a null set) [18, 20].

It is still not known if for almost all $t \in \mathcal{T}$ the map in the family $g_t = tx^\ell - t + 1$ ($\ell > 1$, $t \in [0, 2]$) has a *facim*. It is not known if this can be described in topological terms.

The *facim* μ of an S-unimodal map is ergodic, in particular the frequency of visits of a metrically typical point describes the density of the *facim* [1].

6. Hyperbolicity

As we said while analyzing the linear map, hyperbolicity means that the derivative shrinks or grows exponentially. In general, global and uniform hyperbolicity is nice because there is a definite rate of attraction for the time flowing forward or backward. On compact manifolds uniform (hyperbolic) expansion yields good invariant measure which instead of a deterministic description provides a stochastic one.

However, for maps with a critical point and an expanding fixed point there is no hope for a nice uniform hyperbolicity. There is an expanding part and a contracting part, and in one dimension there is no place to avoid their interference. In this case we must deal with typical points.

For a typical point in case $\mathcal{A}$ the Lyapunov exponent is negative

$$\limsup \log |Df^n(x)|)/n < 0$$

(this is an equivalent condition) [25]. There is a *facim* iff the Lyapunov exponent is positive on the set of positive Lebesgue measure (so in some subcases of $\mathcal{T}$) [14]. Otherwise a typical exponent is zero.

7. Non uniform hyperbolicity — equivalent conditions

Although we have a classification for typical orbits some sets are of special dynamical interest: periodic points, forward and backward images of the critical point.

The attracting case, when the exponent is negative, is well understood. The following statements are equivalent (for S-unimodal maps) (and widely known):

- f has a strictly attracting periodic orbit; there exists a $p = f^n(p)$ with $|Df^n(p)| < 1$.
- The orbit of p attracts the forward critical orbit.
- The forward critical orbit has negative Lyapunov exponent.
- Typical points (in both senses) have negative exponents.
- The orbit of p attracts the typical points.

This situation is invariant under topological conjugacies within the class.

The repelling case is more complicated, nevertheless the following theorem is true

- f has a *facim* iff for (Lebesgue-) almost every point its Lyapunov exponent is positive.

This situation is not invariant under topological conjugacies. For the maps with an acim typical points have orbits dense in a finite union of intervals (but the converse is not always true).

It is interesting that positive Lyapunov exponent on some special sets may imply a *facim*. In particular, the critical value forms such a set. The condition $|Df^n(f(c))| > K\lambda^n$ is called a (forward) Collet-Eckmann (CE) condition [4]. It is equivalent to uniform hyperbolicity on the set of periodic points, to the existence of a *facim* with an exponential decay of correlations, to the exponential decay of the intervals of monotonicity. The invariant measure of CE maps has strong mixing properties (the iterates become more and more independent).

The CE condition is a topological invariant, but there are maps with a *facim* which are not CE. In the quadratic family the parameters of CE maps have positive Lebesgue measure but it is unknown if they have a full measure in the set $\mathcal{T}$ or even in its subset of maps with *facim*.

We summarize this in following equivalent statements.

1. The Lyapunov exponent of the forward orbit of the critical value is positive, i.e. there are $K > 0, \lambda > 1$ with $|Df^n(f(c))| > K\lambda^n$ for all $n > 0$, where$f'(c) = 0$ defines the unique critical point of f. This is called the Collet-Eckmann condition.

2. The Lyapunov exponent of the backward orbit of the critical point is positive i.e. $|Df^n(z)| > K\lambda^n$ whenever $f^n(z) = c$.

3. The periodic points are uniformely hyperbolic i.e. $|Df^n(p)| > K\lambda^n$ whenever $f^n(p) = p$.

4. f has a finite absolutely continuous invariant measure which has an exponential decay of correlations.

5. There exist $N > 0$ and $\delta > 0$ such that for every x there exists an increasing sequence of positive integers n_j of positive frequency ($n_j < Pj$ for some P) with

$$\mathrm{Card}\{i : 0 \le i < n_j, c \in \mathrm{Comp}_{f^i(x)} f^{-(n_j - i)}(f^{n_j}(x) - \delta, f^{n_j}(x) + \delta)\} \le N \ .$$

This situation is topologically invariant (see the last condition), so not all maps with an acim are Collet-Eckmann maps. For the reference: 1)⇒2) [23], 2)⇔3) [22], 1)⇒4) [4, 30, 15], 4)⇒3), 3)⇒1) [25], 1)⇒5), 5)⇒3) [24].

8. References

The results presented above in a very rough and inprecise way are due to many mathematicians. The best reference to one dimensional dynamics is the most recent book by de Melo and van Strien [6] (W. de Melo and S. van Strien, One–Dimensional Dynamics, Springer Verlag, Berlin Heidelberg New York (1993)) for beginners I can recommend also the book by P.Collet and J.-P.Eckmann [3]. For general references in dynamical systems see [7, 28]

Some of the new results mentioned above will be published soon. M. Lyubich [18] gave a topological condition and half of the estimates to prove that the case $\mathcal{T}$ in quadratic family has typically a *facim*, the other half was filled by M. Martens [20]. D. Sands [25] found the way to prove the equivalence of the Collet-Eckmann condition and uniform hyperbolicity on periodic orbits. F. Przytycki [26, 24] gave the topological condition equivalent to the Collet-Eckmann condition. I am very happy that I had the opportunity to cooperate with Marco, Duncan and Feliks in these projects.

References

[1] A. M. Blokh and M. Y. Lyubich. Measurable dynamics of S-unimodal maps of the interval. *Ann. Sci. École Norm. Sup.*, 24(5):545–573, 1991.

[2] H. Bruin, G. Keller, T. Nowicki, and S. van Strien. Wild Cantor attractors exist. *Annals of Math.*, **143**:97–130, 1996.

[3] P. Collet and J.-P. Eckmann. *Iterated Maps of the Interval as Dynamical Systems.* Progress in Physics 1. Birkhäuser, Boston, 1980.

[4] P. Collet and J.-P. Eckmann. Positive Liapunov exponents and absolute continuity for maps of the interval. *Ergodic Theory Dynam. Syst.*, 3:13–46, 1983.

[5] W. de Melo and S. van Strien. A structure theorem in one dimensional dynamics. *Ann. of Math.*, 129:519–546, 1989.

[6] W. de Melo and S. van Strien. *One-dimensional Dynamics.* Springer-Verlag, Berlin, Heidelberg, New York, 1993.

[7] R. L. Devaney. *An Introduction to Chaotic Dynamical Systems.* Addison–Wesley, 1987.

[8] M. Feigenbaum. Quantitative universality for a class of non-linear transformations. *J.Stat.Phys.*, **19,21**:25–52,669–709, 1978/79.

[9] Graczyk J. and Świątek G. Generic hyperbolicity in the logistic family. *Ann. of Math.* **146** (1997), 1–52.

[10] J. Guckenheimer. Sensitive dependence to initial conditions for one dimensional maps. *Commun. Math. Phys.*, 70:133–160, 1979.

[11] F. Hofbauer and G. Keller. Quadratic maps without asymptotic measure. *Commun. Math. Phys.*, 127:319–337, 1990.

[12] M. Jakobson. Absolutely continous invariant measures for one-parameter families of one-dimensional maps. *Communications in Mathematical Physics*, **81**:39–88, 1981.

[13] S. Johnson. Singular measures without restrictive intervals. *Communications in Mathematical Physics*, **110**:185–190, 1987.

[14] G. Keller. Exponents, attractors and Hopf decomposition for interval maps. *Ergodic Theory & Dynamical Systems*, **10**:717–744, 1990.

[15] G. Keller and T. Nowicki. Spectral theory, zeta functions and the distribution of periodic points for Collet-Eckmann maps. *Comm. Math. Phys.*, 149(1):31–69, 1992.

[16] Kozlovski O. Structural stability in one-dimensional dynamics. PhD thesis, University of Amsterdam.

[17] K. Krzyżewski and W. Szlenk. On invariant measures for expanding differentiable mappings. *Studia Math.*, XXXIII:83–92, 1969.

[18] M. Lyubich. Dynamics of quadratic polynomials iii: Parapuzzle and SBR measures. Preprint, Stony Brook, IMS, 1996.

[19] M. Lyubich. Almost every real quadratic map is either regular or stochastic. Preprint, Stony Brook, JMS, 8, 1997.

[20] M. Martens and T. Nowicki. Invariant measures for typical quadratic maps. Preprint, Stony Brook, IMS, 1996.

[21] J. Milnor and W. Thurston. On iterated maps of the interval. In *Dynamical Systems: Proceedings 1986-87*, pages 465–563. Springer-Verlag, 1988. Lecture Notes in Mathematics 1342.

[22] T. Nowicki. On some dynamical properties of S-unimodal maps. *Fundamenta Mathematicae*, **126**:27–43, 1985.

[23] T. Nowicki. A positive Liapunov exponent for the critical value of an S-unimodal mapping implies uniform hyperbolicity. *Ergodic Theory & Dynamical Systems*, **8**:425–435, 1988.

[24] T. Nowicki and F. Przytycki. Topological invariance of the Collet-Eckmann property for S-unimodal maps. Fund. Mat. (155), pages 33–43, 1998.

[25] T. Nowicki and D. Sands. Non-uniform hyperbolicity and universal bounds for S-unimodal maps. To be published in Inv. Math. 1998.

[26] F. Przytycki and S. Rohde. Porosity of Collet-Eckmann Julia sets. Preprint, 1996.

[27] G. Świątek. Hyperbolicity is dense in the real quadratic family. Preprint, Stony Brook, IMS, 1992.

[28] W. Szlenk. *Introduction to the theory of smooth dynamical systems.* PWN, Warsaw, 1981.

[29] J.-C. Yoccoz. C^1-conjugaison des difféomorphismes du cercle. In J. Palis, editor, *Geometric Dynamics*, pages 814–827, Rio de Janeiro 1981, 1983. Springer-Verlag. Lecture Notes in Mathematics 1007.

[30] L.-S. Young. Decay of correlations for certain quadratic maps. *Communications in Mathematical Physics*, **146**:123–138, 1992.

Progress in Mathematics, Vol. 169,

Metastability and Typical Exit Paths in Stochastic Dynamics

E. Olivieri[(1)], E. Scoppola[(2)]

(1) Dipartimento di Matematica, II Università di Roma
Tor Vergata, Via della Ricerca Scientifica, I-00133 Roma, Italy
e-mail: olivieri@mat.utovrm.it

(2) Dipartimento di Matematica, Università di Roma Tre.
Largo S.L. Murialdo 1, I-00146 Roma, Italy
e-mail: scoppola@mat.uniroma3.it

Abstract. In this paper we review and discuss results on metastability. We consider ergodic aperiodic Markov chains with exponentially small transition probabilities and we give a complete description of the typical tube of trajectories during the first excursion outside a general domain Q.

1. Introduction

In this review we want to describe some recent results in the theory of metastability from the point of view of mathematical physics and probability theory.

Mathematically, we will deal with the study of some large deviation problems for families of Markov chains, having finite state space and transition probabilities decaying exponentially fast in a large external parameter β: the so called Freidlin-Wentzell regime.

From physical point of view we analyze a central problem arising in the dynamical description of phase transitions: the decay from metastable to stable equilibrium.

Phase transitions (like, for instance, liquid-vapour for a fluid or positive-negative magnetized phase for ferromagnetic systems) are, no doubt, the most important phenomena in statistical mechanics. As it is well known, all equilibrium aspects are satisfactorily taken into account in the framework of the "Gibbs paradigm" (see, for instance [Ru], [Si], [Ge]). On the other hand, a general theory of non-equilibrium statistical mechanics is still lacking and this is one of the main challenges for contemporary fundamental physics.

Certainly metastability and in particular its peculiar dynamical aspects, play a fundamental role in non-equilibrium statistical mechanics.

Let us rapidly introduce the subject.

Metastability is a relevant phenomenon for thermodynamic systems close to a first order phase transition.

Suppose we start from a given pure equilibrium phase in a suitable region of the phase diagram, and change the thermodynamic parameters to values corresponding to a different equilibrium phase. In particular experimental situations, the system, instead of undergoing a phase transition, can still remain in a "wrong" equilibrium, far from the "true" one but close to what the equilibrium would be at the other side of the transition. This apparent equilibrium, often called "metastable state", persists until an external perturbation or some spontaneous fluctuation leads the system to the stable equilibrium.

Examples of metastable states are supersaturated vapours and solutions, supercooled liquids and ferromagnets with magnetization opposite to the field. We will concentrate on this last example. In this case: starting from a ferromagnet in the presence of a small negative external magnetic field $h < 0$, a metastable state can be obtained by suddenly switching this field to a small positive value. Our system remains with negative magnetization (opposite to the actual value of h) for a time which is long if the external field is small.

We will not provide here a complete list of references on metastability; we want only to mention the paper by O. Penrose and J.L. Lebowitz [PL1], (see also [PL2]), where for the first time the theory of the metastable states has a rigorous formulation, and the paper by M. Cassandro, A. Galves, E. Olivieri and M.E. Vares [CGOV] where the so called "pathwise approach to metastability" has been introduced. This approach turns out to be the starting point of the recent developments on the theory of metastability.

We summarize, in what follows, the mathematical description of metastability, based on the pathwise approach, for the simple case of ferromagnetic spin systems on a lattice.

We will give a simple description of our systems by means of the standard Ising model. Consider a spin variable $\sigma(i)$, assuming values $+1$ or -1, associated to each site i in a box $\Lambda \subset \mathbf{Z}^d$.

To any configuration $\sigma \in \{-1,+1\}^\Lambda$ we associate the energy:

$$H_\Lambda(\sigma) = -\frac{1}{2} \sum_{i,j\in\Lambda,|i-j|=1} \sigma(i)\sigma(j) - \frac{h}{2} \sum_{i\in\Lambda} \sigma(i) \tag{1.2}$$

with $h > 0$ and we assume, for instance, periodic boundary conditions on the boundary of Λ.

The Gibbs measure associated to this Hamiltonian, describing the equilibrium properties of our system, is given by

$$\mu_\Lambda(\sigma) = \frac{e^{-\beta H_\Lambda(\sigma)}}{Z_\Lambda} \tag{1.2}$$

where Z_Λ is the normalization factor called partition function:

$$Z_\Lambda = \sum_{\sigma\in\{-1,+1\}^\Lambda} e^{-\beta H_\Lambda(\sigma)}$$

and $\beta > 0$ represents the reciprocal of the absolute temperature.

It is possible to show (see [Ru], [Si]) that, for $h > 0$ $(h < 0)$, in the thermodynamic limit $\Lambda \to \mathbf{Z}^d$ we get, for every $\beta > 0$, a unique Gibbs state with positive (negative) magnetization.

We will be interested in the regime for which $h > 0$ sufficiently small is given, Λ sufficiently large, but finite is given and β tends to infinity. In these conditions the Gibbs measure will be concentrated on the unique ground state for the energy H.

This configuration of minimal energy is $+\mathbf{1}$, in which $\sigma(i) = +1$ for all i. The configuration $-\mathbf{1}$ ($\sigma(i) = -1$ for all i) is only a local minimum for the Hamiltonian H_Λ (remember $h > 0$) and it is naturally related to metastability.

Let us now define $E(l) \equiv H_\Lambda(\sigma_{(l)}) - H_\Lambda(-\mathbf{1})$, where $\sigma_{(l)}$ is the configuration in which the plus spins are exactly the sites internal to a square of side l drawn in the dual lattice $\mathbf{Z}^d + (\frac{1}{2}, \frac{1}{2})$.

We have: $E(l) = 4l - hl^2$ which is maximum for $l = \frac{2}{h}$. This means that even though an arbitrarily small non-vanishing magnetic field determines the phase, its effects become relevant only on a sufficiently large scale ($l \geq l_c(h) \sim \frac{2}{h}$), since only on these large scales the volume energy dominates the surface energy.

A stochastic dynamics with discrete time is introduced in our system by means of a Markov chain on $S = \{-1, +1\}^\Lambda$ with transition probabilities $P(\sigma, \sigma')$ satisfying the detailed balance condition:

$$\mu_\Lambda(\sigma)P(\sigma, \sigma') = \mu_\Lambda(\sigma')P(\sigma', \sigma) \tag{1.3}$$

and the ergodicity condition

$$\forall \sigma,\, \eta \quad \exists t \in \mathbf{N} \quad \text{such that} \qquad P^t(\sigma, \eta) > 0 \tag{1.4}$$

where $P^t(.,.)$ is the t-step transition probability.

We will denote by σ_t the process at time t starting at σ at time 0.

By using the ergodic theorem for Markov chains it is easy to prove that the t-step transition probabilities $P^t(\sigma, \sigma')$ converge, as t goes to infinity, to the unique invariant measure of the chain. Due to the detailed balance condition (1.3), this turns out to be the Gibbs measure $\mu_\Lambda(\sigma')$. This means that the mean value of each observable f with respect to the Gibbs measure μ_Λ, $\mu_\Lambda(f) = \sum_\sigma f(\sigma)\mu_\Lambda(\sigma)$, can be approximately evaluated as the mean of the observable on the process, $Ef(\sigma_t)$, at large t.

Markov chains satisfying conditions (1.3), (1.4) are actually used for Monte Carlo simulations. An explicit construction of a Markov chain satisfying the previous hypotheses can be given by the Glauber-Metropolis algorithm which is defined as follows: for any $\sigma \in S$ and any i in Λ let

$$\Delta_i H(\sigma) \;=\; H_\Lambda(\sigma^i) - H_\Lambda(\sigma) \tag{1.5}$$

with

$$\sigma^i(j) = \begin{cases} \sigma(j) & \text{if } i \neq j \\ -\sigma(j) & \text{if } i = j. \end{cases} \tag{1.6}$$

We consider the following transition probabilities: if $\sigma \neq \sigma'$

$$P(\sigma, \sigma') = \begin{cases} 0 & \text{if} \quad \sigma' \neq \sigma^i \text{ for all } i \in \Lambda \\ \frac{1}{|\Lambda|}\exp\{-\beta(\Delta_i H(\sigma) \vee 0)\} & \text{if} \quad \sigma' = \sigma^i \text{ for some } i \end{cases} \tag{1.7}$$

$P(\sigma, \sigma)$ is obtained by normalization.

We remark that the dynamics given by (1.7) is not conservative in the sense that the total magnetization is not a conserved quantity. Conservative dynamics (like e.g. Kawasaki dynamics) are not considered in the present paper; the study of metastability in the conservative case is more complicated and still open.

This kind of dynamics are often called stochastic Ising models. Suppose we now analyze the typical paths of our Markov chain defined by (1.7) starting from $-\mathbf{1}$ in the above described Freidlin-Wentzell regime (Λ large, h small fixed, β very large). For instance we may use a computer simulation and perform a large number of independent runs (see [ToMi]). What we see is that, during the evolution, in the sea of minus spins small droplets of pluses appear, but shrinking and disappearing before they are able to become large.

Only after a very long time, under the effect of a large fluctuation, a cluster sufficiently large will appear in the sea. Its behavior is then completely different and it grows without hesitations.

In order to understand this behavior of typical trajectories, let us compare the shrinking and the growing probabilities for a connected cluster of spins $+1$ in the sea of -1's in the two dimensional case.

First of all each cluster of spins $+1$ becomes rectangular in a finite time, independent of β, with a probability of order one following a sequence of transitions with $\Delta_i H(\sigma) < 0$. Indeed the rectangle is the only shape of a cluster in which there are no spins with a number of nearest neighbor opposite spins larger than 2 and there are no spins -1 for which this number is equal to 2.

Starting from a rectangular cluster of $+1$'s, the cut of a corner has a cost $\Delta_i H(\sigma) = h$. Thus the loss of a row of the rectangle of length l has a cost $h(l-1)$ (the last $+1$ in the row is lost with cost 0); similarly for a column. On the other hand the cost of the addition of a new line to the rectangular cluster is $\Delta_i H(\sigma) = 2 - h$.

This means that if the minimal size l of the cluster is such that $h(l-1) > 2-h$, i.e. $l \geq l_c = [\frac{2}{h}] + 1$, the cluster tends to grow, while if $l < l_c$ the cluster tends to shrink.

This heuristic argument has been developed in a rigorous way and metastability has been studied in detail for the 2-d standard stochastic Ising model by R. Schonmann and J. Neves in [NSch1], [NSch2] and [Sch1] (see also [Sch2], [Sch3]).

In [MOS1] metastability has been analyzed in the case of a cluster stochastic dynamics (Swendsen-Wang dynamics).

R. Kotecky and E. Olivieri in [KO1], [KO2] considered different hamiltonians. In particular they studied anisotropic 2-d Ising model and isotropic next nearest neighbor interactions 2-d Ising model and they proved that decay from the metastable to the stable situations typically follows a non-equilibrium path given by a sequence of "non-Wulff" configurations. Other 2-d lattice models have been studied in [CiO], [NaO], [Pe].

Recently metastability in the three dimensional Ising model has been studied by G. Ben Arous and R. Cerf in [BACe]. See also [N].

We also want to mention the papers by R. Schonmann [Sch2], [Sch3] where the more difficult problem of the limit $h \to 0$ for fixed temperature is discussed.

A complete result in this direction has recently been obtained by R. Schonmann and S. Shlosman in [SchSh1].

Dynamics like the one we have described above (in a continuous time version) can also be condidered in the infinite volume situation. This gives rise to one of the most relevant examples of stochastic system of interacting particles (see [Li]). Very interesting results have been recently obtained by P. Dehghanpour and R. Schonmann in the limit $\beta \to \infty$ [DeSch1,2]. See also [Mo]. For the stochastic Ising models in infinite volume we have non-ergodicity of the process on the coexistence region: $h = 0, \beta > \beta_c$, ($\beta_c \equiv$ the reciprocal of the critical temperature). On the other hand, in the complementary uniqueness region: $h \neq 0$ and/or $\beta < \beta_c$, one expects good ergodic properties. Very interesting and deep results have been obtained in this field in particular by Holley (see, for example, the celebrated papers [Ho1,2]). The methods that have been developed handle the region far away from the coexistence line. Near this coexistence line it turns out to be quite difficult to prove, for instance, exponential approach to equilibrium. Consider, in particular, the region $\beta > \beta_c, h$ small but non-vanishing. Nucleation phenomena and the behavior of the system on a scale $l_c(h) = \frac{2}{|h|}$ (going to infinity as h tends to zero) are certainly to be taken into account. In other words the local relaxation from metastability to stability has implications in the global approach to equilibrium. In [MOS2,3], by using results on the behavior of the typical trajectories in a finite volume regime, the authors were able to prove strong exponential ergodicity for $\beta \gg \beta_c$ and $|h| \neq 0$ arbitrarily small in infinite volume. This paper represented a breakthrough in the understanding of the uniqueness regime that now is almost complete ([MO1], [MO2], [MOSch], [SchSh], [CesM]).

Problems somehow related to metastability arise in many different areas of science: biology ([EGJLa]), sociology ([GlHu]), linguistics ([CG], [NiBe]), paleontology ([NewCoKi])... (see also [GSchm]).

In the present paper we want to collect and discuss results on metastability in the limit $\beta \to \infty$ from a general point of view.

This means that we will consider ergodic aperiodic Markov chains X_t with finite state space S and with transition probabilities satisfying:

Property $\mathcal{P}$ *If x and y are communicating states, i.e. $x \neq y$ and $P(x,y) > 0$, then*

$$\exp(-\Delta(x,y)\beta - \gamma\beta) \leq P(x,y) \leq \exp(-\Delta(x,y)\beta + \gamma\beta) \tag{1.8}$$

where $\Delta(.,.)$ is a non-negative function on the set of pairs of communicating states and $\gamma \to 0$ as $\beta \to \infty$.

This kind of Markov chains appear in many models, not all related to physical problems.

As it has been shown by Freidlin and Wentzell [FW], this kind of Markov chains are also useful to study the asymptotic properties of diffusion processes describing small random perturbations of dynamical systems.

Similar Markov chains in a non-stationary situation are also considered in the theory of simulated annealing and optimization ([Ca1,2], [T2], [Ce]).

Let us now start the discussion of the problem of the first exit from a general domain $Q \subset S$.

We will denote by P_x the probability distribution of the process X_t starting from the state x at $t = 0$ and by E_x the corresponding expectation. Moreover, given any set of states $Q \subset S$, we will denote by τ_Q the first hitting time to Q:

$$\tau_Q \equiv \min\{t > 0 : X_t \in Q\} \tag{1.9}$$

For any set $Q \subset S$ we will denote by $Q^c \equiv S \backslash Q$ the complement of Q.

Many results on the first exit problem have been obtained by Freidlin and Wentzel [FW]. In particular they found estimates for the average exit time $E_x \tau_{Q^c}$ and the typical point, on the boundary of Q, reached during the first excursion outside Q, i.e. estimates on $P_x(X_{\tau_{Q^c}} = y)$. Our aim is to provide a complete description of the typical behavior of the Markov chain X_t up to the time τ_{Q^c}, for any set $Q \subset S$, and for sufficiently large β.

We are looking for the minimal set of exiting trajectories such that the sum of the corresponding probabilities is of order one as β goes to infinity.

A state x such that for any $y \neq x$ the quantity $\Delta(x, y)$ appearing in (1.8) is strictly positive is called a *stable state*. For example, in the case of the above defined Metropolis dynamics any strict local minimum for the energy function H is a stable state.

Consider the case of a domain Q completely attracted by a unique stable state $z_0 \in Q$; namely every "null cost" trajectory starting in Q ends in z_0 where the cost of a trajectory $x_0, x_1, \dots, x_n$ is given by $\Delta(x_0, x_1) + \Delta(x_1, x_2) + \dots + \Delta(x_{n-1}, x_n)$. This situation has been analyzed in detail by Freidlin and Wentzel (see [FW]). It turns out that the typical trajectories during the first exit from Q i.e. in the interval of time $[\theta_{\{z_0\}}, \tau_{Q^c}]$ where, for $D \subset Q$

$$\theta_D = \max\{t < \tau_{Q^c}; \, X_t \in D\} \tag{1.10}$$

are suitable sequences "against the drift" emerging from z_0 and leading outside Q without hesitations in a finite time independent of β.

The case of a general Q containing several stable states is much more complicated. In this general case the last escape takes place by visiting a suitable sequence of stable states $z_1, \dots, z_n$ and spending suitable random times inside certain domains $Q_1, \dots, Q_n$ (*permanence sets*) containing $z_1, \dots, z_n$ respectively. This means that in order to define our typical exiting tube we have to consider *resistance times* associated to stable states. These can be considered as a sort of *temporal entropy* that turns out to be necessary to give rise to an efficient escape mechanism. Neglecting these random times, during the escape, would lead to a mechanism extremely depressed in probability.

These permanence sets are union of suitable sets called *cycles* (see Definition 2.2 below). The characteristic property of a cycle A is that the process spends exponentially long time visiting all states in A before leaving it.

Cycles have been introduced by Freidlin and Wentzell, under some particular hypotheses, and generalization of this definition can be found in [OS2] and in [T1],

[ChiCho1] where the cycle decomposition is used in the framework of simulated annealing.

If we consider reversible Markov chains, i.e. if we add to property $\mathcal{P}$ the hypothesis of detailed balance condition (1.3), the problem of the definition of cycles and of the determination of the typical tube is easier and can be reduced to the discussion of the energy landscape. In particular the sequence of sets defining the exiting tube can be obtained by solving a well defined sequence of variational problems (minimax problems) relative to the energy function H. It turns out that these variational problems constitute the only model-dependent work to be done to study the first exit from a general domain in the reversible case (see [OS1]).

In the general non-reversible case, when only property $\mathcal{P}$ is verified, the cycles can be defined, following Freidlin and Wentzell, by means of a graphical technique.

As far as the problem of the determination of the typical exiting tube is concerned, one can adopt two different strategies.

Following Catoni and Cerf [CaCe] (see also [ChiCho2]), given a subset $D \subset Q$ and the cycle decomposition of the set $Q \setminus D$, one can associate to each path of cycles the probability that the process visits exactly this sequence of cycles in the time interval $[\theta_D, \tau_{Q^c}]$. The determination of the typical exiting tube can thus be reduced to the determination of the cycle paths of large probability.

A different approach to the problem is introduced in [OS2], where the analysis of the typical exiting paths is obtained by means of the renormalization procedure introduced in [S1].

In that paper the long time behavior of the chain X_t was studied by means of the construction of a sequence of *renormalized Markov chains* $X_t^{(1)}, X_t^{(2)}, \dots, X_t^{(j)}$, ... whose state spaces $S^{(1)}, S^{(2)}, \dots, S^{(j)}, \dots$ were composed by stable states of increasing stability. These chains provide a rougher and rougher description of our stochastic evolution adapted to the analysis of phenomena taking place in increasing scales of time (exponential in β).

If $S \equiv S^{(0)}$ is the original state space then $S^{(1)}$ is just the set of stable states for the original Markov chain X_t; $X_t^{(1)}$ is suitably defined on $S^{(1)}$. $S^{(2)}$ is the set of stable states for $X_t^{(1)}$ and so on. In [S1] with this construction of renormalized chains, some results on the typical long time behavior of the original chain X_t were easily obtained. In fact to each exponentially long path of the chain X_t, a short path of a chain $X_t^{(N)}$ was associated, with a suitable renormalization index N.

Given $Q \subset S$ we define its boundary:

$$\partial Q = \{x \notin Q : \exists\, y \in Q : P(y,x) > 0\}$$

In order to simplify the description of the first exit from the domain Q of a Markov chain X_t we can consider a new auxiliary chain X_t^Q on the state space $Q \cup \partial Q$ which is equivalent to X_t up to its first exit from Q, but with almost absorbing states in ∂Q. More precisely we define the following transition probabilities:

$$P^Q(x,y) = P(x,y) \qquad \text{for any } x \in Q \tag{1.11}$$

$$P^Q(x,y) = e^{-\beta\Delta(\partial Q)} P(x,y) \qquad \text{for } x \in \partial Q,\ y \neq x \tag{1.12}$$

where $\Delta(\partial Q) > \sum_{y,z \in Q} \Delta(y,z)$.

To this chain is applied in [OS2] the renormalization procedure by introducing the sequence of renormalized chains $X_t^{(1)}, \dots, X_t^{(i)}, \dots$ and the corresponding sequence of state spaces $S^{(1)}, \dots, S^{(i)}, \dots$. We warn the reader of an abuse of notation: in what follows we will only consider the chain P^Q but, to simplify notation, we will omit the superscript Q.

Since $\Delta(\partial Q) > \sum_{y,z \in Q} \Delta(y,z)$ it is immediate to show that there exists a step N of the renormalization iteration such that in Q there are only unstable states with respect to the corresponding chain $X_t^{(N)}$. More precisely let $N = N(Q)$ be the level such that the $(N+1)$-the renormalized Markov chain does not contain states inside Q:

$$N(Q) = \min\{n : S^{(n+1)} \cap Q = \emptyset\} \tag{1.13}$$

This means that the first excursion outside Q for the chain $X_t^{(N)}$ is a sort of "descent" along the drift.

A first rough approximation of the typical tube of escape from Q is thus given by the set of typical trajectories followed during the first excursion outside Q by the chain $X_t^{(N)}$.

It remains the question of "reading" the result in terms of the paths followed on the original scale of time by our original chain X_t.

This problem of analyzing the set of trajectories of the original chain X_t corresponding to a given trajectory of the chain $X_t^{(N)}$, is solved in [OS2]. To any state x^N of $S^{(N)}$ a suitable set $Q_{x^N} \subset S$, a sort of generalized cycle is assigned. This represents the set where the original process X_t typically remains in the interval of time corresponding to a jump of the chain $X_t^{(N)}$. By iterating this idea the exiting tube of paths can be easily determined.

In the remaining part of the present paper we will summarize the results proved in [OS1],[OS2] under some additional assumption that greatly simplify the exposition.

The paper is organized as follows. In section 2 we recall the construction of the exiting tube in the reversible case and we state the main results proved in [OS1]. In sections 3 and 4 we will consider non reversible chains but under a very strong additional hypothesis that allows us to give a simple presentation of the results on the first exit.

2. The exit problem in the reversible case

In this section we will consider ergodic aperiodic Markov chains with transition probabilities satisfying the following:

Hypothesis M *There exists a function $H : S \to \mathbf{R}^+$ such that*

$$P(x,y) \;=\; q(x,y)\exp(-\beta[H(y) - H(x)]_+) \tag{2.1}$$

where $q(x,y) \;=\; q(y,x)$ and $(a)_+$ is the positive part ($:= a \vee 0$) of the real number a.

The above choice corresponds to a *Metropolis Markov chain* (see (1.7)) which is *reversible* in the sense of (1.3) with

$$\mu(x) \propto \exp(-\beta H(x)) \tag{2.2}$$

Since the symmetric matrix q is independent of β we immediately get, from (2.1), that Propery $\mathcal{P}$ is satisfied.

Hypothesis ND

$$H(x) \neq H(y) \quad \forall x \neq y \tag{2.3}$$

This hypothesis of *non-degeneracy* of the energy function $H(x)$ is introduced only to simplify the exposition. It is not needed and more general reversible Markov chains satisfying only the detailed balance condition (1.3) can be considered (see [OS1]) without any additional difficulty.

We observe that we can obtain the validity of (2.3) starting from any Metropolis Markov chain with an arbitrarily small change in the energy landscape.

Following the ideas presented in the introduction, we first define the cycles and we discuss their properties.

In order to construct the set of typical trajectories exiting from a cycle A (see Definition 2.2 below) we will consider in section 2.2 the first descent from any point y_0 in A to the "bottom" $F(A)$ of A. As already observed by R. Schonmann (see [Sch1]), the problem of the typical tube of trajectories during the first excursion outside a cycle A turns out to be simply related, via a time reversal transformation, to the typical tube followed by the process during the first descent to $F(A)$. Using this idea in section 2.3 we will define the typical tube of trajectories exiting from A.

2.1 Cycles and their properties A *path* ω is a sequence $\omega := x_1, \dots, x_N$, $N \in \mathbf{N}$, with x_j, x_{j+1}, $j = 1, \dots, N-1$, communicating states (i.e. $P(x_j, x_{j+1}) > 0$). We often write $\omega : x \to y$ to denote a path joining x to y.

A set $Q \subset S$ is *connected* if $\forall\, x, x' \in Q$ there exists a path $\omega : x \to x'$ all contained in Q.

We denote by M the set of "stable" states namely the local minima of the energy: $x \in M$ iff $H(x) \leq \min_{y \in \partial\{x\}} H(y)$.

We say that a state x is *downhill connected* to a state y if there exists a *down-hill path* $\omega = (x_0 = x, x_1, \dots, x_k = y)$ with $H(x_{i+1}) \leq H(x_i)$, $i = 0, , \dots, k-1$.

It is useful to conventionally extend every down-hill path up to a final stable state $y \in M$.

On the other side a state x is *uphill connected* to a state y if there exists a path $\omega = (x_0 = x, x_1, \dots, x_k = y)$ with $H(x_{i+1}) \geq H(x_i)$, $i = 0, , \dots, k-1$.

Notice that the validity of Hypothesis ND does not prevent, in general, the existence of many different down-hill (or up-hill) paths emerging from the same point y.

Given $Q \subset S$ we denote by $F = F(Q)$ the minimum set of the energy on Q:

$$F(Q) := \{y \in Q : \min_{x \in Q} H(x) = H(y)\} \tag{2.4}$$

and by $U = U(Q)$ the minimum set of the energy on its boundary:

$$U(Q) \equiv F(\partial Q) := \{z \in \partial Q : \min_{x \in \partial Q} H(x) = H(z)\} \tag{2.5}$$

We notice that by Hypothesis ND both $U(Q)$ and $F(Q)$ contain only one state.

Definition 2.1 *For each pair of states $x, y \in S$ we define their minimal saddle point $\mathcal{S}(x,y)$ as follows:*

Let

$$\bar{H}_{x,y} := \min_{\omega:x\to y} \max_{z\in\omega} H(z).$$

Then $\mathcal{S}(x,y)$ is uniquely defined by

$$H(\mathcal{S}(x,y)) = \bar{H}_{x,y} \tag{2.6}$$

This definition is well-posed because of the validity of Hypothesis ND.

By definition we have that $\mathcal{S}(x,y) = \mathcal{S}(y,x)$ $\forall x,y \in S$. Notice that it may happen that $x = \mathcal{S}(y,x)$ or $y = \mathcal{S}(y,x)$.

Given $x \in S$, $Q \subset S$, $x \notin Q$,we define $\mathcal{S}(x,Q)$ by:

$$\min_{w\in Q} H(\mathcal{S}(x,w)) = H(\mathcal{S}(x,Q)) \tag{2.7}$$

Definition 2.2 *A connected set A which satisfies:*

$$\max_{x\in A} H(x) = \bar{H} < H(U(A))$$

is called cycle.

It is easy to see that, under the reversibility hypothesis for our Markov chain, the above definition is strictly related to the one given by Freidlin-Wentzell (see [FW] pag. 198).

In the following we will recall the main properties of cycles (for proofs see [OS1]).

Proposition 2.1 *Given a state $x \in S$ and a real number c the set of all the states connected to x by paths with energy always below c either coincides with S or it is a cycle A with*

$$H(U(A)) \geq c$$

Moreover, given two cycles A_1, A_2, either

1) $A_1 \cap A_2 = \emptyset$

or

2) $A_1 \subset A_2$ *or, viceversa,* $A_2 \subset A_1$

Proposition 2.2 *If A is a cycle then:*

i) *for each $x, y, z \in A$ and $w \notin A$:*

$$H(\mathcal{S}(x,y)) < H(\mathcal{S}(z,w))$$

ii) *for any $x \in A$ the set $\mathcal{S}(x, A^c)$ does not depend on x and coincides with $U(A)$*

Proposition 2.3 *Given a cycle A*

i) *for all* $\epsilon > 0$ *there exist* $\beta_0 > 0$ *and* $k > 0$ *such that for any* $\beta > \beta_0$ *and* $\forall x \in A$

$$P_x(\exp(\beta\,[\,H(U(A)) - H(F(A))\ -\varepsilon\,]) < \tau_{\partial A} < \exp(\beta\,[\,H(U(A)) - H(F(A))\ +\varepsilon\,])\,) \geq 1 - e^{-k\beta}$$

ii) *there exist* $\delta > 0$, $\beta_0 > 0$ *and* $k' > 0$ *such that for all* $\beta > \beta_0$ *and* $\forall x, x' \in A$*:*

$$P_x(\tau_{x'} < \tau_{\partial A}\ ;\ \tau_{x'} < \exp(\beta[H(U(A)) - H(F(A)) - \delta]) \geq 1 - e^{-k'\beta}$$

iii) $\forall x \in A$, $\forall \epsilon > 0$, $y \in \partial A$ *and* β *sufficiently large:*

$$P_x(\,X_{\tau_{\partial A}} = y) \geq \exp(-\beta[H(y) - H(U(A)) + \varepsilon]\,)$$

The property described by point ii) is usually called the *recurrence property* of cycles. It is the main property characterizing cycles and it means that with large probability each state in the cycle is visited by the process before the exit.

A consequence of this property is the asymptotically exponential distribution of the first exit time. More precisely let A be a given cycle. Given a point $x \in F(A)$ let the time $T_\beta = T_\beta(x)$ be defined by

$$P_x(\tau_{\partial A} > T_\beta(x)) \;=\; e^{-1}. \tag{2.8}$$

One has that T_β does not depend on $x \in A$, in the sense of logarithmic equivalence; namely we have $\forall x, y \in A$:

$$\lim_{\beta\to\infty}\frac{1}{\beta}\log\left[\frac{T_\beta(x)}{T_\beta(y)}\right] = 0 \qquad\qquad \lim_{\beta\to\infty}\frac{1}{\beta}\log[T_\beta(x)] \;=\; H(U(A)) - H(F(A)) \tag{2.9}$$

Moreover the asymptotic distribution, (in the sense of the most probable behavior), of the first exit time from A, does not depend on $x \in A$ in the sense of logarithmic equivalence. Namely, $\forall x, y \in A$, $\varepsilon > 0$:

$$\lim_{\beta\to\infty} P_x(T_\beta(y)e^{-\varepsilon\beta} < \tau_{\partial A} < T_\beta(y)e^{+\varepsilon\beta}) \;=\; 1 \tag{2.10}$$

Proposition 2.4 *Let* $T^*_\beta = T_\beta(x^*)$ *where* x^* *is a particular point in A chosen once for all. Then* $\forall\, x \in A$, $\forall\, s \in \mathbf{R}^+$:

$$\lim_{\beta\to\infty} P_x(\frac{\tau_{\partial A}}{T^*_\beta} > s\,) = e^{-s}. \tag{2.11}$$

Moreover, for every $x \in A$ *and* $\varepsilon > 0$, *if* $\mathbf{E}_x$ *denotes average over the trajectories of the process starting, at* $t = 0$, *from* x, *we have:*

$$\exp(\beta\,[H(U(A)) - H(F(A)) - \varepsilon]) < \mathbf{E}_x(\tau_{\partial A}) < \exp(\beta[H(U(A)) - H(F(A)) + \varepsilon] \tag{2.12}$$

2.2 The permanence sets Q_i and the standard cascade We now want to construct the set of typical trajectories starting from an arbitrary state $y_0 \in A$ and descending to $F(A)$. Given y_0 let us consider a downhill path ω_1 starting from y_0. We stress that this path need not be unique. This means that the whole construction we are defining must be repeated for each path.

Let x_1 be the stable state in ω_1. If such a point x_1 is $F(A)$ (y_0 belongs to the "basin of attraction" of $F(A)$) the *cascade* is simply given by $y_0, \omega_1, F(A)$.

Let us now suppose that $x_1 \neq F(A)$. Let H_1 be the energy of the saddle between x_1 and $F(A)$:

$$H(\mathcal{S}(x_1, F(A))) = H_1 \tag{2.13}$$

and let $C(x_1, H_1)$ be the cycle containing x_1 with states of energy stricly less than H_1.

The first *permanence set* Q_1 is the maximal connected set, containing x_1, of points x such that

$$H(\mathcal{S}(x, F(A))) = H_1$$

The permanence set Q_1 turns out to be the union of the cycle $C(x_1, H_1)$, the saddle $\mathcal{S}(x_1, F(A))$ and possibly other cycles contained in A and having $\mathcal{S}(x_1, F(A))$ as common minimal saddle in their boundaries. The above definition of permanence set can be used also when Hypothesis ND does not hold.

Let y_1 be a point in ∂Q_1 such that $\mathcal{S}(y_1, F(A)) < H_1$. We note that there is at least a point in ∂Q_1 with this property since, otherwise, by definition of Q_1, (2.13) does not hold.

We can iterate the procedure by defining a downhill path ω_2 starting from y_1 and lying outside Q_1. Denote by x_2 the final stable state in ω_2 and by H_2 the energy of the saddle between x_2 and $F(A)$. By definition $H_2 < H_1$. As before we construct new sets Q_2 (depending on the choice of y_1 and of the path ω_2 starting from y_1).

If we continue in this way and recursively construct a sequence of the form: $y_2, \omega_3, Q_3 y_3, \ldots$, we necessarily end up with a y_{N-1}, ω_N, for some finite N, with $x_N = F(A)$.

We set $Q_N := F(A)$.

Suppose first that a particular choice is made of $y_0, \omega_1, y_1, \omega_2, \ldots, y_{N-1}, \omega_N$, compatible with the above construction.

Then, automatically, $x_2, \ldots, x_{N-1}, x_N, Q_1, \ldots, Q_{N-1}, Q_N$ are given.

Let

$$\mathcal{T}(y_0, \omega_1, y_1, \omega_2, \ldots, y_{N-1}, \omega_N) = y_0 \cup \omega_1 \cup Q_1 \cup \omega_2, \ldots, Q_{N-1} \cup \omega_N \cup Q_N. \tag{2.14}$$

Any sequence like $\mathcal{T}(y_0, \omega_1, y_1, \omega_2, \ldots, y_{N-1}, \omega_N)$, obtained via the above construction, will be called a *standard cascade*; it can be visualized as a sequence of falls and sets of "communicating" lakes corresponding to cycles.

Notice that, by construction:

$$H(y_{j-1}) > H(y_j), \quad H(F(Q_j)) \leq H(y_{j-1}),\ j = 1, \ldots N$$

and $F(Q_1), \ldots, F(Q_{N-1})$ are strictly higher, in energy, than $F(A)$.

With the above definition of standard cascade we can state the result on typical paths for the first descent to the bottom of the cycle A.

Theorem 2.1 *Given a cycle A, for every $y_0 \in A$,*

i)

$$\exists\, \delta > 0 \ \text{ such that } \lim_{\beta\to\infty} P_{y_0}(\tau_{F(A)} < \exp(\beta[H(U(A)) - H(F(A)) - \delta]) = 1$$

ii)

$$\lim_{\beta\to\infty} P_{y_0}(\forall\, t \le \tau_{F(A)} \ : \ x_t \in \mathcal{T}(y_0, \omega_1, y_1, \omega_2, \dots, y_{N-1}, \omega_N)$$

$$\text{for some } y_0, \omega_1, y_1, \omega_2, \dots, y_{N-1}, \omega_N) \ = \ 1,$$

iii) *Moreover, with probability $\to$ 1 as $\beta \to \infty$, there exists a sequence $y_0, \omega_1, y_1, \omega_2, \dots, y_{N-1}, \omega_N$ defining a standard cascade, such that our process starting at $t = 0$ from y_0, between $t = 0$ and $t = \tau_{F(A)}$, after having followed the initial downhill path ω_1, visits, sequentially, the sets Q_1, Q_2, Q_{N-1} exiting from Q_j through y_j and then following the path ω_{j+1} before entering Q_{j+1}.*
For every $\varepsilon > 0$ with a probability tending to one as $\beta \to \infty$ the process spends inside each Q_j a time which can be estimated from above and from below by $\exp(\beta[H(U(Q_j)) - H(F(Q_j)) \pm \varepsilon]), j = 1, \dots, N-1$.
Finally: before exiting from Q_j it can perform an arbitrary sequence of passages through the cycles belonging to Q_j with the typical behavior described in Proposition 2.3.

2.3 By reversibility from the typical descent to the typical exit Following Schonmann ([Sch1]), we first give some simple general definitions.

Given $x, y \in S$, we denote by $\Omega^*(x,y)$ the set of all paths ω starting from x, visiting y at some finite time t and never visiting x and y in between:

$$\Omega^*(x,y) := \{\omega = x_1, \dots, x_t \text{ for some } t \ : \ x_1 = x,\ x_t = y,\ ;\ x_2, \dots, x_{t-1} \neq x, y\} \tag{2.15}$$

We denote by R the time reversal operator defined on finite paths:

$$\forall\, \omega := (x_1, \dots, x_t) \ : \ R\omega := \bar{\omega} := (x_t, \dots, x_1) \tag{2.16}$$

We naturally define, for every set of paths Δ,

$$R\Delta = \{\bar{\omega} = R\omega \ ;\ \omega \in \Delta\}.$$

Let us call τ_{xy} the last time our process visits the state x before touching, for the first time, y, namely:

$$\tau_{xy} \ := \ \max\{t < \tau_y : X_t = x\} \tag{2.17}$$

Given a finite path $\tilde{\omega} = \tilde{x}_1, \dots, \tilde{x}_t$, we say that our process $\{X_t\}_{t>0}$ *starts as $\tilde{\omega}$* if $X_1 = \tilde{x}_1, \dots, X_t = \tilde{x}_t$.

For any $x, y \in S$ we define a measure on the (infinite) paths $\omega = x_1, \dots, x_t, \dots$ starting at x ($x_1 := x$), as follows:

if $\omega \notin \Omega^*(x,y)$ then $\rho(\omega) = 0$

if $\omega \in \Omega^*(x,y)$ we set $\tilde{\omega}_i = \omega_i$ for all $i < \inf\{t > 0;\ \omega_t = y\}$, that is $\tilde{\omega}$ is the finite path given by the first segment of ω before hitting y. Then

$$\rho(\omega) = P_x(\ \{X_t\}_{t\geq 0} \text{ starts as } \tilde{\omega}\ |\omega \in \Omega^*(x,y))$$

and by $\bar{\rho}(\omega)$ the measure on the paths $\omega = x_1, \dots, x_t, \dots$ given by:

$$\bar{\rho}(\omega) = P_y(\ \{X_t\}_{t\geq 0} \ \text{ starts as } \ R\tilde{\omega}\ |\omega \in\ \Omega^*(y,x)\ ), \ \text{ if } \ \omega\ \in\ \Omega^*(y,x)$$

$$\bar{\rho}(\omega) = 0 \ \text{ otherwise.}$$

In [Sch1] it is proven that, for every $x, y\ \in\ S$, every $\Lambda\ \subset\ \Omega^*(x,y)$,

$$P_x(\{X_t\}_{t\in[\tau_{xy},\tau_y]}\ \in\ \Lambda) = \rho(\Lambda) = \bar{\rho}(R\Lambda) = P_y(\{X_t\}_{t\ \in[\tau_{yx},\tau_x]}\ \in\ R\Lambda). \quad (2.18)$$

Now we are able to state our main result about the typical trajectories realizing the escape from a cycle A.

Theorem 2.2 *Let*

$$\theta_{F(A)} = \max\{t < \tau_{A^c} : X_t\ = F(A).\}$$

Call $\bar{\partial}^- A$ the set of all points $x\ \in A$ uphill connected to $U(A)$. Then:

i)

$$P_{F(A)}(X_{\tau_{A^c}}\ = U(A)) \rightarrow\ 1 \text{ as } \ \beta\ \rightarrow\ \ \infty \quad (2.19)$$

ii)

$$P_{F(A)}(\exists y_0 \in \bar{\partial}^- A, \mathcal{T}(y_0, \omega_1, y_1, \omega_2, \dots, y_{N-1}, \omega_N)\ :$$
$$X_t \in R\mathcal{T}(y_0, \omega_1, y_1, \omega_2, \dots, y_{N-1}, \omega_N) \forall t \in [\theta_{F(A)}, \tau_{A^c} - 1]) \rightarrow 1 \quad \text{as} \quad \beta \rightarrow \infty$$

iii)

During the first excursion from $F(A)$ to A^c, conditioning to $X_{\tau_{A^c}-1} = y$ (for some $y\ \in \bar{\partial}^- A$) and to follow a particular "anticascade" $R\mathcal{T}(y_0\ = y, \omega_1, y_1, \omega_2, \dots, y_{N-1}, \omega_N)$, with probability tending to one as $\beta\ \rightarrow \infty$ the process, between $\theta_{F(A)}$ and τ_y visits the sequence $Q_{N-1}, \dots, Q_1$, spending in each set a time which can be estimated from above and from below by $\exp(\beta[H(U(Q_j)) - H(F(Q_j)) \pm \varepsilon]), j = 1, \dots, N-1$.

3. The renormalization procedure and the cycle decomposition

In the remaining part of this paper we will consider generic Markov chains (i.e. not necessarily reversible) but we will add to property $\mathcal{P}$ the following very particular hypothesis which greately simplifies the exposition of results and proofs. The reader can find in [OS2] the general results.

Hypothesis RI *There is at most a couple of states x, y; $x \neq y$ such that $\Delta(x,y) = 0$ and moreover every non-trivial linear combination with integer coefficients of the not vanishing quantities $\Delta(x,y)$ is different from zero.*

Remark The rational independence condition in RI is even more strict than the analogous condition of non-degeneracy ND of the reversible case. The condition that for a unique pair of states x, y one has $\Delta(x,y) = 0$ is included in RI because it is conserved (and in any case produced) by the renormalization procedure.

Again we notice that with an arbitrarily small change of the $\Delta(x,y)$, any Markov chain satisfying property $\mathcal{P}$ can be modified to satisfy Hypothesis RI.

Given a chain X_t on S let $\Phi(S) \equiv \{\{\phi_i\}_{i\in\mathbf{N}} : \phi_i \in S\}$ be the set of paths. Following the theory of large deviations, developed in [FW], we define, for each $t \in \mathbf{N}$, a functional $I_{[0,t]}$ on $\Phi(S)$ associating to each path $\phi \in \Phi(S)$ the value

$$I_{[0,t]}(\phi) \equiv \sum_{i=0}^{t-1} \Delta(\phi_i, \phi_{i+1}) \tag{3.1}$$

where for $x \neq y$, $P(x,y) > 0$, the function $\Delta(x,y)$ has been defined in (1.8) and we set $\Delta(x,x) = 0$ for each $x \in S$ and $\Delta(x,y) = \infty$ if $P(x,y) = 0$. This functional is the *cost function* of each path ϕ in the sense of the following:

Lemma 3.1 *Let ϕ be a given path starting from x at time 0; then, for $t \in \mathbf{N}$*

i)

$$P_x(X_s = \phi_s \quad \forall s \in [0,t]) \leq e^{-I_{[0,t]}(\phi)\beta + \gamma t\beta}$$

where γ is the quantity introduced in (1.8)

ii) if ϕ is such that $\phi_s \neq \phi_{s+1}$ for any $s \in [0,t]$ then we have also a lower bound:

$$P_x(X_s = \phi_s \quad \forall s \in [0,t]) \geq e^{-I_{[0,t]}(\phi)\beta - \gamma t\beta}$$

iii) for any constant $I_0 > 0$, for any $\alpha > 0$ sufficiently small, for any $t < e^{\alpha\beta}$ and for any sufficiently large β

$$\sup_x P_x(I_{[0,t]}(X_\cdot) \geq I_0) \leq e^{-I_0\beta + \epsilon\beta}$$

where $\epsilon \to 0$ as $\beta \to \infty$.

For each pair of states x, y we define

$$V(x,y) \equiv \inf_{t,\phi:\, \phi_0 = x,\, \phi_t = y} I_{[0,t]}(\phi). \tag{3.2}$$

We say that x is a *stable state* if and only if

$$\text{for any} \quad y \neq x \quad V(x,y) > 0 \tag{3.3}$$

i.e. if each path leaving from x has a positive cost. We will denote by M the set of stable states.

We note that by Hypothesis RI we have that there is at most an unstable state, i.e.$|S \backslash M| \leq 1$. Moreover again Hypothesis RI allows us to exclude the existence of equivalent states, i.e. of states x, y such that $V(x,y) = V(y,x) = 0$, and thus we do not have to discuss the problem in terms of equivalence classes as it has been done in [OS2].

3.1 Construction of the renormalized chain It is immediately proven, using lemma 3.1, that the process spends, with large probability, relatively very short times in the unstable state. This suggests that, if we look at the process X_t on a sufficiently large time scale, then it can be described in terms of transitions between states in M. In this way only the behavior of the process on small times is neglected.

Indeed we can consider the less stable states in M and we can define a time scale t_1 corresponding to this smallest stability:

$$t_1 \equiv [e^{V_1\beta+\delta\beta}] \tag{3.4}$$

where

$$V_1 \equiv \min_{x\in M,\, y\in S,\, x\neq y} V(x,y) \tag{3.5}$$

and $\delta = \delta(\beta) := 2\gamma(\beta)|S|$ (see (1.8)), goes to zero as β tends to infinity.

We can then construct a new Markov chain $X_t^{(1)}$ with state space M, corresponding to the original process with a rescaling time t_1, by defining a sequence of stopping times $\zeta_1, \zeta_2, \ldots, \zeta_n, \ldots$ such that $\zeta_{n+1} - \zeta_n$ is of order t_1 with large probability and X_{ζ_n} belongs to M.

More precisely we define the sequence of stopping times:

$$\zeta_0 \equiv \min\{t \geq 0 : \ X_t \in M\}$$

and for each $n \geq 1$:

$$\begin{aligned} &\sigma_n \equiv \min\{t > \zeta_{n-1} : \ X_t \neq X_{\zeta_{n-1}}\} \\ &\tau_n \equiv \min\{t \geq \sigma_n : \ X_t \in M\} \\ &\zeta_n = \begin{cases} \zeta_{n-1} + t_1 & \text{if } \sigma_n - \zeta_{n-1} > t_1 \\ \tau_n & \text{if } \sigma_n - \zeta_{n-1} \leq t_1 \end{cases} \end{aligned} \tag{3.6}$$

It is easy to see that the sequence $X_n^{(1)} = X_{\zeta_n}$ is a homogeneous Markov chain on the state space $S^{(1)} = M$. For any pair of states $x, y \in M$ we denote by $P^{(1)}(x,y)$ the transition probabilities of the chain $X_n^{(1)}$. It is possible to prove that these transition probabilities satisfy the assumptions (property $\mathcal{P}$ and Hypothesis RI) verified by the original chain X_t.

More precisely for any $x, y \in M, \quad x \neq y$:

$$\begin{aligned} \exp\{-\Delta^{(1)}(x,y)\beta - \gamma'\beta\} \leq P^{(1)}(x,y) &\equiv P(X_{\zeta_n} = y |\, X_{\zeta_{n-1}} = x) \\ &\leq \exp\{-\Delta^{(1)}(x,y)\beta + \gamma'\beta\} \end{aligned} \tag{3.7}$$

The quantities $\Delta^{(1)}(x,y)$ are defined by:

$$\Delta^{(1)}(x,y) = \bar{\Delta}(x,y) - V_1 \tag{3.8}$$

where $\bar{\Delta}(x,y) = [\Delta(x,y) \wedge (\Delta(x,\hat{y}^0) + \Delta(\hat{y}^0,y))]$ if $\hat{y}^0 := S \backslash S^{(1)} \neq \emptyset$ and $\bar{\Delta}(x,y) = \Delta(x,y)$ if $\hat{y}^0 = \emptyset$, and $\gamma' \to 0$ as $\beta \to \infty$ (see [S1]).

Hypothesis RI is also satisfied by the chain $X_t^{(1)}$, since the new quantities $\Delta^{(1)}(x,y)$ turn out to be linear combination of the old quantities $\Delta(.,.)$ with integer coefficients, and there exists at most one unstable state x for the new chain $X_t^{(1)}$.

Thus we have a new chain $X_t^{(1)}$ on the state space $S^{(1)}$, to which we can again apply the same analysis, by defining new stable states, a time scale t_2, a corresponding chain $X_t^{(2)}$ and so on.

This procedure of renormalization can be accomplished for more general Markov chains satisfying only property $\mathcal{P}$. In our case, under Hypothesis RI, at most one unstable state can be removed going from $S^{(n)}$ to $S^{(n+1)}$, i.e. $|S^{(n)}\backslash S^{(n+1)}| \leq 1$. It is also easy to prove that $|S^{(n+2)}| < |S^{(n)}|$.

We will denote by $\hat{y}^n$ the unstable state of $S^{(n)}$, if any, i.e. $\hat{y}^n = S^{(n)}\backslash S^{(n+1)}$. Thus we have: $S^{(n)} = \cup_{i\geq n}\hat{y}^i$.

We recall now the iteration scheme introduced in [S1] (see also [S2]).

Notation: The superscript $^{(k)}$ will denote the various quantities referring to the k-th chain $X_t^{(k)}$; e.g. $\tau_Q^{(k)} = \min\{t : X_t^{(k)} \in Q\}, \quad Q \subset S^{(k)}$.

For any $k \geq 1$ we define the following quantities:
for any $\phi : \mathbf{N} \to S^{(k)}$:

$$I_{[o,t]}^{(k)}(\phi) = \sum_{i=0}^{t-1} \Delta^{(k)}(\phi_i, \phi_{i+1}) \tag{3.9}$$

$$V^{(k)}(x,y) \equiv \min_{t,\phi:\,\phi_0=x,\,\phi_t=y} I_{[o,t]}^{(k)}(\phi) \quad \forall x,y \in S^{(k)} \tag{3.10}$$

$$M^{(k)} = \{x \in S^{(k)} : \forall y \in S^{(k)},\ y \neq x \quad V^{(k)}(x,y) > 0\} \tag{3.11}$$

$$V_{k+1} = \min_{x\in M^{(k)}, y\in S^{(k)}\ x\neq y} V^{(k)}(x,y) \tag{3.12}$$

$$t_{k+1} = e^{V_{k+1}\beta + \delta\beta} \tag{3.13}$$

$$T_1 = t_1$$

$$T_{k+1} = t_1 t_2 \ldots\ldots t_k t_{k+1} \tag{3.14}$$

$$S^{(k+1)} = M^{(k)} \tag{3.15}$$

$$\Delta^{(k+1)}(x,y) = \bar{\Delta}^{(k)}(x,y) - V_{k+1} \quad \forall x,y \in S^{(k+1)} \tag{3.16}$$

where $\bar{\Delta}^{(k)}(x,y) = [\Delta^{(k)}(x,y) \wedge (\Delta^{(k)}(x,\hat{y}^k) + \Delta^{(k)}(\hat{y}^k, y))]$ if $\hat{y}^k \neq \emptyset$ and $\bar{\Delta}^{(k)}(x,y) = \Delta^{(k)}(x,y)$ if $\hat{y}^k = \emptyset$.

The main results proved in [S1] can be summarized as follows:

Theorem 3.1 *Let $W \subset S^{(1)}$ and $B \subset W$. Then for any sufficiently large β and for any $x \in S^{(1)}\backslash W$, $y \in W$ there is a positive γ' depending on γ and tending to zero as $\beta \to \infty$ such that:*

$$\exp\{-\gamma'\beta\}P_x^{(1)}(X_{\tau_W}^{(1)} = y) \leq P_x(X_{\tau_W} = y) \leq \exp\{\gamma'\beta\}P_x^{(1)}(X_{\tau_W}^{(1)} = y)$$

$$\exp\{-\gamma'\beta\}t_1 E_x^{(1)}\tau_W^{(1)} \leq E_x\tau_W \leq \exp\{\gamma'\beta\}t_1 E_x^{(1)}\tau_W^{(1)}$$

$$e^{-\gamma'\beta}\mu^{(1)}(B) \leq \mu(B) \leq e^{\gamma'\beta}\mu^{(1)}(B)$$

for any $B \subset S^{(1)}$ where μ and $\mu^{(1)}$ are the invariant measures of the chains X_t and $X_t^{(1)}$ respectively. Moreover for any $A \in S \backslash M$

$$\mu(A) \leq e^{-V_1\beta+\gamma'\beta}$$

3.2 Tube of paths corresponding to a renormalized path By using the previous construction of renormalized chains we can immediately define a correspondence between trajectories of chains at different levels of the iteration.

More precisely, for any integer n, to any path ϕ of the Markov chain X_t, $\phi \in \Phi(S)$, we can associate a path $\phi^{(n)}$ of the Markov chain $X_t^{(n)}$, $\phi^{(n)} \in \Phi(S^{(n)})$. This is in a sense a projection of the path ϕ in the smaller space $\Phi(S^{(n)})$. On the other hand to any path $\phi^{(n)} \in \Phi(S^{(n)})$ we want to associate a set, say a tube, of paths $\phi \in \Phi(S)$ having projection $\phi^{(n)}$.

Definition 3.1 For each path $\phi \in \Phi(S)$ we evaluate the sequence of stopping times ζ_n and we define a path $\phi^{(1)} \in \Phi(S^{(1)})$ by defining $\phi_i^{(1)} = \phi_{\zeta_i}$.

Using the same construction we can thus define a sequence of trajectories $\{\phi_i^{(n)}\}_{i\in\mathbf{N}}$ in the spaces $\Phi(S^{(n)})$ with $n = 2, 3, \ldots$.

On the other side, to any given sequence of states $\psi^{(n)}$ in $S^{(n)}$ of length T: $\psi_i^{(n)}$, $i \leq T$ we can associate a tube of trajectories in $\Phi(S)$ as

$$\mathcal{T}(\psi^{(n)}) = \{\phi \in \Phi(S) : \phi_i^{(n)} = \psi_i^{(n)} \quad \forall i \leq T\}. \tag{3.17}$$

By construction this application between trajectories in $\Phi(S)$ and trajectories in $\Phi(S^{(n)})$ is such that

$$P^{(n)}(X_s^{(n)} = \phi_s^{(n)}, \forall s \leq T) \asymp P(X_t \in \mathcal{T}(\psi^{(n)})). \tag{3.18}$$

We say here that two quantities A and B are *logarithmically equivalent* and we write $A \asymp B$ if they have the same exponential asymptotic behavior in β namely

$$A \asymp B \quad \text{if and only if} \quad \lim_{\beta\to\infty} \frac{1}{\beta} \log A = \lim_{\beta\to\infty} \frac{1}{\beta} \log B. \tag{3.19}$$

With this application we can also define a sequence of random times Z_k^n corresponding, on the original time scale, to the times k on the time scale of the chain $X^{(n)}$. More precisely, given a path $\phi \in \Phi(S)$, we have defined a sequence of times $\zeta_0, \ldots, \zeta_k, \ldots$ and a path $\phi^{(1)} \in \Phi(S^{(1)})$ associated to it, and, iterating, sequences of times $\zeta_0^{(i)}, \ldots, \zeta_k^{(i)}, \ldots$, and paths $\phi^{(i+1)} \in \Phi(S^{(i+1)})$ for each $i = 0, 1, \ldots$.

Definition 3.2

$$Z_k^n \equiv \zeta_{\zeta^{(1)}_{\zeta^{(2)}_{\cdot^{\cdot^{\zeta_k^{(n-1)}}}}}} \qquad n = 1, 2, \ldots \tag{3.20}$$

This is a random time with respect to the process X_t, corresponding to the time k for the chain $X^{(n)}$.

These times Z_k^n with large probability are of order T_n in the following sense:

Proposition 3.1 *For any $N \geq 1$ and for any sufficiently small positive constant α there exists a positive constant $k = k(\alpha, N)$ such that for any $x \in S^{(N)}$*

$$P_x(T_N e^{-\alpha\beta} \leq Z_1^N \leq T_N e^{\alpha\beta}) \geq 1 - e^{-k\beta}$$

for any sufficiently large β.

Similar results are obtained in [OS2] in the more general case in which Hypothesis RI is not assumed.

In order to complete the characterization of the typical tube of paths in $\Phi(S)$ corresponding to a given path $\phi^{(n)}$, we need to define the cycles. In fact it is possible to associate to each state $x^n \in S^{(n)}$ a cycle Q_{x^n} for the original chain X_t representing the region of the state space S corresponding to the state $x^n \in S^{(n)} \subset S$ under the time rescaling T_n.

3.3 Cycles associated to renormalized states Let us first recall some definitions introduced by Freidlin and Wentzell, in order to discuss the cycle decomposition.

Definition 3.3 (B-graphs): For any set of states $B \subset S$, a B-graph is a graph consisting of arrows $m \to n$ with $m \in B$ and $n \in S$, $m \neq n$ and satisfying the following properties:

1) every state $m \in B$ is the initial point of exactly one arrow,
2) there are no closed cycles in the graph.

Condition 2) can be replaced by:

2′) for any state $m \in B$ there exists a sequence of arrows leading from it to some point $n \in B^c$.

In other words a graph is a forest of trees with roots in B^c and with branches given by arrows directed to the root (i.e. the set B^c is the target of the sequences of arrows).

The set of B-graphs is denoted by $G(B)$; for any graph $g \in G(B)$ we define $\pi(g) = \prod_{m \to n \in g} P(m, n)$.

Warning: Our notation is opposite to that used in [FW], where a W-graph was a graph with target W, i.e. a W^c-graph in our notation.

By using this graphic formalism Freidlin and Wentzel in [FW] prove the following lemmas.

Lemma 3.2 (FW) *The invariant measure of the Markov chain X_t: $\mu(i)$, $i \in S$ is given by:*

$$\mu(i) = \frac{q_i}{\sum_{j \in S} q_j}$$

where

$$q_i = \sum_{g \in G\{S \setminus i\}} \pi(g). \tag{3.21}$$

Lemma 3.3 (FW) *For any $B \subset S$ let τ_B be the first hitting time to B, then for any $j \in B$*

$$P_i(X_{\tau_B} = j) = \frac{\sum_{g \in G_{ij}(B^c)} \pi(g)}{\sum_{g \in G(B^c)} \pi(g)} \tag{3.22}$$

where for any $i \in B^c$ and $j \in B$ we denote by $G_{ij}(B^c)$ the set of B^c-graphs in which the sequence of arrows leading from i into B ends at the point j (i.e. i belongs to the tree with root j).

Lemma 3.4 (FW)

$$\begin{aligned} E_i\tau_B &= \frac{\sum_{g\in G(B^c\setminus\{i\})}\pi(g)+\sum_{j\in B^c,\, j\neq i}\sum_{g\in G_{ij}(B\setminus\{j\})}\pi(g)}{\sum_{g\in G(B^c)}\pi(g)} \\ &= \frac{\sum_{g\in G(i\not\to B)}\pi(g)}{\sum_{g\in G(B^c)}\pi(g)} \end{aligned} \tag{3.23}$$

where $G(i \not\to B)$ is the set of graphs (without closed loops) containing $|B^c|-1$ arrows $m \to n$ each one emerging from a different point $m \in B^c$ and with $n \in S$, $m \neq n$ and not containing chains of arrows leading from i into B.

Lemma 3.2 is easily proved by showing that the quantities q_i satisfy the stationarity equation, Lemmas 3.3 and 3.4 can be proved by induction over the number of states contained in B. (See [FW], pg. 179, 182).

By using property $\mathcal{P}$ these results can be reformulated as follows (see [FW]). Consider the problem of the first exit of our chain X_t from a domain $Q \subset S$ and let τ_{Q^c} be the first exit time from the domain Q.

Proposition 3.2 (FW) *For any $\delta > 0$ there exists $\beta_o > 0$ such that for any $\beta > \beta_0$:*

$$e^{-\beta(W_Q(x,y)-W_Q+\delta)} \le P_x(X_{\tau_{Q^c}} = y) \le e^{-\beta(W_Q(x,y)-W_Q-\delta)} \tag{3.24}$$

for any $x \in Q$ and $y \in Q^c$, with

$$W_Q(x,y) = \min_{g\in G_{xy}(Q)} W(g) \tag{3.25}$$

$$W_Q = \min_{g\in G(Q)} W(g) \tag{3.26}$$

$$W(g) \equiv \sum_{m\to n\in g} \Delta(m,n). \tag{3.27}$$

Moreover for any $x \in Q$:

$$\lim_{\beta\to\infty}\frac{1}{\beta}\log E_x\tau_{Q^c} = W_Q - M_Q(x) \tag{3.28}$$

where

$$M_Q(x) = \min_{g\in G(x\not\to Q^c)} W(g). \tag{3.29}$$

For each $Q \subset S$ let $g_Q^* \in G(Q)$ be a Q-graph minimizing the quantity W_Q defined in (3.26).

We note that by Hypothesis RI for any set $Q \subset S$ there exists a unique graph g_Q^*. Indeed if there were two different minimizing graphs $(g_Q^*)_1$ and $(g_Q^*)_2$, denoting by $(g^*_Q)'_1$ the arrows contained in $(g^*_Q)_1$ not contained in $(g^*_Q)_2$ (and the analog for $(g^*_Q)'_2$) it would be

$$0 = W((g^*_Q)_1) - W(g^*_Q)_2)) = W((g^*_Q)'_1) - W((g^*_Q)'_2)$$

against Hypothesis RI.

For any set Q and $x \in Q$ let

$$R_Q(x) = \{y \in Q^c : \exists x \to \cdots \to y \in g^*_Q\}.$$

Definition of cycles of rank zero Single states in S are called cycles of rank zero or 0-cycles.

Definition of cycles of rank one Consider a set $C = \{x_1, \dots, x_n\}$ contained in S with $n \geq 2$ such that $x_{i+1} = R_{\{x_i\}}(x_i)$ for each $i = 1, \dots, n-1$ and $x_1 = R_{\{x_n\}}(x_n)$. Such a set is called 1-cycle or cycle of rank one.

Every single state not belonging to one of these 1-cycles constitutes by itself a 1-cycle.

Thus we can decompose the state space in 1-cycles: $S = \cup_i C_i$.

If C is a 1-cycle we have that

$$|R_C(x)| = 1 \quad \text{and} \quad R_C(x) = R_C(y) = R_C \quad \forall x, y \in C. \tag{3.30}$$

This property is immediately verified under our Hypothesis RI. This means that graphs minimizing the exit from a cycle are made by a unique tree.

We can then iterate our construction by defining:

Definition of cycles of rank k Consider a set $C = \{C_1, \dots, C_n\}$ contained in S where C_i are $k-1$ cycles and $i = 1, \dots, n$; $n \geq 2$ such that $R_{C_i} \in C_{i+1}$ for each $i = 1, \dots, n-1$ and $R_{C_n} \in C_1$. Such a set is called k-cycle or cycle of rank k.

Every $(k-1)$-cycle not belonging to one of these k-cycles constitutes by itself a k-cycle.

Thus we can decompose the state space in k-cycles.

If C is a k-cycle we can iteratively verify that

$$|R_C(x)| = 1 \quad \text{and} \quad R_C(x) = R_C(y) = R_C \quad \forall x, y \in C. \tag{3.31}$$

The main properties of cycles are summarized in the following.

Proposition 3.3 *For any set C with $|C| > 1$, the following are equivalent:*

i) *C is a cycle (of some rank)*

ii) *there exists $K > 0$ such that for every $x \in C$, $B \subset C$, $x \notin B$ and β sufficiently large we have:*

$$P_x(X_{\tau_{C^c \cup B}} \notin B) < e^{-K\beta} \tag{3.32}$$

iii) *$E_x \tau_{C^c}$ is independent of $x \in C$ in the sense of logarithmic equivalence and for any $B \subset C$ and for any $x \in B$ and $z \in C$:*

$$\lim_{\beta \to \infty} \frac{1}{\beta} \log E_x \tau_{B^c} < \lim_{\beta \to \infty} \frac{1}{\beta} \log E_z \tau_{C^c} \tag{3.33}$$

As in the reversible case, a consequence of the recurrence property of cycles (3.32) is the exponential distribution of their first exit times.

Let us now define the cycle associated to a renormalized state. For any $x^n \in S^{(n)}$ let Q_{x^n} be the maximum cycle such that

$$Q_{x^n} \cap S^{(n-1)} = x^n.$$

We will show that this cycle is the permanence set associated to x^n, i.e. the set of states visited by the chain X_t in the interval of time corresponding to a transition $x^n \to y^n$ of the chain $X_t^{(n)}$. More precisely; let Z_t^{n-1} be defined by (3.20), let

$$\sigma^{(n-1)} := \tau^{(n-1)}_{S^{(n-1)}\setminus x^n}$$

and for any x^n, $y^n \in S^{(n)}$ let $\phi^{(n-1)}(x^n, y^n)$ be the path of states in $S^{(n-1)}$ going from x^n to y^n without visiting other states in $S^{(n)}$ and minimizing the quantity $\bar{\Delta}^{(n-1)}(x^n, y^n)$.

By our hypothesis RI we know that this path is given by:

$$\phi^{(n-1)}(x^n, y^n) = \begin{cases} x^n, y^n & \text{if } \bar{\Delta}^{(n-1)}(x^n, y^n) = \Delta^{(n-1)}(x^n, y^n) \\ x^n, \hat{y}^{n-1}, y^n & \text{otherwise} \end{cases} \tag{3.34}$$

Theorem 3.2 *For any sufficiently small $a > 0$ (see proposition 3.1), we define:*

$$A_1 := \{Z^{n-1}_{\sigma^{(n-1)}-1} \in [T_n e^{-a\beta}, T_n e^{a\beta}]\}$$

$$A_2 := \{Z^{n-1}_{\sigma^{(n-1)}-1} < \tau_{(Q_{x^n})^c} \le Z^{n-1}_{\sigma^{(n-1)}}\}$$

$$A_3 := \{X_t \in \mathcal{T}(\phi^{(n-1)}(x^n, y^n)) \quad \forall t > Z^{n-1}_{\sigma^{(n-1)}-1}\}$$

$$G := \{X_0^{(n)} = x^n, \quad X_1^{(n)} = y^n\}.$$

There exists a positive constant K, depending on a but independent of β, such that for any sufficiently large β we have:

$$P_{x^n}(A_1 \cap A_2 \cap A_3 | G) \ge 1 - e^{-K\beta}$$

Moreover if $x^n = \hat{y}^n$ and if we define

$$A_4 := \{\forall y \in Q_{x^n} \quad \exists t < Z^{n-1}_{\sigma^{(n-1)}-1} \text{ such that } X_t = y\}$$

then we have, for any sufficiently large β

$$P_{x^n}(A_1 \cap A_2 \cap A_3 \cap A_4 | G) \ge 1 - e^{-K\beta}.$$

Sketch of the proof It is sufficient to show that $P_{x^n}(A_i^c|G)$ are exponentially small for $i = 1, \dots, 4$.

Suppose first that $P_{x^n}(G) \asymp 1$.

The estimate of $P_{x^n}(A_1^c|G)$ follows from Proposition 3.1.

To estimate $P_{x^n}(A_2^c|G)$ we first note that, since $Q_{x^n} \cap S^{(n-1)} = x^n$, we have $\tau_{(Q_{x^n})^c} \le Z^{n-1}_{\sigma(n-1)}$. On the other hand

$$P_{x^n}(\tau_{Q^c_{x^n}} < Z^{n-1}_{\sigma(n-1)-1}) \le P_{x^n}(X_{\tau_{Q^c_{x^n}}} \notin R_{Q_{x^n}}) + P_{R_{Q_{x^n}}}(X_{\tau_{S(n-1)}} = x^n). \quad (3.35)$$

The first term in the r.h.s. of (3.35) is estimated by proposition 3.2. The second one can be estimated by considering the minimum cycle containing $R_{Q_{x^n}}$ with non empty intersection with $S^{(n-1)}$. This cycle does not contain x^n, by the maximality of Q_{x^n}, and thus by proposition 3.3, this second term is also exponentially small.

The estimate of $P_{x^n}(A_3^c|G)$ can be obtained by using (3.18) and by applying lemma 3.1 to the chain $X_t^{(n-1)}$; the last probability $P_{x^n}(A_4^c|G)$ is estimated by proposition 3.3.

The case that $P_{x^n}(G)$ is exponentially small is more interesting and more difficult to treat. The reader can find a detailed proof in [OS2].

4. The exit problem in the general case

Let us come now to the problem of the determination of the tube of exit from an arbitrary set $Q \subset S$. We will use the renormalization procedure and the results of the previous section to define the tube of typical exiting paths in terms of a sequence of permanence sets.

As discussed in the introduction, we can apply the renormalization procedure to the auxiliary chain X_t^Q with almost absorbing states in ∂Q (see (1.11), (1.12)) up to the step $N = N(Q)$ of the iteration, such that in Q there is only the unstable state $\hat{y}^N$.

Due to our hypothesis RI there exists a unique transition $\hat{y}^N, z^N$ for the chain $X^{(N)}$ exiting from $\hat{y}^N$ with $\Delta^{(N)}(\hat{y}^N, z^N) = 0$.

By lemma 3.1 we have the following inequality:

$$P_{\hat{y}^N}(X_1^{(N)} = z^N) > 1 - e^{-k\beta} \quad (4.1)$$

This transition $\hat{y}^N, z^N$ can be analyzed in terms of paths of the chain $X^{(N-1)}$.

We can proceed as follows: by applying Theorem 3.2 we can associate to this transition a cycle $Q_{\hat{y}^N}$ and the path $\phi^{(N-1)}(\hat{y}^N, z^N)$ in $\Phi(S^{(N-1)})$ defined by (3.34).

To each transition of this path at level $N-1$ we can associate, again by Theorem 3.2, a cycle $Q_{\phi_j^{N-1}}$ and a new path.

By iterating this argument we have that the transition $\hat{y}^N, z^N$ is described by a sequence of cycles:

$$Q_{\hat{y}^{n_0}}, Q_{\hat{y}^{n_1}}, \dots, Q_{\hat{y}^{n_L}}.$$

Let us analyze the sequence $\hat{y}^{n_0}, \dots, \hat{y}^{n_L}$ arising from the above iteration.

We start with the sequence $\hat{y}^N, z^N$. At the first step we consider transitions at level $N-1$ by substituting to the previous sequence the sequence $\phi^{(N-1)}(\hat{y}^N, z^N)$.

At the k-th step of the iteration we start by a sequence of transitions of the chain at level $N-k+1$, $\hat{\mathbf{y}}^{\mathbf{N-k+1}} = \{\hat{y}^{n_i}\}_i$ with $n_i \geq N-k+1$ and we substitute to each couple of consecutive states $\hat{y}^{n_i}$, $\hat{y}^{n_{i+1}}$ the sequence of transitions between states belonging to $S^{(N-k)}$ given by $\phi^{(N-k)}(\hat{y}^{n_i}, \hat{y}^{n_{i+1}})$. We obtain in this way a new sequence $\hat{\mathbf{y}}^{(\mathbf{N-k})}$ that we call the *refinement* of $\hat{\mathbf{y}}^{\mathbf{N-k+1}}$. This sequence corresponds to the "typical exit path" for the chain $X_t^{(N-k)}$.

The final sequence $\hat{\mathbf{y}}^{\mathbf{0}} = \hat{y}^{n_0}, \ldots, \hat{y}^{n_L}$, with $n_0 = N$, arising from the above iteration will be the refinement of the refinement ... of the refinement of $\hat{y}^N, z^N$.

In order to characterize the typical exit path in the time interval $(\theta_{\hat{y}^N}, \tau_{Q^c}]$ we cancel the first term in $\hat{\mathbf{y}}^{\mathbf{0}}$ corresponding to the first interval of time $[0, \theta_{\hat{y}^N}]$, and we obtain:

$$\hat{\mathbf{y}} := \hat{y}^{n_1}, \ldots, \hat{y}^{n_L}.$$

To this path we can associate the corresponding sequence of permanence sets:

$$\mathcal{T}_{\hat{\mathbf{y}}} := Q_{\hat{y}^{n_1}}, Q_{\hat{y}^{n_1}}, \ldots, Q_{\hat{y}^{n_L}}.$$

The result on the typical exiting tube can thus be stated as follows:

Theorem 4.1 *For each positive a there exists a positive constant K such that for each sufficiently large β we have:*

$$P_{\hat{y}^N}(\{X_t\}_{t\in(\theta_{\hat{y}^N}, \tau_{Q^c}]} \in \mathcal{T}_{\hat{\mathbf{y}}}) \geq 1 - e^{-K\beta}.$$

Moreover with probability going to one as $\beta \to \infty$ the process visits sequentially the sequence of sets given by $\mathcal{T}_{\hat{\mathbf{y}}}$, spending inside each cycle $Q_{\hat{y}^{n_i}}$ a time which can be estimated from above and from below by $T_{n_i} e^{\pm a\beta}$ and visiting all the states inside the cycle before the exit from it.

The proof of this theorem can be obtained by using Theorem 3.2.

Let us now make some remarks on Theorem 4.1.

Given the sequence $\hat{\mathbf{y}}$ consider each element $\hat{y}^{n_j}$ belonging to a portion of the sequence where n_j is monotonocally decreasing, i.e. such that $n_j < n_{j-1}$ and $Q_{\hat{y}^{n_j}} \subset Q_{\hat{y}^{n_{j-1}}}$. This means that the information given by the event:

$$\tau_{Q^c_{\hat{y}^{n_j}}} \asymp T_{n_j} \tag{4.2}$$

can be considered as a "negligible correction" to the statement:

$$\tau_{Q^c_{\hat{y}^{n_{j-1}}}} \asymp T_{n_{j-1}}$$

Then from any sequence $\hat{\mathbf{y}}$ we can extract a new sequence $\hat{\mathbf{y}}'$ in which this kind of monotonically decreasing subsequences have been removed.

More precisely:

Definition 4.1 Let $\{\hat{\mathbf{y}}''\}_i$ be a maximal segment of $\hat{\mathbf{y}}$

$$\{\hat{\mathbf{y}}''\}_i := \{\hat{y}^{n_{l_i}}, \hat{y}^{n_{l_i+1}}, \ldots, \hat{y}^{n_{m_i}}\}$$

such that for any $j \in [l_i, m_i]$ we have $Q_{\hat{y}^{n_j}} \subseteq Q_{\hat{y}^{n_{k_i}}}$ for some $k_i \in [l_i, m_i]$.

Let $\hat{\mathbf{y}}'$ be the sequence obtained by $\hat{\mathbf{y}}$ by replacing each segment $\{\hat{\mathbf{y}}''\}_i$ with the unique term $\hat{y}^{n_{k_i}}$.

This means that by considering the sequence $\hat{\mathbf{y}}'$ instead of $\hat{\mathbf{y}}$, we have removed all "neglegible corrections", like (4.2).

With theorem 4.1 we have described in full detail the exit; with the refinement procedure we have obtained at the same time two different kinds of information. First of all, we have obtained what we call *primary information* (or *level one approximation of the tube of exit*): the family of ordered sequences of different permanence sets visited by the process (namely the set of standard descents $Q_{\hat{\mathbf{y}}'}$) and the typical times spent by our process inside them. The *secondary information* (or level 2 approximation of the tube of exit) concerns further details about the history of the process inside each cycle that it visits when it stays inside some Q_i in $Q_{\mathbf{y}'}$ corresponding to the segments $\{\hat{\mathbf{y}}''\}_i$. It will describe the first descent to the bottom of the cycles and the first excursion outside them.

Up to now we have considered the starting point $\hat{y}^N$. As a final remark we want to note that a general starting point can be considered by distinguishing a first interval of time in which the process falls to the bottom $\hat{y}^N$, and a second, much longer interval of time in which the process exits from Q following the description obtained above, since we can now consider $\hat{y}^N$ as a starting point.

The first transient consisting in the descent to the bottom of Q can be analyzed with the same procedure, by considering refining paths.(See [OS2] for more details).

Acknowledgements. This work has been partially supported by the grant CHRX-CT93-0411 of the Commission of European Communities.

References

[BACe] G. Ben Arous, R. Cerf, "Metastability of the three dimensional Ising model on a torus at very low temperature", Preprint 1995.

[CG] M. Cassandro, A. Galves, "Language acquisition and change in generalized GW model", Preprint MOL4 1995.

[CGOV] M. Cassandro, A. Galves, E. Olivieri, M.E. Vares, "Metastable behavior of stochastic dynamics: a pathwise approach", J. Stat. Phys. **35**, 603–634 (1984).

[Ca1] O. Catoni, "Rough large deviations estimates for simulated annealing. Application to exponential schedules", Ann. Prob. **20**, 1109–1146 (1992).

[Ca2] O. Catoni, "Sharp large deviations estimates for simulated annealing algorithms", Ann. Inst. H. Poincaré **27 no. 3**, 291–382 (1991).

[CaCe] O. Catoni, R. Cerf, "The exit path of a Markov chain with rare transitions", Preprint 1995.

[Ce] R. Cerf, "Une théorie asymptotique des algorithmes génétiques" These Universite Montpellier II (1994).

[CesM] F. Cesi, F. Martinelli, "On the layering transition of an SOS surface interacting with a wall. I. Equilibrium results" J. Stat. Phys. **82**, 823 (1996).

[ChiCho1] T.S. Chiang, Y. Chow, "A limit theorem for a class of inhomogeneous Markov processes", Ann. Prob. **17**, 1483–1502 (1989).

[ChiCho2] T.S. Chiang, Y. Chow, "On the Exit Problem from a Cycle of Simulated Annealing Processes with application – A Backward Equation Approach", Tech. Rept. Inst. of Math. Academia Sinica, 1995.

[CiO] E. Cirillo, E. Olivieri,"Metastability and nucleation for the Blume-Capel model: different mechanisms of transition" Journ. Stat. Phys. **83**, 473–571 (1996).

[DeSch1] P. Dehghanpour, R.H. Schonmann, "Metropolis dynamics relaxation via nucleation and growth", preprint.

[DeSch2] P. Dehghanpour, R.H. Schonmann, "A nucleation-and-growth model", Probab. Th. Rel. Fields, to appear.

[EGJLa] V.R. Eston, A. Galves, C.M. Jacobi, R. Langevin, "Dominance switch between two interacting species and metastability" Atas do II Simposio dos Ecosistemas da costa sul Brasileira, CACIESP, Sau Paulo (1988).

[FW] M.I. Freidlin, A.D. Wentzell, "Random Perturbations of Dynamical Systems", Springer-Verlag (1984).

[GSchm] A. Galves, B. Schmitt, "Occurrence of rare events for mixing dynamical systems" Ann. Inst. H. Poincaré Phys. Theor. **52**, 267 (1990).

[Ge] H.O. Georgii, "Gibbs measures and Phase Transitions", Walter de Gruyter (de Gruyter Studies in Mathematics Vol. 9), Berlin, New York (1988).

[GlHu] S. Glance, B. Huberman, "The dynamics of social dilemmas", Scient. Amer. March 76 (1994).

[Ho1] R. Holley, "Possible rates of convergence in finite range, attractive spin systems", Contemp. Math. **41**, 215 (1985).

[Ho2] R. Holley, "Rapid convergence to equilibrium in one dimensional stochastic Ising models", Ann. Prob. **13**, 72–89 (1985).

[KO1] R. Kotecky, E. Olivieri, "Droplet dynamics for asymmetric Ising model" Journ. Stat. Phys. **70** , 1121–1148, (1993).

[KO2] R. Kotecky, E. Olivieri, "Shapes of growing droplets – a model of escape from a matastable phase" Journ. Stat. Phys. **75** , 409–507, (1994).

[Li] T.M. Liggett, Interacting particle systems, Springer Verlag (1985).

[MO] F. Manzo, E. Olivieri, "Relaxation patterns for competing metastable states: a nucleation and growth model", preprint.

[MO1] F. Martinelli, E. Olivieri, "Approach to equilibrium of Glauber dynamics in the one phase region. I: the attractive case", Comm. Math. Phys. **161**, 447–486 (1994).

[MO2] F. Martinelli, E. Olivieri, "Approach to equilibrium of Glauber dynamics in the one phase region II: the general case", Comm. Math. Phys. **161**, 487–514 (1994).

[MOSch] F. Martinelli, E. Olivieri, R.H. Schonmann, "For 2-d lattice spin systems weak mixing implies strong mixing", Comm. Math. Phys. **165**, 33–47 (1994).

[MOS1] F. Martinelli, E. Olivieri, E. Scoppola, "On the Swendsen and Wang dynamics II: Critical droplets and homogeneous nucleation at low temperature for the two dimensional Ising model", J. Stat. Phys. **62**, 135–159 (1991).

[MOS2] F. Martinelli, E. Olivieri, E. Scoppola, "Metastability and exponential approach to equilibrium for low temperature stochastic Ising models", J. Stat. Phys. **61**, 1105 (1990).

[MOS3] F. Martinelli, E. Olivieri, E. Scoppola, "On the Swendsen and Wang dynamics I: Exponential convergence to equilibrium", J. Stat. Phys. **62**, 117–134 (1991).

[N] E.J. Neves, "A discrete variational problem related to Ising droplets at low temperature", J. Stat. Phys. **80**, 103–123 (1995).

[NaO] F.R. Nardi, E. Olivieri,"Low temperature stochastic dynamics for an Ising model with alternating field". Markov Proc. Rel. Field **2**, 117–166 (1996).

[NSch1] E.J. Neves, R.H. Schonmann, "Behaviour of droplets for a class of Glauber dynamics at very low temperatures" Comm. Math. Phys. **137**, 209 (1991).

[NSch2] E.J. Neves, R.H. Schonmann, "Critical Droplets and Metastability for a Glauber Dynamics at Very Low Temperatures", Prob. Theor. Rel. Fields **91**, 331 (1992).

[NewCoKi] C. Newman, J.E. Cohen, C. Kipnis, "Neo-darwinian evolution implies punctuated equilibria", Nature **315**, 400 (1985).

[NiBe] P. Niyogi, R.C. Berwick, 'Formalizing Triggers: a Learning model for finite spaces', Preprint MIT 1993.

[OS1] E. Olivieri, E. Scoppola, "Markov chains with exponentially small transition probabilities: First exit problem from a general domain – I. The reversible case" JSP **79**, 613 (1995).

[OS2] E. Olivieri, E. Scoppola, "Markov chains with exponentially small transition probabilities: First exit problem from a general domain – II The general case". JSP **84**, 987–1041 (1996).

[Pe] C. Peixoto, "Metastable Behavior of Low Temperature Glauber dynamics with stirring", J. Stat. Phys. **80**, 1165–1184 (1995).

[PL1] O. Penrose, J.L. Lebowitz, "Rigorous treatment of metastable states in van der Waals theory", J. Stat. Phys. **3**, 211–236 (1971).

[PL2] O. Penrose, J.L. Lebowitz, "Towards a rigorous molecular theory of metastability" In Fluctuation Phenomena (second ed.) E.W. Montroll, J.L. Lebowitz ed., North Holland Publ. (1987).

[Ru] D. Ruelle, Statistical Mechanics, Benjamin Inc. (1969).

[Sch1] R. Schonmann, "The pattern of escape from metastability of a stochastic Ising model", Comm. Math. Phys. **147**, 231–240 (1992).

[Sch2] R. Schonmann, "Slow droplet driven relaxation of stochastic Ising models in the vicinity of the phase coexistence region", Comm. Math. Phys. **161**, 1–49 (1994).

[Sch3] R. Schonmann, "Theorems and conjectures on the droplet driven relaxation of stochastic Ising model" in Probability and Phase Transition, Ed. G. Grimmett, NATO ASI Series, Kluwer Ac. Publ. pg. 265–301 (1994).

[SchSh1] R. Schonmann, S. Shlosman, "Wulff droplets and the metastable relaxation of kinetic Ising models", preprint.

[SchSh2] R. Schonmann, S. Shlosman, "Complete Analiticity for 2D Ising completed", Comm. Math. Phys. **170**, 453 (1995).

[S1] E. Scoppola, "Renormalization group for Markov chains and application to metastability", Jour. Stat. Phys. **73**, 83 (1993).

[S2] E. Scoppola, "Metastability for Markov chains: a general procedure based on renormalization group ideas" in Probability and Phase Transition, Ed. G. Grimmett, NATO ASI Series, Kluwer Ac. Publ. (1994).

[Si] Ya.G. Sinai, "Theory of Phase Transition: Rigorous Results", Akademiai Kiado, Budapest (1982).

[ToMi] H. Tomita, S. Miyashita, "Statistical properties of the relaxation process of metastable states in the kinetic Ising model", Phys. Rev. B. Condensed Matter **46**, 8886–8893 (1992).

[T1] A. Touvé, "Cycle Decompositions and Simulated Annealing". SIAM J. Control Optim. To appear.

[T2] A. Trouvé, "Rough large deviation estimates for the optimal convergence speed exponent of generalized simulated annealing algorithms", Ann. Inst. H. Poincaré. To appear.

Progress in Mathematics, Vol. 169, © 1998 Birkhäuser Verlag Basel/Switzerland

Rationality Problems for Group Varieties

VLADIMIR P. PLATONOV

Department of Pure Mathematics
University of Waterloo
Waterloo, Ontario, Canada N2L 3G1

Introduction

The rationality problems for group varieties have a long history. These problems are especially interesting for semisimple group varieties. Let G be a semisimple algebraic group defined over a field K. It is well known that the algebraic variety G is unirational over K, i.e., the field $K(G)$ of rational functions on G is contained in a purely transcendental extension of K (see [DG], Exp.XIV]).

Rationality problems. When is G rational over K ? That is, when is $K(G)$ a purely transcendental extension of K? Equivalently, when is the variety G birationally isomorphic over K to an affine space?
When is G stably rational over K? That is, when does there exist an affine K-space $\mathbb{A}_K^m$ such that $G \times \mathbb{A}_K^m$ is K-rational?

If G is K-split (i.e. there exists a K-split maximal torus T), then G is K-rational. This follows immediately from the Bruhat decomposition:
G is birationally isomorphic over K to the product variety $U \times T \times U$ where U is the unipotent radical of a K-split Borel subgroup of G. In particular, over an algebraically closed field K, the variety G is K-rational.

Since all algebraic K-tori of dimension at most two are rational over any K, and the variety of maximal tori of G is K-rational (see [PR], §2.4 and [V], Th.4.74), it follows that G is K-rational if rank $G \leq 2$.

In general (for arbitrary G and K), these problems are difficult. For a long time, the following problem was open:

Is the variety G rational if G is simply connected?

Many years ago, I obtained the following result (see [$P1$]):

Let $G(K) = SL_n(D)$ where D is a division algebra with center K. Then, for any m which is not square-free, there exists a division algebra D of degree m over $K = \mathbb{Q}(x_1, \ldots, x_n)$, or $\mathbb{Q}_p(x_1, \ldots, x_n), n \geq 2$, such that G is not stably rational over K (here $\mathbb{Q}$ and $\mathbb{Q}_p$ are, respectively, the fields of rational and p-adic numbers, and $x_1, \ldots, x_n$ are independent variables). In fact, I proved even more:

$G(K) = SL_n(D)$ does not have the weak approximation property.

(It is easy to prove that any smooth stably rational variety over K has the weak approximation property (see [PR], Ch.7)). The crucial role in this proof was played by the non-triviality of $SK_1(D \otimes_K F)$ for some extension F of K.

I would like to remark here that the strategies used to prove non-rationality of a given variety are very different from those used to prove rationality. A proof of non-rationality is usually connected with some new invariants (for instance, $SK_1(D)$). On the other hand, a proof of rationality of an algebraic variety involves a direct construction depending on that particular variety.

1. Spinor varieties

Let f be a nondegenerate quadratic form on the n-dimensional vector space V over a field K of characteristic $\neq 2$. It is clear that the variety $SO(f)$ is K-rational: if $L(SO(f))$ denotes the Lie algebra of $SO(f)$, then $x \mapsto (1-x)(1+x)^{-1}$; $x \in L(SO(f))$ gives us a suitable parametrization of $SO(f)$. The group $SO(f)$ has the covering group Spin(f), the spinor variety:

$$1 \to \{\pm 1\} \to Spin(f) \to SO(f) \to 1$$

The next results were inspired by the following natural question posed by P. Deligne (1978):

Is $Spin(f)$ rational over K (in particular, over $K = \mathbb{R}$)?

First of all, the following statement was proven in $[P2]$:

Proposition 1. *If a quadratic form f is K-isotropic, then $Spin(f)$ is K-rational. In particular, if K is either an algebraic extension of $\mathbb{Q}_p$ or a a totally imaginary number field or a global function field, $Spin(f)$ is K-rational.*

Since all groups of rank ≤ 2 are K-rational, the minimal n for which, a priori, there is a possibility of $Spin(f)$ being non-rational over K, is $n = 6$. The next result was relatively surprising:

Theorem 1 ($[P3]$). *Let $n = 4m + 2, n \geq 6$; and let $x_1, \ldots, x_{n-2}$ be algebraically independent over some field L. If $f(y_1, \ldots, y_n) = y_1^2 + x_1 y_2^2 + \cdots + x_{n-2} y_{n-1}^2 + x_1 x_2 \cdots x_{n-2} y_n^2$, then the spinor variety $Spin(f)$ is not stably rational over $K = L(x_1, \ldots, x_{n-2})$.*

Let me recall the notion of R-equivalence.

Definition. *Two points $a, b \in G(K)$ are called strictly R-equivalent if there is a rational map $f : \mathbb{A}^1 \to G$ over K such that $f(0) = a, f(1) = b$. The points a, b are called R-equivalent, if there exists a chain of finitely many points $a = a_0, a_1, \cdots, a_m = b$ in $G(K)$ such that a_i and a_{i+1} are strictly R-equivalent for $0 \leq 1 \leq m - 1$.*

If G is stably rational over K, then any point of $G(K)$ is R-equivalent to 1 (see $[CS]$). I used in the proof of theorem 1 the following i-**effect**:

The center of $Spin(f)$ is $Z = < i >, i^2 = -1$ and i is not R-equivalent to 1.

Hence the spinor variety is not stably rational over K. However, for canonical spinor varieties, we have the following result:

Theorem 2 ($[PC1], [P3]$). *If $f(y) = y_1^2 + \cdots + y_n^2$, then $Spin(f)$ is rational over an arbitrary field K.*

Corollary. *Every real spinor variety $Spin(f)$ is rational over $\mathbb{R}$.*

The proof of theorem 2 is based on the following realization of $Spin(f)$:

Proposition 2 ([$PC1$], [$P3$]). *Let* $f(y) = a_1y_1^2 + \cdots + a_ny_n^2$, $0 \neq a_i \in K$, $i = 1, \ldots, n$. *Then* $Spin(f)$ *is birationally isomorphic over* K *to the hypersurface* M_f *in* $\mathbb{A}^{\frac{n(n-1)}{2}+1}$ *defined by*

$$\alpha z_1^2(a_1 + a_2z_2^2)(a_1 + a_2z_3^2 + a_3z_4^2)\cdots(a_1 + \cdots + a_nz^2_{\frac{n(n-1)}{2}+1}) = 1$$

$$\text{where} \quad \alpha = \begin{cases} 1 & \text{if } n \text{ is odd} \\ a_1 & \text{if } n \text{ is even} \end{cases}$$

In the general case the question of K-rationality of hypersurfaces over K, as in proposition 2, is closely related to the following natural problem about the composition of quadratic forms. Let f_1 and f_2 be quadratic forms in the independent variables x_i and y_j. When is the product f_1f_2 birationally equivalent to a quadratic form Ψ in the variables z_k? In this formulation, the problem was first posed in [$PC1$]. For canonical quadratic forms, our hypersurface is defined by a product of canonical quadratic forms, and one can construct a birational decomposition. For Pfister forms, the corresponding result follows from proposition 2:

Proposition 3. *Let* f *be a Pfister quadratic form. Then* $Spin(f)$ *is rational over an arbitrary field* K.

For an algebraic number field K, which is not totally imaginary, the question of rationality of $Spin(f)$ is still open (for $\mathbb{Q}$, it was proved in [$C1$]). Some adjoint groups of type D_n for odd n are not stably rational even over $\mathbb{Q}$ ([$M2$]).

2. Rationality problem over local and global fields

After the above discussion, it becomes clear that it is possible to deal with the question of rationality of an arbitrary almost simple group only for "good" fields. We have, first of all, the problem over local fields. The solution of this problem was given recently in [$PC2$]:

Theorem A. *Let* K *be a finite extension of* $\mathbb{Q}_p$ *and let* G *be an almost simple algebraic group defined over* K. *If* G *is not of type* A_n, $n \geq 3$, *then* G *is a rational variety over* K.

If G is of type $A_n, n \geq 3$, then G is not necessarily rational over $\mathbb{Q}_p$. A.Merkurjev proved (see [$M1$]): *If* $G(\mathbb{Q}_p) = SL_1(D)$ *where* D *is a division algebra of degree* $4m$ *over* $\mathbb{Q}_p$, *then* G *is not rational over* $\mathbb{Q}_p$. In particular, there exists a $\mathbb{Q}_p$-form G of type D_3 such that G is not rational over $\mathbb{Q}_p$ (recall that $D_3 = A_3$).

In other words, theorem A is surprising – it proves rationality for $D_n, n > 3$, whereas a group of type D_3 need not be rational.

Corollary. *Let* G *be a semisimple algebraic group which is either simply connected, or adjoint without almost simple components of type* A_n. *Then* G *is* K*-rational.*

An intermediate semisimple group (i.e. one which is neither simply connected nor adjoint) G is not necessarily K-rational. There exist even quasisplit semisimple K-groups G which are not K-rational. To describe such an example, consider $\alpha_1, \alpha_2, \alpha_3 \in K^*$ which are independent modulo $(K^*)^2$. Let $f_l = \sum_{i=1}^n x_i^2 -$

$\sum_{j=1}^n y_j^2 + u^2 - \alpha_l v^2$, $l = 1,2,3$, be $(2n+2)$-dimensional quadratic forms over K. Then (see [$PC3$]):

Proposition 4. *$(SO(f_1) \times SO(f_2) \times SO(f_3))/\langle(-1,-1,-1)\rangle$ is not a K-rational group.*

Another example of a non-rational semisimple group was constructed by Ph. Gille. In fact, the first example of a non-rational semisimple algebraic group which is neither simply connected nor adjoint was constructed by J.-P. Serre (see [S], P. 156–159).

For an algebraic number field K the rationality problem is more difficult. I noted already that even for the spinor varieties we do not have the complete solution yet. On the other hand, the situation is quite different for totally imaginary number fields, where we have the following result (see [$PC3$]):

Theorem B. *If G is an almost simple algebraic group defined over a totally imaginary number field K, and if G is not of type A_n, $n \geq 3$, then G is a rational variety over K.*

The proofs of theorems A and B are similar, and are of a much more general nature. In fact, we prove that *an almost simple group with given Tits index (see [T]), which is admissible over local and totally imaginary global fields, is rational over an arbitrary field K*. A description of these results is contained in the next section.

§ 3 Rationality problem over arbitrary fields

Let G be an almost simple algebraic group defined over an arbitrary field K. If S is a maximal K-split torus in G, then the generalised Bruhat decomposition (see [BT]) implies that the variety $G \approx U \times Z_G(S) \times U$ over K, where we use $\approx$ to denote birational isomorphic varieties and U is the unipotent radical of a minimal parabolic K-subgroup of G. Since U is K-rational, it suffices to show that the variety $Z_G(S)$ is K-rational. But, the bundle $Z_G(S) \to Z_G(S)/S$ is locally trivial, so $Z_G(S) \approx Z_G(S)/S \times S$. If we prove that in our situation $Z_G(S)/S$ is K-rational, then G will also be K-rational. If S' is the central torus of $Z_G(S)$, then $Z_G(S) = \Delta S'$, where $\Delta = [Z_G(S), Z_G(S)]$ is the semisimple K-anisotropic kernel of G. There are two possibilities for S':

(i) $S' = S$ (this is true for any inner type group),

(ii) $S' \neq S$, and G is an outer type group.

Usually one can reduce our consideration to the case when $H^1(F, S') = 1$ for any extension F of K, and we get

$$Z_G(S) \approx (Z_G(S)/S') \times S' \approx (\Delta/\Delta \cap S') \times S',$$

hence the variety G is K-rational if $\Delta/\Delta \cap S'$ is. For our investigation, it is essential to know the intersection $\Delta/\Delta \cap S'$ in an explicit form. Let T be a maximal torus of G defined over K and containing S. Let $\Sigma = R(T,G)$ be the root system of G relative to T, and $\Pi = \{\alpha_1, \dots, \alpha_n\}$ be a system of simple roots. If

$\{X_\alpha, \alpha \in \Sigma; H_{\alpha_1}, \dots, H_{\alpha_n}\}$ is a Chevalley basis of the Lie algebra of G, then G is generated by the subgroups $G_\alpha = \langle x_{\pm\alpha}(t) : t \in \overline{K}\rangle$, and the torus T is generated by $\langle h_{\alpha_i}(t_i)\rangle = T \cap G_{\alpha_i}$, $i = 1, \dots, n$. In particular, $T = \langle h_{\alpha_1}(t_1)\rangle \times \cdots \times \langle h_{\alpha_n}(t_n)\rangle$ if G is simply connected. Moreover, $h_\alpha(t)x_\beta(u)h_\alpha(t)^{-1} = x_\beta(t^{\langle\beta,\alpha\rangle}u)$, where $\langle\beta,\alpha\rangle = 2\frac{(\beta,\alpha)}{(\alpha,\alpha)}$ (see [St]). Then the torus S is defined as the identity component of the variety defined by the following system of equations:

$\alpha_i(t) = 1$ *for all non-distinguished vertices of the Tits K-index,*

$\beta(t) = \sigma^*(\beta)(t)$ *for all distinguished vertices $\beta \in \Pi$, and all $\sigma \in \Gamma = Gal(K^{sep}/K)$.*

Here $t \in T$, σ^* is the natural action of Γ on Π. Recall that for inner type groups, we have $\sigma^*(\beta) = \beta$ for all $\beta \in \Pi$.
We assume from now on that $\mathrm{Char}(K) \neq 2$. Most of the results formulated below are contained in [$PC3$].

Let G be a simply connected group of type A_n. If G is of inner type, then $G(K) = SL_n(D)$, where D a division algebra with center K. It is easy to see that $Z_G(S) \approx SL_1(D) \times \mathbb{A}^t$ over K; hence $SL_n(D)$ is K-rational if $SL_1(D)$ is K-rational. For instance, $SL_n(D)$ is K-rational if K is arbitrary and the degree of D is at most 3.

Let now G be a group of outer type A_n. Then G is a special unitary group $SU_{(n+1)/d}(D,h)$, where D is a central division algebra of degree d over a quadratic extension L of K, with an involution σ of the second kind such that $K = \{x \in L : x^\sigma = x\}$, and h is a nondegenerate hermitian form (relative to σ) of index r. We have:

Proposition 5. *The variety $SU_{(n+1)/d}(D,h) \approx SU_1(D,\tau) \times \mathbb{A}^m_K$ for some involution τ of the second kind.*

Corollary. *If $d \leq 3$, then the variety $SU_{(n+1)/d}(D,h)$ is K-rational.*

The rationality of K-isotropic groups of type B_n follows from proposition 1. If G is a simply connected group of type C_n, then $G = SU_{2n/d}(D,h)$, where D is a central division algebra of degree d over K, h is a nondegenerate skew-hermitian form of index r, relative to an involution σ of the first kind such that $dim\ D^\sigma = \frac{d(d+1)}{2}$. The variety $SU_{2n/d}(D,h)$ is always K-rational (can use the generalised Cayley correspondence). For the adjoint group $PSU_{2n/d}(D,h)$, we have:

Proposition 6. *The variety $PSU_{2n/d}(D,h) \approx PSU(D,h_{an}) \times \mathbb{A}^m$, where h_{an} is the anisotropic part of h. In particular, $PSU_{2n/d}(D,h)$ is K-rational when $n - rd \leq 2$.*

The case when G is a group of classical type D_n is one of the most interesting cases. Then G is $SU_{2n/d}(D,h)$, where D is of degree 2^a over K, and h is a nondegenerate hermitian form of index r, relative to an involution σ of the first kind such that $dim\ D^\sigma = \frac{d(d+1)}{2}$. We will identify the center of $SL_1(D)$ with μ_d. The result is:

Theorem C. *Let G be a K-isotropic almost simple group of classical type D_n. If G is not K-rational, then $G \approx SL_1(D)/\mu_{d/2} \times \mathbb{A}^m$ or $G \approx PSU(D,h_{an}) \times \mathbb{A}^m$ for some m. In particular, G is K-rational for $d \leq 4$ and $n - rd \leq 3$.*

A short description of the idea of the proof is contained in $[PC2]$. A result that is useful in the proof of theorem C as well as in dealing with exceptional groups, is:

Proposition 7. *Consider an algebraic group $H = H_1 \times \cdots \times H_n$, where H_i are K-groups of type A_1. Then for any central subgroup C, the variety H/C is K-rational.*

Proposition 8. *If G is K-isotropic, of type F_4, then it is K-rational.*
The proof of this is easy: if G is not K-split, then its K-rank is 1, and so its anisotropic kernel Δ has type B_3; $\Delta/\Delta \cap S$, for a maximal K-split torus S, is isomorphic to $SO(f)$ for some quadratic form f.

For other exceptional groups, one can use the following general result:

Theorem D. *Let G be an almost simple K-group, and suppose the anisotropic kernel Δ has the property that any almost simple component of Δ has rank at most 2. Then the variety G is K-rational.*

For classical groups, this theorem follows from propositions 1,5,6 and theorem C. For exceptional groups, we use the classification of Tits indexes (see $[T]$). Any group of type G_2 is K-rational (as its rank is 2). If G has type F_4 or E_8, and is as in theorem D, it is K-split. If G is of type E_7, and is as in theorem D, and is not K-split, then it is of type $E_{7,4}^9$ (see $[T]$). Then the rationality follows from proposition 7.

For the groups of type E_6, the proof of theorem D is more complicated. We have two Tits indexes that are not K-quasisplit: ${}^1E_{6,2}^{16}$ and ${}^2E_{6,2}^{16''}$. The groups of type ${}^3D_4$ and ${}^6D_4$ are K-quasisplit (hence K-rational) under the conditions of theorem D.

On the proofs of theorems A and B. Now let G be an almost simple algebraic K-group over a local field or a totally imaginary algebraic number field K. Then we can use the following two fundamental facts (see $[PR, Ch.6]$):

(i) If G is not of type A_n, then G is K-isotropic;
(ii) If G is simply connected, then $H^1(K, G) = 1$.

Any central division algebra D over K, which has an involution of the first kind, is a quaternion algebra. Then the proofs of the theorems A and B follow for the groups of type B_n, C_n or D_n, from theorem C and propositions 1,6 and 7. The statement (ii) implies that any K-group G of type E_8, F_4 or G_2 is K-split, therefore, K-rational.

If G is of type E_6 or E_7, we can apply theorem D. Finally, for G of type ${}^3D_4$ or ${}^6D_4$, we have the following general result.

Proposition 9. *If G is a K-isotropic group of type ${}^3D_4$ or ${}^6D_4$ over an arbitrary field K, then G is K-rational.*

J.-P. Serre recently posed the rationality question for semisimple group varieties over $\mathbb{C}(x, y)$. It is possible to prove the following result for classical groups over (more generally) any C_2-field.

Theorem E. *If G is a group of type B_n, C_n or D_n over a C_2-field K, then it is K-rational.*

I believe that this result can be extended to all exceptional groups over C_2-fields. For groups of type F_4, it is true because these groups are split over C_2-fields (see [S]). But, for groups of types $E_i, i = 6, 7, 8$, we need to have a classification of these groups over C_2-fields.

4. Open problems

The following very natural problems about rationality of group varieties are still open:

1. *The problem of rationality for almost simple groups over number fields that are not totally imaginary. Even for the spinor varieties Spin(f), we do not have a complete answer.*
2. *Conjecture ([P3]): Every real algebraic group is rational over $\mathbb{R}$. (Even for classical groups only stable rationality is known – see [C2]).*
3. *The rationality problem over C_2-fields (J.-P. Serre) or, more generally, over fields of cohomological dimension 2.*
4. *The rationality problem for $SL_1(D)$, where D is a division algebra of prime degree over an arbitrary field (this was formulated as a conjecture in [P1]).*

References

[BT] A. Borel & J. Tits, Groupes reductifs, Publ.Math.IHES, 27(1965), 55–151.

[C1] V.I. Chernousov, On the rationality of spinor varieties over the field of rational numbers, Dokl. AN BSSR, 25(1981), 293–296.

[C2] V.I. Chernousov, The group of multipliers of the canonical quadratic form and the stable rationality of the variety PSO, Mat. Zametki, 55(1994), 114–119.

[CS] J.-L. Colliot-Thelene & J.-J. Sansuc, La R-equivalence sur les tores, Ann.Sci. ENS, 10(1977), 175–230.

[DG] M. Demazure et A. Grothendieck, Schemas en groupes, Springer Lecture Notes, v. 152, 1970.

[M1] A.S. Merkurjev, Generic element in SK_1 for simple algebras, K-theory, 7(1993), 1–3.

[M2] A.S. Merkurjev, R-equivalence and rationality problem for semisimple adjoint classical groups, Publ. Math. IHES (to appear).

[P1] V.P. Platonov, Algebraic groups and reduced K-theory, Proc. Internat. Congress Math., Helsinki, v. 1(1980), 311–317.

[P2] V.P. Platonov, On the problem of rationality of spinor varieties, Soviet Math. Dokl., 20(1979), 1027–1031.

[P3] V.P. Platonov, Birational properties of spinor varieties, Proc. Steklov Inst. Math., No. 3(1983), 173–182.

[PC1] V.P. Platonov & V.I. Chernousov, On the rationality of canonical spinor varieties, Soviet Math. Dokl., 21(1980), 830–834.

[PC2] V.P. Platonov & V.I. Chernousov, The rationality problem for group varieties, C.R. Acad. Sci. Paris, 322(1996), 245–250.

[PC3] V.P. Platonov & V.I. Chernousov, The rationality problem for semisimple group varieties, Universität Bielefeld, 37(1997), pp. 26.

[PR] V.P. Platonov & A.S. Rapinchuk, Algebraic groups and number theory, Academic Press, N.Y., 1994, pp. 614.

[S] J.-P. Serre, Cohomologie Galoisienne, Springer, 1994, pp. 181.

[St] R. Steinberg, Lectures on Chevalley groups, Yale University, 1967.

[T] J. Tits, Classification of algebraic semisimple groups, Proc. Symp. Pure Math., 9(1966), 32–62.

[V] V. Voskresenskii, Algebraic tori, Nauka, Moscow, 1977, pp. 223.

Added in proof

We obtained in [PC3] similar results about rationality over global function fields.

Progress in Mathematics, Vol. 169,

Precise Measurements in Symplectic Topology

LEONID POLTEROVICH*

School of Mathematical Sciences
Tel Aviv University
Tel Aviv 69978, Israel
(polterov@math.tau.ac.il)

What is a measurement in differential geometry? Our intuition is based on the Riemannian background where two kinds of quantities appear:

- *local*, like angle or curvature;
- *global*, like the distance between two points, where one should solve a variational problem – to find the shortest path.

The lack of local invariants in symplectic geometry leads to the fact that non-trivial measurements arise there as the solutions of variational problems of quite a complicated nature.

Below we describe an approach to such problems which allows us to get precise answers in a number of interesting situations, and discuss applications to geometry and dynamics. I start with two examples. (See [McD-S2, H-Z] for preliminaries.)

1. The symplectic packing problem

The problem goes back to Gromov [G1]. Let (M, Ω) be a symplectic manifold. A *packing* is a symplectic embedding into M of k disjoint standard balls of equal radii. (Here we assume that the balls lie in the standard Hermitian vector space, and carry the induced symplectic structure).

How do we find the maximal possible volume which can be filled out in M by a symplectic packing? Define a number

$$v_k(M, \Omega) = \sup_f \frac{\text{Vol}(\text{Image } f)}{\text{Vol}(M)},$$

where f runs over all symplectic packings by k balls. Let us emphasize that the common radius of the balls is a variable of this variational problem: the goal is to increase it as far as possible.

The table in Fig. 1 gives the precise values of v_k, $k \leq 9$ for the standard symplectic $\mathbf{CP}^2$ (McDuff-Polterovich [McD-P]).

*revised version

k	1	2	3	4	5	6	7	8	9
$v_k(CP^2)$	1	$\frac{1}{2}$	$\frac{3}{4}$	1	$\frac{20}{25}$	$\frac{24}{25}$	$\frac{63}{64}$	$\frac{288}{289}$	1

Fig. 1

For $k = 2, 3, 5, 6, 7, 8$ the value of v_k is strictly less than 1 and hence we have obstructions to the full filling. In contrast with this, $v_k = 1$ holds for all $k \geq 10$ (P. Biran [Bi]).

2. Symplectic K-area

Let $P \to S^2$ be a symplectic fibration with a simply connected fiber (M, Ω). In other words, each fiber of P carries a symplectic structure, which depends smoothly on the base point and is symplectomorphic to Ω. K-area (P) is an invariant which measures the deviation of P from the flat fibration. This notion in a slightly different setting was introduced recently by Gromov [G2]. It is defined as follows (see [G-L-S], [McD-S2] for preliminaries about connections on symplectic fibrations).

Fix an area form, say τ, on S^2 with total area 1.

A connection ν on P is called *symplectic* if the parallel transport preserves the symplectic structure on fibers. Its curvature ρ^ν is a 2-form on the base which takes values in the Lie algebra of the symplectomorphism group of the fiber, that is in the space of Hamiltonian functions with zero mean. In particular one can write $\rho^\nu = F\tau$ for a real valued function F on P. Define the norm $||\rho^\nu||$ as $\max_P |F|$.

Set

$$\text{K-area}(P) = (\inf_\nu ||\rho^\nu||)^{-1},$$

where ν runs over all symplectic connections on P. This means that we look for the smallest possible curvature norm for a symplectic connection on P.

Interesting examples of symplectic fibrations come from vector bundles over S^2. Let $E \to S^2$ be a rank n complex vector bundle endowed with a Hermitian metric. Then its projectivization $P = \mathbf{P}(E)$ is a symplectic fibration with the fiber $(\mathbf{CP}^{n-1}, \Omega_{st})$. Here, and below, Ω_{st} denotes the standard Fubini-Studi form normalized in such a way that the area of a projective line equals 1.

Theorem 2.A (see [P1]). *Suppose that* $c_1(E) = q \in \{1, \ldots, n-1\}$. *Then*

$$\text{K-area}(P) = \frac{n}{max(q, n-q)}.$$

In fact, there exists a connection with the minimal curvature norm which comes from a certain holomorphic structure on E.

3. Deformations of symplectic forms in cohomology

We have seen two variational problems of quite different natures which admit precise solutions. Interestingly enough, both of them can be reduced to *a deformation problem* of the following type.

Let (P, σ) be a symplectic manifold. Let $\Gamma \subset P$ be a symplectic submanifold, that is the restriction of σ to Γ is non-degenerate. Consider a ray $l \subset H^2(P, \mathbf{R})$ with the origin at $[\sigma]$.

Deformation Problem 3.A How far one can deform σ through symplectic forms on P whose restriction to Γ is non-degenerate in such a way that the cohomology class moves in the direction of l?

Example 3.B (symplectic blow up). Let (M, Ω) be a closed *Kähler* symplectic manifold, and let $p : P \to M$ be the complex blow up of M at k points. Note that $H^2(P, \mathbf{R})$ admits a natural decomposition

$$H^2(P, \mathbf{R}) = p^* H^2(M, \mathbf{R}) \oplus \mathrm{Span}_{\mathbf{R}}\{e_1, \dots, e_k\},$$

where e_i, $i = 1, \dots, k$ are classes Poincare dual to the exceptional divisors. Set $b = p^*[\Omega]$ and $c = -(e_1 + \cdots + e_k)$. The symplectic blow up construction (see [McD-P], [McD-S2]) ensures that for $\epsilon > 0$ small enough the class $b + \epsilon c$ can be represented by a canonical (up to isotopy) symplectic form σ on P whose restriction to each exceptional divisor is a multiple of the standard form on the projective space. We are going to study deformations of σ. Consider a ray

$$l = \{b + tc,\ t \geq \epsilon\}$$

in $H^2(P, \mathbf{R})$. Let Γ be the union of the exceptional divisors.

The deformation problem in this situation is formulated as follows. One has to find the maximal possible T such that there exists a smooth family of symplectic forms ω_t, $t \in [\epsilon; T)$ on P so that $\omega_\epsilon = \sigma$, $\omega_t|_\Gamma$ is non-degenerate and

$$[\omega_t] = b + tc$$

for all $t \in [\epsilon; T)$.

This problem is closely related to symplectic packings. Namely, one can show [McD-P] that

$$v_k(M, \Omega) = \frac{kT^n}{n!\mathrm{Vol}(M)},$$

where $n = \dim_{\mathbf{C}} M$.

Example 3.C (weak coupling). Let $p : P \to S^2$ be a symplectic fibration with a closed simply connected fiber (M, Ω). Denote by a the generator of $H^2(S^2, \mathbf{Z})$, and set $b = p^* a$. Let $c \in H^2(P, \mathbf{R})$ be the unique class whose restriction on fibers coincides with $[\Omega]$ and whose top power vanishes. The weak coupling construction due to Guillemin, Lerman and Sternberg [G-L-S, McD-S2] ensures that for $\epsilon > 0$ small enough the class $b+\epsilon c$ is represented by a canonical (up to isotopy) symplectic form σ on P whose restriction to each fiber is a multiple of Ω. Consider a ray

$$l = \{b + tc,\ t \geq \epsilon\}$$

in $H^2(P, \mathbf{R})$ and choose a fiber Γ of p.

The deformation problem in this situation is formulated exactly as in the example 3.B above. It is closely related to the symplectic K-area. Namely, one can show [P1] that if T is the solution of the deformation problem, then

$$\text{K-area}(P) \leq T.$$

4. Methods of almost complex geometry

The deformation problem 3.A is a symplectic counterpart of the classical question of complex geometry about the structure of the Kähler cone of a complex manifold. Therefore it is natural to attack this problem using Gromov's theory of pseudo-holomorphic curves [G1, McD-S1] which provides us with a powerful extension of algebraic-geometric methods for symplectic purposes.

Let (P, σ) be a symplectic manifold. Denote by X the connected component of the space of all symplectic forms on P which contains σ. Consider a fibration $Y \to X$ defined as follows. The fiber over a form $\omega \in X$ is the space of all almost complex structures J on P which are *compatible* with ω, that is the expression $\omega(\xi, J\eta)$, $\xi, \eta \in TP$ is a Riemannian metric on P. One can show that each fiber is contractible.

Choose an almost complex structure J on P compatible with a symplectic form ω. A *pseudo-holomorphic curve* (or J-curve) is a mapping, say ϕ, of a Riemann surface (Σ, i) into P which satisfies the Cauchy-Riemann equation $d\phi \circ i = J \circ d\phi$. Note that if a J-curve ϕ is non-constant then ω takes a strictly positive value on the homology class represented by ϕ.

Let us return to the deformation problem 3.A. Consider a path $\{\sigma_t\}_{t\in[0;1]}$ of symplectic forms on P starting at σ. Choose a lift $\{J_t\}$ of this path to Y. Assume that a class $A \in H_2(P)$ is represented by a non-constant J_0-holomorphic curve ϕ. Now the following argument can be applied. Namely, using Gromov-Witten invariants [R, McD-S1] one can show in many interesting situations that *the curve ϕ must persist along every generic deformation.* In particular, this means that the class A is also represented by a J_1-holomorphic curve. As a consequence we get an a priori estimate $([\sigma_1], A) > 0$, which provides us with an obstruction to deformations of the symplectic form σ in cohomology.

Example 4.A An important example of a persisting pseudo-holomorphic curve is an *exceptional* 2-sphere in a symplectic 4-manifold, that is an embedded pseudo-holomorphic sphere with the self-intersection number equal to -1. In view of the discussion above, every exceptional sphere gives a formal obstruction to deformation of a symplectic form in cohomology. Consider for instance the blow up of $\mathbf{CP}^2$ at k points where $2 \leq k \leq 8$. In this situation the complete list of exceptional spheres is known (e.g. for $k = 2$ it includes exceptional divisors together with the proper transform of the line on the plane which passes through the blown up points). It follows from a version of the Nakai criterion that the corresponding list of obstructions is complete, that is all non-trivial obstructions to deformations of symplectic forms come from the exceptional spheres. Using this fact and the argument in 3.B above, it is not hard to obtain the precise solutions of the packing problem given in Fig. 1.

What is the role played by the submanifold Γ in the formulation of the deformation problem 3.A? The previous example shows that in a low dimensional situation one may solve an even more general deformation problem where one does not assume non-degeneracy of symplectic forms on some submanifold of P. However, such an additional assumption is essential for proving persistence of pseudo-holomorphic curves in higher dimensions.

We conclude this section with a remark that the interaction between Algebraic Geometry and Symplectic Topology turns out to be useful for both fields. In a recent paper [La] one can find a beautiful application of symplectic packings to a purely algebraic-geometric problem on the Seshadri constants of abelian varieties.

5. Measurements on the group of Hamiltonian diffeomorphisms

There exists a canonical notion of *distance* in Symplectic Geometry. The distance is defined not on a symplectic manifold itself, but on its group of symmetries – on the infinite-dimensional group of all Hamiltonian diffeomorphisms. This discovery due to H. Hofer yielded many interesting developments [H-Z, McD-S2]. In this section we give an application of our approach to a computation of the *length spectrum* in Hofer's geometry.

Let Ham be the (infinite-dimensional!) group of all Hamiltonian diffeomorphisms of (M, Ω). If M is simply connected, this is just the identity component of the group of all symplectomorphisms. Given a path $\{f_t\}$ in Ham ($t \in [0;1]$), define its *length* as follows. Let $F : M \times S^1 \to \mathbf{R}$ be the associated Hamiltonian function normalized so that the mean value of F_t over M vanishes for all $t \in S^1$. Then $\text{length}\{f_t\} = \max|F|$. With this language the distance between two Hamiltonian diffeomorphisms is just the minimal possible length of a path between them.

For an element $\gamma \in \pi_1(\text{Ham})$ define its norm as

$$||\gamma|| = \inf\ \text{length}\{f_t\},$$

where the infimum is taken over all loops $\{f_t\}$ with $f_0 = f_1 = \mathbb{1}$ representing γ. The next result gives a precise calculation of this norm for projective spaces.

Theorem 5.A (see [P1]). *Let* $M = (\mathbf{CP}^n, \Omega_{st})$. *Let* $\gamma \in \pi_1(\text{Ham})$ *be the element represented by the loop*

$$(z_0 : z_1 : \cdots : z_n) \to (e^{2\pi it} z_0 : z_1 : \cdots : z_n).$$

Then

$$||\gamma^q|| = \frac{\max(q, n+1-q)}{n+1},$$

for all $q \in \{1, \dots, n\}$.

There exists a loop of the minimal length which is given by a one-parametric subgroup of

$$PU(n+1) \subset \text{Ham}.$$

For the 2-sphere ($n = 1$) this result was proved earlier in [L-McD].

The proof of Theorem 5.A is based on the following estimate 5.B which establishes an interrelation between Hofer's geometry and symplectic K-area. Loops of Hamiltonian diffeomorphisms give rise to symplectic fibrations over S^2. It turns out that the K-area of these fibrations is closely related to metrical properties of loops in the framework of Hofer's geometry.

Let $\{f_t\}, t \in S^1$ be a loop of Hamiltonian diffeomorphisms of (M, Ω) which represents an element $\gamma \in \pi_1(\text{Ham})$. Let D_+ and D_- be two copies of the disc D^2

bounded by S^1. Consider a map $\Psi : M \times S^1 \to M \times S^1$ given by $(z,t) \to (f_t z, t)$. Define now a new manifold

$$P_\gamma = (M \times D_-) \cup_\Psi (M \times D_+).$$

It is clear that P_γ has the canonical structure of a symplectic fibration over S^2 which depends only on the element γ.

Theorem 5.B (see [P1]). *The following inequality holds:*

$$||\gamma|| \geq \frac{1}{\text{K-area}(P_\gamma)}.$$

6. Applications to classical mechanics

Symplectic measurements are used in various problems of classical mechanics. We illustrate this by showing a link between mechanics and Hofer's geometry, and present a result (see Theorem 6.A below) which is based on the approach discussed above. The underlying philosophy is as follows. Mechanics studies the behavior of a Hamiltonian flow f_t on a manifold M. In other words, we want to get information about a complicated tangle of trajectories. In Hofer's geometry the picture is in some sense simpler. Our flow can be interpreted as a *single* curve $\{f_t\}$ on the group Ham of all Hamiltonian diffeomorphisms of M. One can suggest that the geometric properties of this curve (say, its deviation from a minimal geodesic) reflect some dynamical properties of the flow. A rigorous result in this direction concerning the destruction of KAM tori in a specific model was obtained in [B-P2]. This result was recently generalized by K.F. Siburg [Si] who found an interesting interrelation between Hofer's geometry and Mather's minimal measures.

Another example is Arnold's empirical principle [A] that flows which twist in the positive direction correspond to essentially long paths on the group. This principle was confirmed in certain situations by Eliashberg-Ratiu [E-R] and by Bialy-Polterovich [B-P1] under the assumption that the symplectic form is exact. Here the relation between length and twisting can already be seen on the level of the Lie algebra of the group Ham. In the non-exact case, some global argument is needed. A result in this direction was obtained recently in [P2]. Its precise formulation is as follows.

Let (M, Ω) be a closed simply connected *monotone* symplectic manifold, that is $[\Omega] = \kappa c_1(TM)$ for some $\kappa > 0$. As before, denote by Ham the group of Hamiltonian diffeomorphisms of M. Define a natural homomorphism

$$I : \pi_1(\text{Ham}) \to \mathbf{R}$$

which is built up from two well-known ingredients – the symplectic action and the Maslov index – as follows.

Let $\{f_t\}$ be a loop of Hamiltonian diffeomorphisms of M with $f_0 = f_1 = \mathbb{1}$, which is generated by a normalized 1-periodic Hamiltonian function $F : M \times S^1 \to \mathbf{R}$. Take a point $z \in M$, and consider its orbit $x(t) = f_t z$. The map $x : S^1 \to M$ can be extended to a map $\bar{x} : D^2 \to M$. Trivializing the bundle $\bar{x}^*(TM)$, one can define in the standard way the Maslov index $\mu(\bar{x})$ of the linearized flow $\{f_{t*}\}_{t\in[0;1]}$ along the orbit. Let

$$\mathcal{A}(\bar{x}) = \int_{D^2} \bar{x}^*\Omega - \int_0^1 F(x(t),t)dt$$

be the symplectic action. It is not hard to check that the quantity

$$I = \mathcal{A}(\bar{x}) - \frac{\kappa}{2}\mu(\bar{x})$$

depends only on the homotopy class of the loop $\{f_t\}$ in $\pi_1(\text{Ham})$. Moreover, the mapping $I : \pi_1(\text{Ham}) \to \mathbf{R}$ is in fact *a homomorphism.*

Notice that the symplectic action and the Maslov class were used in [E-R] and [B-P] respectively in order to describe the twisting. Therefore it is natural to suggest that the homomorphism I is a natural "global" measure of twisting of Hamiltonian loops on monotone symplectic manifolds. Our next result confirms Arnold's principle in this setting.

Theorem 6.A (see[P2]). *Let (M,Ω) be a monotone closed simply connected symplectic manifold. Then for all $\gamma \in \pi_1(\text{Ham})$ the following inequality holds:*

$$||\gamma|| \geq |I(\gamma)|.$$

The proof of this theorem goes along the following lines. Applying Theorem 5.B we reduce the statement to a deformation problem as in Example 3.C above. Then one can get the needed estimate using persisting pseudo-holomorphic curves as sketched in Section 4. The existence of such curves follows from a recent result due to P. Seidel [Se] on the canonical action of $\pi_1(\text{Ham})$ on the quantum cohomology ring of M.

Acknowledgments. I thank Paul Biran, Viktor L. Ginzburg, Dusa McDuff and Karl Friedrich Siburg for their help in the preparation of this paper.

References

[A] V.I. Arnold, First steps in Symplectic Topology, *Russ. Mat. Surveys* **41** (1986), 1–21.

[B-P1] M. Bialy and L. Polterovich, Optical Hamiltonian functions, in *Geometry in PDE*, A. Prastaro and Th.M. Rassias eds., World Sc. Publ. Co. (1994), pp. 32–50.

[B-P2] M. Bialy and L. Polterovich, Invariant tori and Symplectic Topology, *Amer. Math. Soc. Transl.* (2) **171** (1996), 23–33.

[Bi] P. Biran, Symplectic packing in dimension 4, *Geom. and Funct. Analysis*, **7** (1997), 420–437.

[E-R] Y. Eliashberg and T. Ratiu, The diameter of the symplectomorphism group is infinite, *Invent. Math.* **103** (1991), 327–340.

[G1] M. Gromov, Pseudo-holomorphic curves in symplectic manifolds, *Invent. Math.* **82** (1985), 307–347.

[G2] M. Gromov, Positive curvature, macroscopic dimension, spectral gaps and higher signatures, in *Functional Analysis on the eve of the 21st century*, S. Gindikin, J. Lepowsky, R. Wilson eds., Birkhäuser, 1996.

[G-L-S] V. Guillemin, E. Lerman and S. Sternberg, *Symplectic fibrations and multiplicity diagrams*, Cambridge University Press, 1996.

[H-Z] H. Hofer and E. Zehnder, *Symplectic invariants and Hamiltonian dynamics*, Birkhäuser, 1994.

[La] R. Lazarsfeld, Lengths of periods and the Seshadri constants of abelian varieties, *Math. Research Letters* **3** (1996), 439–447.

[L-McD] F. Lalonde and D. McDuff, Hofer's L^∞-geometry: energy and stability of flows II, *Invent. Math.* **122** (1995), 35–69.

[McD-P] D. McDuff and L. Polterovich. Symplectic packings and algebraic geometry, *Invent. Math.* **115** (1994), 405–429.

[McD-S1] D. McDuff and D. Salamon, *J-holomorphic curves and quantum cohomology*, AMS, 1994.

[McD-S2] D. McDuff and D. Salamon, *Introduction to Symplectic Topology*, Clarendon Press, Oxford, 1995.

[P1] L. Polterovich, Gromov's K-area and symplectic rigidity, *Geom. and Funct. Analysis* **6** (1996), 726–739.

[P2] L. Polterovich, Hamiltonian loops and Arnold's principle, Amer. Math. Soc. Transl. (2) **180** (1997), 181–187.

[R] Y. Ruan, Topological sigma model and Donaldson type invariants in Gromov theory, *Duke Math. Journ.* **83** (1996), 461–500.

[Se] P. Seidel, π_1 of symplectic automorphisms groups and invertibles in quantum homology rings, *Geom. and Funct. Anal.* **7** (1997), 1046–1095.

[Si] K.F. Siburg, Action minimizing measures and the geometry of the Hamiltonian diffeomorphism group, to appear in *Duke. Math. Journ.*

Note added in proofs: More information about interrelations between Hofer's geometry, K-area and coupling can be found in a paper

L. Polterovich, Symplectic aspects of the first eigenvalue,
to appear in *Crelle's Journal.*

Progress in Mathematics, Vol. 169, © 1998 Birkhäuser Verlag Basel/Switzerland

Nonlinear Partial Differential Equations, Birkhoff Normal Forms, and KAM Theory

JÜRGEN PÖSCHEL

Universität Stuttgart, Mathematisches Institut A,
Pfaffenwaldring 57, D–70569 Stuttgart
poschel@mathematik.uni-stuttgart.de

Introduction

The purpose of this talk is twofold. First, I want to give another example of how tools and techniques which are well proven in the world of finite dimensional dynamical systems may be applied in the world of infinite dimensional evolution equations. In this case the tool is the so-called Birkhoff normal form of Hamiltonian mechanics, which allows Hamiltonian systems near an equilibrium to be viewed as small perturbations of integrable systems. In the infinite dimensional world, such a normal form allows us to view certain nonlinear evolution equations not only as small perturbations of integrable *partial* differential equation, but also as small perturbations of infinite dimensional, integrable *ordinary* differential equations. In addition, the calculations involved are rather elementary.

Second, such normal forms enable one to apply, in a rather effortless way, an infinite dimensional extension of the classical KAM theory and thus establish the existence of large families of time-quasi-periodic solutions which are linearly stable. This may help to explain the numerical observations of recurrent, non-ergodic behaviour for those evolution equations. Here, we continue work of Sergej Kuksin, which he described at the first ECM in Paris in 1992.

1. Nonlinear Schrödinger and wave equations

As a first example, consider the nonlinear Schrödinger equation

$$\mathrm{i}u_t = u_{xx} - mu - f(|u|^2)u \tag{1}$$

on the bounded interval $[0, \pi]$ with Dirichlet boundary conditions

$$u(0,t) = 0 = u(\pi, t), \qquad -\infty < t < \infty.$$

Here, m is a real parameter, and f is a real analytic function around 0 in $\mathbb{C}$ such that

$$f(|u|^2)u = a\,|u|^2\,u + \cdots, \qquad a \neq 0.$$

So after rescaling u we have

$$\mathrm{i}u_t = u_{xx} - mu \mp |u|^2\,u + O_5(u).$$

This equation is Hamiltonian. As the phase space, take $H_0^1([0,\pi])$, the Sobolev space of all complex valued L^2-functions on $[0,\pi]$ with an L^2-derivative and vanishing boundary values. As the Hamiltonian, take

$$H = \frac{1}{2}\langle Au, u\rangle + \frac{1}{2}\int_0^\pi F(|u|^2)\,\mathrm{d}x,$$

where $A = -\mathrm{d}^2/\mathrm{d}x^2 + m$, F is the primitive of f with, say, $F(0) = 0$, and $\langle u, v\rangle = \mathrm{Re}\int_0^\pi u\bar{v}\,\mathrm{d}x$. Then (1) is equivalent to

$$\dot{u} = \mathrm{i}\nabla H(u),$$

where the gradient is taken with respect to $\langle\cdot,\cdot\rangle$. The underlying symplectic structure is $[u,\tilde{u}] = \langle \mathrm{i}u, \tilde{u}\rangle$.

As a second example, consider the nonlinear wave equation

$$u_{tt} = u_{xx} - mu - f(u) \tag{2}$$

on the same bounded interval $[0,\pi]$ with Dirichlet boundary conditions. In this case, $m > 0$, and f is a real analytic, *odd* function such that

$$f(u) = au^3 + O_5(u), \qquad a \neq 0.$$

This class of equations comprises the sine-Gordon, the sinh-Gordon and the ϕ^4-equation, given by

$$mu + f(u) = \begin{cases} \sin u \\ \sinh u \\ u + u^3 \end{cases},$$

respectively, as well as small perturbations of them of order five and more.

In this case, the phase space may be taken as $H_0^1([0,\pi]) \times L^2([0,\pi])$ with real coordinates u and $v = u_t$. As the Hamiltonian, take

$$H = \frac{1}{2}\langle v, v\rangle + \frac{1}{2}\langle Au, u\rangle + \int_0^\pi F(u)\,\mathrm{d}x,$$

where A and F are as above, and $\langle\cdot,\cdot\rangle$ is the standard L^2-product. Then (2) is equivalent to

$$\dot{u} = \frac{\partial H}{\partial v} = v, \qquad \dot{v} = -\frac{\partial H}{\partial u} = -Au - f(u).$$

The underlying symplectic structure is $[(u,v),(\tilde{u},\tilde{v})] = \langle u,\tilde{v}\rangle - \langle \tilde{u}, v\rangle$.

We are going to study solutions of these equations of small amplitude. In a first approximation this allows us to consider the nonlinear terms as small perturbations of the linear Schrödinger and the Klein-Gordon equation,

$$\mathrm{i}u_t = u_{xx} - mu \qquad \text{and} \qquad u_{tt} = u_{xx} - mu, \tag{3}$$

respectively. These are, of course, completely understood. Every solution is the superposition of the harmonic oscillations of their basic modes $\phi_k = \sin kx$, $k \geq 1$, which oscillate with *fixed* frequencies

$$\lambda_k = k^2 + m \qquad \text{and} \qquad \lambda_k = \sqrt{k^2 + m},$$

and arbitrary amplitudes. For instance,

$$u(x,t) = \sum_{k\geq 1} a_k e^{\mathrm{i}\lambda_k t}\phi_k(x), \qquad \sum_{k\geq 1} k^2 |a_k|^2 < \infty,$$

is the general solution in the Schrödinger case. The combined motion of the modes is periodic, quasi-periodic or almost-periodic, depending on whether one, finitely many, or infinitely many modes are excited. Moreover, the whole phase space is completely filled by these types of solutions. That is, the linear systems are completely integrable.

Upon restoring the nonlinear terms this foliation will not persist in its entirety, due to the long range coupling effects of the nonlinearities, and resonances, or near resonances, among the different modes. This applies even to the families of time-periodic solutions. One may only hope to prove the persistence of a *large portion* of it in the regime of small amplitude solutions, using some extension of KAM theory to infinite dimensional systems. Integrable linear systems, however, are totally *degenerate* from the point of view of KAM theory, as they show no frequency amplitude modulation whatsoever – all the frequencies are fixed. Hence, KAM theory is not applicable at this stage.

This difficulty may be circumvented by replacing the scalar parameter m by a potential $Q(x,\xi)$, which depends on sufficiently many external parameters $\xi = (\xi_1, \ldots, \xi_n)$. They allow the frequencies of the basic modes to be varied and may thus replace the usual nondegeneracy condition of KAM theory. As a result one obtains large Cantor sets of parameters for which there exist quasi-periodic solutions of small amplitude. This approach was taken by Wayne [35] and in some cases by Kuksin – see for example [22, 24, 25]. However, these Cantor sets do not include any open interval of constant potentials $Q \equiv m$ due to infinitely many nonresonance conditions imposed on the frequencies λ_k. Also, the number of parameters needed grows with the number of independent frequencies of the quasi-periodic solutions considered.

But linear systems are not the only integrable approximations that are available. There are also *nonlinear* integrable pde. In the case of the Schrödinger equation this is the Zakharov-Shabat equation

$$\mathrm{i}u_t = u_{xx} - mu - |u|^2 u$$

on the real line with periodic boundary conditions [38]. In the case of the wave equation this is the sine-Gordon or the sinh-Gordon equation with periodic boundary conditions on the real line. So it is natural to consider (1) and (2) as perturbations of such nonlinear integrable pde's. This approach was taken by Kuksin and his collaborators – see [3, 6].

This approach, however, entails a lot of technical difficulties. First, a detailed knowledge of the solutions of these integrable pde is required, for example through explicit representations by theta-functions, to obtain the variational equations of motion along unperturbed quasi-periodic solutions and to bring these into constant coefficient form. Second, these coefficients have to satisfy certain nondegeneracy conditions in order to apply KAM, and this is quite difficult to verify. All this can be done, but the amount of work is formidable – see also [3, 4, 5, 6].

It is the purpose of this talk to describe an alternative approach via nonlinear integrable *ode*, which by comparison is short and elementary. This approach was first proposed in [29]. For the Schrödinger equation this was carried out by Kuksin and the author [28], and for the wave equation by the author [30]. The first step is to rewrite the nonlinear pde as an infinite dimensional system of ode's by introducing infinitely many coordinates. To simplify notation, we now focus our attention on the nonlinearities $\mp |u|^2 u$ and $\mp u^3$, respectively, since terms of higher order will not make any difference.

Using Fourier's classical approach we make the *ansatz* for the Schrödinger equation

$$u = \sum_{k\geq 1} q_k(t)\phi_k(x).$$

The time dependent coefficients are taken from some Hilbert space $\ell^s_{\mathbb{C}}$, $s \geq 1$, of complex sequences $q = (q_1, q_2, \ldots)$ with $\sum_{k\geq 1} k^{2s} |q_k|^2 < \infty$. One then obtains a Hamiltonian system on $\ell^s_{\mathbb{C}}$ with the Hamiltonian

$$\begin{aligned} H &= \Lambda + G \\ &= \frac{1}{2}\sum_{k\geq 1} \lambda_k |q_k|^2 \pm \frac{1}{4}\sum_{i,j,k,l} G_{ijkl} q_i q_j \bar{q}_k \bar{q}_l, \end{aligned} \tag{4}$$

where

$$\lambda_k = k^2 + m, \qquad G_{ijkl} = \int_0^\pi \phi_i\phi_j\phi_k\phi_l \, \mathrm{d}x,$$

and equations of motion

$$\dot{q}_k = 2\mathrm{i}\frac{\partial H}{\partial \bar{q}_k} = \mathrm{i}\lambda_k q_k + \ldots, \qquad k \geq 1.$$

These are the classical Hamiltonian equations of motion for the real and imaginary parts of $q_k = x_k + \mathrm{i}y_k$ written in complex notation. The underlying symplectic structure is $\frac{\mathrm{i}}{2}\sum_k dq_k \wedge d\bar{q}_k$.

The quadratic term Λ describes the linear integrable Schrödinger equation and gives rise to a linear Hamiltonian vector field which is *unbounded* of order 2. The fourth order term G is not integrable, but gives rise to a *bounded* vector field on $\ell^s_{\mathbb{C}}$ of order 0.

The same approach to the wave equation requires only a little more notation. Here we write

$$u = \sum_{k\geq 1} \frac{q_k(t)}{\sqrt{\lambda_k}}\phi_k(x), \qquad v = u_t = \sum_{k\geq 1} \sqrt{\lambda_k} p_k(t)\phi(x),$$

taking $(q, p) = (q_1, q_2, \ldots, p_1, p_2, \ldots)$ from some real Hilbert space $\ell^s \times \ell^s$ with $s \geq \frac{1}{2}$. One obtains the Hamiltonian

$$\begin{aligned} H &= \Lambda + G \\ &= \frac{1}{2}\sum_{k\geq 1} \lambda_k(p_k^2 + q_k^2) \pm \frac{1}{4}\sum_{i,j,k,l} G_{ijkl} q_i q_j q_k q_l, \end{aligned} \tag{5}$$

where

$$\lambda_k = \sqrt{k^2 + m}, \qquad G_{ijkl} = \frac{1}{\sqrt{\lambda_i \lambda_j \lambda_k \lambda_l}} \int_0^\pi \phi_i \phi_j \phi_k \phi_l \,\mathrm{d}x,$$

and equations of motion

$$\dot{q}_k = \frac{\partial H}{\partial p_k} = \lambda_k p_k, \qquad \dot{p}_k = -\frac{\partial H}{\partial q_k} = -\lambda_k q_k - \frac{\partial G}{\partial q_k}.$$

The underlying symplectic structure is $\sum_k dq_k \wedge dp_k$. In this case the linear integrable vector field is unbounded of order 1 on $\ell^s \times \ell^s$, whereas the nonlinear vector field is bounded of order -1.

In both cases we are now dealing with a Hamiltonian system of infinitely many degrees of freedom near an *elliptic* equilibrium located at the origin. In the classical theory, the standard tool to investigate such systems is their *Birkhoff normal form*, or its generalizations. Remarkably, such a normal form is available here without any further assumption. Moreover, its coefficients are easily determined.

To this end one first notices that in both cases the fourth order coefficients satisfy

$$G_{ijkl} = 0 \quad \Leftrightarrow \quad i \pm j \pm k \pm l \neq 0,$$

for any combination of plus and minus signs. Next, in the Schrödinger case, one has the following elementary nonresonance of order four among the frequencies at the equilibrium.

LEMMA 1.1 ([28]) *If $i \pm j \pm k \pm l = 0$, but $\{i,j\} \neq \{k,l\}$, then for $\lambda_k = k^2 + m$ one has*

$$\lambda_i + \lambda_j - \lambda_k - \lambda_l = i^2 + j^2 - k^2 - l^2 \neq 0.$$

This allows us to define a convergent normalizing coordinate transformation in exactly the same way as in the classical case by comparison of coefficients, since these are the only divisors that show up [12]. Thus, from the Hamiltonian (4) we can remove all terms of order four which are not integrable. The upshot is the following result.

THEOREM 1.2 ([28]) *There exists a real analytic, symplectic coordinate transformation Φ in a neighbourhood of the origin in $\ell^s_{\mathbb{C}}$ that for all m takes the Hamiltonian (4) of the nonlinear Schrödinger equation into its Birkhoff normal form of order four,*

$$H \circ \Phi = \frac{1}{2} \sum_{k \geq 1} \lambda_k \left|q_k\right|^2 + \frac{1}{2} \sum_{k,l \geq 1} Q_{kl} \left|q_k\right|^2 \left|q_l\right|^2 + O_6(q),$$

with uniquely determined coefficients

$$Q_{kl} = \frac{2 - \delta_{kl}}{2} G_{klkl} = \frac{4 - \delta_{kl}}{4\pi}.$$

The order of the corresponding nonlinear vector fields is still 0.

Thus, we obtain an infinite dimensional Hamiltonian system which is completely integrable up to order four in the classical sense.

The corresponding result for the wave equation is somewhat less complete due to asymptotic resonances among its frequencies λ_k. Here, the following estimate holds.

LEMMA 1.3 ([30]) *If i, j, k, l are positive integers, no two of which are equal such that $i \pm j \pm k \pm l = 0$, then for $\lambda_k = \sqrt{k^2 + m}$ one has*

$$|\lambda_i \pm \lambda_j \pm \lambda_k \pm \lambda_l| \geq \frac{cm}{\sqrt{(n^2+m)^3}}, \qquad n = \min(i, j, k, l),$$

for arbitrary combinations of plus and minus signs with some absolute constant c. Hence the left-hand side is uniformly bounded away from zero on every compact m-interval in $(0, \infty)$.

As a consequence, we obtain the following partial normal form result. Let $z_k = \frac{1}{\sqrt{2}}(q_k + \mathrm{i}p_k)$ and $\bar{z}_k = \frac{1}{\sqrt{2}}(q_k - \mathrm{i}p_k)$. Then $z \in \ell^s_{\mathbb{C}}$.

THEOREM 1.4 ([30]) *For each $n \geq 1$ and $m > 0$ there exists a real analytic, symplectic coordinate transformation Φ_n in some neighbourhood of the origin in $\ell^s_{\mathbb{C}}$ that takes the Hamiltonian (5) into*

$$H \circ \Phi_n = \sum_{k \geq 1} \lambda_k |z_k|^2 + \frac{1}{2} \sum_{\min(k,l) \leq n} Q_{kl} |z_k|^2 |z_l|^2 + O_4(z') + O_6(z),$$

with uniquely determined coefficients

$$Q_{kl} = 12(2 - \delta_{kl}) G_{klkl} = \frac{6}{\pi} \cdot \frac{4 - \delta_{kl}}{\lambda_k \lambda_l}$$

and $z' = (z_{n+1}, z_{n+2}, \ldots)$. The order of the associated nonlinear vector fields is still -1.

Moreover, the neighbourhood can be chosen uniformly for every compact m-interval in $(0, \infty)$, and the dependence of Φ_n on m is real analytic.

The first two terms of the normal form describe again an infinite dimensional integrable Hamiltonian system. Note that the interaction of the first n modes with *all* modes is normalized, but not the interaction of the higher modes with each other. Incidentally, it is possible to produce a complete normal form as in Theorem 1.2, but only at the expense of increasing the order of the nonlinear vector fields by 1. To such a normal form, the KAM theorem of section 4 below is *not* applicable.

We are now in a position to study the fate of quasi-periodic solutions, where only a finite number of modes is excited. Their analogue in the pde approach are the so-called finite-gap-solutions. Suppose for simplicity that we want to excite the first n modes, $1 \leq n < \infty$. Introduce action-angle-coordinates for them by writing

$$q_k = z_k = \sqrt{I_k}\, e^{\mathrm{i}\varphi_k}, \qquad 1 \leq k \leq n.$$

In both cases we obtain a Hamiltonian $H = H_0 + H_1$ in coordinates (φ, I, z') in $\mathbb{T}^n \times \mathbb{R}^n \times \ell^s_{\mathbb{C}}$, where

$$H_0 = \sum_{1\le k\le n} \lambda_k I_k + \frac{1}{2} \sum_{1\le k,l\le n} Q_{kl} I_k I_l + \sum_{k\ge n+1} \lambda_k \left|z_k\right|^2 + \sum_{1\le k\le n<l} Q_{kl} I_k \left|z_l\right|^2 \quad (6)$$

and

$$H_1 = O_4(z') + O_6(z).$$

H_0 describes a *nonlinear* integrable Hamiltonian system with equations of motion

$$\begin{aligned} \dot\varphi_k &= \lambda_k + \sum_{1\le l\le n} Q_{kl} I_l + \dots, \\ \dot I_k &= 0, \\ \dot z_k &= \mathrm{i}\lambda_k z_k + \dots. \end{aligned}$$

There is an n-parameter family of invariant tori

$$\mathbb{T}^n \times \{I\} \times \{0\}, \qquad I \in \mathbb{R}^n_+,$$

carrying linear flows with amplitude-dependent frequencies, along which the variational equations of motion have constant coefficients Q_{kl} that are known explicitly. Note that this family is *different* from the analogous family for the quadratic Hamiltonian $\sum_k \lambda_k \left|z_k\right|^2$, where there is *no* frequency modulation.

To this Hamiltonian H_0 an infinite dimensional version of the KAM theorem given by Kuksin [25] or the author [31] can be applied. The nondegeneracy conditions for H_0 are easily checked, and H_1 may be regarded as a small perturbation of H_0, when the amplitudes are sufficiently small. As a result, for each n one obtains a Cantor family of time-quasi-periodic solutions of the nonlinear Schrödinger and wave equations with n independent frequencies, filling n-dimensional invariant tori in phase space which are linearly stable. The relative $2n$-dimensional Hausdorff measure of all these tori approaches 1 as one approaches the equilibrium $u \equiv 0$. For a more detailed and complete statement of the conclusions we refer to [28, 30].

To sum up, the key ingredients of this approach are the following. To get started the linear operator involved – in this case $-\mathrm{d}^2/\mathrm{d}x^2 + m$ with Dirichlet boundary conditions – has to have a pure point spectrum with a complete set of eigenfunctions. The point eigenvalues have to avoid certain low order resonances so that the transformation into Birkhoff normal form is defined. This, for example, makes it necessary that the cubic nonlinearities

$$\pm \left|u\right|^2 u \qquad \text{and} \qquad \pm u^3$$

do not depend explicitly on x; otherwise, there were non-integrable resonant terms which can not be transformed away for all parameter values of m. Their sign, however, is completely irrelevant. So it is not necessary to distinguish between the focusing and the defocusing cases.

The nonresonance conditions are reminiscent of Lyapunov's center theorem in the finite-dimensional theory: if, at an elliptic equilibrium of a real analytic Hamiltonian with characteristic frequencies $\lambda_1, \dots, \lambda_n$, one has

$$\frac{\lambda_k}{\lambda_n} \notin \mathbb{Z}, \qquad 1 \le k \le n-1,$$

then there exists a disc through the equilibrium filled with periodic solutions with frequencies λ approaching λ_n as they approach the equilibrium. The nonresonance conditions are necessary: given a resonance there are nonlinearities so that such a family of periodic solutions does not exist [33].

For infinite dimensional systems there are similar results concerning the *non-persistence of breathers* under generic perturbations. See for example [16] and [34], respectively, for recent results concerning the sine-Gordon equation and certain classes of nonlinear wave equations. Very loosely speaking, the resonance occurs between the point eigenvalue of the unperturbed periodic solution – the breather – and the continuous spectrum of the unperturbed operator. These results indicate that to require a pure point spectrum not in low order resonance is not a technical shortcoming, but essential for persistence results of the kind above.

Some further restrictions arise from the requirements of the KAM theorem itself. First, the characteristic frequencies λ_k of the equilibrium have to grow at least linearly:

$$\lambda_k \sim k^d, \qquad d \geq 1.$$

This restricts the existence results essentially to problems with a 1-dimensional x-space. Second, the frequencies also have to be *simple*. This restricts one essentially to Sturm-Liouville boundary conditions, and excludes periodic boundary conditions, where eigenvalues are asymptotically double.

This restriction is more of a technical nature. Recently, Craig & Wayne [14] extended the Lyapunov center theorem to infinite dimensions and constructed Cantor discs of time-periodic solutions of a nonlinear wave equation with periodic boundary conditions. They used a Lyapunov-Schmidt reduction scheme together with small divisor estimates in the spirit of KAM theory. Their approach was then extended considerably by Bourgain, who obtained not only quasi-periodic solutions for Schrödinger equations [9], but also periodic and quasi-periodic solutions for some *two*-dimensional Schrödinger and wave equation [7, 8, 11]. With this technique, double eigenvalues are permitted. On the other hand, one does *not* obtain the linear stability of the solutions so constructed.

The results described so far concern the existence of quasi-periodic solutions filling finite-dimensional invariant tori in an infinite dimensional phase space. Nothing is known, however, about the existence of *almost-periodic* solutions. On one hand, the generic solution of the linear equations (3) are almost-periodic, with infinitely many excited modes and infinitely many independent frequencies. On the other hand, the nonlinearities effect a strong, long range coupling among all of them. This effect is beyond the control of the current techniques.

There are, however, existence results for simplified problems. Bourgain [10] considered the Schrödinger equation

$$\mathrm{i}u_t = u_{xx} - V(x)u - |u|^2 u$$

on $[0, \pi]$ with Dirichlet boundary conditions, depending on some analytic potential V. Given an almost-periodic solution of the linear equation with *very* rapidly decreasing amplitudes and nonresonant frequencies, he showed that the potential V may be modified so that this solution persists for the nonlinear equation.

Thus, the potential serves as an infinite dimensional parameter, which has to be chosen properly for each initial choice of amplitudes. This result is obtained by iterating the Lyapunov-Schmidt reduction introduced by Craig & Wayne.

A similar result was obtained independently by the author for the equation

$$\mathrm{i}u_t = u_{xx} - V(x)u - N(u),$$

where

$$N(u) = \Psi\big(f(|\Psi u|^2)\Psi u\big),$$

with $\Psi\colon u \mapsto \psi * u$ a convolution operator which is smoothing of order greater than 1/4. This smoothing makes it possible to *iterate* the KAM theorem about the existence of quasi-periodic solutions. As a result, one obtains for – in a suitable sense – almost all potentials V a set of almost periodic solutions, which – again in a suitable sense – has density one at the origin. See [32] for more details.

2. Perturbed KdV equations

Now we consider small Hamiltonian perturbations of the best known nonlinear integrable pde, the KdV equation. That is, we consider the equation

$$u_t = u_{xxx} - 6uu_x + \varepsilon\partial_x f(u) \tag{7}$$

with periodic boundary conditions,

$$u(x+1,t) = u(x,t),$$

where f is some real analytic function on the real line. For simplicity of exposition, we may restrict ourselves to the case of zero mean solutions,

$$[u] = \int_0^1 u(x,t)\,\mathrm{d}x = 0,$$

although the general case presents no additional difficulties.

The KdV equation belongs to a whole family of integrable pde. For instance, we may also consider perturbations of the second KdV equation,

$$u_t = u_{xxxxx} - 10uu_{xxx} - 20u_xu_{xx} + 30u^2u_x + \varepsilon\partial_x f(u), \tag{8}$$

subject to the same boundary and mean value conditions. We will consider both equations simultaneously.

These equations are Hamiltonian. As a phase space take the Sobolev space $H_0^s(S^1)$, $s \geq 1$, of real valued functions on $\mathbb{R}$ with period 1 and mean value zero, endowed with the Poisson structure due to Gardner,

$$\{F,G\} = \int_0^1 \frac{\partial F}{\partial u}\frac{\mathrm{d}}{\mathrm{d}x}\frac{\partial G}{\partial u}\,\mathrm{d}x.$$

The Hamiltonians corresponding to the above equations are then

$$H^1(u) = \int_0^1 \big(\tfrac{1}{2}u_x^2 + u^3 + \varepsilon F(u)\big)\,\mathrm{d}x$$

and

$$H^2(u) = \int_0^1 \left(\frac{1}{2}u_{xx}^2 + 5uu_x^2 + \frac{5}{2}u^4 + \varepsilon F(u)\right) \mathrm{d}x,$$

where F is the primitive of f with $F(0) = 0$. The above evolution equations are then equivalent to

$$\dot{u} = \frac{\mathrm{d}}{\mathrm{d}x}\frac{\partial H^\iota}{\partial u}$$

for $\iota = 1$ and $\iota = 2$, respectively.

Note that for the first KdV equation, the linear operator is unbounded of order 3, while the nonlinear operator is unbounded of order 1. For the second KdV equation, these numbers are 5 and 3, respectively. So, in contrast to the nonlinear Schrödinger and wave equations considered in section 1, the nonlinearities are also given by *unbounded* operators. This causes some complications in the KAM theory, as we will indicate below.

Results for the perturbed KdV equation were first obtained by Kuksin [23, 27]. He showed that the so-called n-gap solutions persist for sufficiently small ε, where the bound on ε depends on the solution considered, in particular its number of gaps. To apply the KAM theory, however, he had to go a long way. First, the Its-Matveev formula for periodic solutions and the Zakharov-Akhiezer-Ψ-functions were needed to obtain the variational equations of motion along n-gap solutions and to put them into constant coefficient form. Then, further investigations based on a beautiful result of Krichever [21] were needed to verify the nondegeneracy and nonresonance conditions required by the KAM theorem.

The use of Birkhoff normal forms provides an alternative, elementary approach to the same results, at least for *small* solutions. The following presents joint work of Kappeler and the author [20]. Write

$$u = \sum_{k \neq 0} \gamma_k q_k e^{2\pi \mathrm{i} kx} = \Psi q \tag{9}$$

with

$$\gamma_k = \sqrt{2\pi\,|k|}, \qquad \bar{q}_k = q_{-k},$$

and $q = (q_1, q_2, \ldots) \in \ell_{\mathrm{C}}^{s+1/2}$. The KdV Hamiltonian for $\varepsilon = 0$ in these new coordinates is

$$\begin{aligned} H^1(q) &= \Lambda + G \\ &= \sum_{k \geq 1} (2\pi k)^3 \,|q_k|^2 + \sum_{k+l+m=0} \gamma_k\gamma_l\gamma_m q_k q_l q_m. \end{aligned}$$

The equations of motion are the classical ones in complex coordinates:

$$\dot{q}_k = \mathrm{i}\frac{\partial H^1}{\partial \bar{q}_k}, \qquad k \geq 1.$$

The second KdV Hamiltonian is also easy to write down. It just contains more terms and will be given below in normal form.

To transform the first Hamiltonian into its Birkhoff normal form, two coordinate transformations are required, one to eliminate the cubic term, and one to normalize the resulting fourth order term. Both calculations are elementary. The pertinent nonresonance conditions are satisfied, since first, no three cubics add up to zero, and second, an analogue of Lemma 1.1 for four cubics holds. The result is the following.

THEOREM 2.1 ([20]) *There exists a real analytic, symplectic coordinate transformation* Φ *in a neighbourhood of the origin in* $\ell_{\mathbb{C}}^{s+1/2}$, *which transforms the first KdV Hamiltonian into*

$$H^1 \circ \Phi = \sum_{k\geq 1} (2\pi k)^3 \left|q_k\right|^2 - 3 \sum_{k\geq 1} \left|q_k\right|^4 + O_5(q)$$

and preserves the order of the nonlinear vector fields. This transformation satisfies $\Phi(0) = 0$, $d_0\Phi = \mathrm{id}$.

Note that the quartic term is in diagonal form.

The Hamiltonians of the KdV family are all in involution with each other. As a consequence, this transformation Φ not only normalizes the first KdV Hamiltonian up to order four, but indeed simultaneously normalizes *every other* KdV Hamiltonian up to order four as well. This considerably simplifies the calculations. To find the normal form of the second KdV Hamiltonian, we only have to calculate *one* Poisson bracket, since the coefficients of Φ are known already. Compare also Ito [18] for a sharper version in finite dimensions.

THEOREM 2.2 ([20]) *The transformation* Φ *of the above theorem transforms every Hamiltonian of the KdV family into its Birkhoff normal form up to order four, preserving the order of the nonlinear vector fields. In particular, for the second KdV Hamiltonian one finds*

$$H^2 \circ \Phi = \sum_{k\geq 1} (2\pi k)^5 \left|q_k\right|^2 + \sum_{k,l\geq 1} Q_{kl} \left|q_k\right|^2 \left|q_l\right|^2 + O_5(q)$$

with uniquely determined coefficients

$$Q_{kl} = 20\pi^2 (6 - 7\delta_{kl})\, kl.$$

These normal forms are sufficient to prove the persistence of quasi-periodic solutions of sufficiently small amplitude under small Hamiltonian perturbations of the equations as given in (7) and (8). Moreover, if the perturbing terms are of order three or more in u, no extra parameter ε is needed to make the perturbing terms small. The nonresonance and nondegeneracy conditions are easily checked with the explicit coefficients at hand, so that the KAM theorem as given below can be applied in the usual way.

More is needed, however, to study large solutions as well. Here, the global action-angle coordinates constructed by Kappeler *et al.* [1, 2, 19] provide a very convenient handle. In our context, their result can be formulated as follows.

THEOREM 2.3 ([1, 2, 19]) *There exists a real analytic, symplectic diffeomorphism*

$$\Omega\colon\ \ell_{\mathbb{C}}^{3/2} \to H_0^1(S^1), \quad q \mapsto u,$$

with a real analytic inverse, such that the KdV Hamiltonian (7) is transformed into a function of the amplitudes $|q|^2 = (|q_1|^2, \ldots)$ *alone:*

$$H^1 \circ \Omega = \hat{H}^1(|q|^2).$$

The same holds for every other Hamiltonian in the KdV family.

This transformation satisfies

$$\Omega(0) = 0, \qquad d_0\Omega = \Psi,$$

where Ψ *is the transformation in* (9). *Moreover, the restriction of* Ω *to each subspace* $\ell_{\mathbb{C}}^{s+1/2}$, $s \geq 1$, *is a bianalytic diffeomorphism between* $\ell_{\mathbb{C}}^{s+1/2}$ *and* $H_0^s(S^1)$.

We may view Ω as a global transformation taking H^1 into a complete global Birkhoff normal form, whereas our simple transformation $\Psi \circ \Phi$ is a local transformation into a local Birkhoff normal form of order four only. Both transformations, however, have the *same* linearization at the origin. The uniqueness of the Birkhoff normal form then implies that the two normal forms must agree up to order four. In other words, the local results provides us with the first terms of the Taylor series expansion of the globally integrable Hamiltonians. In the following we drop the $\hat{}$.

THEOREM 2.4 ([20]) Ω *transforms the first and second KdV Hamiltonians into the globally integrable Hamiltonians*

$$H^\iota(|q|^2) = \sum_{k\geq 1} \lambda_k^\iota \, |q_k|^2 + \frac{1}{2} \sum_{k,l\geq 1} Q_{kl}^\iota \, |q_k|^2 \, |q_l|^2 + \ldots,$$

where

$$\lambda_k^1 = (2\pi k)^3, \qquad Q_{kl}^1 = -6\delta_{kl},$$

and

$$\lambda_k^2 = (2\pi k)^5, \qquad Q_{kl}^2 = 40\pi^2(6 - 7\delta_{kl})kl.$$

Now we are in a position to study perturbations of large n-gap solutions. Introducing the usual action-angle coordinates

$$q_k = \sqrt{I_k}\, e^{\mathrm{i}\varphi_k}, \qquad 1 \leq k \leq n,$$

we obtain an integrable system with Hamiltonian

$$H^\iota = \sum_{1\leq k\leq n} \lambda_k^\iota I_k + \frac{1}{2} \sum_{1\leq k,l\leq n} Q_{kl}^\iota I_k I_l + \sum_{k\geq n+1} \lambda_k^\iota \, |q_k|^2 + \sum_{1\leq k\leq n<l} Q_{kl}^\iota I_k \, |q_l|^2 + \cdots \tag{10}$$

and equations of motion

$$\dot{\varphi}_k = \omega_k^\iota, \qquad \dot{I}_k = 0, \qquad \dot{q}_k = \mathrm{i}\omega_k^\iota q_k,$$

where the frequencies ω_k^ι are real analytic functions of I and $|q'|^2 = (|q_{n+1}|^2, \dots)$ with expansion

$$\omega_k^\iota(I, |q'|^2) = \lambda_k^\iota + \sum_{1\le l\le n} Q_{kl}^\iota I_l + \cdots .$$

The dots stand for terms of higher order in I and at least first order in $|q'|^2$.

Thus, the variational equations along each invariant torus $\mathbb{T}^n \times \{I\} \times \{0\}$ have constant coefficients, which are real analytic functions of the position I of the torus. The relevant nonresonance and nondegeneracy conditions are easily checked at $I = 0$. Hence, by analyticity and local finiteness, they are satisfied on a dense open subset of the phase space. We may thus apply KAM theory to obtain the persistence of almost all finite gap solutions of the first and second KdV equation under sufficiently small Hamiltonian perturbations of the equations as given in (7) and (8).

Of course, the admissible size of the perturbations depends on the solution considered. For a fixed size of ε, a Cantor set of quasi-periodic solutions is obtained, which asymptotically fills the phase space as ε tends to zero.

To sum up, in the Hamiltonian perturbation theory of the KdV family, the Birkhoff normal form serves two purposes. First, it provides an easy way to apply KAM theory to small solutions. Second – and this is probably more important – in connection with the global coordinates of Kappeler *et al.* it provides a sufficient control over the variational equations to verify the nonresonance and nondegeneracy conditions almost everywhere. So for this purpose the deep and intricate results of Krichever are not needed.

3. Water waves

So far we have considered equations which could be understood as small perturbations of some nonlinear integrable pde. Now I want to describe another application of the Birkhoff normal form where there is seemingly no integrable pde in the background. This application is due to Zakharov [37] and Craig & Worfolk [15, 13], and I will merely report their results.

Consider the motion of an incompressible, irrotational ideal fluid in a two-dimensional domain of infinite depth,

$$S(\eta) = \{\, (x, y) : 0 < x < 2\pi, \ \ -\infty < y < \eta(x, t) \,\} .$$

There is a velocity potential φ, so that on this domain the velocities are

$$u = \nabla\varphi, \qquad \Delta\varphi = 0. \tag{11}$$

The boundary condition are

$$\varphi_y \to 0 \qquad \text{as} \qquad y \to -\infty$$

at the bottom and periodic at the lateral sides,

$$\eta(x+2\pi,t)=\eta(x,t), \qquad \varphi(x+2\pi,t)=\varphi(x,t).$$

The nonlinearity and time dependence of the problem are introduced by the boundary condition at the free top surface,

$$\eta_t=\varphi_y-\eta_x\varphi_x, \qquad \varphi_t=-g\eta-\frac{1}{2}\langle\nabla\varphi,\nabla\varphi\rangle,$$

where g is the gravitational constant.

These equations can be written in Hamiltonian form. The Hamiltonian is the total energy of the system. The appropriate choice of canonically conjugate variables was given by Zakharov [36] as

$$\eta(x) \qquad \text{and} \qquad \xi(x)=\varphi(x,\eta(x)),$$

the free surface and the trace of the potential on it. This trace together with the boundary conditions at the bottom and at the sides uniquely determines the harmonic potential φ on the domain $S(\eta)$. The Hamiltonian then turns out to be

$$\begin{aligned} H(\xi,\eta) &= K+V \\ &= \frac{1}{2}\int_0^{2\pi} \xi G(\eta)\xi\,\mathrm{d}x+\frac{1}{2}\int_0^{2\pi} g\eta^2\,\mathrm{d}x. \end{aligned}$$

Here, $G(\eta)$ is the Dirichlet-Neumann operator which associates with ξ the normal derivative of the harmonic solution φ with $\varphi(x,\eta(x))=\xi(x)$. The equations of motion are now

$$\dot\eta=\frac{\partial H}{\partial\xi}=G(\eta)\xi, \qquad \dot\xi=-\frac{\partial H}{\partial\eta}=-g\eta-\frac{\partial K}{\partial\eta}.$$

In the following we may restrict ourselves to zero mean functions,

$$[\eta]=0, \qquad [\xi]=0,$$

since these means are conserved quantities.

The Dirichlet-Neumann operator G is analytic in η, so it admits an expansion $G=G_0+G_1+\dots$ into homogeneous terms in η. Consequently, there is a similar expansion of the Hamiltonian H,

$$H=H_2+H_3+\cdots,$$

where H_j is homogeneous of degree j in ξ and η. Thus, 0 is an equilibrium, which agrees well with the fact that the flat surface is a steady state of our system. In particular,

$$H_2=\frac{1}{2}\int_0^{2\pi}(\xi G_0\xi+g\eta^2)\,\mathrm{d}x.$$

Moreover, at infinite depth one simply has $G_0 = -\mathrm{i}\partial_x$. Expanding ξ and η into Fourier series with complex coefficients ξ_k and η_k and no constant terms, one obtains

$$H_2 = \sum_{k\geq 1}(k\,|\xi_k|^2 + g\,|\eta_k|^2).$$

Thus, 0 is an elliptic equilibrium with characteristic frequencies

$$\omega_k = \sqrt{gk}, \qquad k \geq 1.$$

This is the well-known frequency dispersion for linear water waves.

Now consider the next terms of the Hamiltonian H. They can be written down explicitly, either by hand or by computer, but their exact form is not relevant here. As before, the cubic term H_3 contains only monomials in ξ_k and η_k whose indices add up to zero. They turn out to be nonresonant, since similar to Lemma 1.1 one has

$$\omega_k \pm \omega_l \pm \omega_m \neq 0 \qquad \text{if} \qquad k \pm l \pm m = 0.$$

So formally they can be transformed away. Craig & Worfolk [15] showed that this formal transformation is indeed convergent.

THEOREM 3.1 ([15]) *There exists a real analytic, symplectic coordinate transformation between small neighbourhoods of the origin in some suitably defined Hilbert space of exponentially decreasing sequences of Fourier coefficients such that*

$$H \circ \Phi = H_2 + \tilde{H}_4 + \tilde{H}_5 + \cdots.$$

That is, the third order term is completely removed.

The fourth order term, however, is more troublesome. There are not only the classical, *benign* resonances of order four of the form

$$\omega_k + \omega_l - \omega_k - \omega_l = 0,$$

which are not harmful, since they give rise to *integrable* terms in the normal form. There are also *non-benign* resonances such as

$$\omega_1 - \omega_4 - \omega_4 + \omega_9 = 0.$$

Indeed, all the non-benign resonances are easily classified – they are of this form with indices

$$pm^2, \quad p(m+1)^2, \quad pm^2(m+1)^2, \quad p(m^2+m+1)^2,$$

with positive integers p and m. In the original Hamiltonian H the corresponding coefficients are *not* all zero. But then, a little miracle happens – which was first observed by Zakharov [37] and reported in Dyachenko & Zakharov [17].

THEOREM 3.2 ([17, 15]) *In the transformed Hamiltonian $H \circ \Phi$ of the preceding theorem* all *the non-benign coefficients of $\tilde{H}_4$* vanish. *Hence, there exists a formal symplectic coordinate transformation Ψ that takes $H \circ \Phi$ into an integrable Birkhoff normal form up to order four.*

It has not been shown yet, however, that this formal transformation is convergent in some suitable sense.

If a statement is true for $n = 3$ and $n = 4$, it is customary to conjecture that it holds for all n. Therefore, Dyachenko and Zakharov [17] conjectured that the Birkhoff normal form of H is integrable to every order. Craig & Worfolk [15] showed, however, that this is already false at $n = 5$. The transformed Hamiltonian contains nonvanishing resonant terms of order five, which can not be transformed away. This makes it very likely that the whole Hamiltonian is not integrable.

To sum up, the Hamiltonian of deep water waves has the following properties. Due to some unexpected cancellation of terms, there are formal symplectic coordinates around the equilibrium, in which the Hamiltonian is in Birkhoff normal form up to order four. So, modulo their analyticity, the system may be viewed as a small perturbation of an infinite dimensional system of nonlinear integrable ode's. This integrability can not be pushed to higher order, due to resonant terms of order five.

Unfortunately, the present KAM theories do *not* apply to this Hamiltonian, even when it is proven to be analytic. Its characteristic frequencies are

$$\omega_k \sim \sqrt{k},$$

whereas for the KAM theories,

$$\omega_k \sim k$$

is the limiting case.

4. The basic KAM theorem

For the sake of completeness and reference we include a version of the infinite dimensional KAM theorem that is applicable to the Hamiltonians in section 1 and 2.

To study perturbations of the Hamiltonians (6) and (10) it is convenient to introduce local coordinates

$$\varphi = x, \qquad I = \xi + y$$

around each torus $\mathbb{T}^n \times \{\xi\} \times \{0\}$. We obtain a *family* of Hamiltonians

$$H_0 = \sum_{1 \le k \le n} \omega_k(\xi) y_k + \sum_{k \ge n+1} \omega_k \left|q_k\right|^2 + \cdots$$

with

$$\omega_k(\xi) = \lambda_k + \sum_{1 \le l \le n} Q_{kl} \xi_l,$$

for which we want to prove the persistence of the invariant torus $\mathbb{T}^n \times \{0\} \times \{0\}$.

It turns out that the terms denoted by the dots may also be regarded as perturbations. So, we now consider a slightly more general family of simple Hamiltonians

$$H_0 = \sum_{1 \le k \le n} \omega_k(\xi) y_k + \sum_{k \ge n+1} \Omega_k(\xi) \left|q_k\right|^2$$

on the phase space $\mathbb{T}^n \times \mathbb{R}^n \times \ell^s_{\mathbb{C}}$, where the frequencies $\omega = (\omega_1, \dots, \omega_n)$ and $\Omega = (\Omega_{n+1}, \Omega_{n+2}, \dots)$ depend on n parameters

$$\xi \in \Pi \subset \mathbb{R}^n.$$

The parameter domain Π may be any *closed* subset of $\mathbb{R}^n$ of positive Lebesgue measure. For instance, Π may be a Cantor set.

For each parameter value ξ, there is an invariant torus $\mathbb{T}^n \times \{0\} \times \{0\}$ with linear flow, which is determined by the "internal" frequencies $\omega(\xi)$. In its normal space described by the q-coordinates, the origin is an elliptic equilibrium with "external" frequencies $\Omega(\xi)$. The aim is to prove the persistence of a large portion of this family under small perturbations $H = H_0 + H_1$ of H_0. To this end the following assumptions are made.

A – *Spectral Asymptotics.* There exists $d > 1$ and $\beta > 1$ such that

$$\Omega_k(\xi) = k^d + \ldots + \tilde{\Omega}_k(\xi),$$

where the dots stand for fixed lower order terms not depending on ξ, and where the weighted functions

$$\frac{\tilde{\Omega}_k(\xi)}{k^{d-\beta}}$$

are *uniformly* Lipschitz on Π.

We assume here that $d > 1$. The case $d = 1$ may also be handled – see [25, 31] – but is somewhat more involved and not relevant here. We therefore omit it for the sake of brevity.

B – *Nondegeneracy.* The map

$$\xi \mapsto \omega(\xi)$$

is a homeomorphism between Π and its image which is Lipschitz in both directions. Moreover, for all $(k, l) \in \mathbb{Z}^n \times \mathbb{Z}^\infty$ with $1 \leq \sum_i |l_i| \leq 2$ the resonance sets

$$R_{kl} = \{\xi \in \Pi : \langle k, \omega(\xi) \rangle + \langle l, \Omega(\xi) \rangle = 0\}$$

have Lebesgue measure zero.

We note that assumption A implies that R_{kl} is *empty* for *almost all* k and l, so this assumption together with assumption A concerns only finitely many k and l.

C – *Regularity.* The perturbation H_1 is real analytic in the space coordinates and Lipschitz in the parameters, and for each ξ its Hamiltonian vector field is a map

$$X_{H_1} \colon \mathbb{T}^n \times \mathbb{R}^n \times \ell^s_{\mathbb{C}} \supset D \to \mathbb{T}^n \times \mathbb{R}^n \times \ell^{s-\sigma}_{\mathbb{C}}$$

with

$$\sigma < d - 1,$$

where D is some neighbourhood of $\mathbb{T}^n \times \{0\} \times \{0\}$.

To formulate a smallness condition for the perturbations we introduce the complex domains

$$D(w,r)\colon \quad |\mathrm{Im}\, x| < w,\ |y| < r^2,\ \|q\|_s < r$$

and the weighted vector field norms

$$\|X_H\|_r = |\partial_y H| + r^{-2}\,|\partial_x H| + r^{-1}\,\|\partial_{\bar q} H\|_{s-\sigma}\,.$$

Then, $\|X_H\|_{r,D(w,r)\times\Pi}$ denotes the supremum of $\|X_H\|_r$ over the set $D(w,r)\times\Pi$. Similarly, $\|X_H\|^L_{r,D(w,r)\times\Pi}$ denotes the associated Lipschitz semi-norm with respect to the parameters ξ over Π.

THEOREM 4.1 ([20, 25, 27, 31]) *Suppose $H = H_0 + H_1$ satisfies assumptions A, B and C. If*

$$\varepsilon = \|X_{H_1}\|_{r,D(w,r)\times\Pi} + \|X_{H_1}\|^L_{r,D(w,r)\times\Pi}$$

is sufficiently small, then there exists

(a) *a Cantor set $\Pi' \subset \Pi$ such that the Lebesgue measure of $\Pi\backslash\Pi'$ tends to zero with ε;*

(b) *a Lipschitz continuous family of real analytic, symplectic coordinate transformations*

$$\Psi\colon \mathbb{T}^n \times \mathbb{R}^n \times \ell^s_{\mathbb{C}} \times \Pi' \to \mathbb{T}^n \times \mathbb{R}^n \times \ell^s_{\mathbb{C}}$$

close to the identity; and

(c) *perturbed frequencies ω' and Ω' close to the unperturbed ones, which are Lipschitz on Π', such that*

$$H \circ \Psi = \sum_{1\le k\le n} \omega'_k(\xi) y_k + \sum_{k\ge n+1} \Omega'_k(\xi)\,|q_k|^2 + \cdots,$$

where the dots stand for higher order terms in y and q. Hence, for each $\xi \in \Pi'$ the perturbed Hamiltonian admits a linearly stable invariant n-torus with internal frequencies $\omega'(\xi)$ and external frequencies $\Omega'(\xi)$.

For a more precise version of this theorem see [31] and [20]. Note that its "output" is of the same form as its "input". This makes it possible to iterate the theorem to construct almost-periodic solutions for certain Hamiltonians depending on infinitely many parameters [32].

The first versions of this KAM theorem [25, 31] had to impose the additional hypotheses

$$\sigma \le 0,$$

that is, the perturbing vector fields had to be *bounded* as operators. This was sufficient for perturbing Schrödinger and wave equations, but not sufficient for the KdV equations, where the operators are *unbounded.* Strictly speaking, the proof of Kuksin's result [23] contained a gap, which he filled only recently in [27]. It required a very delicate estimate of a linear small divisor equation with *large, variable* coefficients. For another version, see also [20].

References

[1] D. Bättig, A.M. Bloch, J.-C. Guillot, & T. Kappeler, On the symplectic structure of the phase space for periodic KdV, Toda, and defocusing NLS. *Duke Math. J.* **79** (1995), 549–604.

[2] D. Bättig, T. Kappeler, & B. Mityagin, On the Korteweg-de Vries equation: Convergent Birkhoff normal form. J. Funct. Anal. **140** (1996), 335–358.

[3] R.F. Bikbaev & S.B. Kuksin, A periodic boundary value problem for the Sine-Gordon equation, small Hamiltonian perturbations of it, and KAM-deformations of finite-gap tori. *Algebra i Analis* **4** (1992) [Russian]. English translation in *St. Petersburg Math. J.* **4** (1993), 439–468.

[4] R.F. Bikbaev & S.B. Kuksin, On the parametrization of finite-gap solutions by frequency vector and wave-number vector and a theorem of I. Krichever. *Lett. Math. Phys.* **28** (1993), 115–122.

[5] A. Bobenko & S.B. Kuksin, Finite-gap periodic solutions of the KdV equation are non-degenerate. *Physics Letters* **161** (1991), 274–276.

[6] A. Bobenko & S.B. Kuksin, The nonlinear Klein-Gordon equation on an interval as a perturbed sine-Gordon equation. *Comment. Math. Helv.* **70** (1995), 63–112.

[7] J. Bourgain, Periodic solutions of hamiltonian perturbations of linear Schrödinger equations in higher dimensions. Preprint (1994).

[8] J. Bourgain, Quasi-periodic solutions of hamiltonian perturbations of 2D linear Schrödinger equations. Preprint IHES (1994).

[9] J. Bourgain, Construction of quasi-periodic solutions for hamiltonian perturbations of linear equations and applications to nonlinear pde. *Intern. Math. Res. Not.* **11** (1994), 475–497.

[10] J. Bourgain, Construction of approximative and quasi-periodic solutions of perturbed linear Schrödinger and wave equations. Preprint (1995).

[11] J. Bourgain, Construction of periodic solutions of nonlinear wave equations in higher dimension. *Geom. Funct. Anal.* **5** (1995), 629–639.

[12] R.C. Churchill, M. Kummer, & D.L. Rod, On averaging, reduction and symmetry in Hamiltonian systems. *J. Diff. Equations* **49** (1983), 359–414.

[13] W. Craig, Birkhoff normal forms for water waves. *Contemporary Mathematics* **200** (1996), 57–74.

[14] W. Craig & C.E. Wayne, Newton's method and periodic solutions of nonlinear wave equations. *Commun. Pure Appl. Math.* **46** (1993), 1409–1498.

[15] W. Craig & P.A. Worfolk, An integrable normal form for water waves at infinite depth. *Physica D* **84** (1995), 513–531.

[16] J. Denzler, Nonpersistence of breather families for the perturbed sine-Gordon equation. *Commun. Math. Phys.* **158** (1993), 397–430.

[17] A.I. Dyachenko & V.E. Zakharov, Is free-surface hydrodynamics an integrable system? *Phys. Lett. A* **190** (1994), 144–148.

[18] H. Ito, Convergence of Birkhoff normal forms for integrable systems. *Comment. Math. Helvetici* **64** (1989), 412–461.

[19] T. Kappeler, Fibration of the phase space for the Korteweg-de Vries equation. *Ann. Inst. Fourier* **41** (1991), 539–575.

[20] T. Kappeler & J. Pöschel, Perturbations of KdV equations. In preparation.

[21] I. Krichever, "Hessians" of integrals of the Korteweg-de Vries equation and perturbations of finite-gap solutions. *Funktional. Anal. i Prilozhen.* **21** (1987), 22–37. English translation in *Soviet Math. Dokl.* **27** (1983).

[22] S.B. KUKSIN, Hamiltonian perturbations of infinite dimensional linear systems with an imaginary spectrum. *Funktsional. Anal. Prilozhen.* **21** (1987), 22–37 [Russian]. English translation in *Funct. Anal. Appl.* **21** (1987), 192–205.

[23] S.B. KUKSIN, Perturbation theory for quasiperiodic solutions of infinite-dimensional Hamiltonian systems, and its application to the Korteweg-de Vries equation. *Matem. Sbornik* **136** (1988) [Russian]. English translation in *Math. USSR Sbornik* **64** (1989), 397–413.

[24] S.B. KUKSIN, Conservative perturbations of infinite dimensional linear systems depending on a vector parameter. *Funct. Anal. Appl.* **23** (1989), 62–63.

[25] S.B. KUKSIN, *Nearly integrable infinite-dimensional Hamiltonian systems.* Lecture Notes in Mathematics 1556, Springer, 1993.

[26] S.B. KUKSIN, KAM-theory for partial differential equations. In *Proceedings First European Congress of Mathematics* (Paris 1992), Birkhäuser, Basel, 1994, 123–157.

[27] S.B. KUKSIN, A KAM theorem for equations of the Korteweg-de Vries type. Preprint (1996).

[28] S.B. KUKSIN & J. PÖSCHEL, Invariant Cantor manifolds of quasi-periodic oscillations for a nonlinear Schrödinger equation. *Ann. Math.* **143** (1996), 149–179.

[29] J. PÖSCHEL, On elliptic lower dimensional tori in Hamiltonian systems. *Math. Z.* **202** (1989), 559–608.

[30] J. PÖSCHEL, Quasi-periodic solutions for a nonlinear wave equation. *Comment. Math. Helv.* **71** (1996), 269–296.

[31] J. PÖSCHEL, A KAM-theorem for some nonlinear partial differential equations. *Ann. Sc. Norm. Sup Pisa* **23** (1996), 119–148.

[32] J. PÖSCHEL, On the construction of almost periodic solutions for a nonlinear Schrödinger equation. In *Proceedings NATO ASI, S'Agaro* 1995 (to appear).

[33] C.L. SIEGEL & J. MOSER, *Lectures on Celestial Mechanics.* Grundlehren 187, Springer, Berlin, 1971.

[34] I.M. SIGAL, Non-linear wave and Schrödinger equations I. Instability of periodic and quasiperiodic solutions. *Commun. Math. Phys.* **153** (1993), 297–320.

[35] C.E. WAYNE, Periodic and quasi-periodic solutions of nonlinear wave equations via KAM theory. *Commun. Math. Phys.* **127** (1990), 479–528.

[36] V.E. ZAKHAROV, Stability of periodic waves of finite amplitude on the surface of deep fluid. *J. Appl. Mech. Phys.* **2** (1968), 190–194.

[37] V.E. ZAKHAROV, Space-time spectra of sea waves in the weak turbulence approximation. Seminar talk at the 'Workshop on waves in the Ocean', MSRI, Berkeley (1994).

[38] V.E. ZAKHAROV & A.B. SHABAT, Exact theory of two-dimensional self-focusing and one-dimensional self-modulation of waves in nonlinear media. *Zh. Eksp. Teor. Fiz* **61** (1971), 118–134 [Russian]. English translation in *Soviet Physics JETP* **34** (1972), 62–69.

Progress in Mathematics, Vol. 169, © 1998 Birkhäuser Verlag Basel/Switzerland

Group Enumeration and Where It Leads Us

L. Pyber

Mathematical Institute of the Hungarian Academy of Sciences,
Budapest P.O.Box 127, H–1364. Hungary
e-mail: pyber@math-inst.hu

1. Introduction

Let $g(n)$ be the number of (isomorphism classes of) groups of order n. One of the most intriguing problems of asymptotic group theory is to obtain sharp estimates for $g(n)$ (see section 2 for the history of the problem).

Recently it was proved [Py1] that $g(n) \leq n^{(\frac{2}{27}+o(1))(\log_2 n)^2}$ as $n \to \infty$ and it was already known from G. Higman's previous work [Hi1] that $\frac{2}{27}$ is the best constant in such an estimate.

C. C. Sims [Si] has obtained sharp upper bounds for $g(n)$ in the case when n is a prime-power. The basic approach of the proof in the general case is to obtain a "Reduction Theorem". Namely one has to show that the number of groups of order n with given Sylow p-subgroups is much smaller than the bound required for $g(n)$. This is made more explicit in section 3 where we give further examples of such reduction theorems.

The proof of the Reduction Theorem draws attention to enumeration type problems for finite permutation groups and linear groups. Recent results in this field are reviewed in section 4, where we also give an outline of the proof of the Reduction Theorem.

These results can be applied to estimate the subgroup growth of certain residually finite groups [Lu1], [LPSh], [PSh1]. More significantly they form the basis of investigations concerning the maximal subgroup growth of profinite groups.

Note that in recent work A. Mann and A. Shalev [MSh] have discovered a beautiful connection between maximal subgroup growth and the distribution of finite generating sets in profinite groups.

These developments are surveyed in sections 5 and 6.

By a well-known result of J. D. Dixon [Di] a random pair of elements of the symmetric group $\mathrm{Sym}(n)$ generates $\mathrm{Sym}(n)$ or $\mathrm{Alt}(n)$ with probability 1 as $n \to \infty$. During the last decade various extensions and analogues of Dixon's Theorem have appeared. These results form the basis of applications of probabilistic ideas in the solutions of some classical problems concerning residual properties of free groups, the modular group, etc. In section 7 we give an account of applications of the probabilistic method in group theory.

In the final section we consider the following problem: which finitely generated profinite groups G have a dense free (discrete) subgroup? Particular cases of this problem are related to maximal subgroup growth, residual properties of free groups and to identities for pro-p groups.

For a more detailed account of some of the topics considered here see [Lu2], [MS1], [Py2] and [Py3].

2. Group enumeration

The study of $g(n)$ was started in 1895 by O. Hölder [Hö] who established a beautiful formula for $g(n)$ in the case when n is squarefree. Let $\nu_p(n)$ denote the number of prime-divisors of n which are $\equiv 1 (\bmod p)$.

THEOREM 2.1. *For n squarefree*

$$g(n) = \sum_{d/n} \prod_{p/d} \left(\frac{p^{\nu_p(n/d)} - 1}{p-1} \right)$$

where the inner product runs over the prime divisors p of d.

There is a more general expression for counting groups G of order n with all their Sylow subgroups cyclic [MM]. This implies in particular that the number of such groups is at most $\frac{1}{2}n$.

It seems, however, unlikely that an explicit formula for $g(n)$ will be obtained in the case where n is a prime-power. Work of G. Higman [Hi1] and of C. C. Sims [Si] in the early 1960-s gave a surprising asymptotic formula for the number of p-groups.

THEOREM 2.2. *For p a fixed prime*

$$g(p^\alpha) = p^{\left(\frac{2}{27}+o(1)\right)\alpha^3} \text{ as } \alpha \to \infty.$$

Somewhat later P. M. Neumann [Ne2] proved that if the number of simple groups of order n is not outrageous then

$$g(n) \le n^{c\lambda^2}$$

where $\lambda = \lambda(n)$ denotes the number of prime-factors of n, counting multiplicities.

We now know from the Classification Theorem of finite simple groups that there are at most two finite simple groups of the same order.

Improving the argument of Neumann, D. F. Holt [Ho] proved that

$$g(n) \le n^{\frac{1}{6}\lambda^2+\lambda}.$$

Neumann and McIver [MN] conjectured that the upper bound in 2.2 extends to the case of general n. This conjecture was confirmed in [Py1].

THEOREM 2.3.

$$g(n) \le n^{(\frac{2}{27}+o(1))\mu^2} \quad \text{as } \mu \to \infty$$

where $\mu = \mu(n)$ denotes the maximal exponent of a prime in the prime decomposition of n (Note that $\mu(n) \le \lambda(n) \le \log_2 n$.).

While this result is in some sense sharp it still leaves room for improvements. For example it would be interesting to find an explicit function $f(n)$ such that $f(n) \le g(n) \le f(n)^c$ for some constant c.

Let us denote the number of groups (nilpotent groups) of order at most n by $g^*(n)$ ($g^*_{\mathrm{nilp}}(n)$).

Conjecture 2.4.

$$\lim_{n\to\infty} \frac{g^*_{\mathrm{nilp}}(n)}{g^*(n)} = 1.$$

That is we conjecture that almost all finite groups are nilpotent.

A lot of work has been done on counting groups of various restricted types. Improvements of the Higman–Sims Theorem appear in [Bl] and [NS]. For applications of the methods of Higman and Sims in the context of Lie algebras and finite rings see [KP], [Ne1] and [NS].

It is shown in [MN] that the number of d-generated groups of order p^α is at most $p^{\frac{1}{2}(d+1)\alpha^2+o(\alpha)}$. It would be of interest to prove that the number of d-generated groups of order n is at most $n^{c\log n}$ for some constant c.

For number-theoretic investigations concerning the number of abelian groups and groups of squarefree order see [Li2], [MS2] and the references therein.

3. Reduction theorems

As indicated earlier, Theorem 2.3 is a direct consequence of 2.2 and the following result [Py1].

REDUCTION THEOREM 3.1. *The number of groups of order n with given Sylow p-subgroups is at most*

$$n^{100\mu(n)}.$$

Denote by $s(G)$ the number of subgroups of a finite group G. The proof of 3.1 uses a result of similar flavour for $s(G)$ [BPSh] (which is proved in [Py1] in the relevant special cases). For a prime p let G_p denote the Sylow p-subgroup of the group G.

THEOREM 3.2.

$$s(G) \le |G|^{c+1} \prod_p s(G_p) \quad \text{for some } c > 0.$$

By a beautiful result of Aschbacher and Guralnick [AG] every finite group G can be generated by a solvable subgroup S and a single element g of G. On the other hand it is proved in [BPSh] that the number of maximal solvable subgroups of G is at most $|G|^c$. Theorem 3.2 is an easy consequence of these results.

There is some evidence [BP] that in fact we can take here $c = 1$.

The proof of 3.1 uses yet another reduction theorem [Gu], [Lu3].

THEOREM 3.3. *If every Sylow subgroup of G can be generated by d elements then G can be generated by $d+1$ elements.*

It is clear that such reduction theorems are extremely useful tools in obtaining sharp asymptotic results for various classes of groups.

We end this section by mentioning three elegant conjectures concerning invariants of finite groups.

Conjecture 3.4. [Ne3]

$$|\operatorname{Aut}(G)| \text{ divides } |G/Z(G)| \prod_p |\operatorname{Aut}(G_p)|.$$

Denote by $k(G)$ the number of conjugacy classes of G and by $b(G)$ the maximal size of a conjugacy class.

Conjecture 3.5. [NV]

$$\prod_p b(G_p) \leq b(G).$$

Conjecture 3.6.

$$k(G) \leq \prod_p k(G_p).$$

We believe that a proof of these conjectures would lead to new insights concerning the structure of a general finite group.

4. Enumerating permutation groups

We start this section with an outline of the proof of the Reduction Theorem for solvable groups.

Recall that $F(G)$, the Fitting subgroup of G, is the maximal nilpotent normal subgroup of G. The most significant fact about $F(G)$ is that for solvable groups $C_G(F(G))$ is contained in $F(G)$, hence $Z(F(G)) = C_G(F(G))$. This implies that the quotient group $G/Z(F(G))$ has a natural embedding into $\operatorname{Aut}(F(G))$.

It is easy to see that the number of possibilities for $F(G)$ for a solvable group G with given Sylow p-subgroups is small (less than $n^{c\mu(n)}$).

$\operatorname{Aut}(F(G))$ is a relatively small group with a transparent structure. Using some delicate arguments one can show that the number of possibilities for $G/Z(F(G))$ is also small. Finally, using the theory of group extensions one can show that there are only a small number of possibilities for glueing $F(G)$ and $G/Z(F(G))$ together.

For arbitrary groups the proof follows a similar pattern, this time using the so-called generalized Fitting subgroup $F^*(G)$.

Similar ideas play an important role in the proof of the following result [PSh2].

THEOREM 4.1. *The number of primitive permutation groups of degree n (up to permutation equivalence) is at most*

$$n^{c\mu(n)} \text{ for some constant } c.$$

It is also proved in [PSh2] that for some $\varepsilon > 0$ and for infinitely many n $\mathrm{Sym}(n)$ has at least $n^{\varepsilon \log n/\log\log n}$ pairwise non-isomorphic primitive solvable subgroups, that is 4.1 is close to best possible.

Motivated by important applications (see section 5, in particular Proposition 5.2) we propose the following conjecture [PSh2].

Conjecture 4.2. The number of r-generated primitive permutation groups of degree n (up to permutation equivalence) is at most n^c where c depends on r.

We call a (not necessarily finite) group $\mathrm{Alt}(d)$*-free* if it does not involve the alternating group $\mathrm{Alt}(d)$ as a section. This class is a natural extension of the class of solvable groups.

Various hard algorithmic problems can be solved in polynomial time for $\mathrm{Alt}(d)$-free permutation groups (see [KL2]). The reason for this is that, as proved by Babai, Cameron and Pálfy [BCP], primitive $\mathrm{Alt}(d)$-free groups of degree n have size less than $n^{c(d)}$. This makes it possible to devise efficient divide and conquer algorithms.

THEOREM 4.3. [BPSh] *The number of maximal* $\mathrm{Alt}(d)$*-free primitive permutation groups of degree n (up to permutation equivalence) is at most*

$$n^c \text{ where } c \text{ depends on } d.$$

Combining 4.3 with the result of Babai, Cameron and Pálfy we obtain that Conjecture 4.2 holds for $\mathrm{Alt}(d)$-free groups. This fact will be used in section 5.

The total number of permutation groups, $s(\mathrm{Sym}(n))$, is much larger then the number of primitive groups.

PROPOSITION 4.4. [Py2]

$$2^{\left(\frac{1}{16}+o(1)\right)n^2} \le s(\mathrm{Sym}(n)) \le 2^{cn^2}.$$

We conjecture that the lower bound is sharp. Using the ideas of section 3 it follows that in order to prove this conjecture one only has to consider solvable groups.

5. Positively finitely generated groups

Recall that a profinite group is an inverse limit of finite groups, hence it is a compact topological group. Such a group G (and its cartesian powers G^k) can be viewed as a probability space with respect to the normalized Haar measure.

The group G is termed positively finitely generated (PFG) if for some k the measure $P(G,k)$ of the set of k-tuples generating G is positive in G^k.

This concept first arose in the context of field arithmetic. Answering a question posed in [FJ] Kantor and Lubotzky [KL1] showed that $\widehat{F_d}$, the free profinite group of rank d, is not PFG.

It was also noted in [KL1] that free abelian groups are PFG. Many more examples are given in [Ma1], in particular finitely generated pro-(finite solvable) groups are shown to be PFG as is the profinite completion of $\mathrm{PSL}(d, \mathbb{Z})$ for $d \geq 3$.

These results were unified in [BPSh].

THEOREM 5.1. *Let G be a finitely generated profinite group which is not PFG. Then every finite alternating group (and therefore every finite group) can be obtained as a quotient of an open subgroup of G.*

How does one prove such a result? If a k-tuple does not generate G then it is contained in some maximal subgroup M of G. Hence if G has few maximal subgroups then it is PFG. More precisely this is true if $m_n(G)$, the number of (open) maximal subgroups of index n is bounded by n^c for some constant c. Groups with this property are said to have polynomial maximal subgroup growth.

Let $N_{\text{prim}}(r, n)$ denote the number of (conjugacy classes of) r-generated primitive subgroups of $\mathrm{Sym}(n)$. A maximal subgroup of index n in G determines a primitive subgroup of $\mathrm{Sym}(n)$ in a natural way. Let $M_G(n)$ denote the maximal order of such primitive images of G.

A simple counting argument yields the following basic observation.

PROPOSITION 5.2. *Let G be an r-generated group. Then*

$$m_n(G) \leq n \cdot N_{\text{prim}}(r, n) \cdot M_G(n)^r.$$

Now if G is $\mathrm{Alt}(d)$-free then by the Babai–Cameron–Pálfy Theorem [BCP] $M_G(n)$ is bounded by a polynomial in n.

Moreover one can substitute $N_{\text{prim}}(r, n)$ in the above expression by the number of (conjugacy classes of) r-generated $\mathrm{Alt}(d)$-free primitive subgroups of $\mathrm{Sym}(n)$. As noted after 4.3 this function is also bounded by a polynomial in n. Hence G has polynomial maximal subgroup growth and therefore it is PFG.

By a very surprising result of Mann and Shalev [MSh] if a group is PFG then indeed it has few maximal subgroups.

THEOREM 5.3. *A profinite group is PFG if and only if it has polynomial maximal subgroup growth.*

The proof of this theorem relies on the following deep consequence of the Classification Theorem [MSh].

THEOREM 5.4. *There is a constant c such that $m_n(G) \leq cn^2$ for all finite almost simple groups G and for all positive integers n.*

This estimate also plays a crucial role in the proof of Theorem 4.1.

In view of the above results it would be most interesting to characterize PFG groups. The converse of 5.2 does not hold; examples of PFG groups which do involve every finite group as an upper section are given in [KL1] and [Bh1]. For example, using a sharp version of Dixon's Theorem [Ba] Kantor and Lubotzky show that the cartesian product of all finite alternating groups (of degree ≥ 5) is generated by 3 random elements with positive probability.

The following problem is due to Mann [Ma1].

Problem 5.5. Let H be an open subgroup of the PFG group G. Is it a PFG group?

Indeed, it is not even known whether there exist virtually free PFG groups.

6. Maximal subgroup growth

An easy consequence of Dixon's Theorem is that almost all index n subgroups of the free group F_d are maximal. Together with what is known about the subgroup growth of F_d this implies the following [Li1].

THEOREM 6.1.

$$m_n(F_d) \sim n(n!)^{d-1} \text{ as } n \to \infty.$$

Note that if G is a d-generated discrete or profinite group then $m_n(G) \leq m_n(F_d)$.

Confirming a conjecture of Lubotzky, T. Müller and the present author obtained the following analogue of Dixon's Theorem [MP].

THEOREM 6.2. *Let G_1 and G_2 be two fixed non-trivial finite groups not both of order* 2. *Then random copies of G_1 and G_2 generate* Alt(n) *or* Sym(n) *with probability* 1 *as $n \to \infty$.*

This implies that almost all subgroups of the free product $G_1 * G_2$ are maximal. This is true in particular for the modular group $\mathrm{PSL}(2, \mathbb{Z})$.

Motivated by the previous examples we make the following definition. We call a (discrete) group Γ a *Dixon group* if for almost all index n subgroups the image of Γ with respect to the natural homomorphism into Sym(n) contains Alt(n).

Problem 6.3. Which groups are Dixon groups?

To start with one should consider the following.

Problem 6.4. Which 1-relator groups have infinitely many alternating or symmetric groups as quotients?

Let us return now to our discussion of maximal subgroup growth.

DEFINITION. Let f be a function. The group G has maximal subgroup growth of type f if there exists positive constants c and b such that

$$m_n(G) \leq f(n)^c \text{ for all sufficiently large } n,$$

$$m_n(G) \geq f(n)^b \text{ for infinitely many } n.$$

Müller [Mü] gave asymptotic formulas for the subgroup growth of free products Γ of finite groups. By 6.2 these asymptotic formulas are also valid for $m_n(\Gamma)$. In particular it follows that such a group Γ has maximal subgroup growth of type n^n (unless $\Gamma = C_2 * C_2$).

By considering suitable cartesian products of alternating groups Mann [Ma1] proved that there is a continuous range of possible types of maximal subgroup growth.

Theorem 6.5. *Let f be a non-decreasing function with $n \leq f(n) \leq n^n$ for all n. Then there exists a 2-generated profinite group whose maximal subgroup growth is of type f.*

Using Theorem 4.1 one can investigate the properties of groups of intermediate maximal subgroup growth [PSh2].

Theorem 6.6. *Let G be a finitely generated profinite group. Then*

(i) *either $m_n(G) \leq n^{c \log n}$ for some $c > 0$ and for all n or* $\mathrm{Alt}(n)$ *occurs as an upper composition factor of G for infinitely many n.*
(ii) *either $m_n(G) \leq n^{c\sqrt{n}}$ for some $c > 0$ and for all n, or for infinitely many n G has a quotient which is isomorphic to* $\mathrm{Sym}(n)$ *or* $\mathrm{Alt}(n)$.

It follows that if G has very large maximal subgroup growth then it has an infinite PFG quotient. It would be interesting to characterize groups with this property.

Let us end this section with some more philosophical remarks.

The subgroup growth of a residually finite group can be thought of as an asymptotic invariant which helps us to classify such groups. More complicated groups tend to have larger subgroup growth. Maximal subgroup growth is not really useful for distinguishing groups. Investigations concerning maximal subgroup growth, however, form the basis of a probabilistic theory of arbitrary finitely generated profinite groups (not just PFG groups). See section 8 for additional justification of this philosophy.

7. Probabilistic methods

Since Erdős started solving combinatorial problems by means of random constructions, ideas related to randomness have become important in many branches of discrete mathematics. Such ideas seem to be useful in group theory as well. In particular randomized algorithms and questions concerning random generation of simple groups have been much investigated in the past few years.

Kantor and Lubotzky [KL1] obtained an analogue of Dixon's Theorem for classical groups and for the exceptional groups of small rank. Liebeck and Shalev [LSh1] finished the proof of the following.

Theorem 7.1. *Let S be a finite simple group and let G be an almost simple group with $S \leq G \leq \mathrm{Aut}(S)$. If $P(G)$ is the probability that two randomly chosen elements of G generate a subgroup containing S then $P(G) \to 1$ as $|S| \to \infty$.*

While Dixon's proof is elementary, the proofs in [KL1] and [LSh1] rely on the Classification Theorem plus detailed knowledge of maximal subgroups of simple groups of Lie type.

Liebeck and Shalev tackled a classical problem using the probabilistic method [LSh2].

Theorem 7.2. *Let $S \neq PSp_4(q)$ be a finite simple classical group. Then the probability that a random involution and a randomly chosen element of order 3 generate S tends to 1 as $|S| \to \infty$.*

If $S = PSp_4(p^k)$ where p is a prime exceeding 3, then the above probability tends to 1/2 as $|S| \to \infty$.

This implies that all but finitely many of the finite simple classical groups different from $PSp_4(q)$ ($q = 2^k$ or 3^k) are homomorphic images of the modular group $\mathrm{PSL}(2,\mathbb{Z})$. In [LSh2] it is also shown that for $p = 2$ or 3 $PSp_4(p^k)$ is not $(2,3)$ generated.

Guralnick and Kantor [GK] using a similar approach proved that sufficiently large simple groups S have the 3/2 generation property, i.e. each $g \in S$, $g \neq 1$, generates S with some $h \in S$.

T. Weigel recently settled a classical conjecture of Magnus concerning residual properties of free goups. He proved that if $\mathcal{S}$ is any infinite series of non-abelian finite simple groups then the intersection of all normal subgroups N of F_2 such that the quotient group F_2/N is in $\mathcal{S}$ is trivial [We1], [We2], [We3].

Weigel's Theorem is equivalent to the following:

If $w(x,y)$ is a non-reducible word in x, x^{-1}, y, y^{-1} then for some $S \in \mathcal{S}$ and a pair of elements g,h of S we have

(i) g and h generate S,
(ii) $w(g,h) \neq 1$.

THEOREM 7.3. [DPSSh] *If $\mathcal{S}$ and $w(x,y)$ are as above, then for $S \in \mathcal{S}$ and a random pair of elements g,h we have $w(g,h) \neq 1$ with probability 1 as $|S| \to \infty$.*

Together with 7.1 this shows that a random pair of elements satisfies both (i) and (ii) above with high probability, i.e. these two results together imply Weigel's Theorem.

A different kind of application of the probabilistic method appears in [Ol]. By a classical observation of G. A. Miller a group of order p^α (p prime) contains an abelian subgroup of order at least $p^{\sqrt{2\alpha}}$. By considering certain random p-groups with central elementary abelian Frattini subgroups, Olshanskii proved that there exists p-groups of order p^α without abelian subgroups of order greater than $p^{\sqrt{8\alpha}}$.

The above groups are essentially the same as the groups used by Higman [Hi1] to obtain the lower bound on the number of p-groups.

It would be of interest to develop a systematic theory of such random Higman groups. For some results in this direction see [BGH], [BGP], [Go], [Hi2], [HP1], [Lu5], [Ma2], and [Ma3].

8. Dense free subgroups of profinite groups

A subgroup Γ of a profinite group G is called dense in G if the topological closure of Γ in G coincides with G.

M. Bhattacharjee considered certain profinite groups W which can be obtained as inverse limits of iterated wreath products of finite alternating groups. She proved [Bh1] that a random pair of elements generates such a group W (as a topological group) with positive probability. Furthermore, it follows from the results in [Bh2] that a random pair of elements generates a free discrete subgroup of W with probability 1. These results together give a probabilistic proof of the fact that W has a dense F_2 subgroup.

It can be shown that Weigel's Theorem is equivalent to the following.

THEOREM 8.1. *If G is a 2-generated profinite group with arbitrarily large non-abelian simple quotients then G has a dense F_2 subgroup.*

Moreover, the probabilistic proof indicated in section 7 can be rephrased in terms of profinite groups. As a byproduct we obtain the following.

THEOREM 8.2. *If G is as in Theorem 8.1 then the probability that a random pair of elements generates a (discrete) F_2 subgroup is* 1.

Analogous results hold for profinite groups with maximal subgroup growth exceeding $n^{c\sqrt{n}}$ (this follows from 6.6 (ii)).

More significantly, the following is true [HP2].

THEOREM 8.3. *Let Γ be a finitely generated linear group and $G = \hat{\Gamma}$ its profinite completion. Then either Γ is solvable-by-finite or G has an open subgroup H such that H has a dense free subgroup F (of finite rank).*

There is a completely different source of profinite groups having dense free subgroups.

Extending an observation from [WZ] one can show [LP] that a d-generated pro-p-group has a dense F_d subgroup exactly if it satisfies no coset identity (a group law on the cosets of an open subgroup).

As proved by Wilson and Zelmanov [WZ] if a pro-p-group G satisfies a coset identity then a certain Lie algebra associated with G satisfies an identity.

The above results suggest a number of open problems [LP]. Let G be a finitely generated profinite group.

Problem 8.4. Suppose G has arbitrarily large alternating sections. Does G have a dense free subgroup of finite rank?

Problem 8.5. Suppose G has an F_2 subgroup. Does G also have a dense free subgroup of finite rank?

Problem 8.6. Suppose G has a dense free subgroup of finite rank. Is it true that a random pair of elements of G generates an F_2 subgroup with probability 1?

References

[AG] M. Aschbacher and R. M. Guralnick, *Solvable generation of groups and Sylow subgroups of the lower central series*, J. Algebra, **77** (1982), 189–201.

[Ba] L. Babai, *The probability of generating the symmetric group*, J. Comb. Theory (A), **52** (1989), 148–153.

[BCP] L. Babai, P. J. Cameron and P. P. Pálfy, *On the orders of primitive groups with restricted nonabelian composition factors*, J. Algebra, **79** (1982), 161–168.

[BGH] J. Buhler, R. Gupta and J. Harris, *Isotropic subspaces for skewforms and maximal abelian subgroups of p-groups*, J. Algebra, **108** (1987), 269–279.

[BGP] L. Babai, A. J. Goodman and L. Pyber, *On faithful permutation representations of small degree*, Comm. in Algebra, **21** (1993), 1587–1602.

[Bh1] M. Bhattacharjee, *The probability of generating certain profinite groups by two elements*, Israel J. Math., **86** (1994), 311–330.

[Bh2] M. Bhattacharjee, *Free subgroups in certain profinite groups*, in preparation.

[Bl] S. R. Blackburn, *Group enumeration*, D. Phil. thesis, Oxford, 1992.

[BP] A. V. Borovik and L. Pyber, *On the number of maximal solvable subgroups of a finite group*, in preparation.

[BPSh] A. V. Borovik, L. Pyber and A. Shalev, *Maximal subgroups in finite and profinite groups* Trans. Amer. Math. Soc., **348** (1996), 3745–3761.

[Di] J. D. Dixon, *The probability of generating the symmetric group*, Math. Z. **110** (1969), 199–205.

[DPSSh] J. D. Dixon, L. Pyber, Á. Seress and A. Shalev, *Residual properties of free groups and probabilistic methods*, in preparation.

[FJ] M. D. Fried and M. Jarden, *Field Arithmetic*, Springer, Berlin, 1986.

[GK] R. M. Guralnick and W. M. Kantor, *Probabilistic generation of finite simple groups*, in preparation.

[Go] A. J. Goodman, *The edge-orbit conjecture of Babai*, J. Comb. Theory (B), **57** (1993), 26–35.

[Gu] R. M. Guralnick, *On the number of generators of a finite group*, Arch. Math., **53** (1989), 521–523.

[Hi1] G. Higman, *Enumerating p-groups, I: Inequalities*, Proc. London Math. Soc. (3), **10** (1960), 24–30.

[Hi2] G. Higman, *Enumerating p-groups, II: Problems whose solution is PORC*, Proc. London Math. Soc. (3), **10** (1960), 562–582.

[Ho] D. F. Holt, *Enumerating perfect groups*, J. London Math. Soc., **39** (1989), 67–78.

[Hö] O. Hölder, *Die Gruppen mit quadratfreier Ordnungszahl*, Nachr. König. Ges. Wiss. Göttingen Math.-Phys. Kl. (1895), 211–229.

[HP1] H. W. Henn and S. Priddy, *p-Nilpotence, classifying space indecomposability, and other properties of almost all finite groups*, Comment. Math. Helv., **69** (1994), 335–350.

[HP2] E. Hrushovski and L. Pyber, unpublished.

[KL1] W. M. Kantor and A. Lubotzky, *The probability of generating a finite classical group*, Geom. Dedicata, **36** (1990), 67–87.

[KL2] W. M. Kantor and E. M. Luks, *Computing in quotient groups*, Proc. 22nd ACM STOC (1990), 524–534.

[KP] R. L. Kruse and D. T. Price, *Enumerating finite rings*, J. London Math. Soc., (2), **2** (1970), 149–159.

[Li1] V. A. Liskovets, *Towards the enumeration of subgroups of the free group*, Dokl. AN BSSR, **15** (1971), 6–9, (in Russian).

[Li2] H.-Q. Liu, *On the number of abelian groups of a given order*, Acta Arith. **59** (1991), 261–277.

[LP] L. Lévai and L. Pyber, *Dense free subgroups in profinite groups*, in preparation.

[LPSh] A. Lubotzky, L. Pyber and A. Shalev, *Discrete groups of slow subgroup growth*, Israel J. Math. (Amitsur Volume), **96** (1996), 399–418.

[LSh1] M. W. Liebeck and A. Shalev, *The probability of generating a finite simple group*, Geom. Dedicata, **56** (1995), 103–113.

[LSh2] M. W. Liebeck and A. Shalev, *Classical groups, probabilistic methods, and the $(2,3)$-generation problem*, Annals of Math., **144** (1996), 77–125.

[Lu1] A. Lubotzky, *Subgroup growth and congruence subgroups*, Invent. Math., **119** (1995), 267–195.

[Lu2] A. Lubotzky, *Subgroup growth*, in Proc. ICM Zürich '94, Vol.1, Birkhäuser, Basel, 1995, pp. 309–317.

[Lu3] A. Lucchini, *A bound for the number of generators of a finite group*, Arch. Math., **53** (1989), 313–317.

[Lu4] A. Lucchini, *Some profinite groups containing a dense free subgroup*, in preparation.

[Lu5] E. M. Luks, *What is a typical nilpotent Lie algebra?* in Computers in nonassociative rings and algebras, (eds. R. E. Beck and B. Kolman) Acad. Press, 1977, pp. 189–207.

[Ma1] A. Mann, *Positively finitely generated groups*, Forum Math., **8** (1996), 429–459.

[Ma2] G. Zh. Mantashyan, *The number of generators of finite p-groups and nilpotent groups without torsion and dimensions of associative rings and Lie algebras*, Matematika, **6** (1988), 178–186, (in Russian), Zbl. Math., **744** 20034.

[Ma3] U. Martin, *Almost all p-groups have automorphism group a p-group*, Bull. Amer. Math. Soc., **15** (1986), 78–82.

[MM] M. R. Murty and V. K. Murty, *On groups of squarefree order*, Math. Annalen, **267** (1989), 299–304.

[MN] A. McIver and P. M. Neumann, *Enumerating finite groups*, Quart. J. Math. Oxford, **38** (1987), 473–488.

[MP] T. Müller and L. Pyber, unpublished.

[MS1] A. Mann and D. Segal, *Subgroup growth: some current developments*, in Infinite groups 94 (eds. de Giovanni and Nevell) Walter de Gruyter, Berlin-New York, 1995, pp. 179–197.

[MS2] M. R. Murty and S. Srinivasan, *On the number of groups of squarefree order*, Can. Math. Bull., **30** (1987), 87–92.

[MSh] A. Mann and A. Shalev, *Simple groups, maximal subgroups and probabilistic aspects of profinite groups*, Israel J. Math. (Amitsur Volume), **96** (1996), 449–468.

[Mü] T. Müller, *Subgroup growth of free products*, Invent. Math., to appear.

[Ne1] Yu. A. Neretin, *An estimate of the number of parameters defining an n-dimensional algebra*, Math. USSR Izvestiya, **30** (1988), 283–294.

[Ne2] P. M. Neumann, *An enumeration theorem for finite groups*, Quart. J. Math. Oxford, **20** (1969), 395–401.

[Ne3] P. M. Neumann, *Proof of a conjecture by Garrett Birkhoff and Philip Hall on the automorphisms of a finite group*, Bull. London. Soc., **27** (1995), 222–224.

[NS] M. F. Newman and C. Seeley, *Modality for nilpotent Lie algebras*, in preparation.

[NV] P. M. Neumann and M. R. Vaughan-Lee, *An essay on BFC groups*, Proc. London Math. Soc. (3), **35** (1977), 213–237.

[Ol] A. Yu. Olshanskii, *The number of generators and orders of abelian subgroups of finite p-groups*, Math. Notes **23** (1978), 183–185.

[PSh1] L. Pyber and A. Shalev, *Groups with super-exponential subgroup growth*, Combinatorica, **16** (1996), 527–533.

[PSh2] L. Pyber and A. Shalev, *Asymptotic results for primitive permutation groups*, J. Algebra, **188** (1997), 103–124.

[Py1] L. Pyber, *Enumerating finite groups of given order*, Annals of Math., **137** (1993), 203–220.

[Py2] L. Pyber, *Asymptotic results for permutation groups*, in Groups and Computation, Proc. Workshop New Brunswick 1991, DIMACS Series in Discrete Mathematics and Theoretical Computer Science, Vol.11 (eds. L. Finkelstein and W. M. Kantor) 1993, pp. 197–219.

[Py3] L. Pyber, *Asymptotic results for simple groups and some applications*, in Groups and Computation, Proc. Workshop New Brunswick 1995, DIMACS Series in Discrete Mathematics and Theoretical Computer Science Vol. 128, (eds. L. Finkelstein and W. M. Kantor), 1997, 309–327.

[Si] C. C. Sims, *Enumerating p-groups*, Proc London Math. Soc., **15** (1965), 151–166.

[We1] T. S. Weigel, *Residual properties of free groups*, J. Algebra, **160** (1993), 16–41.

[We2] T. S. Weigel, *Residual properties of free groups, II.*, Comm. Algebra, **20** (1992) 1395–1425.

[We3] T. S. Weigel, *Residual properties of free groups, III.*, Israel J. Math., **77** (1992) 65–81.

[WZ] J. S. Wilson and E. Zelmanov, *Identities for Lie algebras of pro-p groups*, J. of Pure and Applied Algebra, **81** (1992), 103–109.

Progress in Mathematics, Vol. 169, © 1998 Birkhäuser Verlag Basel/Switzerland

Studying Dynamical Systems with Algebraic Tools

NÁNDOR SIMÁNYI*

Bolyai Institute of Mathematics
The University of Szeged
Szeged, Aradi Vértanuk t. 1.
H-6720 Hungary

The proper mathematical formulation of Ludwig Boltzmann's ergodic hypothesis, which has been the source of so much interest and discussion in the last hundred years, is still not clear. For systems of elastic hard balls on a torus, however, Yakov Sinai, in 1963, [Sin(1963)] gave a stronger, and at the same time mathematically rigorous, version of Boltzmann's hypothesis: *the system of an arbitrarily fixed number N of identical elastic hard balls moving in the ν-torus $\mathbb{T}_L^\nu = \mathbb{R}^\nu/L\cdot\mathbb{Z}^\nu$ ($\nu \geq 2$) is ergodic – of course, on the submanifold of the phase space specified by the trivial conservation laws.* Boltzmann used his ergodic hypothesis when laying the foundations of statistical physics, and its various forms are still intensively used in modern statistical physics. The meaning of Sinai's hypothesis for the theory of dynamical systems, which was partially based on the physical arguments of Krylov's 1942 thesis (cf. [K(1979)] and its afterword written by Ya. G. Sinai, [Sin(1979)]), is stressed by the fact that the interaction of elastic hard balls defines the only physical system of an arbitrary number of particles in arbitrary dimension whose dynamical behaviour has been so far at least guessed – except for the completely integrable system of harmonic oscillators. (As to the history of Boltzmann's hypothesis, see the recent work [Sz(1996)].)

The aim of the present survey is to give the reader a sketchy review of the newly developed algebraic methods that have been successfully used in the proof of a weak version of the ergodic hypothesis for hard ball systems, which we call the weak version of the *Boltzmann-Sinai Ergodic Hypothesis.* Our formulation is more general than Sinai's, in that we consider balls with different masses rather than those with necessarily identical masses, though our theroem is only established for vectors of outer geometric parameters $(\vec{m}; L) = (m_1, \dots, m_N; L) \in \mathbb{R}^{N+1}$ not belonging to a countable union of proper analytic submanifolds.

Let us first specify the model and formulate the theorem.

Assume that, in general, a system of $N (\geq 2)$ balls, identified as $1, 2, \dots, N$, of masses $m_1, \dots, m_N (> 0)$ and radius $r > 0$ are given in $\mathbb{T}_L^\nu = \mathbb{R}^\nu/L \cdot \mathbb{Z}^\nu$, the

*) Research supported by the Hungarian National Foundation for Scientific Research, grant No. OTKA 16425. E-mail: simanyi@math.u-szeged.hu

ν-dimensional unit L-torus ($\nu \geq 2$). Denote the phase point of the ith ball by $(q_i, v_i) \in \mathbb{T}_L^\nu \times \mathbb{R}^\nu$. A priori, the configuration space $\tilde{\mathbf{Q}}$ of the N balls is a subset of $\mathbb{T}_L^{N\cdot\nu}$: from $\mathbb{T}_L^{N\cdot\nu}$ we cut out $\binom{N}{2}$ cylindric scatterers:

$$\tilde{C}_{i,j} = \left\{Q = (q_1, \ldots, q_N) \in \mathbb{T}_L^{N\cdot\nu} : \|q_i - q_j\| < 2r\right\},$$

$1 \leq i < j \leq N$, or in other words $\widetilde{\mathbf{Q}} := \mathbb{T}_L^{N\nu} \setminus \bigcup_{1\leq i<j\leq N} \tilde{C}_{i,j}$. The energy $H = \frac{1}{2}\sum_1^N m_i v_i^2$ and the total momentum $P = \sum_1^N m_i v_i$ are first integrals of motion. Thus, without loss of generality, we can assume that $H = \frac{1}{2}$, $P = 0$. (If $P \neq 0$, then the system has an additional conditionally periodic or periodic motion.) Now, for these values of H and P, we define our dynamical system.

In earlier works (cf. [S-Sz(1995)-II] and the references therein), where the masses were identical, one could and – to obtain ergodicity – one had to fix a center of mass. For different masses – as observed in [S-W(1989)] – this is generically not possible and we should follow a different approach.

The equivalence relation Ψ over $\widetilde{\mathbf{Q}}$, defined by $Q \sim_\Psi Q^*$ *if and only if there exists an* $a \in \mathbb{T}_L^\nu$ *such that for every* $i \in [1, N]$ $\quad q_i^* = q_i + a$, allows us to introduce $\mathbf{Q} := \widetilde{\mathbf{Q}}/\Psi$ (the equivalence relation means, in other words, that for the internal coordinates $q_i - q_j = q_i^* - q_j^*$ holds for every $i, j \in [1, N]$). The set $\mathbf{Q}$, a compact, flat Riemannian manifold with boundary will actually be the configuration space of our system, whereas its phase space will be the ellipsoid bundle $\mathbf{M} := \mathbf{Q} \times \mathcal{E}$, where $\mathcal{E}$ denotes the ellipsoid $\sum_{i=1}^N m_i(dq_i)^2 = 1, \quad \sum_{i=1}^N m_i dq_i = 0$. Clearly, $d := \dim \mathbf{Q} = N\nu - \nu$, and $\dim \mathcal{E} = d - 1$. The well-known Liouville measure μ is invariant with respect to the time evolution $S^{\mathbb{R}} := \{S^t : t \in \mathbb{R}\}$ of our dynamical system defined by elastic collisions of the balls of positive masses $m_1, \ldots, m_N$ and their uniform free motion.

The dynamical system $\left(\mathbf{M}, S^{\mathbb{R}}, \mu\right)_{\vec{m},L}$ is called *the standard billiard ball system with the vector* $(\vec{m}; L)$ *of outer geometric parameters.*

Denote by $R_0 := R_0(N, \nu, L)$ the interval of those values of $r > 0$, for which the interior $\mathrm{Int}\mathbf{M} = \mathrm{Int}\mathbf{Q} \times \mathcal{E}$ of the phase space of the standard billiard ball flow is connected.

The main result of Domokos Szász and myself [S-Sz(1996)] is the following

THEOREM. *For* $N, \nu \geq 2$ *and* $r \in R_0$ *the standard billiard ball system* $\left(\mathbf{M}, \{S^{\mathbb{R}}\}, \mu\right)_{\vec{m},L}$ *has nonzero relevant Lyapunov exponents* μ*-almost everywhere and, consequently, it has (at most countably many) open ergodic components so that the restriction of the billiard flow to each ergodic component is a K-mixing flow, provided that the vector* $(\vec{m}; L) \in \mathbb{R}_+^{N+1}$ *of outer geometric parameters does not belong to a countable union of proper analytic submanifolds of exceptional vectors* $(\vec{m}; L)$.

Note 1

In general, if $r \notin R_0$, then $\mathrm{Int}\mathbf{M}$ decomposes into a finite number of connected components. Then our methods provide that the standard billiard ball flow has nonzero relevant Lyapunov exponents (almost everywhere) on each of these connected components whenever the vector $(\vec{m}; L) \in \mathbb{R}_+^{N+1}$ does not belong to a countable union of proper analytic submanifolds of exceptional vectors $(\vec{m}; L)$.

Note 2

As to the basic results (and history) concerning the Boltzmann and Sinai hypotheses we refer to the recent survey [Sz(1996)]. The approach of [S-Sz(1996)] uses the following basic results: the Chernov-Sinai Fundamental Theorem (the Theorem on Local Ergodicity) for Semi-Dispersing Billiards [(S-Ch(1987)] in its form proved in [K-S-Sz(1990)] on the one hand, and my theorems on the Uniform Avoiding of Balls and the appropriate generalization of my Connecting Path Formula on the other hand (see [Sim(1992)], where, in fact, the Boltzmann-Sinai Ergodic Hypothesis was already established for the case when $\nu \geq N$).

In the first part of this review the reader will be reminded of the basic strategy for establishing global ergodicity of semi-dispersing billiards as was suggested in our series of works with András Krámli and Domokos Szász (see the references), and then we will point out the new elements of the method of the present work.

The basic notion in the theory of semi-dispersing billiards is that of the *sufficiency* of a phase point or, equivalently, of its trajectory. The conceptual importance of sufficiency can be explained as follows: in a suitably small neighborhood of a (typical) phase point of a dispersing billiard the system is *hyperbolic*, i.e. its relevant Lyapunov exponents are not zero. For a semi-dispersing billiard the same property holds for sufficient points only. Physically speaking, a phase point is sufficient if its trajectory encounters in its history all possible degrees of freedom of the system. Then the theorem on local ergodicity for semi-dispersing billiards (see the main result of [S-Ch (1987)]) says that – under certain conditions – a suitably small neighborhood of a sufficient phase point does belong to one ergodic component. As a consequence, if, for instance, almost every phase point of a semi-dispersing billiard is sufficient, then this property implies that the ergodic components are open and, therefore, their number is countable. To obtain, however, the (global) ergodicity of the flow, a more stringent property of the subset of non-sufficient points is needed. For that purpose, in our work with A. Krámli and D. Szász [K-S-Sz(1989)], the topological property of *slimness*, closely related to that of *topological codimension two*, was suggested and used fruitfully in later works.

Definition. *We say that a subset of a smooth manifold M is* **slim** *if it can be covered by a countable union of codimension two (i.e. at least two), closed subsets of μ-measure zero, where μ is a smooth measure on M.*

Remark

The phrase *dimension* in this review always means the topological or Hurewicz dimension of a separable metric space, see for instance [H-W(1941)].

It has long been an accepted idea among experts that the right way of proving the ergodicity of hard ball systems is an induction on the number N of balls.

According to our strategy initiated in [K-S-Sz(1991)] and explained in the introductions of [K-S-Sz(1992)] and [Sim(1992)-I] , there are three major parts in such an induction, once a combinatorial property, called the *richness* of the symbolic collision sequence of a trajectory, has been suitably defined.

(1) The "geometric-algebraic considerations" on the codimension of the manifolds describing the non-sufficient trajectory segments with a combinatorially rich symbolic collision structure;
(2) Proof of slimness (negligibility) of the set of phase points with a combinatorially non-rich collision sequence by using the K-property of less than N hard balls in the torus;
(3) Checking the Chernov-Sinai Ansatz – an important necessary condition for the proof of the Theorem on Local Ergodicity, see Condition (A) before Theorem 5 in [S-Ch(1987)] or Condition 3.1 in [K-S-Sz(1990)]. This step also uses the K-property of less than N balls and step (1) for the system of N hard balls in the torus.

We emphasize here that in order to prove the above theorem (i.e. that the relevant Lyapunov exponents are nonzero almost everywhere) it is enough to obtain one codimension in (1) and to prove zero Liouville measure instead of slimness in (2).

As a consequence of the methods of [Sim(1992)], the main difficulty in our general case is to settle the geometric-algebraic part (1), and here we needed to use some algebraic tools, see below. To do so, we have introduced *varying masses*, and can claim nonzero relevant Lyapunov exponents for typical mass vectors only. Our novel approach of playing with arbitrary masses actually necessitates the *complexification* of the dynamics. The complexification is required because of the inductive derivation of the Key Lemma (see below), so that one does not have to worry about the existence of orbit segments with a prescribed symbolic collision sequence: they do exist, thanks to the algebraic closedness of the ground field $\mathbb{C}$.

The neutral subspace. Linearity vs. hyperbolicity

Let us consider a trajectory segment

$$S^{[a,b]}(x) = \left\{q_i(t),\, v_i(t)\big|\; i = 1, \ldots, N;\; a \le t \le b\right\}$$

of the standard billiard flow $\left(\mathbf{M}, S^{\mathbb{R}}, \mu\right)_{\vec{m};L}$ with the symbolic collision sequence $\Sigma = (\sigma_1, \ldots, \sigma_n)$, where the symbol $\sigma_k = (i_k, j_k)$ $(1 \le i_k < j_k \le N)$ means that the k-th collision takes place between the i_k-th and j_k-th particles. We are interested in studying those infinitesimal variations

$$\{\delta q_i(t),\, \delta v_i(t)|\; i = 1, \ldots, N;\; a \le t \le b\}$$

$$\sum_{i=1}^{N} m_i \delta q_i(t) = \sum_{i=1}^{N} m_i \delta v_i(t) = \sum_{i=1}^{N} m_i \langle v_i(t), \delta v_i(t)\rangle = 0$$

of this trajectory segment $S^{[a,b]}(x)$ (i.e. Jacobi vector fields along this orbit segment) for which *absolutely no scattering effect* of our cylindric billiard can be observed or, equivalently, *no exponential divergence of the trajectories* can be detected along this infinitesimal variation or, equivalently again, $\delta v_i(t) = 0$ for $i = 1, \ldots, N$ and $a \leq t \leq b$. Such parallel, or "neutral" Jacobi vector fields are uniquely determined by the initial variation of the positions

$$\delta q(a) = (\delta q_1(a), \ldots, \delta q_N(a)).$$

These vectors constitute the so-called *neutral vector space* $\mathcal{N}\left(S^{[a,b]}(x)\right)$.

DEFINITION. *The trajectory segment $S^{[a,b]}(x)$ is called sufficient, or hyperbolic, iff* $\dim \mathcal{N}\left(S^{[a,b]}(x)\right) = 1$, *i.e. the neutral space only contains the flow direction. (Which direction necessarily belongs to the neutral space.)*

The key feature of the neutral space $\mathcal{N}\left(S^{[a,b]}(x)\right)$ is that apparently "everything" is linear on it, especially the variation $-\alpha_k = \delta t_k$ of the time of the collision σ_k is a linear function of the neutral vector $\delta q(a) = (\delta q_1(a), \ldots, \delta q_N(a))$. The name of the linear functional $\alpha_k : \mathcal{N}\left(S^{[a,b]}(x)\right) \to \mathbb{R}$ is the *advance* of the collision σ_k.

It is easy to see that the above defined sufficiency is equivalent to the equality of all advance functionals: $\alpha_1 = \cdots = \alpha_n$.

The advantage of sufficiency can be clearly seen from

PROPOSITION. *Every sufficient phase point $x_0 \in \text{Int}\mathbf{M}$ (with at most one singularity on its orbit) has an open neighborhood $x_0 \in U_0 \subset \mathbf{M}$ in $\mathbf{M}$ with the following property:*

The tangent space $\mathcal{T}_x\mathbf{M}$ of the phase space $\mathbf{M}$ at almost every point $x \in U_0$ splits into a direct sum of vector spaces

$$\mathcal{T}_x\mathbf{M} = E_s(x) \oplus E_u(x) \oplus E_0(x)$$

so that for every nonzero vector $v \in E_s(x)$ we have

$$\lim_{t\to\infty} t^{-1} \log \left\|(DS^t)(v)\right\| < 0,$$

whereas for every nonzero vector $v \in E_u(x)$ one has

$$\lim_{t\to-\infty} t^{-1} \log \left\|(DS^t)(v)\right\| > 0,$$

and, finally, $\dim E_0(x) = 1$. *(The subspace $E_0(x)$ merely contains the flow direction. The above statement is the precise notion of having nonzero relevant Lyapunov exponents in U_0.)*

Furthermore, the neighborhood U_0 belongs to a single ergodic component of the billiard flow.

This result is essentially the celebrated Theorem on Local Ergodicity by N.I. Chernov and Ya.G. Sinai, [S-Ch(1987)]. It says – besides local ergodicity – that the flow behaves in U_0 in a hyperbolic manner, or there is (locally) an exponential divergence of nearby trajectories similar to what we can detect in a geodesic flow on a compact, negatively curved manifold.

It is a direct consequence of the Connecting Path Formula (CPF, Lemma 2.9 of [Sim(1992)-II]) that we can compute the neutral space $\mathcal{N}\left(S^{[a,b]}(x)\right)$ by solving a homogeneous system of linear equations

$$\sum_{k=1}^{n} \alpha_k \Gamma_{ik} = 0, \; i = 1, 2, \ldots, n + P_\Sigma - N, \qquad (*)$$

where each equation in (*) is a ν-dimensional vector equation, the coefficients $\Gamma_{ik} \in \mathbb{R}^\nu$ are certain linear combinations of relative velocities $v_i^k - v_j^k$ (the coefficients of those linear combinations are just fractional linear expressions of the masses) and, finally, P_Σ denotes the number of connected components of the collision graph of Σ. Now the sufficiency (hyperbolicity) of the trajectory segment $S^{[a,b]}(x)$ can be characterized in terms of (*) as follows: the system (*) has a one-dimensional (minimum dimension!) solution set.

Combinatorial richness of the symbolic sequence

The fundamental notion (to appear in [S-Sz(1996)]) guaranteeing the (almost sure) sufficiency of $S^{[a,b]}(x)$ is the *combinatorial richness* of the symbolic collision sequence $\Sigma = (\sigma_1, \ldots, \sigma_n)$ of $S^{[a,b]}(x)$:

DEFINITION. *We say that* $\Sigma = (\sigma_1, \ldots, \sigma_n)$ *is* **rich** *if* Σ *contains at least* $C(N)$ *consecutive, connected collision subgraphs, where the natural numbers* $C(N)$ $(N \geq 2)$ *can be defined by the following recursion:* $C(2) = 1$, $C(N) = \frac{N}{2} \cdot \max\{C(N-1), 3\}$ $(N \geq 3)$, *that is,* $C(2) = 1$, $C(N) = \frac{3N!}{2^{N-1}}$ *for* $N \geq 3$.

We note that, as has been proved in [K-S-Sz(1991)], for $N = 3$ the value $C(3) = 2$ already does the job.

The measure-theoretic part of the inductive step (of the proof of full hyperbolicity, that is, that the relevant Lyapunov exponents are almost everywhere nonzero) is the following easy lemma:

LEMMA. *Assume that all hard ball systems with masses* $m_{i_1}, m_{i_2}, \ldots, m_{i_l}$ $(1 \leq i_1 < i_2 < \cdots < i_l \leq N, \, l < N)$ *have nonzero (relevant) Lyapunov exponents almost everywhere, i.e. they are fully hyperbolic. Then for* $\mu_{\vec{m};L}$*-almost every phase point* $x \in \mathbf{M}(m_1, \ldots, m_N; L)$ *the orbit* $S^{\mathbb{R}}x$ *has infinitely many consecutive, connected collision subgraphs.*

In a strengthened version of the above lemma (Theorem 5.1 of [Sim(1992)-I]) the notion of zero measure is replaced by the concept of slimness. That result proved to be crucial in the proof of the ergodicity of standard hard ball systems with $N \leq \nu$.

The missing element of the "puzzle" is then to prove that the combinatorial richness defined above actually implies that the trajectory segment is almost surely hyperbolic.

Polynomial equations determining the dynamics

Let us now fix the symbolic collision sequence $\Sigma = (\sigma_1, \dots, \sigma_n)$ and explore the algebraic relationship between the data $\{q_i^{k-1}, v_i^{k-1} | i = 1, \dots, N\}$ and $\{q_i^k, v_i^k | i = 1, \dots, N\}$, $k = 1, \dots, n$. The symbol $v_i^k = v_i^+(t_k) \in \mathbb{R}^\nu$ shall denote the velocity $\dot{q}_i$ of the i-th ball right *after* the k-th collision $\sigma_k = \{i_k, j_k\}$ $(1 \le i_k < j_k \le N)$ occuring at time t_k, $k = 1, \dots, n$. Of course, there is no need to deal with the velocities right *before* collisions, since $v_i^-(t_k) = v_i^+(t_{k-1}) = v_i^{k-1}$. As usual, $q_i^k \in \mathbb{T}_L^\nu = \mathbb{R}^\nu / L \cdot \mathbb{Z}^\nu$ denotes the position of the center of the i-th ball at the moment t_k of the k-th collision. By definition, the data $\{q_i^0, v_i^0 | i = 1, \dots, N\}$ correspond to the initial (non-collisioñ) phase point x_0. We also set $t_0 = 0$.

Since we would like to carry out arithmetic operations on these data, the periodic positions $q_i^k \in \mathbb{T}_L^\nu$ are not suitable for this purpose. Therefore, instead of studying the genuine orbit segments $S^{[0,b]} x_0$, we will deal with their Euclidean liftings.

Proposition. *Let $S^{[0,b]} x_0 = \{q_i(t), v_i(t) | 0 \le t \le b\}$ be an orbit segment as above, and assume that certain pre-images (Euclidean liftings) $\tilde{q}_i(0) = \tilde{q}_i^0 \in \mathbb{R}^\nu$ of the positions $q_i(0) \in \mathbb{T}_L^\nu = \mathbb{R}^\nu / L \cdot \mathbb{Z}^\nu$ are given. Then there is a uniquely defined, continuous, Euclidean lifting $\{\tilde{q}_i(t) \in \mathbb{R}^\nu | 0 \le t \le b\}$ of the given orbit segment which is an extension of the initial lifting $\{\tilde{q}_i^0 | i = 1, \dots, N\}$. Moreover, for every collision σ_k there exists a uniquely defined integer vector $a_k \in \mathbb{Z}^\nu$ – named the adjustment vector of σ_k – such that*

$$\left\| \tilde{q}_{i_k}^k - \tilde{q}_{j_k}^k - L \cdot a_k \right\|^2 - 4r^2 = 0.$$

We shall call the orbit segment $\tilde{\gamma} = \{\tilde{q}_i^k, v_i^k | i = 1, \dots, N;\ k = 0, \dots, n\}$ a **lifted orbit segment with the system of adjustment vectors** $\mathcal{A} = (a_1, \dots, a_n) \in \mathbb{Z}^{n\nu}$.

The next result establishes the already mentioned polynomial relationships between the kinetic data

$$\{\tilde{q}_i^{k-1}, v_i^{k-1} | i = 1, \dots, N\}$$

and

$$\{\tilde{q}_i^k, v_i^k | i = 1, \dots, N\},$$

$k = 1, \dots, n$.

Proposition. *Using all notations and notions from above, one has the following polynomial relations between the kinetic data at σ_{k-1} and σ_k, $k = 1, \dots, n$. In order to simplify the notations, we write down those equations as if σ_k was $\{1, 2\}$:*

$$v_i^k = v_i^{k-1}, \quad i \notin \{1, 2\}, \tag{A}$$

$$m_1 v_1^k + m_2 v_2^k = m_1 v_1^{k-1} + m_2 v_2^{k-1} \tag{B}$$

(conservation of the momentum),

$$v_1^k - v_2^k = v_1^{k-1} - v_2^{k-1} - \frac{1}{2r^2}\left\langle v_1^{k-1} - v_2^{k-1}; \tilde{q}_1^k - \tilde{q}_2^k - L \cdot a_k \right\rangle \cdot \left(\tilde{q}_1^k - \tilde{q}_2^k - L \cdot a_k\right) \tag{C}$$

(reflection of the relative velocity determined by the elastic collision),

$$\tilde{q}_i^k = \tilde{q}_i^{k-1} + \tau_k v_i^{k-1}, \quad i = 1, \ldots, N, \tag{D}$$

where the time slot $\tau_k = t_k - t_{k-1}$ ($t_0 := 0$) is determined by the quadratic equation

$$\left\| \tilde{q}_1^{k-1} - \tilde{q}_2^{k-1} + \tau_k \left(v_1^{k-1} - v_2^{k-1}\right) - L \cdot a_k \right\|^2 = 4r^2. \tag{E}$$

Remark

Observe that the new velocities v_1^k and v_2^k can be computed (by using the above equations) as follows:

$$v_1^k = v_1^{k-1} - \frac{\mu_2}{2r^2}\left\langle v_1^{k-1} - v_2^{k-1}; \tilde{q}_1^k - \tilde{q}_2^k - L \cdot a_k \right\rangle \cdot \left(\tilde{q}_1^k - \tilde{q}_2^k - L \cdot a_k\right),$$
$$v_2^k = v_2^{k-1} + \frac{\mu_1}{2r^2}\left\langle v_1^{k-1} - v_2^{k-1}; \tilde{q}_1^k - \tilde{q}_2^k - L \cdot a_k \right\rangle \cdot \left(\tilde{q}_1^k - \tilde{q}_2^k - L \cdot a_k\right),$$

where $\mu_1 = m_1/(m_1 + m_2)$, $\mu_2 = 1 - \mu_1$. We also note that the equations above even extend analytically to the case when one of the two masses, say m_1, is equal to zero:

$$v_1^k = v_1^{k-1} - \frac{1}{2r^2}\left\langle v_1^{k-1} - v_2^{k-1}; \tilde{q}_1^k - \tilde{q}_2^k - L \cdot a_k \right\rangle \cdot \left(\tilde{q}_1^k - \tilde{q}_2^k - L \cdot a_k\right),$$
$$v_2^k = v_2^{k-1}.$$

We would like to emphasize here that the symbol $\langle\, . ;\, . \,\rangle$ denotes the Euclidean inner product of ν-dimensional real vectors and $\|\, . \,\|$ is the corresponding norm.

Then we *complexify* the whole system by considering all quantities $\tilde{q}_i^k \in \mathbb{C}^\nu$, $v_i^k \in \mathbb{C}^\nu$, m_i, $L \in \mathbb{C}$ as complex variables interrelated by the above polynomial equations (A-E). The equations (A-E) enable us to express each kinetic variable $\left(\tilde{q}_i^k\right)_j$ or $\left(v_i^k\right)_j$ ($i = 1, \ldots, N$, $k = 0, \ldots, n$, $j = 1, \ldots, \nu$, the subscript j denotes the j-th component of a ν-vector) as an algebraic function $f(\vec{x})$ of the initial kinetic variables

$$\vec{x} = \left\{ (\tilde{q}_i^0)_j, (v_i^0)_j, m_i, L \,|\, i = 1, \ldots, N; j = 1, \ldots, \nu \right\}.$$

As mentioned earlier, the sufficiency (hyperbolicity) of the trajectory segment $S^{[0,b]}(x)$ is equivalent to the fact that the homogeneous linear system (*) has a one-dimensional solution set. This assertion, along with the previous statement and some elementary linear algebra, yields that – in terms of the initial variables $\vec{x} \in \mathbb{C}^{(2\nu+1)N+1}$ – the hyperbolicity of $S^{[0,b]}(x)$ is equivalent to the simultaneous vanishing of finitely many algebraic functions:

$$f_1(\vec{x}) = f_2(\vec{x}) = \cdots = f_s(\vec{x}) = 0. \tag{**}$$

The inductive step $N-1 \longmapsto N$

The crucial combinatorial lemma, making the induction available, is

LEMMA. *Let $N \geq 3$, and suppose that the symbolic collision sequence $\Sigma = (\sigma_1, \dots, \sigma_n)$ for N particles is $C(N)$-rich. Then we can find a particle, say the one with label N, and two indices $1 \leq p < q \leq n$ such that*

(i) $N \in \sigma_p \cap \sigma_q$,
(ii) $N \notin \bigcup_{j=p+1}^{q-1} \sigma_j$,
(iii) $\sigma_p = \sigma_q \Longrightarrow \exists j\; p < j < q \,\&\, \sigma_p \cap \sigma_j \neq \emptyset$,

and

(iv) *Σ' is $C(N-1)$-rich on the vertex set $\{1, \dots, N-1\}$.*

Here we denote by Σ' the symbolic sequence that can be obtained from Σ by discarding all edges containing N.

Having this result and (**) at hand, we can carry out the substitution $m_N = 0$ in the equations (**). This means studying the orbit segments with an "infinitely light" N-th particle having zero mass. By using the induction hypothesis and some smart craftmanship, it can be shown that the Σ-trajectories are typically hyperbolic under the side condition $m_N = 0$. Then it is easy to conclude that in the generic case $m_N \neq 0$, too, the typical Σ-trajectory segments are sufficient.

Some open problems

1. To prove global ergodicity requires either

(i) a significant refinement of the algebraic methods sketched above, so that codimension-two exceptional manifolds can be obtained instead of just proper submanifolds,

or

(ii) some further "dynamical" ideas by showing that codimension-one, analytic submanifolds of the phase space cannot cut the phase space into several (more than one) open ergodic components.

(The second direction seems more promising.)

2. Box billiards, i.e. the motion of elastic balls in certain "box-like" containers, especially in rectangular boxes. Are such billiard flows ergodic? How to prove ergodicity? What kind of algebraic or dynamical methods ought to be developed in order to get a right approach?

3. The Main Conjecture of Cylindric Billiards.
The research on cylindric billiards conducted jointly with D. Szász in [S-Sz(1994)] suggests the following conjecture:

Conjecture on Cylindric Billiards. *Consider a cylindric billiard with circular cylindric scatterers. (In the case of a ball system, circularity is equivalent to having equal masses. Of course, as usual, we assume the connectedness of the interior of the configuration space.) Denote by G the transformation group on the sphere $\mathcal{E}$ of compound velocities generated by all rotations around the constituent spaces of the cylinders. Then this cylindric billiard is ergodic if and only if G acts transitively on $\mathcal{E}$.*

References

[B-S(1973)] L.A. Bunimovich, Ya.G. Sinai, *The fundamental theorem of the theory of scattering billiards*, Math. USSR-Sb. **19**, 407–423 (1973).

[C-F-S(1982)] I.P. Cornfeld, S.V. Fomin, Ya.G. Sinai, *Ergodic Theory*, Berlin, Heidelberg, New York: Springer 1982.

[C-H(1996)] N.I. Chernov, C. Haskell, *Nonuniformly hyperbolic K-systems are Bernoulli*, Ergodic Theory and Dynamical Systems **16**, 19–44 (1996).

[G-O(1974)] G. Gallavotti, D. Ornstein, *Billiards and Bernoulli schemes*, Commun. Math. Phys. **38**, 83–101 (1974).

[H-W(1941)] W. Hurewicz, H. Wallman, *Dimension Theory*, Princeton, 1941.

[K(1979)] N.S. Krylov, *The Processes of Relaxation of Statistical Systems and the Criterion of Mechanical Instability*, Thesis, Moscow, (1942); *Republished in English by Princeton University Press*, Princeton N. J. (1979).

[K-S-Sz(1989)] A. Krámli, N. Simányi, D. Szász, *Ergodic Properties of Semi–Dispersing Billiards I. Two Cylindric Scatterers in the 3–D Torus*, Nonlinearity **2**, 311–326 (1989).

[K-S-Sz(1990)] A. Krámli, N. Simányi, D. Szász, *A "Transversal" Fundamental Theorem for Semi-Dispersing Billiards*, Commun. Math. Phys. **129**, 535–560 (1990).

[K-S-Sz(1991)] A. Krámli, N. Simányi, D. Szász, *The K–Property of Three Billiard Balls*, Annals of Mathematics **133**, 37–72 (1991).

[K-S-Sz(1992)] A. Krámli, N. Simányi, D. Szász, *The K–Property of Four Billiard Balls*, Commun. Math. Phys. **144**, 107–148 (1992).

[O-W(1996)] D. Ornstein, B. Weiss, *On the Bernoulli Nature of Systems with Some Hyperbolic Structure*, Ergodic Theory and Dynamical Systems (accepted for publication), pp. 23 (1996).

[Sin(1963)] Ya.G. Sinai, *On the Foundation of the Ergodic Hypothesis for a Dynamical System of Statistical Mechanics*, Soviet Math. Dokl. **4**, 1818–1822 (1963).

[Sin(1970)] Ya.G. Sinai, *Dynamical Systems with Elastic Reflections*, Russian Math. Surveys **25:2**, 137–189 (1970).

[Sin(1979)] Ya.G. Sinai, *Development of Krylov's Ideas*, Afterword to N.S. Krylov's "Works on the foundations of statistical physics", Princeton University Press (1979).

[Sim(1992)] N. Simányi, *The K-property of N billiard balls I*, Invent. Math. **108**, 521–548 (1992); *II.*, ibidem **110**, 151–172 (1992).

[S-Ch(1987)] Ya.G. Sinai, N.I. Chernov, *Ergodic properties of certain systems of 2–D discs and 3–D balls*, Russian Math. Surveys **(3) 42**, 181–207 (1987).

[S-Sz(1994)] N. Simányi, D. Szász, *The K-property of 4-D Billiards with Non-orthogonal Cylindric Scatterers*, J. Stat. Phys. **76, Nos. 1/2**, 587–604 (1994).

[S-Sz(1995)-I] N. Simányi, D. Szász, *The Boltzmann-Sinai ergodic hypothesis for hard ball systems*, Manuscript.

[S-Sz(1995)-II] N. Simányi, D. Szász, *The K-property of Hamiltonian systems with restricted hard ball interactions*, Mathematical Research Letters **2, No. 6**, 751–770 (1995).

[S-Sz(1996)] N. Simányi, D. Szász, *Hard Ball Systems Are Completely Hyperbolic*, Submitted to Annals of Mathematics.

[S-W(1989)] N. Simányi, M. Wojtkowski, *Two-particle billiard system with arbitrary mass ratio*, Ergod. Th. & Dynam. Sys. **9**, 165–171 (1989).

[Sz(1993)] D. Szász, *Ergodicity of classical billiard balls*, Physica A **194**, 86–92 (1993).

[Sz(1994)] D. Szász, *The K-property of 'Orthogonal' Cylindric Billiards*, Commun. Math. Phys. **160**, 581–597 (1994).

[Sz(1996)] D. Szász, *Boltzmann's Ergodic Hypothesis, a Conjecture for Centuries?*, Studia Sci. Math. Hung **31**, 299–322 (1996).

Progress in Mathematics, Vol. 169, © 1998 Birkhäuser Verlag Basel/Switzerland

Mathematical Results on the Structure of Large Atoms

J.P. Solovej

Department of Mathematics
University of Copenhagen
Universitetsparken 5
DK-2100 Copenhagen, Denmark
e-mail: solovej@math.ku.dk

Abstract. One of the qualitative features of the periodic table is that all atoms have essentially the same size. More precisely, in each group of the periodic table the atomic radius is nearly constant. I shall in my lecture present a mathematical result, that, to some extent, explains this very important feature. In the spirit of the congress I will show how many branches of mathematics need to be united to understand the structure of large atoms. Among the tools I shall discuss are generalized Sobolev inequalities, semi-classical estimates, variational methods, and non-linear elliptic PDE.

1. Introduction

In this lecture it is my intention to give a very elementary presentation of some of the mathematics used in studying the structure of atoms. I shall mainly be concerned with the energy and size of atoms. One of the remarkable features is that the size of atoms varies quite *periodically* through the periodic table. We are so used to thinking of the periodic table as being periodic that this statement hardly sounds striking. It should however be quite surprising that for instance Lithium and Uranium have essentially the same size. After all Lithium is made up of only 3 electrons whereas Uranium consists of 92 electrons. The explanation behind this periodicity in the size of atoms is due to a very delicate balance between the forces governing the structure of atoms. This balance is thus of great importance in understanding the basic properties of the world around us.

Although, we are far from having a complete mathematical understanding of the structure of atoms and molecules we have made substantial progress over the past 30 years. It is my hope to illustrate how many branches of mathematics must be combined in order to describe the basic features of atoms.

2. Basics of the mathematics of atoms

Consider an atom consisting of N negatively charged electrons (charge $-e$) and a nucleus of positive charge Ze. In this lecture I shall discuss the ground state of such an atom (mainly the neutral case $N = Z$). I shall focus on *the ground state density and energy.* The ground state density is a non-negative function $\rho \in L^1(\mathbb{R}^3)$ describing the density of electrons, thus $\int \rho = N$.

The ground state energy $E(N,Z)<0$ is the *total* binding energy of the atom. The goal is to analyze the size of atoms. For any $\nu>0$ we define the *radius to the last* ν *electrons* $\mathcal{R}_\nu(N,Z)$ by

$$\int_{|x|>\mathcal{R}_\nu(N,Z)} \rho(x)dx=\nu.$$

It is of course not true that $\mathcal{R}_\nu(Z,Z)$ is strictly speaking periodic in Z. A weaker, but mathematically more tractable, aspect of the 'periodicity' is the fact that $\mathcal{R}_\nu(Z,Z)$ remains bounded above and below as Z tends to infinity.

It is true that in nature Z remains bounded (less than about 100) and discussing large Z is therefore purely academic.

Let me, however, emphasize that the interest here is really not on large Z. Allowing Z to be large is used in order to be able to make qualitative mathematical statements, rather than just numerical calculations, about the function $Z\mapsto\mathcal{R}_\nu(Z,Z)$.

I shall introduce three different models describing the atom. In increasing complexity these are *the Thomas-Fermi model* (TF), *the Hartree-Fock model* (HF), and the *Schrödinger model.*

As is probably well known the Schrödinger model gives an extremely accurate description of atoms. Unfortunately, this model is too complex for many both theoretical and numerical considerations.

The HF and TF models can be considered as approximations to the Schrödinger model. The HF model and various generalizations of it have been widely used in chemistry over the past 70 years. In Schrödinger theory it is still an open problem to prove adequate bounds on $\mathcal{R}_\nu(Z,Z)$.

In HF theory I have recently (see Theorem 1 below) been able to obtain such bounds [10]. These bounds also prove the *ionization conjecture* for the Hartree-Fock model, namely that ionization energies are bounded independently of Z. I shall not discuss ionization energies here. (Some years ago I had [9] results about simplified versions of the HF model.)

I shall next introduce the basic ingredients of the Schrödinger and HF models. I shall return to TF theory in Sect. 6. below.

2.1. Schrödinger theory In the Schrödinger model the ground state of the atom is described by a many-body wave function

$$\Psi\in\bigotimes^N L^2(\mathbb{R}^3;\mathbb{C}^2)=L^2(\mathbb{R}^{3N};\mathbb{C}^{2^N}).$$

(The function takes values in $\mathbb{C}^{2^N}$ because each electron has two possible spin states.) The ground state density is given in terms of Ψ by

$$\rho(x)= \tag{1}$$
$$\sum_{i=1}^N\int_{\mathbb{R}^{3(N-1)}}\|\Psi(x_1,\ \ldots,x_{i-1},x,x_{i+1},\ldots,x_N)\|^2_{\mathbb{C}^{2^N}}dx_1\cdots dx_{i-1}dx_{i+1}\cdots dx_N.$$

Thus if $\|\Psi\|=1$ then $\int\rho=N$.

There is an additional feature which is of extreme importance. Namely the *Pauli exclusion principle.* The fact that two electrons cannot occupy the same state. Mathematically it is expressed by the fact that one only considers antisymmetric wave functions

$$\Psi \in \bigwedge^{N} L^2(\mathbb{R}^3; \mathbb{C}^2). \tag{2}$$

In order to define the energy we introduce the (unbounded) *energy operator*

$$H_{N,Z} = \sum_{i=1}^{N} \left(-\frac{\hbar^2}{2m_e}\Delta_i - Ze|x_i|^{-1} \right) + \sum_{1\leq i<j\leq N} e^2|x_i - x_j|^{-1}.$$

For simplicity I shall henceforth use atomic units, where Planck's constant $\hbar$, the electron mass m_e and the unit charge e all are equal to 1. The term $-\Delta_i$ represents the operator for the kinetic energy of electron number i. The second term $-Z|x_i|^{-1}$ is the attractive energy between the nucleus and electron i and finally the term $|x_i - x_j|^{-1}$ is the repulsive energy between electron i and electron j. All these terms act diagonally on the $\mathbb{C}^2$ (spin) components. The ground state energy is then

$$E(N,Z) = \inf_{\Psi\neq 0} \frac{(H_{N,Z}\Psi, \Psi)}{\|\Psi\|^2}. \tag{3}$$

The infimum is over functions of the form (2). Here we use the convention that $(H_{N,Z}\Psi, \Psi) = \infty$ unless all of the (non-negative) terms

$$(-\Delta_i\Psi, \Psi) = \|\nabla_i\Psi\|^2, \quad (|x_i|^{-1}\Psi, \Psi), \quad (|x_i - x_j|^{-1}\Psi, \Psi)$$

are finite. All of these terms are finite if Ψ is in H^1, the first Sobolev space. The function $\Delta_i\Psi$ makes sense only if Ψ is in H^2, but by an integration by parts $(-\Delta_i\Psi, \Psi) = \|\nabla_i\Psi\|^2$ and the right side makes sense as long as Ψ is just in H^1.

We say that the atom has a ground state if the infimum in (3) is attained. A minimizer Ψ, which is then an eigenfunction of $H_{N,Z}$, is called a *ground state wave function* and the corresponding density ρ is *a ground state density.* (Note the minimizer need not be unique.)

2.2. Hartree-Fock (HF) Theory In Hartree-Fock theory one simply limits the allowed wave functions. The Hartree-Fock energy $E^{\mathrm{HF}}(N,Z)$ is defined as in (3), but instead of minimizing over all functions of the form (2) one minmizes only over simple wedge products

$$\Psi = \phi_1 \wedge \ldots \wedge \phi_N, \tag{4}$$

where $\phi_1, \ldots, \phi_N \in L^2(\mathbb{R}^3; \mathbb{C}^2)$ are often referred to as one-particle orbitals.

As before the atom is said to have an HF ground state if the infimum is attained and a minimizer is called an *HF ground state wave function.* Of course an HF ground state wave function is *not* in general an eigenfunction of $H_{N,Z}$, but there is another Euler-Lagrange equation corresponding to the HF minimization. It can be expressed as follows.

Let ρ^{HF} be the density of the HF minimizer Ψ^{HF} (defined as in (1) with Ψ replaced by Ψ^{HF}).

Consider the *mean field potential*

$$\phi^{\mathrm{HF}}(x) = Z|x|^{-1} - \int \rho^{\mathrm{HF}}(y)|x-y|^{-1}dy, \tag{5}$$

which is the mean field felt by a test particle. Then the N one-particle orbitals defining Ψ^{HF} span the N-dimensional eigenspace corresponding to the lowest eigenvalues of the *mean field operator*

$$H_{N,Z}^{\mathrm{HF}} = -\tfrac{1}{2}\Delta - \phi^{\mathrm{HF}} - \mathcal{K}. \tag{6}$$

Here $\mathcal{K}$ is the *exchange operator* which (among other things) makes up for the fact that ϕ^{HF} includes self-interaction. The exchange operator is somewhat difficult to express and I shall not discuss it further here.

The fact that Ψ^{HF} corresponds to the 'lowest' eigenfunctions of the mean field operator $H_{N,Z}^{\mathrm{HF}}$ is often referred to as *self-consistency.*

One should observe that whereas Schrödinger theory is a *linear* theory the HF model is a *non-linear* variational problem. The HF variational problem was studied in [5], where it was also proved that minimizers exist if $N < Z+1$.

Here is the main result that I would like to explain in this lecture.

THEOREM 1 **(Main theorem)** *There exist universal constants $\nu, C_\pm > 0$ (in particular independent of Z) such that for Z large enough*

$$C_- \leq \mathcal{R}_\nu^{\mathrm{HF}}(Z,Z) \leq C_+.$$

Here $\mathcal{R}_\nu^{\mathrm{HF}}(Z,Z)$ is the radius to the last ν electrons for the neutral HF atom.

This theorem follows easily from the results announced in [10].

3. The Sobolev inequality

As an elementary illustration I shall here use the classical Sobolev inequality $\int_{\mathbb{R}^3}|\nabla\phi|^2 \geq C_{\mathrm{s}}(\int_{\mathbb{R}^3}|\phi|^6)^{1/3}$ to estimate the energy of atoms.

In fact, for any $0 < V \in L^{5/2}(\mathbb{R}^3)$ we have using Sobolev's and Hölder's inequalities

$$\begin{aligned}
\left(\left(-\tfrac{1}{2}\Delta - V\right)\phi,\phi\right) &= \tfrac{1}{2}\int|\nabla\phi|^2 - \int V|\phi|^2 \\
&\geq \tfrac{1}{2}C_{\mathrm{s}}\left(\int|\phi|^6\right)^{1/3} - \left(\int V^{5/2}\int|\phi|^2\right)^{2/5}\left(\int|\phi|^6\right)^{1/5} \\
&\geq -\frac{2}{5}\left(\frac{6}{5C_{\mathrm{s}}}\right)^{3/2}\left(\int V^{5/2}\right)\left(\int|\phi|^2\right).
\end{aligned}$$

Thus the lowest eigenvalue of the operator $-\frac{1}{2}\Delta - V$ is bounded below by $-\frac{2}{5}\left(\frac{6}{5C_s}\right)^{3/2}\int V^{5/2}$. For normalized N-particle wave functions Ψ we clearly have

$$\left(\sum_{i=1}^{N}\left(-\tfrac{1}{2}\Delta_i - V(x_i)\right)\Psi,\Psi\right) \geq -\frac{2}{5}\left(\frac{6}{5C_s}\right)^{3/2} N\int V^{5/2}. \tag{7}$$

Let us now ignore both the electron-electron repulsion and the Pauli exclusion principle we then find for atoms

$$\begin{aligned}\left(\sum_{i=1}^{N}\left(-\tfrac{1}{2}\Delta_i - \frac{Z}{|x_i|}\right)\Psi,\Psi\right) &\geq -(\text{const.})NZ^{5/2}\int\limits_{|x|<R}|x|^{-5/2}dx - NZR^{-1}\\ &= -(\text{const.})NZ^2,\end{aligned} \tag{8}$$

where we used that $Z|x|^{-1} \leq ZR^{-1}$ for $|x| > R$ and we picked $R = (\text{const.})Z^{-1}$.

A slightly more careful analysis[1] shows that if we ignore both the electron-electron repulsion and the Pauli exclusion principle the radius of the atom is indeed $\sim Z^{-1}$, i.e., not at all bounded independently of Z. In fact, this is even true if one includes the electron-electron repulsion (but still ignores the Pauli principle).

Unfortunately, I do not know a good reference for this last fact, but I believe it can be easily derived from the analysis in [1] or [8].

4. The Lieb-Thirring inequality

In order to include the Pauli exclusion principle in the above considerations one needs a generalization of the Sobolev inequality to antisymmetric functions. The following generalized Sobolev inequality was proved by Lieb and Thirring [7]. For normalized functions Ψ satisfying (2) we have

$$\sum_{i=1}^{N}\int\limits_{\mathbb{R}^{3N}}\|\nabla_i\Psi(x_1,\ldots,x_N)\|^2dx_1\ldots dx_N \geq C_{\text{LT}}\int\limits_{\mathbb{R}^3}(\rho(x))^{5/3}dx,$$

where ρ is the density defined in (1). For normalized Ψ satisfying (2) and $0 < V \in L^{5/2}(\mathbb{R}^3)$ we now get

$$\begin{aligned}\left(\sum_{i}\left(-\tfrac{1}{2}\Delta_i - V(x_i)\right)\Psi,\Psi\right) &\geq \tfrac{1}{2}C_{\text{LT}}\int\rho^{5/3} - \int V\rho\\ &\geq -\frac{2}{5}\left(\frac{6}{5C_{\text{LT}}}\right)^{3/2}\int V^{5/2}.\end{aligned} \tag{9}$$

Note that in contrast to (7), which holds for all functions Ψ, the estimate (9), which holds only for antisymmetric Ψ, has no N-dependence.

---

[1] The analysis can indeed be done exactly, since this is simply the standard hydrogen atom. I am choosing this different approach for the sake of illustration

If we choose Ψ of the form (4) with the one-particle orbitals $\phi_1, \ldots, \phi_N$ being eigenfunctions corresponding to negative eigenvalues of $-\frac{1}{2}\Delta - V$ we find that the sum of the negative eigenvalues of $-\frac{1}{2}\Delta - V$ is bounded below by $-\frac{2}{5}\left(\frac{6}{5C_{\mathrm{LT}}}\right)^{3/2}$ $\int V^{5/2}$. Compare this with the fact that the Sobolev inequality gave a similar estimate only on the lowest eigenvalue.

Let us now improve (8) by including the Pauli principle.

If Ψ is a normalized *antisymmetric* wave function we find

$$\begin{aligned}\left(\sum_{i=1}^{N}\left(-\tfrac{1}{2}\Delta_i - \frac{Z}{|x_i|}\right)\Psi, \Psi\right) &\geq -(\text{const.})Z^{5/2}\int\limits_{|x|<R}|x|^{-5/2}dx - NZR^{-1} \\ &= -(\text{const.})N^{1/3}Z^2, \end{aligned} \tag{10}$$

where we this time choose $R = (\text{const.})N^{2/3}Z^{-1}$. Again a more careful analysis shows that this is indeed the radius of the atom. Thus if one includes the Pauli principle but ignores the electron-electron repulsion a neutral atom ($N = Z$) has radius $\sim Z^{-1/3}$.

5. Weyl type semiclassical asymptotics

Up to now I have discussed only estimates. I shall now address the exact balance between the nuclear attraction on the one hand and the electron-electron repulsion and the Pauli principle on the other hand.

The first step is to understand that the estimates discussed so far are actually in some sense exact. In fact, we shall now see that these estimates can indeed be understood from a semiclassical point of view.

It is well-known from the classical results of Weyl [12] that the large eigenvalue asymptotics can be obtained from classical phase space estimates.

In our context the classical equivalent of the sum of the negative eigenvalues of $-\frac{1}{2}\Delta - V$ is

$$2(2\pi)^{-3}\iint\limits_{\frac{1}{2}p^2 - V(x)<0}\left(\tfrac{1}{2}p^2 - V(x)\right)dpdx = -\frac{4\sqrt{2}}{15\pi^2}\int V^{5/2}. \tag{11}$$

The classical equivalent of the density (of the many-particle wave function with orbitals corresponding to all eigenfunctions with negative eigenvalues) is

$$\rho^{\text{classical}}(x) = 2(2\pi)^{-3}\int\limits_{\frac{1}{2}p^2 - V(x)<0} 1\, dp = \frac{2\sqrt{2}}{3\pi^2}V(x)^{3/2}. \tag{12}$$

The factor 2 in front of these classical expressions is due to the spin degeneracy.

The equivalent of Weyl's classical eigenvalue asymptotics can in this context be formulated as follows.

THEOREM 2 **(Weyl asymptotics)** *Consider* $0 < V \in L^{5/2}(\mathbb{R}^3)$ *and* $\lambda > 0$. *Let* $e_1(\lambda), e_2(\lambda), \ldots < 0$ *and* $u_1, u_2, \ldots \in L^2(\mathbb{R}^3; \mathbb{C}^2)$ *denote the negative eigenvalues and corresponding eigenfunctions of the operator* $-\frac{1}{2}\Delta - \lambda V$. *Then*

$$\begin{aligned}\lim_{\lambda\to\infty} \lambda^{-5/2} \sum_i e_i(\lambda) &= -\frac{4\sqrt{2}}{15\pi^2} \int V^{5/2} \\ \lim_{\lambda\to\infty} \lambda^{-3/2} \sum_i \|u_i(x)\|^2 &= \frac{2\sqrt{2}}{3\pi^2} V(x)^{3/2},\end{aligned}$$

where the last limit is in the sense of distributions.

One famous (unproven) conjecture in the subject states that the inequality in (9) is correct if the right side is replaced by the right side of (11). This would then, in view of the Weyl asymptotics Theorem, be the sharp form of the inequality.

Turning again to atoms it should now not be surprising that the limit of many particles, e.g., $N = Z \to \infty$ can be treated using the semiclassical approximation. This is Thomas-Fermi (TF) theory.

6. Thomas-Fermi (TF) theory

Recall the Hartree-Fock mean field consistency. Namely that the HF mean field potential ϕ^{HF} can be expressed, as in (5), in terms of the density ρ^{HF} of the 'first' N eigenfunctions of the mean field operator (6). Let us now ignore the exchange operator $\mathcal{K}$ and also assume that the number of negative eigenvalues of $-\frac{1}{2}\Delta - \phi^{\mathrm{HF}}$ is almost Z for large Z. (Both of these statements can be made into rigorous estimates, but this is beyond the scope of this simple presentation.) We can then express the self-consistency for a neutral atom ($N = Z$) in the semiclassical approximation. It simply states that the density ρ and the potential ϕ should be related by

$$\phi(x) = Z|x|^{-1} - \int \rho(y)|x-y|^{-1} dy \tag{13}$$

$$\rho(x) = \frac{2\sqrt{2}}{3\pi^2} \phi(x)^{3/2}. \tag{14}$$

Note that this set of equations expresses both the electrostatic attraction and repulsion (13) and the Pauli-principle (14).

It was shown by Lieb and Simon in [6] (see also [4]) that there exists a unique pair of positive functions (the Thomas-Fermi solutions) $(\rho^{\mathrm{TF}}, \phi^{\mathrm{TF}})$ solving the equations (13) and (14). The existence and uniqueness was proved by variational techniques. In fact, the TF density is the minimizer of the TF functional

$$\begin{aligned}\mathcal{E}(\rho) = {} & \tfrac{3}{10}(3\pi^2)^{2/3} \int \rho(x)^{5/3} dx - \int Z|x|^{-1}\rho(x) dx \\ & + \tfrac{1}{2} \iint \rho(x)|x-y|^{-1}\rho(y) dx dy.\end{aligned}$$

One of the important results also proved in [6] is that the minimal value of $\mathcal{E}(\rho)$ indeed approximates both the Schrödinger energy $E(Z,Z)$ and the HF energy $E^{\rm HF}(Z,Z)$ to leading order as $Z\to\infty$.

It is often more convenient to eliminate $\rho^{\rm TF}$ from the equations (13) and (14). This can be done by observing that (13) implies that $(4\pi)^{-1}\Delta\phi=\rho-Z\delta_0$ (where δ_0 is the Dirac delta measure at 0). One then arrives at the *Thomas-Fermi equation* for $\phi^{\rm TF}$

$$\Delta\phi^{\rm TF}=\frac{8\sqrt{2}}{3\pi}\phi^{\rm TF}(x)^{3/2}-4\pi Z\delta_0. \tag{15}$$

This non-linear elliptic PDE also has a unique positive solution. The Thomas-Fermi potential has the following scaling behavior

$$\phi^{\rm TF}(x)=Z^{4/3}\phi_1(Z^{1/3}x), \tag{16}$$

where ϕ_1 is a function independent of Z. This scaling seems to indicate that the atomic radius should indeed behave like $Z^{-1/3}$. This is true for the bulk of the electrons, but not if we ask for where the last few electrons are.

The universal behavior of the size of atoms can, in the semiclassical TF approximation, indeed be explained from the TF equation (15).

In fact, it follows that the solution has the Z independent asymptotic behavior $3^4 2^{-3}\pi^2|x|^{-4}$ for large $|x|$. Note that this asymptotic form obeys the scaling in (16). The asymptotic form was discovered by Sommerfeld [11] in 1932.

Moreover, for all $x\neq 0$

$$\lim_{Z\to\infty}\phi^{\rm TF}(x)=3^4 2^{-3}\pi^2|x|^{-4}. \tag{17}$$

In fact, the Sommerfeld asymptotics is a very robust property of solutions to TF-like equations.

The following version (see [10]) can be proved by fairly simple comparison arguments. Similar estimates were proved by Brezis and Lieb in [2].

THEOREM 3 **(Sommerfeld estimate)** *Assume $\phi\geq 0$ is continuous on $\{|x|\geq r\}$ and satisfies the TF differential equation*

$$\Delta\phi(x)=\frac{8\sqrt{2}}{3\pi}\phi(x)^{3/2}, \tag{18}$$

for $|x|>r$. Let

$$a(r)=\sup_{|x|=r}\left[\left(\frac{\phi^{(x)}}{3^4 2^{-3}\pi^2|x|^{-4}}\right)^{\frac12}-1\right]\quad \textit{and}\quad A(r)=\sup_{|x|=r}[\phi(x)-3^4 2^{-3}\pi^2|x|^{-4}].$$

Then

$$\left(1+a(r)\frac{r^{\tau-4}}{|x|^{\tau-4}}\right)^{-2}-1\leq\phi(x)-3^4 2^{-3}\pi^2|x|^{-4}\leq A(r)r^{\tau}|x|^{-\tau}, \tag{19}$$

where $\tau=(1+\sqrt{73})/2\approx 4.8$.

Note that the asymptotic expression $3^4 2^{-3}\pi^2|x|^{-4}$, in fact, solves the TF equation (18).

Using the Sommerfeld asymptotics and the equations (13) and (14) it is fairly straightforward to prove the equivalent of Theorem 1 in the TF model.

We thus see how the delicate balance involving electrostatics and the Pauli principle, at least on the semiclassical level, explains the universal behavior of atoms.

Using this to arrive at the same conclusion for HF theory is what I shall turn to next.

As already mentioned it is still an open problem to do this in Schrödinger theory.

7. Comparing HF to TF

In this section I shall explain how the Sommerfeld asymptotics holds also in HF theory. The main theorem, Theorem 1, is not a direct consequence of the Sommerfeld asymptotics, but can be proved using similar ideas (see [10]). I concentrate on the Sommerfeld asymptotics because I find it conceptually more important.

THEOREM 4 **(TF approximation to HF)** *There exist universal constants $\varepsilon_1, D > 0$ such that for $|x| < D$*

$$|\phi^{\mathrm{HF}}(x) - \phi^{\mathrm{TF}}(x)| \leq (\phi^{\mathrm{TF}}(x))^{1-\varepsilon_1}. \tag{20}$$

It can be easily seen from (17) that $\phi^{\mathrm{TF}}(x)$ becomes large for $|x| < D$ if Z is large enough and D is small enough. Hence $(\phi^{\mathrm{TF}}(x))^{1-\varepsilon_1}$ is smaller than $\phi^{\mathrm{TF}}(x)$ and (20) really describes the asymptotics of ϕ^{HF} for small x. Note, however, that D is independent of Z as it should be in order to explain the universal behavior of the size of atoms.

It may be surprising that we here give the asymptotics of $\phi^{\mathrm{HF}}(x)$ near $|x| = 0$ whereas the Sommerfeld asymptotics for ϕ^{TF} was for large $|x|$. The point to observe is that what is meant by large depends on Z. As we saw in (16) large $|x|$ means large compared to $Z^{-1/3}$.

The idea behind the proof of Theorem 4 is to show that for $|x| < D$

- $\phi^{\mathrm{HF}}(x) > 0$
- ϕ^{HF} approximately satisfies the TF equation (18).

One of the tools in approximating HF by TF is to control the 'Coulomb norm'

$$\|\rho^{\mathrm{HF}} - \rho^{\mathrm{TF}}\|_C = \tfrac{1}{2}\iint (\rho^{\mathrm{HF}}(x) - \rho^{\mathrm{TF}}(x))\,|x-y|^{-1}\,(\rho^{\mathrm{HF}}(y) - \rho^{\mathrm{TF}}(y))\,dxdy$$

That this is a norm (on say $C_0^\infty(\mathbb{R}^3)$) follows from the fact that $|x|^{-1}$ is a function of positive type (its Fourier transform is (const.)$|p|^{-2} > 0$). The Coulomb norm was first used by Fefferman and Seco in [3].

The main problem in using the Coulomb norm is that it is a global quantity. Thus $\|\rho^{\mathrm{HF}} - \rho^{\mathrm{TF}}\|_C$ will depend on Z and cannot be used to control what happens at $|x|$ independent of Z.

This problem is circumvented by an iterative scheme (see [10]).

Assuming that (20) has been established for all $|x| < r$ for some $r > 0$ one proves that $\phi^{\mathrm{HF}}(x)$ for $|x| > r$ can be found from a new variational problem which does not depend (too heavily) on $|x| < r$.

For this new problem the Coulomb norm will have an order which is adequate for controlling $\phi^{\mathrm{HF}}(x)$ for $|x|$ values larger than r. It is this separation of the region $|x| > r$ from the region $|x| < r$ which has not yet been achieved in Schrödinger theory.

References

[1] V. Bach, Ionization energies of bosonic Coulomb systems, *Lett. Math. Phys.*, **21**, 139–149, (1991).

[2] H. Brezis & E.H. Lieb, Long range potentials in Thomas-Fermi theory. *Commun. Math. Phys.*, **65**, 231–246, (1979).

[3] C. Fefferman & L.A. Seco, Asymptotic neutrality of large ions. *Commun. Math. Phys.*, **128**, 109–130, (1990).

[4] E.H. Lieb Thomas-Fermi and related theories of atoms and molecules, *Rev. Modern Phys.*, **53**, 603–641, (1981).

[5] E.H. Lieb & B. Simon, The Hartree-Fock theory for Coulomb systems, *Commun. - Math. Phys.*, **53**, 185–194, (1977).

[6] E.H. Lieb & B. Simon, Thomas-Fermi theory of atoms, molecules and solids, *Adv. in Math.*, **23**, 22-116, (1977).

[7] E.H. Lieb & W.E. Thirring, Inequalities for the moments of the eigenvalues of the Schrödinger Hamiltonian and their relation to Sobolev inequalities, in *Studies in mathematical physics*, (E. Lieb, B. Simon, and A.S. Wightman, Eds.), Princeton Univ. Press, Princeton, New Jersey, 269–330, 1976.

[8] J.P. Solovej, Asymptotics for bosonic atoms. *Lett. Math. Phys.*, **20**, 165–172, 1990.

[9] J.P. Solovej, Proof of the ionization conjecture in a reduced Hartree-Fock model, *Invent. Math.*, **104**, 291–311, (1991).

[10] J.P. Solovej, The size of atoms in Hartree-Fock theory, in *Partial Differential Equations and Mathematical Physics*, Progress in Nonlinear Differential Equations and Their Applications, Vol. **21**, L. Hörmander and A. Melin (Eds.), 321–332, Birkhäuser, (1995).

[11] A. Sommerfeld, Asymptotische Integration der Differentialgleichung des Thomas-Fermischen Atoms, *Z. Phys.*, **78**, 283–308, (1932).

[12] Hermann Weyl, Über die asymptotische Verteilung der Eigenwerte. *Nachrichten der Königlichen Gesellschaft der Wissenschaften zu Göttingen. Mathematisch-Naturwissenschaftliche Klasse*, 110–117, 1911.

Progress in Mathematics, Vol. 169, © 1998 Birkhäuser Verlag Basel/Switzerland

Geography of Irreducible 4-Manifolds

András Stipsicz

Department of Analysis, ELTE TTK, 1088. Múzeum KRT. 6–8
Budapest, Hungary
e-mail: stipsicz@cs.elte.hu

1. Introduction

In the following we intend to give a short overview of recent results concerning simply connected, oriented, closed, smooth 4-manifolds. Throughout this paper we will call a 4-dimensional manifold with the above four properties an *admissible* 4-manifold.

One of the most important classical invariants of an oriented, closed 4-manifold X is its cohomology ring $\mathrm{H}^*(X;\mathbb{Z})$ – which, in the simply connected case, is determined by $\mathrm{H}^2(X;\mathbb{Z})$ and the cup product (intersection form) $Q_X\colon \mathrm{H}^2(X;\mathbb{Z})\times \mathrm{H}^2(X;\mathbb{Z}) \to \mathbb{Z}$,

$$Q_X(a,b) = \langle a \cup b, [X]\rangle.$$

Remark 1.1 Poincaré duality $PD\colon \mathrm{H}^2(X;\mathbb{Z}) \to \mathrm{H}_2(X;\mathbb{Z})$ gives an isomorphism between the second cohomology and homology groups of an oriented, closed 4-manifold. It is easy to prove that elements of $\mathrm{H}_2(X;\mathbb{Z})$ can be represented by embedded surfaces. If Σ_α denotes a surface corresponding to $\alpha \in \mathrm{H}_2(X;\mathbb{Z})$, and $a = PD(\alpha),\ b = PD(\beta) \in \mathrm{H}^2(X;\mathbb{Z})$, then

$$Q_X(a,b) = \#(\Sigma_\alpha \cap \Sigma_\beta).$$

To form this intersection we chose the surfaces Σ_α and Σ_β to meet transversally. The orientation of these surfaces and of X assign ± 1 to every intersection point, and $\#$ denotes the sum of those ± 1's. The above equality explains the name *intersection form*.

By a theorem of Freedman the intersection form Q_X determines an admissible 4-manifold X *up to homeomorphism*:

Theorem 1.2 (Fr) *X_1, X_2 admissible 4-manifolds are homeomorphic iff Q_{X_1} and Q_{X_2} are equivalent as bilinear forms.* □

Theorems of Rohlin, Donaldson and Furuta ([R], [D], [F]) give restrictions on the possible forms arising as intersection forms of admissible 4-manifolds. We summarize these results in the following theorem:

THEOREM 1.3 *If Q_X is the intersection form of an admissible 4-manifold, then Q_X is equivalent either to*

- $n\langle 1\rangle \oplus m\langle -1\rangle$, $n, m \geq 0$, *or to*
- $2kE_8 \oplus lH$, $l \geq 2|k|$, $k \in \mathbb{Z}$. □

Here E_8 denotes the negative definite 8×8 matrix corresponding to the Dynkin diagram of the exceptional Lie-algebra E_8:

$$\begin{pmatrix} -2 & 0 & -1 & 0 & 0 & 0 & 0 & 0 \\ 0 & -2 & 0 & -1 & 0 & 0 & 0 & 0 \\ -1 & 0 & -2 & -1 & 0 & 0 & 0 & 0 \\ 0 & -1 & -1 & -2 & -1 & 0 & 0 & 0 \\ 0 & 0 & 0 & -1 & -2 & -1 & 0 & 0 \\ 0 & 0 & 0 & 0 & -1 & -2 & -1 & 0 \\ 0 & 0 & 0 & 0 & 0 & -1 & -2 & -1 \\ 0 & 0 & 0 & 0 & 0 & 0 & -1 & -2 \end{pmatrix},$$

while $H = \left[\begin{smallmatrix} 0 & 1 \\ 1 & 0 \end{smallmatrix}\right]$. If $Q_X \cong 2kE_8 \oplus lH$, then the admissible 4-manifold X admits a (unique) spin structure, while if $Q_X \cong n\langle 1\rangle \oplus m\langle -1\rangle$, X has no spin structure.

DEFINITION 1.4 $b_+(X)$ *(and $b_-(X)$ resp.)* denotes the dimension of a maximal positive (negative) definite subspace of $(\mathrm{H}_2(X;\mathbb{Z}), Q_X)$. $\sigma(X) = b_+(X) - b_-(X)$ is the signature X. Note that the Euler-characteristic $e(X)$ equals to $b_+(X) + b_-(X) + 2$ in case X is admissible.

Note also that $\sigma(X) = n - m$ if $Q_X \cong n\langle 1\rangle \oplus m\langle -1\rangle$ and $-16k$ if $Q_X \cong 2kE_8 \oplus lH$. In particular the signature $\sigma(X)$, the Euler characteristic $e(X)$ and the type of the manifold (whether it is spin or not) determines Q_X, and so – by Theorem 1.2 – the homeomorphism type of an admissible 4-manifold.

EXAMPLES 1.5 The intersection form of the complex projective plane $\mathbb{CP}^2$ is equivalent to $\langle 1\rangle$. Note that since $\mathbb{CP}^2$ is a complex manifold, it has a canonical orientation. $\overline{\mathbb{CP}}^2$ denotes the same manifold with the opposite orientation; $Q_{\overline{\mathbb{CP}}^2} = \langle -1\rangle$. If $\mathbb{CP}^1$ denotes the complex projective line, then $\mathbb{CP}^1 \times \mathbb{CP}^1$ is an admissible 4-manifold and $Q_{\mathbb{CP}^1\times\mathbb{CP}^1} = \left[\begin{smallmatrix} 0 & 1 \\ 1 & 0 \end{smallmatrix}\right] = H$.

It is somewhat harder to prove that the complex surface

$$S_4 = \{[z_0 : z_1 : z_2 : z_3] \in \mathbb{CP}^3 \mid \sum_{i=0}^{3} z_i^4 = 0\}$$

has intersection form $Q_{S_4} = 2E_8 \oplus 3H$. Since for a connected sum $X = X_1 \# X_2$ the intersection form Q_X is equal to $Q_{X_1} \oplus Q_{X_2}$, a form equivalent to $n\langle 1\rangle \oplus m\langle -1\rangle$ can be realized as the intersection form of an admissible 4-manifold. Likewise if $l \geq 3|k|$, then there is an admissible 4-manifold X with $Q_X \cong 2kE_8 \oplus lH$ ($n\mathbb{CP}^2 \# m\overline{\mathbb{CP}}^2$ and $kS_4 \# (l - 3|k|)H$ will do). It is an open question whether a form equivalent to $2kE_8 \oplus lH$ with $2|k| < l < 3|k|$ can be realized as intersection form of an admissible 4-manifold.

THE $\frac{11}{8}$-CONJECTURE 1.6 *If $Q_X \cong 2kE_8 \oplus lH$ is the intersection form of an admissible spin 4-manifold, then $l \geq 3|k|$.*

Summarizing the above theorems one can see that – modulo the $\frac{11}{8}$-conjecture – the *topological* classification of admissible 4-manifolds is well understood. Our ultimate goal is, however, to study *smooth* properties of smooth 4-manifolds. To achieve this, one needs a finer, *diffeomorphism*-invariant of admissible 4-manifolds. A recently discovered set of invariants, the Seiberg-Witten basic classes, will be discussed in the next section.

2. Basic classes of smooth 4-manifolds

In the following we will always assume that the admissible 4-manifold X has odd $b_+(X)$ and $b_+(X) > 1$. For such a manifold a finite set of cohomology classes, the set of Seiberg-Witten (SW-)basic classes

$$\mathcal{B}_X = \{K_i\}_{i=1}^s \subset \mathrm{H}^2(X;\mathbb{Z})$$

can be defined. Because of the complicated nature of the necessary background material we omit the description of the full and rigorous definition of the set of basic classes, see e.g. [A], [Sa]. Here we only give a very rough outline of the idea of the definition of $\mathcal{B}_X$:

If $K \in \mathrm{H}^2(X;\mathbb{Z})$ satisfies $K(\alpha) \equiv Q_X(\alpha,\alpha) \pmod 2$ for all $\alpha \in \mathrm{H}_2(X;\mathbb{Z})$, then K is a characteristic cohomology element. A characteristic cohomology element of an admissible 4-manifold X determines a spin$_c$ structure – a generalization of the spin structure. Using this spin$_c$ structure, a Riemannian metric g and a self-dual 2-form μ on X, fairly standard constructions, give rise to the pair of Seiberg-Witten equations (see [Sa]). The solution space of these equations, denoted by $\mathcal{M}_K(g,\mu)$, is called the *moduli space*. Using homological arguments one can produce a number $SW_X(K) \in \mathbb{Z}$ from the space $\mathcal{M}_K(g,\mu)$, and – as the notation suggests – in case $b_+(X) > 1$, $SW_X(K)$ will be independent of g and μ. If $K \in \mathrm{H}^2(X;\mathbb{Z})$ is *not* characteristic, then we define $SW_X(K)$ to be zero.

DEFINITION 2.1

$$\mathcal{B}_X = \{K \in \mathrm{H}^2(X;\mathbb{Z}) \mid SW_X(K) \neq 0\}.$$

THEOREM 2.2 *If $f\colon X_1 \to X_2$ is an orientation preserving diffeomorphism, then for every $K \in \mathcal{B}_{X_2}$ we have $f^*(K) \in \mathcal{B}_{X_1}$. Moreover $SW_{X_1}(f^*(K)) = \pm SW_{X_2}(K)$.* □

REMARK 2.3

- In case $b_+(X)$ is even, the homological argument quoted in the above description gives $SW_X(K) = 0$ for all K. This is the reason for our restriction on the parity of $b_+(X)$.
- If $b_+(X) = 1$, the value $SW_X(K)$ associated to the moduli space $\mathcal{M}_K(g,\mu)$ depends on the choices g and μ. The case of $b_+(X) = 1$ is well understood, but we will omit the discussion of this topic here.

- The idea of the definition of SW-basic classes rests on previous work of Donaldson and Kronheimer-Mrowka. Using ASD connections Donaldson defined a polynomial invariant for admissible 4-manifolds with $b_+(X) > 1$ and odd ([D1]). Based on this definition, Kronheimer and Mrowka later proved a structure theorem for the Donaldson-invariants and defined the Kronheimer-Mrowka (KM-)basic classes of admissible 4-manifolds satisfying an additional condition, called the *simple type* condition (see [KM]). Recent work of Pidstrigatch and Tyurin underpin the conjecture, that the set of KM-basic classes and the set of SW-basic classes actually coincide. In this paper we will mostly deal with SW-basic classes – we will formulate some results concerning the KM-basic classes of certain admissible manifolds as well.

- As results in the following sections will show, $\mathcal{B}_X$ is a finer invariant of the smooth 4-manifold X than Q_X. By applying Theorems 3.1, 3.3 and 1.2, one can easily find homeomorphic manifolds having different sets of basic classes, proving that those manifolds are non-diffeomorphic. In fact (a more detailed version of Theorem 3.1) gives infinitely many homeomorphic, but pairwise non-diffeomorphic manifolds – a phenomenon unique in dimension 4.

Let us formulate the two questions we would like to focus on in the rest of the paper.

- **Q1:** How can one determine the set $\mathcal{B}_X$ for a given admissible 4-manifold X?
- **Q2:** What consequences can one derive from the knowledge of $\mathcal{B}_X$ regarding the geometry of X?

From results concerning **Q1** and **Q2** we will deduce partial results concerning the smooth classification of admissible 4-manifolds.

3. Theorems about SW-basic classes

Let us first list some of the most important theorems about the computation of basic classes. First we give statements concerning basic classes of complex surfaces (real 4-dimensional manifolds). We assume that the reader is familiar with the basic notions of the theory of complex surfaces (see e.g. [BPV]).

THEOREM 3.1 *1. If X is a $K3$-surface, then $\mathcal{B}_X = \{0\}$.*

2. If X is a minimal elliptic surface, then $\mathcal{B}_X = \{k \cdot c_1(X) \mid k \in A_X\}$, where A_X is a well-understood subset of $\mathbb{Q} \cap [-1, 1]$. (We do not describe A_X here only to keep this introduction short.)

3. If X is a minimal surface of general type, then $\mathcal{B}_X = \{\pm c_1(X)\}$. □

Note that by the classification of complex surfaces every minimal, simply connected, complex surface with $b_+ > 1$ fits in one of the above categories. Note also, that in all the above cases $K \in \mathcal{B}_X$ implies that $K^2 = 3\sigma(X) + e(X)$.

DEFINITION 3.2 If $K^2 = 3\sigma(X) + 2e(X)$ holds for every class $K \in \mathcal{B}_X$, then the admissible 4-manifold X has (SW-) simple type.

It is an open (and very interesting) question whether all admissible 4-manifolds (with $b_+(X) > 1$) have simple type. No example for violating the simple type condition (with $b_+(X) > 1$) has been found yet. The following theorem describes the computation of $\mathcal{B}_X$ in case X is a connected sum of other manifolds.

THEOREM 3.3

- *If $X = X_1 \# X_2$ and $b_+(X_i) > 0$ $(i = 1, 2)$, then $\mathcal{B}_X = \emptyset$.*
- *If $X = Y \# \overline{\mathbb{CP}}^2$ and Y is of simple type, then*

$$\mathcal{B}_X = \{K_i \pm E \mid K_i \in \mathcal{B}_Y\}.$$

  $E \in \mathrm{H}^2(X; \mathbb{Z})$ is the Poincaré dual of the exceptional sphere in $\overline{\mathbb{CP}}^2$.
- *More generally, if $X = Y \# N$, where Y is of simple type and $b_+(N) = 0$ (so $\mathrm{H}^2(N; \mathbb{Z})$ has an orthogonal basis $\{E_i \in \mathrm{H}^2(N; \mathbb{Z}) \mid i = 1, 2, \ldots, b_2(N)\}$ with $E_i^2 = -1$), then*

$$\mathcal{B}_X = \{K_i \pm E_1 \pm \cdots \pm E_{b_2(N)} \mid K_i \in \mathcal{B}_Y\}. \qquad \square$$

REMARK 3.4 Similar relations can be proved without assuming that Y is of simple type. Without giving the precise formulation we only note here that $\{K_i \pm E_1 \pm \cdots \pm E_{b_2(N)} \mid K_i \in \mathcal{B}_Y\} \subset \mathcal{B}_X$ holds in the general case.

A wide class of admissible 4-manifolds with b_+ odd have an extra structure which can be capitalized on in the computation of $\mathcal{B}_X$. Recall that a 4-manifold X has a symplectic structure if there exists a non-degenerated, closed 2-form ω on X. By choosing a compatible almost complex structure, $c_1(X, \omega) \in \mathrm{H}^2(X; \mathbb{Z})$ first Chern-class can be defined for the symplectic manifold (X, ω). Note that since simply connected complex surfaces are Kähler, the class of symplectic 4-manifolds contains all the admissible complex surfaces. Fundamental results of Taubes give partial information about the set $\mathcal{B}_X$ of a symplectic manifold (X, ω).

THEOREM 3.5 *If (X, ω) is a symplectic manifold with $b_+(X) > 1$, then $\pm c_1(X, \omega) \in \mathcal{B}_X$.* $\square$

REMARK 3.6

- In his seminal paper ([T1]) Taubes actually proves that $SW_X(\pm c_1(X, \omega)) = \pm 1$. The knowledge of the exact value of $SW_X(\pm c_1(X, \omega))$ will be important later on.
- More delicate arguments of Taubes establish a relationship between elements of $\mathcal{B}_X$ and the Gromow invariants of the symplectic manifold (X, ω) – see [T2, T3].

The following general construction – due to Gompf ([G]) – gives many examples of symplectic manifolds.

Assume that X_1, X_2 are symplectic 4-manifolds and $F_i \subset X_i$ $(i = 1, 2)$ symplectic surfaces in them. Assume also that the genera of F_i are equal and $[F_1]^2 + [F_2]^2 = 0$. $\nu(F_i)$ will denote a tubular neighborhood of $F_i \subset X_i$. Because of our assumptions, a diffeomorphism $\Psi\colon F_1 \to F_2$ can be lifted to an orientation reversing diffeomorphism $\varphi\colon \partial(\nu(F_1)) \to \partial(\nu(F_2))$ between the boundaries of the tubular neighborhoods. By choosing such a φ, a new manifold

$$X_\varphi = (X_1 \setminus \nu(F_1)) \cup_\varphi (X_2 \setminus \nu(F_2))$$

can be constructed. In [G] Gompf proves that X_φ has a symplectic structure. X_φ is called the symplectic normal connected sum of X_1 and X_2 along $F = F_1$, and it is denoted by $X_\varphi = X_1 \#_{(F,\varphi)} X_2$. Note that (as the notation suggests) the diffeomorphism type of X_φ might depend on the particular choice of φ.

As a special case of the above construction one can define the following operation (see [S1]). Assume that the 4-manifold X contains a torus $F \subset X$ with $[F]^2 = -1$. Note that since the generic cubic curve in $\mathbb{CP}^2$ is a torus with square 9, $\mathbb{CP}^2 \# 8\overline{\mathbb{CP}}^2$ contains a symplectic torus T with square 1. Apply the above construction of Gompf in case $(X_1, F_1) = (X, F)$ and $(X_2, F_2) = (\mathbb{CP}^2 \# 8\overline{\mathbb{CP}}^2, T)$. If X is a symplectic manifold and $F \subset X$ is its symplectic submanifold, then

$$X_F = X \#_{(F,\varphi)} (\mathbb{CP}^2 \# 8\overline{\mathbb{CP}}^2)$$

is a symplectic manifold. The following theorem was proved for KM-basic classes and used the assumption that F lies in a cusp neighborhood (see [S1]).

THEOREM 3.7 *Assume that X is an admissible 4-manifold, $b_+(X) > 1$ and odd, X has KM-simple type and X contains a (-1)-torus as described above. Then the KM-basic classes of X_F are precisely the classes*

$$\{K_i + Q_X(K_i, PD([F])) \cdot PD([F])\}_{i=1}^s$$

where $\{K_i\}_1^s$ are the basic classes of X. □

REMARK 3.8 Theorem 3.7 has not been proved for SW-basic classes yet. By the conjectured equivalence (cf. Remark 2.3) however one might hope that the same statement holds for SW-basic classes. On the other hand, most of the statements (in particular Theorems 3.3 and 3.9) has been proved earlier for KM-basic classes as well.

Let us finish this section with a typical theorem related to question **Q2**. Theorem 3.9 was proved by Kronheimer and Mrowka ([KM1]); earlier Kronheimer and Mrowka gave a proof for the same statement for KM-basic classes – once X is of simple type.

THEOREM 3.9 *(The generalized adjunction formula) If $\Sigma \subset X$ is an embedded surface with $0 \neq [\Sigma] \in \mathrm{H}_2(X;\mathbb{Z})$, $[\Sigma]^2 \geq 0$ and genus $g(\Sigma)$, then*

$$2g(\Sigma) - 2 \geq [\Sigma]^2 + |K([\Sigma])|$$

for all $K \in \mathcal{B}_X$ basic class of the admissible 4-manifold X ($b_+(X) > 1$ and odd). □

Applying Theorems 3.9 and 3.1, the following corollary (often called the "Thom-conjecture") can be easily proved ([KM1], [MSzT]):

COROLLARY 3.10 *Assume that X is a complex surface with $b_+(X) > 1$, $C \subset X$ is an embedded, connected, smooth, complex curve while $C' \subset X$ is a connected, smooth submanifold of X. Assume also that $[C] = [C'] \in \mathrm{H}_2(X;\mathbb{Z})$ and $[C]^2 \geq 0$. If $g(C)$, $g(C')$ denote the genera of C and C' respectively, then*

$$g(C') \geq g(C). \qquad \square$$

REMARK 3.11 Corollary 3.10 holds for surfaces with $b_+(X) = 1$ as well; in these cases however the proof is more complicated.

4. Irreducible 4-manifolds

In the light of Theorem 3.3 one has to know the basic classes only for 4-manifolds not admitting decomposition as connected sum.

DEFINITION 4.1 A smooth 4-manifold X is *irreducible* if for every decomposition $X = X_1 \# X_2$ either X_1 or X_2 is homeomorphic to the 4-sphere S^4.

Obviously every 4-manifold X admits a connected sum decomposition $X = X_1 \# \cdots \# X_n$ into irreducible pieces. Note, however, that this decomposition is not unique. $\mathbb{CP}^2$, $\overline{\mathbb{CP}}^2$ and $\mathbb{CP}^1 \times \mathbb{CP}^1$ are irreducible 4-manifolds; and since $X = \mathbb{CP}^2 \# 2\overline{\mathbb{CP}}^2$ is diffeomorphic to $(\mathbb{CP}^1 \times \mathbb{CP}^1) \# \overline{\mathbb{CP}}^2$, X has two different decompositions as a connected sum of irreducible manifolds. Despite this non-uniqueness, the understanding of irreducible smooth 4-manifolds would be a major step towards the general classification of 4-manifolds. Here, after a little historical background, we will give some statements concerning the "geography" of irreducible 4-manifolds.

One would hope that irreducibility implies the existence of some extra structure on the admissible 4-manifold X. A conjecture (frequently attributed to Thom) stated that an irreducible 4-manifold X is always diffeomorphic to a complex surface – with one of its orientations. This conjecture was first disproved by Gompf and Mrowka ([GM]) in 1990; they used the construction described in Section 3 to construct interesting new manifolds and the newly developed gauge-theoretic techniques to prove that those new manifolds are, in fact, irreducible and non-complex. Following their footsteps many other examples were found for irreducible, non-complex, admissible 4-manifolds (see e.g. [FS], [Sz], [S1]). Since the operation described in Section 3 is a *symplectic operation*, almost all these irreducible manifolds were in fact symplectic. Results of Taubes about basic classes of symplectic manifolds strengthened the following conjecture: an irreducible 4-manifold (with one of its orientations) always carries a symplectic structure. Although counterexamples to this conjecture with non-trivial fundamental group were found shortly after the introduction of SW-basic classes, no admissible counterexample has been found until very recently. A few weeks ago Z. Szabó described a way of constructing infinitely many admissible, irreducible 4-manifolds which do not carry symplectic structure. An outline of the construction of a symplectic non-complex and a non-symplectic irreducible manifold will be given in the last section. The following observation turns out to be a useful tool in proving irreducibility for a given admissible 4-manifold X.

PROPOSITION 4.2 *If $\mathcal{B}_X \neq \emptyset$ and -4 is not in $\{(K_i - K_j)^2 \mid K_i, K_j \in \mathcal{B}_X\}$, then X is irreducible.*

Proof. By (1) of Theorem 3.3 $\mathcal{B}_X \neq \emptyset$ implies that X cannot admit a decomposition $X = X_1 \# X_2$ with $b_+(X_i) > 0$ $(i = 1, 2)$. (3) of Theorem 3.3 shows, that if $X = X_1 \# N$ with Q_N negative definite $(b_+(N) = 0)$, then $K + E_1 + E_2 + \cdots + E_{b_2(N)}$ and $K - E_1 + E_2 + \cdots + E_{b_2(N)}$ are in $\mathcal{B}_X$. The square of the difference of these two classes is -4, which contradicts our second assumption. So if $X = X_1 \# X_2$, then $b_+(X_2) = 0$ and $b_2(X_2) = 0$, which implies – by Theorem 1.2 –, that X_2 is homeomorphic to S^4, and so X is irreducible. □

REMARK 4.3

- Since blowing up and down are symplectic operations, one can talk about *minimal* symplectic 4-manifolds – just like in the complex analytic category. As a corollary of Taubes' work (see [T2]), one can easily see that a minimal symplectic 4-manifold is, in fact, irreducible.
- One might hope that irreducibility implies $\mathcal{B}_X \neq \emptyset$ (at least with one of the two orientations of X). No admissible counterexample to this hope has yet been found, but there are examples of irreducible manifolds with fundamental group isomorphic to Z_2 and with no basic classes ([W]). In these later cases irreducibility was proved by a covering trick originally due to Kotschick.

The rest of this section is devoted to the introduction of the notion "geography" of irreducible 4-manifolds and the description of some results concerning this geography. Based on algebraic geometric motivations two integers, $c_1^2(X)$ and $\chi(X)$, can be associated to an admissible 4-manifold X with odd $b_+(X)$.

DEFINITION 4.4

$$c_1^2(X) := 3\sigma(X) + 2e(X) \quad \textit{and} \quad \chi(X) := \frac{1}{2}(b_+(X) + 1) = \frac{\sigma(X) + e(X)}{4}.$$

Note that $c_1^2(X)$ and $\chi(X)$ determine $\sigma(X)$ and $e(X)$, and so $b_+(X)$ and $b_-(X)$ of the admissible 4-manifold X. By results discussed in the Introduction this means that $c_1^2(X)$ and $\chi(X)$ determine Q_X up to type (spinnes). Recall that Q_X uniquely determines the homeomorphism class of the admissible 4-manifold X. By geography of irreducible 4-manifolds (or minimal symplectic 4-manifolds) we mean the determination of the set of pairs $(a, b) \in \mathbb{Z} \times \mathbb{Z}$ such that there is an irreducible (minimal symplectic resp.) admissible 4-manifold X such that

$$a = c_1^2(X) \quad \textit{and} \quad b = \chi(X).$$

The name "geography" for such studies is originally due to Persson ([P]); he used it for complex surfaces of general type. A quick review of a few statements concerning the geography problem of complex surfaces will be given in the next theorem. Note, however, that even the solution of the geography problem would not be enough for the complete description of irreducible (or symplectic) admissible 4-manifolds. A complete solution of the classification problem would also contain the description of *all* irreducible (symplectic resp.) admissible 4-manifolds corresponding to $(a, b) \in \mathbb{Z}^2$. Research in this direction is called *botany* (again after Persson).

THEOREM 4.5 *If X is an admissible minimal complex surface, then X is irreducible. By definition $\chi(X) > 0$ always holds. Moreover*

- $c_1^2(X) = 0$ *or*
- $c_1^2(X) > 0$ *and* $2\chi(X) - 6 \leq c_1^2(X) \leq 9\chi(X)$ *(Noether and Bogomolov-Miyaoka-Yau inequalities).* □

Minimal elliptic surfaces correspond to points $(0, b) \in \mathbb{Z}^2$. The region $D = \{(a, b) \in \mathbb{Z}^2 \mid a > 0,\ 2b - 6 \leq a \leq 9b\}$ is not filled completely – although most points in D do correspond to minimal surfaces of general type (for such results see [C], [P], [MT]).

A combination of the construction of Gompf (described in Section 3) and Theorem 3.7 proves the following result – showing that there are points in $\mathbb{Z} \times \mathbb{Z}$ outside of D corresponding to irreducible 4-manifolds (see [S2]).

THEOREM 4.6 *For every pair $(a, b) \in \mathbb{N} \times \mathbb{N}$ with $0 \leq a \leq 2b - 6$ there exists an irreducible 4-manifold X with*

$$a = c_1^2(X) \quad \textit{and} \quad b = \chi(X). \qquad \square$$

REMARK 4.7 Most of the examples given in [S2] are proved to have symplectic structure. Obviously these manifolds are not even homotopy equivalent to minimal complex surfaces. One of these example will be described in the next section. There are other sets of examples (given by Gompf in [G] and by Fintushel-Stern in [FS1]) covering the same region.

Finally let us list some interesting unsolved questions regarding the geography problem.

- None of the above results treats admissible 4-manifolds with even b_+. There are proposed candidates for irreducible admissible 4-manifolds with even b_+ – one such example, constructed from the $K3$-surface, is due to Gompf –, but because of the lack of any sufficiently fine invariant there is no known way to prove irreducibility.
- For all known irreducible examples $c_1^2(X) \geq 0$ is satisfied. In addition, as an application of Taubes' results, $c_1^2(X) \geq 0$ can be proved once X is a minimal symplectic manifold. This means that to find an irreducible admissible example with $c_1^2(X) < 0$ one has to come out with some exotic construction – the non-symplectic examples of Szabó have $c_1^2(X) = 0$. On the other hand, the fact $c_1^2(X) \geq 0$ for all irreducible admissible 4-manifold would imply the $\frac{11}{8}$-conjecture (1.6).
- Another challenging problem is the replacement of the Bogomolov-Miyaoka-Yau inequality $c_1^2(X) \leq 9\chi(X)$ (proved for complex surfaces) for irreducible (and also for symplectic) admissible 4-manifolds.

5. Two examples of irreducible 4-manifolds

In this last section we outline the construction of a symplectic non-complex, and a non-symplectic irreducible admissible 4-manifold. Familiarity with logarithmic transformation will be needed later on.

T^4 denotes the 4-dimensional torus $S^1 \times S^1 \times S^1 \times S^1$. It contains many 2-dimensional tori, for later reference we will specify 4 of them. Choose $p, q, r \in S^1$ three distinct points and take

- $T_{12} = S^1 \times S^1 \times \{p\} \times \{p\}$,
- $T_{34} = \{p\} \times \{p\} \times S^1 \times S^1$,
- $T_{13} = S^1 \times \{q\} \times S^1 \times \{q\}$ and
- $T_{14} = S^1 \times \{r\} \times \{r\} \times S^1$ 2-dimensional tori in T^4.

Note that T_{12} and T_{34} intersect each other in $P = \{p\}\times\{p\}\times\{p\}\times\{p\}$, otherwise all these tori are disjoint. Blow up T^4 in the above point P, and take the proper transforms $\tilde{T}_{12}$ and $\tilde{T}_{34}$ of T_{12} and T_{34}. In this way we have $\tilde{T}_{12}$, $\tilde{T}_{34}$, T_{13}, T_{14} four disjoint tori in the blown up manifold $T^4\#\overline{\mathbb{CP}}^2$. The first two of these four tori have square (-1) while the last two have square 0. Recall that by blowing up a smooth cubic curve in $\mathbb{CP}^2$ eight times, we get a torus $F_1 \subset \mathbb{CP}^2\#8\overline{\mathbb{CP}}^2$ with square 1. One further blow up gives a torus F_0 with square 0 in $\mathbb{CP}^2\#9\overline{\mathbb{CP}}^2$. Now all the components are defined to construct our "master-manifold" X: take first the normal connected sum of $(T^4\#\overline{\mathbb{CP}}^2, \tilde{T}_{12})$ and $(\mathbb{CP}^2\#8\overline{\mathbb{CP}}^2, F_1)$ along the tori indicated. The resulting manifold Y will still contain the torus T_{13} (since $\tilde{T}_{12}$ and T_{13} were disjoint), so form

$$X = (Y, T_{13})\#_F(\mathbb{CP}^2\#9\overline{\mathbb{CP}}^2, F_0) =$$

$$= (\mathbb{CP}^2\#8\overline{\mathbb{CP}}^2, F_1)\#_{F_1}(T^4\#\overline{\mathbb{CP}}^2, (\tilde{T}_{12}, T_{13}))\#_{F_0}(\mathbb{CP}^2\#9\overline{\mathbb{CP}}^2, F_0).$$

The following proposition summarizes the most important properties of X:

PROPOSITION 5.1 *X has a symplectic structure; moreover $\tilde{T}_{34}$ and T_{14} are symplectic submanifolds of X. The fundamental group $\pi_1(X)$ is isomorphic to $\mathbb{Z}$.* □

First let us give the construction of the minimal symplectic (admissible) 4-manifold which is not diffeomorphic to a minimal complex surface.

THEOREM 5.2 *The symplectic normal sum $X_F = (X, \tilde{T}_{34})\#_{F,\varphi}(\mathbb{CP}^2\#8\overline{\mathbb{CP}}^2, F_1)$ is a simply connected, minimal symplectic manifold which is not diffeomorphic to a minimal complex surface.*

Proof. By the symplectic normal sum theorem of Gompf (quoted in Section 3) X_F is obviously symplectic. Applying Theorems 3.3, 3.7 and 3.9 the basic classes of X_F can be determined:

$$\mathcal{B}_{X_F} = \{\pm([\tilde{T}_{12}] + [T_{34}]) \pm [T_{13}]\}.$$

Using Proposition 4.2 one can easily see that X_F is irreducible. Comparing $\mathcal{B}_{X_F}$ with the set of basic classes of complex surfaces (given in Theorem 3.1) it is obvious that X_F is not diffeomorphic to a minimal complex surface. □

REMARK 5.3 Note that a torus R homologous to T_{13} in T^4 (for example $S^1 \times \{q_1\} \times S^1 \times \{q_1\}$ for a point $q_1 \in S^1$ distinct from p, q, r chosen before) can still be found in X_F. Taking the symplectic normal sum $Y = (X_F, R)\#_R(\mathbb{CP}^2\#9\overline{\mathbb{CP}}^2, F_0)$ we get a minimal symplectic 4-manifold not even homotopy equivalent to a minimal complex surface. Again, minimality can be proved by Theorem 3.7 and Proposition 4.2.

The construction (recently announced by Z. Szabó) of an irreducible, admissible 4-manifold with no symplectic structure also starts with the manifold X described above. The 4-manifold W_n is defined as a certain logarithmic transformation of X along the torus T_{14}. Recall that a (topological) logarithmic transformation on a torus L with square 0 in a 4-manifold M means the following operation: delete the tubular neighborhood $\nu(L) \cong D^2 \times T^2$ of L from M and then glue it back via a self-diffeomorphism of $\partial(\nu(L)) = T^3 = S^1 \times S^1 \times S^1$. By choosing an appropriate logarithmic transformation with multiplicity n, Szabó proved the following (see [Sz1]):

THEOREM 5.4 *W_n is an admissible 4-manifold, so in particular $\pi_1(W_n) = 1$. W_n has two basic classes $\pm K$ and*

$$SW_{W_n}(\pm K) = \pm n.$$ □

Since for a symplectic manifold (M, ω) the value of $SW_M(\pm c_1(M, \omega))$ is ± 1 (cf. Remark 3.6), once $n > 1$ the manifold W_n is not symplectic. On the other hand the application of Proposition 4.2 ($\mathcal{B}_{W_n} = \{\pm K\}$) shows that W_n is minimal.

REMARK 5.5 Since T_{14} can be regarded as a submanifold of X_F as well, and since it can be represented by a node in X_F (but not in X!), the multiplicity 2 logarithmic transformation produces a symplectic manifold $X_F(2)$. Note however, that $X_F(2)$ is the normal sum of $(W_2, \tilde{T}_{34})$ and $(\mathbb{CP}^2\#8\overline{\mathbb{CP}}^2, F_1)$. So although W_2 is not symplectic, the normal sum of it with $\mathbb{CP}^2\#8\overline{\mathbb{CP}}^2$ brings the manifold back to the symplectic world.

References

[A] S. Akbulut, *Lectures on Seiberg-Witten invariants*, Turkish J. of Math. **20** (1996), 95–119.

[BPV] W. Barth, C. Peters and A. Van de Ven, *Compact complex surfaces*, Ergebnisse der Mathematik, Springer-Verlag Berlin (1984)

[C] Z. Chen, *On the geography of surfaces – simply connected minimal surfaces with positive index*, Math. Ann. **277**, 141–164.

[D] S.K. Donaldson, *An application of gauge theory to four dimensional topology*, Journal of Differential Geometry **18** (1983), 279–315.

[D1] S.K. Donaldson, *Polynomial invariants for smooth 4-manifolds*, Topology **29** (1990), 257–315.

[Fr] M. Freedman, *The topology of four-dimensional manifolds*, Journal of Differential Geometry **17** (1982), 357–453.

[FS] R. Fintushel and R. Stern, *Surgery in cusp neighborhoods and the geography of irreducible 4-manifolds*, Invent. Math. **117** (1994) 455–523.

[FS1] R. Fintushel and R. Stern, *Rational blowdown of smooth 4-manifolds*, preprint 1995.

[F] M. Furuta, *Monopole equation and the $\frac{11}{8}$-conjecture*, preprint

[G] R. Gompf, *A New Construction of Symplectic Manifolds*, Annals of Math. **142** (1995), 527–595.

[GM] R.E. Gompf and T.S. Mrowka, *Irreducible 4-manifolds need not be complex*, Annals of Math. **138** (1993), 61–111.

[KM] P. Kronheimer and T. Mrowka, *Recurrence Relations and Asymptotics for Four-Manifold Invariants*, to appear in BAMS

[KM1] P.B. Kronheimer and T.S Mrowka, *The genus of embedded surfaces in the projective plane*, Math. Research Letters **1** (1994), 797–808.

[MSzT] J.W. Morgan, Z. Szabó and C.H. Taubes, *A product formula for the Seiberg-Witten invariants and the generalized Thom-Conjecture*, to appear in Journal of Differential Geometry.

[MT] B. Moishezon and M. Teichner, *Simply connected surfaces of positive index*, Invent. Math. **89** (1987), 601–643.

[P] U. Persson, *Chern invariants of surfaces of general type*, Compositio Mathematica **43** (1981), 3–58.

[R] V. Rohlin, *New results in the theory of four dimensional manifolds*, Dokl. Akad. Nauk. SSSR **84** (1952), 221–224.

[S] A. Stipsicz, *Donaldson invariants of cerain symplectic manifolds*, J. reine angew. Math. **465** (1995), 1–10.

[S1] A. Stipsicz, *Donaldson series and (−1)-tori*, J. reine angew. Math. **465** (1995), 133–144.

[S2] A. Stipsicz, *A note on the geography of symplectic manifolds*, Turkish J. of Math. **20** (1996), 135–139.

[Sa] D. Salamon, *Spin geometry and Seiberg-Witten invariants*, in preparation

[Sz] Z. Szabó, *Irreducible four-manifolds with small Euler-characteristics*, Topology **35** (1996), 411–426.

[Sz1] Z. Szabó, *Simply-connected irreducible 4-manifolds with no symplectic structures*, preprint 1996.

[T1] C.H. Taubes, *The Seiberg-Witten invariants and symplectic form*, Math. Research Letters **1** (1994), 809–822.

[T2] C.H. Taubes, *The Seiberg-Witten and the Gromov invariants*, Math. Research Letters **2** (1995), 221–238.

[T3] C.H. Taubes, *$SW \to Gr$, From the Seiberg-Witten equations to pseudoholomorphic curves*, to appear in Jour. Amer. Math. Soc.

[W] S. Wang, *A vanishing theorem for Seiberg-Witten invariants*, Math. Research Letters, **2** (1995).

Progress in Mathematics, Vol. 169, © 1998 Birkhäuser Verlag Basel/Switzerland

Transversals of d-intervals – Comparing three Approaches

GÁBOR TARDOS

Mathematical Institute of the Hungarian Academy of Sciences
Reàltanoda utca 13–15
Budapest, Hungary

1. The problem

In this short note we compare three different methods for solving the same combinatorial problem. We start by describing the problem itself.

Consider any set system F of non-empty sets. By the *matching number* $\nu(F)$ of F we mean the maximal number of pairwise disjoint sets of F. A *transversal* of F is a set S that nontrivially intersects all sets in F. In this case S is said to *cover* F. By the *transversal number* $\tau(F)$ of F we mean the minimum cardinality of a transversal of F.

We clearly have

$$\nu(F) \leq \tau(F)$$

for any set system F, but no inequality holds in full generality in the opposite direction. Finding special classes of set systems for which equality holds or the transversal number can be bounded in terms of the matching number is a central problem in combinatorial duality theory.

The König-Hall theorem gives an example for such a class. Considering bipartite graphs as sets of edges and identifying an edge with the set of its two endpoints one gets that $\tau = \nu$ for finite bipartite graphs. For general (non-bipartite) graphs as set systems one trivially has $\tau \leq 2\nu$ which is tight, as graphs consisting of triangle components show.

Another example where equality holds was noticed by Gallai (see Hajnal and Surányi [5]): For finite sets F of intervals of a line we have $\tau(F) = \nu(F)$.

The class of set systems we consider in this note is closely related to the second example mentioned. Let d be a positive integer. A *d-interval* is a non-empty union of d closed intervals on the line. We are going to compare the matching and transversal numbers of families of d-intervals.

Sometimes it is more natural to assume that the components of the d-intervals are separated, that they lie on distinct lines. Let us fix d distinct parallel lines and define a *separated d-interval* to be the union of d non-empty closed intervals, one on each line. If I is a d-interval and $1 \leq i \leq d$ we say that I^i, the i'th *component* of I, is the interval of I on the i'th line. We remark here that part of the literature uses the term "d-interval" for this and "homogeneous d-interval" for our previous definition.

Let us note here that the assumption of the (separated) d-intervals to be closed is not crucial, it can be replaced by an assumption on the finiteness of the set systems we consider. One of these assumptions is essential though, since for infinite families of open intervals even Gallai's observation fails.

Let us define here the functions we are interested in. For positive integers k and d let let us define $g(d,k)$ to be the maximal transversal number of any family of d-intervals with matching number at most k. Similarly let $f(d,k)$ be the maximal transversal number of any family of separated d-intervals with matching number at most k. In what follows we survey the techniques used in upper bounding these functions. In this notation Gallai's observation can be stated as

$$f(1,k) = g(1,k) = k.$$

We discuss the three methods for bounding f and g in the order they were invented. Separated d-intervals were introduced by Gallai and he posed the problem of finding $f(d,1)$. Gyárfás and Lehel [4] established that for families of (separated) d-intervals the transversal number can be bounded in terms of d and the matching number, i.e. $f(d,k)$ and $g(d,k)$ are finite. The actual bound they got is very weak (and it is left implicit in their paper). Their strictly elementary argument is discussed in the next section. We also mention elementary lower bounds for f and g and a connection between these functions in the same section.

In the third section we discuss the topological approach. Surprisingly it is the topology of higher dimensional spaces that gives the best known upper bound for the transversal number of (separated) d-intervals. It gives tight results for both 2-intervals and separated 2-intervals. This technique was used by Tardos [12] and Kaiser [6].

In the fourth section we discuss the approach of Alon [1] based on the duality of linear programming. The bounds achieved are much better than the ones discussed in the second section but worse than the ones we get from the topological approach by a factor of about 2. The major advantage of this method is simplicity; Alon's paper is less than a page long.

In the last section we mention related results.

2. The elementary approach

Gyárfás and Lehel [4] were the first to prove that for families F of separated d-intervals $\tau(F)$ is bounded in terms of $\nu(F)$ and d. They proved

Theorem 1. [4] *For positive integers d and k we have*

$$f(d+1,k) \leq f(d, k((k+1)^d - 1)) + k.$$

This proves that $f(d,k))$ is always finite but the bound one gets is worse than $k^{d!}$. To illustrate the proof technique we show a similar proof of a somewhat stronger statement in Károlyi and Tardos [8]:

Theorem 2. [8] *For integers $d \geq 2$ and $k \geq 1$ we have*

$$f(d,k) \leq f(d,k-1) + f(d-1,k(d-1)) + (d-1)^2k + 1.$$

Proof. Let F be a family of separated d-intervals with $\nu(F) = k$. For compactness reasons we may assume F is finite. Imagine sliding a point x on the first line from left to right and stopping as soon as we have a set $S \subset F$ of $(d-1)k+1$ d-intervals satisfying that pairwise intersections of members of S are contained in the first line and the first component of each d-interval in S is to the left of x possibly containing x. We may suppose that such point x exists as, otherwise, clearly $\tau(F) \leq f(d-1,k(d-1))$. Let us fix x and S.

We partition $F = F_1 \cup F_2 \cup F_3$ by putting the d-interval $I \in F$ into F_1 if the first component I^1 is strictly to the left of x; we put $I \in F_2$ if $x \in I^1$ and finally we put $x \in F_3$ if I^1 is strictly to the right of x. We can find a transversal of F_1 consisting of $f(d-1,k(d-1))$ points outside the first line. The singleton $\{x\}$ clearly covers F_2. It remains to prove that $\tau(F_3) \leq f(d,k-1)+k(d-1)^2$. For $2 \leq i \leq d$ consider the components $\{I^i | I \in S\}$ and take a set C_i of $k(d-1)$ points on the i'th line separating these $k(d-1)+1$ disjoint intervals. Let F_3' be the part of F_3 not covered by any of the sets C_i. Now each component of a d-interval in F_3' can intersect at most one element of S and the first component intersects none. Thus if we had a family $S' \subset F_3'$ of k pairwise disjoint d-intervals then they would intersect at most $(d-1)k$ members of S, meaning that we could extend S' with an element of S to get $k+1$ pairwise disjoint d-intervals in F. The contradiction shows $\nu(F_3') < k$ and thus $\tau(F_3') \leq f(d,k-1)$. Therefore $\tau(F_3) \leq f(d,k-1)+k(d-1)^2$ as required. □

Using the base cases $f(1,k) = k$ and $f(d,0) = 0$ the result above yields $f(d,k) = O(k^d)$ for any fixed d. This result is still far from optimal; for better bounds see the two following sections.

One can regard separated d-intervals as special cases of d intervals. This yields the trivial connection between f and g:

Lemma 3. *For every d and k we have $f(d,k) \leq g(d,k)$.*

Proof. For any family F of separated d-intervals we construct a family F' of d-intervals with $\tau(F') = \tau(F)$ and $\nu(F') = \nu(F)$. We take a function ϕ that maps the d lines of the separated d-intervals homeomorphically to d disjoint open intervals of the one line of the d-intervals. Now ϕ maps separated d-intervals to d-intervals and we can simply take $F' = \{\phi(I) | I \in F\}$ □

Inequality in the other direction is less obvious. In [4] the finiteness of g was established by showing:

Lemma 4. [4] *For $d \geq 2$ and $k \geq 1$ we have*

$$g(d,k) \leq (g(d-1,k))^{d-1} f(d,k) + \sum_{i=1}^{d-1} (g(d-1,k))^i.$$

A stronger bound appears in [8]:

Lemma 5. [8] *For every d and k we have*

$$g(d,k) \leq f(d, 2d(d-1)k).$$

We include the sketch of the proof of this weak result because it nicely complements the trivial observation of Lemma 3.

Proof sketch: For any family F of d-intervals one can construct a family F' of separated d-intervals such that the d-intervals in F are obtained from the members of F' by perpendicular projection. Clearly $\tau(F) \leq \tau(F')$ in this case. Thus it is enough to prove that $\nu(F') \leq 2d(d-1)\nu(F)$. This is so, since for any pairwise disjoint family S of separated d-intervals the greedy algorithm finds a subset of size $\lceil |S|/(2d(d-1)) \rceil$ of whose projections are also pairwise disjoint. $\square$

We conclude this section with an easy lower bound on f.

Theorem 6. [8] *For positive integers d and k we have*

$$f(d,k) \geq kf(d,1) \geq dk.$$

Proof. We identify the d lines on which our separated d-intervals lie with the line of the real numbers. Let us consider the set F of separated d-intervals I defined by $I^i = [a, b_i]$ for $i = 1, \dots, d$ for all possible values $0 < a < 1$, $a < b_i \leq 1$ for all $i = 1, \dots, d$ and $b_i = 1$ for at least one $i = 1, \dots, d$. It is easy to see that the longest component in any pair of separated d-intervals of F contains an entire component of the other separated d-interval, thus the pair is intersecting and we have $\nu(F) = 1$. One can also see that any finite transversal contains the point 1 in each of the d lines, thus $\tau(F) \geq d$ and we have proved the second inequality.

To prove the first inequality consider a pairwise intersecting family F of separated d-intervals with transversal number $f(d,1)$. We may suppose F covers a finite segment of the d lines and thus we may take the union of k translates of F covering pairwise disjoint segments. This construction clearly multiplies both the matching and the transversal numbers by k. $\square$

Notice that the family constructed in the proof above allows us to obtain the same ordering for any $i = 1, \dots, d$ if we order the separated d-intervals according the left endpoint of their i'th component. It was noticed in [8] that this bound is tight for such families, i.e. this property implies $\tau(F) \leq d\nu(F)$.

As we are going to see in the next section (see Theorem 7), the bound of Theorem 6 is tight for $d = 2$. It is, however, not tight in general. An example of ten separated 3-intervals in [4] shows $f(3,1) \geq 4$ (equality is also established there). Theorem 6 with Lemma 3 shows $g(d,k) \geq dk$. This is not tight even for $d = 2$. An example of six 2-intervals in [4] shows $g(2,1) \geq 3$ (equality is also established there, while $g(2,k) = 3k$ is shown in [6], see Corollary 12).

3. The topological approach

Using the topology of higher dimensional spaces in the investigation of transversals of d-intervals was initiated in [12] where the present author used simplicial complexes to establish the tight bound for separated 2-intervals:

Theorem 7. [12] $f(2,k) = 2k$.

Proof sketch: Let us take a family F of separated 2-intervals. We may suppose F covers a finite part of the two lines, and say that each element of F is contained in the separated 2-interval $[X_0, X_{k+1}] \cup [Y_0, Y_{k+1}]$. Now consider the $2k$-tuples of points $T = (X_1, \ldots, X_k; Y_1, \ldots, Y_k)$ with $X_0 \le X_1 \le \ldots \le X_{k+1}$ and $Y_0 \le Y_1 \le \ldots \le Y_{k+1}$. The space S of these candidate transversals T is the direct product of two k-simplexes. We define bad subsets $B_{i,j} \subset S$ for $0 \le i,j \le k$ by letting $T \in B_{i,j}$ if and only if there exists an element of F contained in $(X_i, X_{i+1}) \cup (Y_j, Y_{j+1})$.

The hard part of the proof is to show the following for any collection of open subsets $B_{i,j} \subseteq S$. If $T \notin B_{i,j}$ holds whenever $X_i = X_{i+1}$ or $Y_j = Y_{j+1}$ (boundary condition) then either the union of all the $B_{i,j}$ does not cover S or else we have a permutation σ of $\{0, \ldots, k\}$ such that the intersection of the sets $B_{i,\sigma(i)}$ is nonempty.

The proof of this claim in [12] is tedious and uses the topology of simplicial complexes. We omit it here.

To apply the above statement we have to show that our bad sets $B_{i,j}$ satisfy the conditions there. The boundary condition is trivial. The sets $B_{i,j}$ are open since the separated d-intervals in F are closed. Therefore we have one of the two possible consequences there.

By the definition of the bad sets $B_{i,j}$ any point $T = (X_1, \ldots, X_k; Y_1, \ldots, Y_k) \in S$ outside all of the bad sets represents a transversal $\{X_1, \ldots, X_k, Y_1, \ldots, Y_k\}$ of F. Thus if $\cup_{0\le i,j\le k} B_{i,j} \ne S$ we have $\tau(F) \le 2k$.

If $T = (X_1, \ldots, X_k; Y_1, \ldots, Y_k) \in \cap_{i=0}^{k} B_{i,\sigma(i)}$ for a permutation σ of $\{0, \ldots, k\}$, then the elements I_i of F contained in $(X_i, X_{i+1}) \cup (Y_j, Y_{j+1})$ are pairwise disjoint for $i = 0, \ldots, k$ thus $\nu(F) \ge k+1$.

For families F of separated 2-intervals with $\nu(F) \le k$ we thus must have $\tau(F) \le 2k$ and therefore $f(2,k) \le 2k$ as claimed. The inequality in the opposite direction comes from Theorem 6. □

Notice the striking similarity of the core topological statement of the proof to Sperner's Lemma. As it was already pointed out in [12] the same proof can bound $f(d,k)$ for arbitrary d if the right analog of the topological statement on the alternatives is proved. This was implicitly done by Kaiser [6]. He also simplified the presentation of the argument and through the use of the Borsuk-Ulam theorem greatly simplified the proof.

We present here Kaiser's upper bound on $g(d,n)$, i.e. the transversal number of d-intervals.

Theorem 8. [6] $g(d,k) \le (d^2 - d + 1)k$.

Proof. Take a family F of d-intervals. We may suppose again that all the d-intervals of F are contained in the unit interval $[0,1]$.

Let us fix $n \ge 1$ and define the n-simplex $S = \{(y_1, \ldots, y_n) | 0 \le y_1 \le \ldots \le y_n \le 1\}$. We think of $y = (y_1, \ldots, y_n) \in S$ to represent the set $\bar{y} = \{y_1, \ldots, y_n\}$ and thus S is the space of transversal-candidates for F of size at most n. We set $V = \{0, 1, \ldots, n\}$. A point $y = (y_1, \ldots, y_n) \in S$ determines the open intervals

$L_i(y) = (y_i, y_{i+1})$ for $i \in V$ where we take $y_0 = 0$ and $y_{d+1} = 1$. (Note that some of these intervals may be empty.) For a set $e \subset V$ we define

$$w_y(e) = \sup_I dist(\bar{y}, I)$$

with I ranging over the d-intervals of F contained in $\cup_{i \in e} L_i(y)$ but intersecting every $L_i(y)$ with $i \in e$. Here *dist* denotes minimum distance between sets on the line.

For $i \in V$ and $y \in S$ we define

$$w_y(i) = \sum_{e \ni i} w_y(e).$$

Observation 9.

(a) *$w_y(e) \geq 0$; furthermore, $w_y(e) = 0$ for every set e of size $|e| > d$ or for $e = \emptyset$.*
(b) *$w_y(e) = 0$ if there exists an index $i \in e$ with $L_i(y) = \emptyset$ and thus $w_y(i) = 0$ if $L_i(y) = \emptyset$.*
(c) *If $w_y(e) = 0$ for every set $e \subset V$ then $\bar{y}$ is a transversal of F.*
(d) *As a function of $y \in S$ all the functions $w_y(e)$ $(e \subset V)$ and thus the functions $w_y(i)$ $(i \in V)$ are continuous.*

Proof. (a) follows from the fact that F is a collection of nonempty d-intervals. Parts (b) and (d) are trivial. For (c) notice that any element of F disjoint from the set represented by y has a positive contribution to the supremum in the definition of $w_y(E)$ for a (unique) set e. Here we have also used that F consists of closed sets. □

The following observation is the heart of this proof. This is the point where topology is used. Note that we did not suppose anything about the family F or the parameter n for the observation.

Lemma 10. *There exists a point $y \in S$ for which $w_y(i)$ is independent of $i \in V$.*

Proof. To be able to use the Borsuk-Ulam theorem we have to shift attention from the simplex S to the sphere $S^n = \{(z_0, \ldots, z_n | \sum z_i^2 = 1\} \subset R^{n+1}$. We do it through the map $g : S^n \to S$ defined by

$$g(z_0, \ldots, z_n) = (t_1, \ldots, t_n),$$

$$t_i = \sum_{j=0}^{i-1} z_i^2, \quad \text{for } i = 1, \ldots, n.$$

Note that $g(z)$ does not depend on the sign of the coordinates of z.

We combine g with the weight functions $w_y(i)$ to define the functions $h_i : S^n \to \mathbb{R}$ for $i = 1, \ldots, n$ by

$$h_i(z) = \text{sign}(z_i) w_{g(z)}(i) - \text{sign}(z_0) w_{g(z)}(0)$$

where $z = (z_0, \ldots, z_n)$.

Notice that by Observation 9 (d) all functions in the above definition are continuous except for the sign function. But at the jump of the sign function $\text{sign}(z_i)$ $(i \in V)$ we have $L_i(g(z)) = \emptyset$ and thus by Observation 9 (b) we have $w_{g(z)}(i) = 0$. Therefore the functions h_i are continuous everywhere.

Notice also the antipodality: $h_i(-z) = -h_i(z)$ follows from $g(z) = g(-z)$. Now we can use the Borsuk-Ulam theorem. It states that n continuous, antipodal, real functions on S^n always have a common zero. (See, e.g. [11].) In particular there exists a $z \in S^n$ for which $h_i(z) = 0$ for $i = 1, \ldots, n$. This clearly implies that for $y = g(z)$ we have $w_y(i) = w_y(0)$ for $i = 1, \ldots, n$ as claimed. □

For the rest of the proof we fix $y \in S$ to be the point claimed in Observation 9 and let x be the common value of $w_y(i)$ for $i \in V$. The connection between the matching and transversal numbers of F will follow from the following alternative. If $x = 0$, then the transversal number is small; if $x \neq 0$, then the matching number is large.

Suppose first that $x = 0$. Then clearly $w_y(e) = 0$ for all $e \subset V$. By Observation 9 (c) this implies $\tau(F) \leq n$.

Let us now suppose that $x > 0$. We define $M = (V, E)$ to be the hypergraph on the vertex set V with the edge set $E = \{e \subset V | w_y(e) > 0\}$.

Let us first observe that $\nu(M) \leq \nu(F)$. Indeed, we can choose a d-interval $I_e \in F$ for every edge $e \in E$ such that $I_e \subset \cup_{i \in e} L_y(i)$. The d-intervals chosen for disjoint edges are clearly disjoint, thus the inequality holds as claimed. Thus it is enough for us to lower bound $\nu(M)$.

A *partial matching* of a hypergraph N is an assignment of non-negative weights to the edges such that the sum of weights on edges containing any particular vertex is at most 1. The *partial matching number* $\nu^*(N)$ is the supremum of the total weights assigned by any partial matching of M. We clearly have $\nu^*(N) \geq \nu(N)$. We will later need inequalities in the other direction.

The function $w_y(e)/x$ clearly forms a partial matching of M. Therefore $\nu^*(M) \geq \sum_{e \in E} w_y(e)/x$. Here, by Observation 9 (a), all edges in E have size at most d and thus $d \sum_{e \in E} w_y(e) \geq \sum_{e \in E} |e| w_y(e) = \sum_{i \in V} \sum_{e \ni i} w_y(e) = (n+1)x$. Thus

$$\nu^*(M) \geq \frac{n+1}{d}.$$

The last task to finish the proof entails connecting $\nu^*(M)$ and $\nu(M)$. The following observation is trivial. For every hypergraph M with edges having at most d elements we have:

$$\nu^*(M) \leq d\nu(M).$$

For the proof notice that if $M' \subset M$ is any maximal matching, then every edge of M contains one of the points in $H = \cup M'$ and thus the weight of any edge in a fractional matching is counted at the vertex weights of vertices in H, thus $\nu^*(M) \leq |H| \leq d|M'|$. Notice that this already implies

$$g(d,k) \leq d^2 k.$$

Slight improvement of the trivial argument above is given by Füredi in [3]:

Lemma 11. [3] *For every hypergraph M with edges having at most d elements we have:* $\nu^*(M) \le (d-1+1/d)\nu(M)$.

Now we have that either $x = 0$ and then $\tau(F) \le n$ or $x \ne 0$ and then $\nu(F) \ge \nu(M) \ge \nu^*(M)/(d-1+1/d) \ge (n+1)/(d^2-d+1)$. Thus with the choice $n = (d^2-d+1)k$ for any family F with $\nu(F) \le k$ $x \ne 0$ is impossible thus we must have $\tau(F) \le n$. This finishes the proof of Theorem 8. □

We further remark here that if $d > 2$ and there is no projective plane of order $d-1$, the statement of Lemma 11 can be made stronger to $\nu^*(M) \le (d-1)\nu(M)$, see [3]. This leads to the tighter bound $g(d,k) \le (d^2-d)k$ for such d. Let us also mention that the bound of Theorem 8 is tight for the $d = 2$ special case:

Corollary 12. [6] $g(2,k) = 3k$

Proof. $g(2,k) \le 3k$ is a special case of Theorem 8. The example in [4] already mentioned at the end of Section 2 shows $g(2,1) \ge 3$ and $g(2,k) \ge kg(2,1)$ is trivial (similarly to the first inequality of Theorem 6). □

We remark that the special case $d = 2$ of Lemma 11 was first proved by Lovász in [9].

Using Lemma 3 one sees that the bound of Theorem 8 also applies for f in place of g. Using the proof technique directly (Kaiser [6]) establishes the slightly better bound:

Theorem 13. [6] $f(d,k) \le (d^2-d)k$.

The proof is similar but one has to use a recent extension of the Borsuk-Ulam theorem by Ramos [10]. This theorem is about the common zeros of functions defined on product of spheres. Notice that Theorem 7 is a special case of this Theorem 13.

4. Using the duality theorem

In this section we survey the short note of Noga Alon [1]. This paper uses the method of Alon and Kleitman [2]; it is based on the duality theorem for linear programming and on Turán's theorem in extremal graph theory. It is worth including the short proof in its entirety.

Theorem 14. [1] $g(d,k) \le 2d^2k-1$.

Proof. Let F be a family of d-intervals with $\nu(F) \le k$. For compactness reasons it is again enough to consider finite families, let $n = |F|$. As there are no $k+1$ pairwise disjoint elements of F Turán's theorem yields that there are more than $n^2/(2k)$ unordered pairs of (not necessarily) distinct intersecting members of F. For any such pair P one finds two ordered triplets (p, I, I') with p being an endpoint in a member of the d-interval I, $p \in I'$ and $\{I, I'\} = P$. There are at most $2dn$ endpoints p, thus one of them is contained in $n/(2dk)$ d-intervals of F.

Let us assign nonnegative rational weights a_I/b to the d-intervals $I \in F$. Applying the above argument to a multiset F' obtained from F via replicating each d-interval I a_I times, one gets a point p that lies on a collection of d-intervals representing more than $1/(2dk)$ fraction of the total weight. We apply the min-max theorem of linear programming to deduce the existence of a multiset M of m points such that each d-interval in F contains more than $m/(2dk)$ of them. Now we select the points in M whose rank in the linear ordering is divisible by $\lfloor m/(2d^2k)\rfloor+1$. Any interval disjoint from these points contains at most $m/(2d^2k)$ points of M, thus none of the d-intervals in F can be disjoint of the less than $2d^2k$ points selected. □

The bound obtained here is worse than Kaiser's bound discussed in the preceding section by a factor approaching 2 for large d.

5. Applications and open problems

It is not surprising that one can use the connection of the matching and transversal numbers of (separated) d-intervals to obtain similar connection for other families of (mostly geometrical) sets. We give one example here. Let S be a closed set in the Euclidean plane. We define an *S-box* to be the intersection of an axis-parallel rectangle with S. In [8] Theorem 7 is used to prove

Theorem 15. [8] *For any family F of S-boxes*

$$\tau(F) \le 4\nu(F)\lfloor \log \nu(F) + 1\rfloor^2.$$

It is perhaps more noteworthy when a topological method similar to that of Section 3 is proved to be applicable in a different setting. An example is the paper of Kaiser and Rabinovich [7] about a Helly-type problem. They consider *convex (n,d)-bodies*, ordered n-tuples of convex sets in R^d. A family F of convex (n,d)-bodies is called *weakly intersecting* if there is an n-tuple $(p_1,\dots,p_n)$ of points in R^d such that for every convex (n,d)-body $(K_1,\dots,K_n)\in F$ one has $p_i\in K_i$ for at least one $i=1,\dots,n$. The family F is called strongly intersecting if there exist a point $p\in R^d$ and an index $1\le i\le n$ with $p\in K_i$ for each convex (n,d)-body $(K_1,\dots,K_n)\in F$.

Theorem 16. [7] *Suppose every subfamily F' of the family F of convex (n,d)-bodies is strongly intersecting if $|F'|\le \lceil \log_2(n+2)\rceil^d$. Then F is weakly intersecting.*

Finding out if there are more combinatorial problems where a similar line of reasoning is fruitful and what these problems are can be an interesting research project.

A survey paper on a simple combinatorial problem of finding the values of the function $f(d,k)$ and $g(d,k)$ must admit that none of these values is known exactly for $d\ge 3$ and $k\ge 1$ except for the single case $f(3,1)=4$ settled in [4]. It seems that the most intriguing question is how to find the order of magnitude for the maximum transversal number of an *intersecting* family of d-intervals. Note that the upper bound for this number $g(d,1)$ is quadratic (Theorem 8) while the lower bound is linear (Theorem 6).

References

[1] N. Alon, Piercing d-intervals, to appear in *Discrete Comput. Geom.*

[2] N. Alon and D. J. Kleitman, Piercing convex sets and the Hadwiger Debrunner (p, q)-problem, *Advences in Mathematics* **96** (1992), 103–112.

[3] Z. Füredi, Maximum degree and fractional matchings in uniform hypergraphs, *Combinatorica* **1**, 2 (1981), 155–162.

[4] A. Gyárfás and J. Lehel, A Helly-type problem in trees, in *Combinatorial Theory and its Applications*, P. Erdős, A. Rényi and V. T. Sós, Eds., North-Holland, Amsterdam 1970, 571–584.

[5] A. Hajnal and J. Surányi, Über die Auflösung von Graphen in vollständige Teilgraphen, *Ann. Univ. Sci. Budapest* (1958), 113–121.

[6] T. Kaiser, Transversals of d-intervals, to appear in *Discrete Comput. Geom.*

[7] T. Kaiser and Y. Rabinovich, Intersection properties of families of convex (n, d)-bodies, to appear in *Discrete Comput. Geom.*

[8] Gy. Károlyi and G. Tardos, On point covers of multiple intervals and axis-parallel rectangles, *Combinatorica*, **16** (2) (1996) 213-222.

[9] L. Lovász, Minimax theorems for hypergraphs, in *Hypergraph Seminar 1972* (C. Berge and D.K. Ray-Chaudhuri eds.), *Lecture Notes* **411**, Springer-Verlag, Berlin, 1972.

[10] E. A. Ramos, Equipartition of mass distributions by hyperplanes, *Discrete Comput. Geom.* **15** (1996) 147–167.

[11] E. H. Spanier, Algebraic Topology, McGraw-Hill, 1966.

[12] G. Tardos, Transversals of 2-intervals, a topological approach, *Combinatorica* **15**, 1 (1995), 123–134.

Progress in Mathematics, Vol. 169, © 1998 Birkhäuser Verlag Basel/Switzerland

Algebras with Involution and Classical Groups

JEAN-PIERRE TIGNOL*

Université catholique de Louvain, B–1348 Louvain-la-Neuve, Belgium

Deep relations between linear algebraic groups over an arbitrary field and central simple algebras with involution can be traced back to two main sources. In [40], Weil[1] shows that the connected component of the identity in the automorphism group of a separable algebra with involution is almost[2] always a semisimple linear algebraic group of adjoint type and that, conversely, almost[3] every semisimple linear algebraic group of adjoint type can be obtained in this way. The groups derived from algebras with involution are called *classical*.

On the other hand, investigating representations of a semisimple linear algebraic group G over an arbitrary field F, Tits proves in [38] that some irreducible representations ρ over a separable closure can be descended to representations of G in the group $\mathrm{GL}(A_\rho)$ of invertible elements in some uniquely determined central simple F-algebra A_ρ. This algebra carries a canonical involution in certain cases.

Although these results have been known and used for more than twenty years, investigations from different sources unexpectedly converged in the last few years to renew interest in this topic. The primary reason for this convergence may be that the theory of linear algebraic groups provides the proper level of generality to make explicit the relationship between various results on central simple algebras and quadratic forms.

Index reduction formulas are a case in point. It is usually difficult to control the behaviour of the Schur index of a central simple algebra under scalar extension. However, Schofield and Van den Bergh [32] found a way to use Quillen's computation of K-groups of Severi-Brauer varieties to determine by how much the index of a central simple algebra is reduced when scalars are extended to the function field of the Severi-Brauer variety of another central simple algebra. This idea of Schofield and Van den Bergh was soon used to derive index reduction formulas for scalar extension to function fields of various classes of varieties X from K-theory information on X: Blanchet [6] considered the case of twisted Grassmannians and Merkurjev [24] the case of quadrics. Tao [35] then considered varieties arising from central simple algebras with involution, which he called *involution varieties*. In a sense, these varieties are hybrids of Severi-Brauer varieties and quadrics, making

*Supported in part by the National Fund for Scientific Research (Belgium).

[1]Weil attributes the idea to Siegel and to "a famous winner of many cocycle races".

[2]Factors of degree 2 with involution of orthogonal type are excluded, since their automorphism group is commutative. There are restrictions on the characteristic of the base field also: see the end of section 1.

[3]The exceptional groups and the groups of trialitarian type D_4 are excluded.

clear that orthogonal[4] involutions on central simple algebras may be thought of as twisted forms of projective quadrics: the involution variety of a central simple algebra A with involution of orthogonal type may be regarded as a quadric hypersurface on the Severi-Brauer variety of A. It was soon realized that all the varieties thus considered, including the transfers of Severi-Brauer varieties and of twisted Grassmannians treated by Saltman [30] by different K-theory techniques, were special cases of varieties of parabolic subgroups in a semisimple linear algebraic group. A general index reduction formula was then set up by Merkurjev, Panin and Wadsworth [26], subsuming all the particular cases found so far and enlightening the occurrence of auxiliary central simple algebras in these formulas: these are the Tits algebras associated with representations of the group. Simultaneously, a similar generalization was achieved by Kersten and Rehmann [11], who showed that various types of generic splitting fields designed for central simple algebras and for quadratic forms are special cases of functions fields of varieties of parabolic subgroups.

Other impetus for the study of algebras with involution was given by Knus, Ojanguren, Parimala and Sridharan in a series of papers[5] investigating quadratic forms of low rank. In [12], [17], quadratic forms of dimension 6 and discriminant 1 are shown to be (generalized) pfaffians of central simple algebras of degree 4 with involution. Generalized pfaffians are used in [18] and [19] to define a discriminant for involutions, a notion which was already implicit (at least for involutions on central simple algebras) in Jacobson's and Tits' construction of Clifford algebras [9], [37]. The case of algebras of degree 4 has been specially fascinating since Albert proved that every central simple algebra of degree 4 with an involution leaving the center elementwise invariant decomposes into a tensor product of two quaternion subalgebras [2] (see also [29]). For a given involution there may not be any decomposition into quaternion subalgebras stable under the involution, however, as was shown by Amitsur, Rowen and Tignol [3]. Knus, Parimala and Sridharan [18], [20] showed that a stable decomposition exists if and only if the discriminant of the involution is trivial. After [18], it became apparent that seemingly unrelated investigations by Lam, Leep and Tignol [21] on the discriminants of subfields of biquaternion algebras could yield significant information on the discriminants of involutions on these algebras; this was made explicit by Knus, Lam, Shapiro and Tignol [15].

With hindsight, the relationship between 6-dimensional quadratic forms and biquaternion algebras, first observed by Albert [1], as well as various results relating orthogonal involutions on biquaternion algebras with quaternion algebras over quadratic extensions, and symplectic involutions on biquaternion algebras with 5-dimensional quadratic forms, can all be seen to derive from the exceptional isomorphisms of classical groups suggested by the equalities of Dynkin diagrams $A_3 = D_3$, $D_2 = A_1 \times A_1$ and $B_2 = C_2$. This variety of approaches to biquaternion algebras was put to remarkable use by Rost in his computation of the reduced Whitehead group of these algebras: see [25], [16].

[4]See section 1 for the definition of an orthogonal involution.

[5]A comprehensive account of these investigations can be found in [13], [14].

The purpose of the rest of this survey is to outline the properties of biquaternion algebras which derive from the exceptional isomorphisms of classical groups. General properties of central simple algebras with involution, their invariants and their automorphism groups, are discussed in the first three sections. A detailed exposition with complete proofs will be found in [16].

1. Algebras with involution

The involutions considered in this paper are ring anti-automorphisms of period 2. The most important examples in relation with algebraic groups are obtained as follows: if b is a nonsingular symmetric or skew-symmetric bilinear form on a finite-dimensional vector space V over a field F of arbitrary characteristic, there is a unique involution σ_b on $\operatorname{End}_F(V)$ such that

$$b\Big(f(x),y\Big) = b\Big(x,\sigma_b(f)(y)\Big)$$

for all $x,y \in V$ and all $f \in \operatorname{End}_F(V)$. The map σ_b is called the *adjoint involution* with respect to the bilinear form b. For $f \in F = F\cdot\operatorname{Id}_V$, we clearly have $\sigma_b(f) = f$. It is easily seen that every involution on $\operatorname{End}_F(V)$ which leaves F elementwise invariant is the adjoint involution with respect to some nonsingular symmetric or skew-symmetric bilinear form on V, uniquely determined up to a factor in $F^\times$ (see [31, §8.7], [16, Ch. I]).

Similarly, if W is a finite-dimensional vector space over a field K and ι is an automorphism of period 2 of K, every involution on $\operatorname{End}_K(W)$ which restricts to ι on K is adjoint to a nonsingular hermitian form on W with respect to ι, uniquely determined up to a factor in the subfield of K elementwise invariant under ι.

This example can be further generalized: every finite-dimensional simple algebra may be represented as $\operatorname{End}_D(U)$ for some finite-dimensional division algebra D and some finite-dimensional right vector space U over D, by a theorem of Wedderburn, and every involution on $\operatorname{End}_D(U)$ is adjoint to a nonsingular hermitian or skew-hermitian form on U with respect to an involution on D. This form is uniquely determined up to a central factor invariant under the involution [31, §8.7]. Therefore, it is equivalent to study involutions on central simple algebras or hermitian and skew-hermitian forms up to a scalar factor over division algebras. The point of view of hermitian spaces affords greater flexibility because the orthogonal sum defines a group structure on Witt-equivalence classes; on the other hand, involutions are better behaved regarding scalar extension: if σ is an involution on an F-algebra A and L/F is any field extension, the involution $\sigma_L = \sigma\otimes\operatorname{Id}_L$ on $A_L = A\otimes_F L$ is clearly related to σ, whereas the relationship between the hermitian forms associated with σ and σ_L is unclear, if only because the algebras associated with A and A_L by the Wedderburn theorem may have different indices.

It may be more suggestive to think of the involutions leaving F elementwise invariant on a central simple F-algebra A as twisted forms of nonsingular symmetric or skew-symmetric bilinear forms up to scalar factors. Under scalar extension to a separable closure F_s the algebra splits as $A\otimes F_s \simeq \operatorname{End}_{F_s}(V)$ for some vector space V over F_s, and every involution on $\mathrm{End}_{F_s}(V)$ is adjoint to a nonsingular symmetric or skew-symmetric bilinear form.

In the case where the center K is not elementwise invariant under the involution, the algebra structure should be considered over the subfield F of invariant elements (of codimension 2 in K), so that the involution extends to scalar extensions. Since $K \otimes_F F_s \simeq F_s \times F_s$, the algebra decomposes under scalar extension to F_s into a direct product of two factors which are exchanged by the involution (and are therefore anti-isomorphic to each other).

Involutions on simple algebras may thus be divided into the following types:

orthogonal: An involution σ on a central simple F-algebra A is called *orthogonal* if

$$(A,\sigma) \otimes F_s \simeq \left(\mathrm{End}_{F_s}(V), \sigma_b\right)$$

for some nonsingular symmetric bilinear form b on a finite-dimensional vector space V over F_s. If $\mathrm{char}\, F = 2$, the form b is also required to be nonalternating, in the sense that there exists $x \in V$ such that $b(x,x) \neq 0$.

symplectic: An involution σ on a central simple F-algebra A is called *symplectic* if

$$(A,\sigma) \otimes F_s \simeq \left(\mathrm{End}_{F_s}(V), \sigma_b\right)$$

for some nonsingular alternating bilinear form b on a finite-dimensional vector space V over F_s. The dimension of V, which is by definition the degree of A, is necessarily even in this case.

unitary: An involution τ on a simple algebra B over a field K which is a separable quadratic extension of a subfield F is called *unitary* if

$$(B,\tau) \otimes_F F_s \simeq \left(\mathrm{End}_{F_s}(V) \times \mathrm{End}_{F_s}(V)^{\mathrm{op}}, \varepsilon\right)$$

for some finite-dimensional vector space V, where $\mathrm{End}_{F_s}(V)^{\mathrm{op}}$ is the opposite algebra to $\mathrm{End}_{F_s}(V)$ and ε is the *exchange involution*, defined by

$$\varepsilon(f, g^{\mathrm{op}}) = (g, f^{\mathrm{op}}).$$

Slightly abusing terminology so that the notion will be stable under scalar extension, we call (B,τ) a *central simple algebra with unitary involution over F* whenever the relation above holds, and call $\dim_{F_s} V$ the *degree* of (B,τ). We thus allow B to be of the form $E \times E^{\mathrm{op}}$ (hence not simple!) for some central simple F-algebra E, with τ the exchange involution. Note that the center of a central simple algebra with unitary involution (B,τ) over F is *not* F, but an étale quadratic extension of F, i.e. either a separable quadratic field extension of F or an algebra isomorphic to $F \times F$.

Involutions of unitary type on a simple algebra can be characterized by the condition that the center is not elementwise invariant. Involutions of orthogonal and symplectic type can also be distinguished "rationally", i.e. without scalar extension: if the degree of the central simple F-algebra A is n (i.e.

$[A:F]=n^2$), extension to a separable closure shows that the dimension of the space $\operatorname{Sym}(A,\sigma)$ of symmetric elements is either $n(n+1)/2$ or $n(n-1)/2$. If char $F\neq 2$, the involution σ is orthogonal if $\dim_F \operatorname{Sym}(A,\sigma)=n(n+1)/2$ and symplectic if $\dim_F \operatorname{Sym}(A,\sigma)=n(n-1)/2$. If char $F=2$, we always have $\dim_F \operatorname{Sym}(A,\sigma)=n(n+1)/2$; the involution σ is orthogonal if there exists a symmetric element s whose reduced trace $\operatorname{Trd}_A(s)$ is not zero and symplectic if $\operatorname{Trd}_A\bigl(\operatorname{Sym}(A,\sigma)\bigr)=\{0\}$.

As pointed out above, orthogonal involutions on a split algebra $\operatorname{End}_F(V)$ are adjoint to symmetric bilinear forms on V. In order to define orthogonal groups however, quadratic forms rather than symmetric bilinear forms are needed. These two notions are in one-to-one correspondence if the characteristic char F is different from 2, but in characteristic 2 they are different. Therefore if char $F=2$ a more precise structure is required. In [37], Tits defines generalized quadratic forms for this purpose. A slightly different solution is used in [16], where *quadratic pairs* on a central simple F-algebra A of characteristic 2 are defined as pairs (σ,f) where σ is a symplectic involution and $f:\operatorname{Sym}(A,\sigma)\to F$ is a linear map such that $f\bigl(x+\sigma(x)\bigr)=\operatorname{Trd}_A(x)$ for all $x\in A$.

To avoid these complications, we shall henceforth assume that the characteristic of the base field F is different from 2.

2. Automorphisms

An automorphism of an algebra with involution (A,σ) is an algebra-automorphism of A which commutes with σ. Inner automorphisms which commute with σ are easily determined: denoting by $\operatorname{Int}(a)$ the automorphism of conjugation by $a\in A^\times$, which carries $x\in A$ to axa^{-1}, we have $\sigma\circ\operatorname{Int}(a)=\operatorname{Int}(\sigma(a)^{-1})$, hence $\operatorname{Int}(a)\in\operatorname{Aut}(A,\sigma)$ if and only if $\sigma(a)a$ is central in A.

For instance, if b is a nonsingular symmetric bilinear form on a finite-dimensional vector space V over a field F, inner automorphisms of $\operatorname{End}_F(V)$ which commute with the adjoint involution σ_b are induced by linear transformations $f\in\operatorname{GL}(V)$ such that $\sigma_b(f)f\in F^\times$. In view of the definition of σ_b, the condition $\sigma_b(f)f=\lambda\in F^\times$ is equivalent to: $b\bigl(f(x),f(y)\bigr)=\lambda\, b(x,y)$ for all $x,y\in V$, which means that f is a *similitude* of (V,b) with multiplier λ.

More generally, the group of similitudes of an algebra with involution (A,σ) with center $Z(A)$ is defined by

$$\operatorname{Sim}(A,\sigma)=\Bigl\{a\in A^\times \mid \sigma(a)a\in Z(A)\Bigr\};$$

the map $a\mapsto\operatorname{Int}(a)$ is then a homomorphism from $\operatorname{Sim}(A,\sigma)$ to $\operatorname{Aut}(A,\sigma)$ which induces an isomorphism from $\operatorname{Sim}(A,\sigma)/Z(A)^\times$ onto the group of inner automorphisms of (A,σ). For $a\in\operatorname{Sim}(A,\sigma)$, the element $\sigma(a)a\in Z(A)^\times$ is called the *multiplier* of the similitude a. The similitudes with multiplier 1 are the *isometries*:

$$\operatorname{Iso}(A,\sigma)=\Bigl\{a\in A^\times \mid \sigma(a)a=1\Bigr\}.$$

To give more precise information, we consider separately the various cases where (A,σ) is central simple.

2.1. Orthogonal case Suppose A is a central simple algebra over a field F of characteristic different from 2, and let σ be an orthogonal involution on A. We then denote

$$\mathrm{Iso}(A,\sigma)=\mathrm{O}(A,\sigma),\qquad \mathrm{Sim}(A,\sigma)=\mathrm{GO}(A,\sigma)$$

and

$$\mathrm{PGO}(A,\sigma)=\mathrm{GO}(A,\sigma)/F^\times.$$

Let also

$$\mathrm{O}^+(A,\sigma)=\Big\{a\in\mathrm{O}(A,\sigma)\mid \mathrm{Nrd}_A(a)=1\Big\},$$

where $\mathrm{Nrd}_A: A^\times\to F^\times$ is the reduced norm map.

Since every automorphism of A is inner, by the Skolem-Noether theorem, the map $a\mapsto \mathrm{Int}(a)$ induces an isomorphism

$$\mathrm{PGO}(A,\sigma)\xrightarrow{\sim}\mathrm{Aut}(A,\sigma).$$

As observed in the example above, in the split case $(\mathrm{End}_F(V),\sigma_b)$ where b is a nonsingular symmetric bilinear form, $\mathrm{O}(\mathrm{End}_F(V),\sigma_b)=\mathrm{O}(V,b)$ is the orthogonal group of the space (V,b). Since every central simple algebra A with orthogonal involution σ becomes isomorphic to $(\mathrm{End}_F(V),\sigma_b)$ after scalar extension, the group $\mathrm{O}(A,\sigma)$ is a form of an orthogonal group.

If the degree $\deg A$ is odd, then A is necessarily split, because the exponent of A is either 1 or 2, since σ determines an isomorphism $A\simeq A^{\mathrm{op}}$, and the index of A is a multiple of the exponent which divides $\deg A$. An easy argument involving the discriminant of bilinear forms shows that the multiplier of every similitude of (A,σ) is a square, hence

$$\mathrm{GO}(A,\sigma)=F^\times\times\mathrm{O}^+(A,\sigma)\qquad\text{and}\qquad \mathrm{PGO}(A,\sigma)=\mathrm{O}^+(A,\sigma).$$

The group $\mathrm{PGO}(A,\sigma)$ is an absolutely simple adjoint group of type B_n, if $\deg A = 2n+1>1$.

If $\deg A$ is even, the groups $\mathrm{GO}(A,\sigma)$ and $\mathrm{PGO}(A,\sigma)$ are not connected. The connected component of the identity in $\mathrm{GO}(A,\sigma)$ is

$$\mathrm{GO}^+(A,\sigma)=\Big\{a\in\mathrm{GO}(A,\sigma)\mid \mathrm{Nrd}_A(a)=(\sigma(a)a)^{\deg A/2}\Big\},$$

and $\mathrm{PGO}^+(A,\sigma)=\mathrm{GO}^+(A,\sigma)/F^\times$ is the connected component of the identity in $\mathrm{PGO}(A,\sigma)=\mathrm{Aut}(A,\sigma)$:

$$\mathrm{PGO}^+(A,\sigma)=\mathrm{Aut}(A,\sigma)^0.$$

This group is an adjoint semisimple group of type D_n if $\deg A=2n\geq 4$; it is absolutely simple if $n>3$. The corresponding simply connected group is the spin group $\mathrm{Spin}(A,\sigma)$ which lies in the (generalized, even) Clifford algebra $C(A,\sigma)$ discussed in section 3.

2.2. Symplectic case Suppose A is a central simple algebra over a field F and σ is a symplectic involution on A. We then denote

$$\mathrm{Iso}(A,\sigma) = \mathrm{Sp}(A,\sigma), \qquad \mathrm{Sim}(A,\sigma) = \mathrm{GSp}(A,\sigma)$$

and

$$\mathrm{PGSp}(A,\sigma) = \mathrm{GSp}(A,\sigma)/F^\times.$$

Since every automorphism of A is inner, by the Skolem-Noether theorem, the map $a \mapsto \mathrm{Int}(a)$ induces an isomorphism

$$\mathrm{PGSp}(A,\sigma) \xrightarrow{\sim} \mathrm{Aut}(A,\sigma).$$

In the split case $(\mathrm{End}_F(V), \sigma_b)$ where b is a nonsingular alternating bilinear form, the same arguments as in the orthogonal case show that $\mathrm{Sp}(\mathrm{End}_F(V), \sigma_b)$ is the group of isometries of (V, b). Since every central simple algebra with symplectic involution (A,σ) becomes isomorphic to $(\mathrm{End}_F(V), \sigma_b)$ after scalar extension, it follows that $\mathrm{Sp}(A,\sigma)$ is a form of a symplectic group.

The groups $\mathrm{Sp}(A,\sigma)$, $\mathrm{GSp}(A,\sigma)$ and $\mathrm{PGSp}(A,\sigma)$ are connected; moreover, the group $\mathrm{Sp}(A,\sigma)$ (resp. $\mathrm{PGSp}(A,\sigma)$) is absolutely almost simple and simply connected (resp. absolutely simple adjoint) of type C_n if $\deg A = 2n$.

2.3. Unitary case Suppose (B,τ) is a central simple algebra with unitary involution over a field F, and let K denote the center of B, which is an étale quadratic extension of F. We denote

$$\mathrm{Iso}(B,\tau) = \mathrm{U}(B,\tau), \qquad \mathrm{Sim}(B,\tau) = \mathrm{GU}(B,\tau)$$

and

$$\mathrm{PGU}(B,\tau) = \mathrm{GU}(B,\tau)/K^\times.$$

Let also

$$\mathrm{SU}(B,\tau) = \left\{ u \in \mathrm{U}(B,\tau) \mid \mathrm{Nrd}_B(u) = 1 \right\}.$$

The algebra B may have automorphisms which do not restrict to the identity on K; those which leave K elementwise invariant and commute with τ form the connected component of the identity in $\mathrm{Aut}(B,\tau)$. The Skolem-Noether theorem shows that they are all inner, as in the preceding cases:

$$\mathrm{PGU}(B,\tau) \xrightarrow{\sim} \mathrm{Aut}(B,\tau)^0.$$

If $B = \mathrm{End}_K(W)$ for some K-vector space W, and $\tau = \sigma_h$ is the adjoint involution with respect to some nonsingular hermitian form h on W, then $\mathrm{U}(B,\tau) = \mathrm{U}(W,h)$ is the unitary group of the hermitian space (W,h).

If $K = F \times F$, then $B = E \times E^{\mathrm{op}}$ for some central simple F-algebra E, and we may assume that $\tau = \varepsilon$ is the exchange involution. In that case

$$\mathrm{U}(B,\tau) = \left\{ (u, (u^{-1})^{\mathrm{op}}) \mid u \in E^\times \right\} \simeq \mathrm{GL}(E),$$

hence

$$\mathrm{SU}(B,\tau) \simeq \mathrm{SL}(E) \qquad \text{and} \qquad \mathrm{PGU}(B,\tau) \simeq \mathrm{PGL}(E).$$

Since every central simple algebra with unitary involution becomes isomorphic to $(\mathrm{End}_F(V) \times \mathrm{End}_F(V)^{\mathrm{op}}, \varepsilon)$ after scalar extension, the group $\mathrm{SU}(B,\tau)$ is a form of a special linear group. It is a simply connected absolutely almost simple group of type A_{n-1} if $\deg(B,\tau) = n \geq 2$; the corresponding adjoint simple group is $\mathrm{PGU}(B,\tau)$.

3. Invariants

Invariants of central simple algebras with involution or, equivalently, of their algebraic groups of automorphisms, can be obtained in various ways. The most obvious invariants are the degree of the algebra and the type of the involution, which are closely related to the rank and the type of the group, as observed in the preceding section. The classification of semisimple linear algebraic groups over an arbitrary field readily provides subtler invariants which Tits [36] combines under the notion of *index*: the action of the absolute Galois group on the Dynkin diagram of the group, the Schur index of the central simple algebra and the Witt index of the hermitian or skew-hermitian form associated to the involution. Additional invariants of central simple algebras with involution can be obtained by considering the various central simple algebras associated by Tits [38] with representations of the simply connected covers of their automorphism groups. (See below for a description of these algebras).

Bayer-Fluckiger and Parimala [4] have shown that these "classical" invariants are sufficient to classify central simple algebras with involution over fields of cohomological dimension at most 2, thereby solving in the affirmative an old conjecture of Serre on the cohomology of simply connected algebraic groups.

Higher cohomological invariants, with values in $H^3(F, \mathbf{Q}/\mathbf{Z}(2))$, have been defined by Rost (unpublished). They have been explicitly determined in special cases only. We refer to Serre's Bourbaki talk [34] for an outline of their definition and to [8] for the case of central simple algebras with unitary involution of degree 3.

Further invariants can be derived from (reduced) trace bilinear forms: every orthogonal or symplectic involution σ on a central simple F-algebra A induces an algebra isomorphism

$$\sigma_* : A \otimes A \xrightarrow{\sim} \mathrm{End}_F(A)$$

such that $\sigma_*(a \otimes b)$ maps $x \in A$ to $ax\sigma(b)$. Under this isomorphism, the involution $\sigma \otimes \sigma$ corresponds to the adjoint involution with respect to the symmetric bilinear form T_σ on A defined by $T_\sigma(x,y) = \mathrm{Trd}_A(\sigma(x)y)$ for $x, y \in A$. This observation was used by Lewis and Tignol [22] to define the *signature* of σ for a given ordering P as $\sqrt{\mathrm{sgn}_P T_\sigma}$. Precise information on σ is not easily derived from T_σ however, since T_σ really determines $\sigma \otimes \sigma$ rather than σ. The discriminant of T_σ is always trivial and its Hasse (-Clifford-Witt) invariant, computed by Saltman (unpublished), by Quéguiner [28] and by Lewis [23], only depends on A and the discriminant of σ (if σ is orthogonal; see below for the definition of the discriminant). The restriction of T_σ to the space $\mathrm{Sym}(A,\sigma)$ of symmetric elements seems to be a more precise invariant; at least its discriminant yields the discriminant of σ [28], [23]. For unitary involutions, this restriction may be used to define signatures, as was shown by Quéguiner [27]. Its Hasse invariant yields the Brauer class of the discriminant algebra of the involution (defined below) [16], [28]. Bayer-Fluckiger and Parimala [5] have also shown that the classical invariants and signatures may be used to classify central simple algebras with involution over fields of virtual cohomological dimension at most 2.

In the rest of this section, we review the classical invariants for the various types of central simple algebras with involution.

3.1. Orthogonal case Let A be a central simple algebra with an orthogonal involution σ over a field F with characteristic different from 2. As observed in the preceding section, central simple algebras of odd degree with orthogonal involution are split. Therefore, if $\deg A$ is odd, the involution σ is adjoint to a nonsingular symmetric bilinear form over an F-vector space of dimension $\deg A$. Multiplying by a suitable scalar, we may assume that this form has trivial discriminant; it is then uniquely determined by the involution. The Witt index of the bilinear form and the even Clifford algebra of the associated quadratic form are invariants of the involution.

Suppose $\deg A$ is even. As observed by Knus, Parimala and Sridharan [19], we have $\operatorname{Nrd}_A(a)\operatorname{Nrd}_A(b)^{-1} \in F^{\times 2}$ for all skew-symmetric units $a, b \in A^\times$. Therefore, the square class of the reduced norm $\operatorname{Nrd}_A(a)$ does not depend on the choice of the skew-symmetric unit $a \in A^\times$, and we may define the *discriminant* of σ by

$$\operatorname{disc} \sigma = (-1)^{\deg A/2} \cdot \operatorname{Nrd}_A(a) \cdot F^{\times 2} \in F^\times / F^{\times 2}$$

for any $a \in A^\times$ such that $\sigma(a) = -a$. For the adjoint involution σ_b with respect to a symmetric bilinear form b, the discriminant $\operatorname{disc} \sigma_b$ is the (signed) discriminant of b. If $\deg A \geq 4$, an element of the absolute Galois group of F acts trivially on the Dynkin diagram of $\mathrm{PGO}^+(A, \sigma)$ if and only if it restricts to the identity on the field $F(\sqrt{\operatorname{disc} \sigma})$; in particular, the action of the absolute Galois group is trivial if and only if $\operatorname{disc} \sigma = 1$.

Every similitude g of a quadratic space (V, q) acts on the even Clifford algebra $C_0(V, q)$: if the multiplier of g is $\lambda \in F^\times$, we let g act on products of vectors $v \cdot w$ by $g(v \cdot w) = \lambda^{-1} g(v) \cdot g(w)$. Since scalars act trivially, this action factors through an action of $\mathrm{PGO}(V, q)$. Using this observation, Jacobson showed in [9] that a (generalized, even) Clifford algebra $C(A, \sigma)$ can be defined by Galois descent for every central simple algebra with orthogonal involution (A, σ), in such a way that

$$C\Big(\operatorname{End}_F(V), \sigma_b\Big) = C_0(V, q)$$

if q is the quadratic form associated with the nonsingular symmetric bilinear form b, and σ_b is the adjoint involution with respect to b. A "rational" definition of the Clifford algebra $C(A, \sigma)$ (or more precisely of the Clifford algebra of a "generalized quadratic form", taking into account the characteristic 2 case) was subsequently proposed by Tits [37]. The center of $C(A, \sigma)$ is the étale quadratic F-algebra

$F[\sqrt{\operatorname{disc}\sigma}]$ (which is the field $F(\sqrt{\operatorname{disc}\sigma})$ if $\operatorname{disc}\sigma \neq 1$ and is isomorphic to $F \times F$ if $\operatorname{disc}\sigma = 1$). The algebra $C(A,\sigma)$ carries a canonical involution $\underline{\sigma}$, which in the split case $C(\operatorname{End}_F(V),\sigma_b) = C_0(V,q)$ maps $v{\cdot}w$ to $w{\cdot}v$ for $v, w \in V$. This involution is orthogonal if $\deg A$ is divisible by 8, and symplectic if $\deg A$ is congruent to 4 modulo 8; it is unitary if $\deg A$ is congruent to 2 modulo 4.

Mimicking the construction of the spin group in the split case (see [31, §9.3]), one may define a spin group $\operatorname{Spin}(A,\sigma) \subset C(A,\sigma)^\times$, which is the simply connected cover of the group $\operatorname{PGO}^+(A,\sigma)$. The Clifford algebra $C(A,\sigma)$ is a Tits algebra associated with some representation of the spin group $\operatorname{Spin}(A,\sigma)$ (extended to $F(\sqrt{\operatorname{disc}\sigma})$) if $\operatorname{disc}\sigma \neq 1$. If $\operatorname{disc}\sigma = 1$, the algebra $C(A,\sigma)$ decomposes into a direct product of two central simple F-algebras

$$C(A,\sigma) = C_+(A,\sigma) \times C_-(A,\sigma),$$

where the components $C_\pm(A,\sigma)$ are the Tits algebras associated with the "half-spin" representations.

3.2. Symplectic case There is no "classical" invariant for symplectic involutions σ on a central simple algebra A besides the Schur index of A and the Witt index of the hermitian or skew-hermitian form associated with σ, since the absolute Galois group acts trivially on the Dynkin diagram C_n and the Tits algebras of representations of $\operatorname{Sp}(A,\sigma)$ are Brauer-equivalent to A if they are not split.

3.3. Unitary case Let (B,τ) be a central simple algebra with unitary involution over a field F, and let K denote the center of B, which is an étale quadratic F-algebra. The action of the absolute Galois group of F on the Dynkin diagram of $\operatorname{PGU}(B,\tau)$ is trivial if $K \simeq F \times F$ and factors through a nontrivial action of the Galois group of K/F if K is a field.

The first case is known as the *inner case*. In this case, we may assume $B = E \times E^{\mathrm{op}}$ for some central simple F-algebra E, and $\tau = \varepsilon$ is the exchange involution. For each integer $i = 1, \dots, \deg E$ one may define a central simple F-algebra $\lambda^i E$ Brauer-equivalent to $E^{\otimes i}$ of degree the binomial coefficient $\binom{\deg E}{i}$, in such a way that for every F-vector space V

$$\lambda^i \operatorname{End}_F(V) = \operatorname{End}_F(\textstyle\bigwedge^i V),$$

where $\bigwedge^i V$ is the i-th exterior power of V. The algebras $\lambda^i E$ are Tits algebras associated with representations of the simply connected group $\operatorname{SU}(B,\tau) = \operatorname{SL}(E)$.

In the case where K is a field, known as the *outer case*, there is a nontrivial Tits algebra only when $\deg(B,\tau)$ is even. If $\deg(B,\tau) = 2n$, the central simple algebra $\lambda^n B$ carries a canonical involution γ, which in the split case $B = \operatorname{End}_K(W)$ is adjoint to the canonical pairing induced by the exterior product

$$\wedge : \textstyle\bigwedge^n W \times \bigwedge^n W \to \bigwedge^{2n} W \ (\simeq K).$$

This canonical involution γ leaves K elementwise invariant, since the exterior product $\wedge$ is K-bilinear. On the other hand, the involution τ induces a unitary involution $\lambda^n\tau$ on $\lambda^n B$, which commutes with γ. Therefore, $\gamma\circ\lambda^n\tau$ is an automorphism of period 2 on $\lambda^n B$ which restricts to the nontrivial automorphism of K/F; it follows that the F-algebra $D(B,\tau)$ of invariant elements under $\gamma\circ\lambda^n\tau$ satisfies

$$D(B,\tau)\otimes_F K\simeq\lambda^n B.$$

The algebra $D(B,\tau)$ is therefore central simple of degree $\binom{2n}{n}$. Moreover, the involutions γ and $\lambda^n\tau$ coincide on $D(B,\tau)$ and define on this algebra a canonical involution $\underline{\tau}$ of orthogonal type if n is even and of symplectic type if n is odd. The algebra $D(B,\tau)$ is a Tits algebra associated with a representation of $\mathrm{SU}(B,\tau)$. It is called the *discriminant algebra* of (B,τ) because in the case where $B=\mathrm{End}_K(W)$ for some K-vector space W and $\tau=\sigma_h$ is the adjoint involution with respect to a hermitian form h with diagonalization $\langle\alpha_1,\dots,\alpha_{2n}\rangle$ (with $\alpha_1,\dots,\alpha_{2n}\in F^\times$), the algebra $D(B,\tau)$ is Brauer-equivalent to the quaternion algebra $\big(K,(-1)^n\alpha_1\cdots\alpha_{2n}\big)_F$.

4. Exceptional isomorphisms

The equalities of Dynkin diagrams $A_1=B_1=C_1$, $B_2=C_2$, $D_2=A_1\times A_1$ and $A_3=D_3$ reflect well-known isomorphisms between algebraic groups of various types: $\mathrm{PGL}_2\simeq\mathrm{O}_3^+\simeq\mathrm{PGSp}_2$, $\mathrm{O}_5^+\simeq\mathrm{PGSp}_4$, $\mathrm{PGO}_4^+\simeq\mathrm{PGL}_2\times\mathrm{PGL}_2$ and $\mathrm{PGL}_4\simeq\mathrm{PGO}_6^+$. The last two isomorphisms may be extended to non-connected groups: if (V,q) is a 4-dimensional hyperbolic quadratic space over a field F, there are 2-dimensional F-vector spaces V_1,V_2 such that $C_0(V,q)\simeq\mathrm{End}_F(V_1)\times\mathrm{End}_F(V_2)$; the action of similitudes on the even Clifford algebra mentioned in the preceding section yields an isomorphism

$$\mathrm{PGO}(V,q)\xrightarrow{\sim}\mathrm{Aut}\Big(\mathrm{End}_F(V_1)\times\mathrm{End}_F(V_2)\Big).$$

Similarly, if (W,q) is a 6-dimensional hyperbolic quadratic space, one may find 3-dimensional vector spaces W_1,W_2 and a decomposition of the even Clifford algebra $C_0(W,q)\simeq\mathrm{End}_F(W_1)\times\mathrm{End}_F(W_2)$. We then get an isomorphism

$$\mathrm{PGO}(W,q)\xrightarrow{\sim}\mathrm{Aut}\Big(\mathrm{End}_F(W_1)\times\mathrm{End}_F(W_2),\tau\Big),$$

where τ is a unitary involution which corresponds to the canonical involution on $C_0(W,q)$.

To derive information on algebras with involution from these isomorphisms, it suffices to compare the nonabelian Galois H^1-cohomology sets of both sides. As observed in section 2, we have $\mathrm{PGO}(V,q)=\mathrm{Aut}(\mathrm{End}_F(V),\sigma_b)$ where σ_b is the adjoint involution with respect to the polar symmetric bilinear form b of q; therefore the general methods of nonabelian cohomology (see [33, Ch. III]) show that the set $H^1(F,\mathrm{PGO}(V,q))$ is in one-to-one correspondence with the isomorphism classes of F-algebras with orthogonal involution of degree $\dim V$. The cohomology sets of the other groups can be similarly interpreted, and the isomorphisms of groups yield canonical bijections between isomorphism classes of central simple algebras with involution of various types.

These bijections can also be obtained in a more direct and explicit way. To formulate the main result, define the following categories (in fact they are groupoids because the maps are isomorphisms):

$\mathcal{A}_1 \times \mathcal{A}_1$ is the category whose objects are quaternion algebras over an étale quadratic F-algebra, and whose maps are F-algebra isomorphisms;

$\mathcal{A}_3$ is the category whose objects are central simple algebras with unitary involution of unitary type of degree 4 over F, and whose maps are F-algebra isomorphisms preserving the involutions;

$\mathcal{D}_n$ $(n = 2, 3)$ is the category whose objects are central simple F-algebras of degree $2n$ with orthogonal involution, and whose maps are F-algebra isomorphisms preserving the involutions.

The Clifford algebra construction defines functors

$$\mathbf{C}: \mathcal{D}_2 \to \mathcal{A}_1 \times \mathcal{A}_1 \qquad \text{and} \qquad \mathbf{C}: \mathcal{D}_3 \to \mathcal{A}_3.$$

Conversely, if $Q \in \mathcal{A}_1 \times \mathcal{A}_1$ has center K, there is a norm (or corestriction) F-algebra $N_{K/F}(Q)$ such that $N_{K/F}(Q) \otimes_F K$ is the tensor product of Q and its conjugate ${}^{\iota}Q$, where ι is the nontrivial automorphism of K/F. In particular, the norm $N_{K/F}(Q)$ is central simple of degree 4. It carries an orthogonal involution $N_{K/F}(\gamma)$, where γ is the quaternion conjugation on Q, and the norm construction defines a functor

$$\mathbf{N}: \mathcal{A}_1 \times \mathcal{A}_1 \to \mathcal{D}_2.$$

Similarly, the discriminant algebra construction yields a functor

$$\mathbf{D}: \mathcal{A}_3 \to \mathcal{D}_3.$$

THEOREM. *The above-defined functors induce equivalences of categories*

$$\mathcal{D}_2 \equiv \mathcal{A}_1 \times \mathcal{A}_1 \qquad \textit{and} \qquad \mathcal{A}_3 \equiv \mathcal{D}_3.$$

Under these equivalences, the discriminant of $(A, \sigma) \in \mathcal{D}_n$ $(n = 2, 3)$ is the square class $\delta \cdot F^{\times 2}$ such that $F[\sqrt{\delta}]$ is the center of the corresponding algebra in $\mathcal{A}_1 \times \mathcal{A}_1$ or $\mathcal{A}_3$. Restricting to involutions of trivial discriminant, we get in the case $\mathcal{D}_2 \equiv \mathcal{A}_1 \times \mathcal{A}_1$:

COROLLARY 1. *The norm and Clifford algebra functors define an equivalence between the groupoid of central simple F-algebras of degree 4 with orthogonal involution of trivial discriminant and the groupoid of pairs of quaternion F-algebras.*

Therefore, for every central simple F-algebra of degree 4 with orthogonal involution of trivial discriminant (A, σ), there is a pair of quaternion F-algebras $\{Q_1, Q_2\}$, uniquely determined up to isomorphism, such that

$$(A, \sigma) \simeq (Q_1, \gamma_1) \otimes (Q_2, \gamma_2)$$

where γ_1, γ_2 are the conjugation involutions on Q_1 and Q_2. The algebras Q_1, Q_2 are such that

$$C(A,\sigma) \simeq Q_1 \times Q_2.$$

We thus recover the Knus–Parimala–Sridharan criterion for the stable decomposability of central simple algebras of degree 4 with orthogonal involution [18], [20].

The corresponding result in the case $\mathcal{D}_3 \equiv \mathcal{A}_3$ is the following:

Corollary 2. *The discriminant and Clifford algebra functors define an equivalence between the groupoid of central simple F-algebras of degree 6 with orthogonal involution of trivial discriminant and the groupoid of F-algebras of the form $E \times E^{\mathrm{op}}$, where E is a central simple F-algebra of degree 4.*

Under this equivalence, an algebra $E \times E^{\mathrm{op}}$ is mapped to $\lambda^2 E$ with its canonical involution γ, as described in the preceding section; conversely, an algebra (A,σ) with orthogonal involution of trivial discriminant is mapped to $C(A,\sigma) = C_+(A,\sigma) \times C_-(A,\sigma)$. Note that $E_1 \times E_1^{\mathrm{op}} \simeq E_2 \times E_2^{\mathrm{op}}$ if and only if $E_1 \simeq E_2$ or E_2^{op}. Moreover, $E \simeq E^{\mathrm{op}}$ if and only if the exponent of E is 1 or 2, if and only if $\lambda^2 E$ is split. Therefore, the equivalence above, further restricted to algebras $E \times E^{\mathrm{op}}$ where E has exponent 1 or 2, yields a one-to-one correspondence between isomorphism classes of central simple F-algebras of degree 4 and exponent 1 or 2, and split central simple algebras of degree 6 with orthogonal involution of trivial discriminant. The isomorphism class of E is mapped to the isomorphism class of $(\lambda^2 E, \gamma)$ and the isomorphism class of (A,σ) is mapped to the isomorphism class of $C_+(A,\sigma) \simeq C_-(A,\sigma)$. Since split central simple algebras with orthogonal involution have the form $(\mathrm{End}_F(V), \sigma_b)$, where b is a nonsingular symmetric bilinear form uniquely determined by the involution up to a scalar factor, the latter set is also in bijection with the similarity classes of quadratic spaces of dimension 6 and trivial discriminant.

Corollary 3. *The map which carries (V,q) to $C_\pm(V,q)$ defines a one-to-one correspondence between similarity classes of 6-dimensional quadratic spaces with trivial discriminant and isomorphism classes of central simple algebras of degree 4 and exponent 2.*

Explicitly, the similarity class of the quadratic form with diagonalization $\langle a_1, b_1, -a_1b_1, -a_2, -b_2, a_2b_2\rangle$ is mapped to the biquaternion algebra $(a_1,b_1)_F \otimes (a_2,b_2)_F$, as is easily verified by a Clifford algebra computation. Since every quadratic form of dimension 6 and trivial discriminant is similar to a form of the type above, it follows that every central simple algebra of degree 4 and exponent 2 is isomorphic to a tensor product of two quaternion algebras, as was first proved by Albert [2]. The quadratic form $\langle a_1, b_1, -a_1b_1, -a_2, -b_2, a_2b_2\rangle$ is now called an *Albert form* of the biquaternion algebra $(a_1,b_1)_F \otimes (a_2,b_2)_F$. The fact that the Albert form is uniquely determined up to similarity by the biquaternion algebra was first shown by Jacobson [10], by Jordan algebra methods. Albert also established a close relation between the Schur index of the biquaternion algebra and the Witt index of the associated quadratic form [1]; this relation can also be derived from the general index theory of semisimple linear algebraic groups and the equivalence $\mathcal{A}_3 \equiv \mathcal{D}_3$.

This equivalence can also be used to get explicit descriptions of the groups of multipliers of similitudes and of spinor norms of an Albert form, completing results of Dieudonné [7] and Van der Waerden [39].

The other isomorphisms of classical groups $B_2 = C_2$ and $A_1 = B_1 = C_1$ also lend themselves to reinterpretations as equivalences of suitably defined groupoids. Details are in [16, Ch. IV].

References

[1] A.A. Albert, *On the Wedderburn norm condition for cyclic algebras,* Bull. Amer. Math. Soc., **37** (1931), 301–312 (pp. 185–196 in: *Collected Mathematical Papers, Part 1,* Amer. Math. Soc., Providence, R.I., 1993).

[2] A.A. Albert, *Normal division algebras of degree four over an algebraic field,* Trans. Amer. Math. Soc., **34** (1932), 363–372 (pp. 279–288 in: *Collected Mathematical Papers, Part 1,* Amer. Math. Soc., Providence, R.I., 1993).

[3] S.A. Amitsur, L.H. Rowen, J.-P. Tignol, *Division algebras of degree 4 and 8 with involution,* Israel J. Math., **33** (1979), 133–148.

[4] E. Bayer-Fluckiger, R. Parimala, *Galois cohomology of the classical groups over fields of cohomological dimension* ≤ 2, Invent. Math., **122** (1995), 195–229.

[5] E. Bayer-Fluckiger, R. Parimala, *Classical groups and Hasse principle,* prépublications de l'équipe de mathématiques de Besançon n° 96/03, to appear.

[6] A. Blanchet, *Function fields of generalized Brauer-Severi varieties,* Comm. Algebra, **19** (1991), 97–118.

[7] J. Dieudonné, *Les extensions quadratiques des corps non commutatifs et leurs applications,* Acta Math., **87** (1952), 175–242 (pp. 296–363 in: *Choix d'Œuvres Mathématiques, tome 2,* Hermann, Paris, 1981).

[8] D. Haile, M.-A. Knus, M. Rost, J.-P. Tignol, *Algebras of odd degree with involution, trace forms and dihedral extensions,* Israel J. Math. **96** (1996), 299–340.

[9] N. Jacobson. *Clifford algebras for algebras with involution of type D,* J. Algebra, **1** (1964), 288–300 (pp. 516–528 in: *Collected Mathematical Papers, vol. 2,* Birkhäuser, Boston, 1989).

[10] N. Jacobson, *Some applications of Jordan norms to involutorial simple associative algebras,* Advances in Math., **48** (1983), 149–165 (pp. 251–267 in: *Collected Mathematical Papers, vol. 3,* Birkhäuser, Boston, 1989).

[11] I. Kersten, U. Rehmann, *Generic splitting of reductive groups,* Tôhoku Math. J., **46** (1994), 35–70.

[12] M.-A. Knus, *Pfaffians and quadratic forms,* Advances in Math., **71** (1988), 1–20.

[13] M.-A. Knus, *Quadratic Forms, Clifford Algebras and Spinors,* Seminários de Matemática, IMECC, Caminas, 1988.

[14] M.-A. Knus, *Quadratic and Hermitian Forms over Rings,* Grundlehren math. Wiss., **294**, Springer-Verlag, Berlin 1991.

[15] M.-A. Knus, T.Y. Lam, D.B. Shapiro, J.-P. Tignol, *Discriminants of involutions on biquaternion algebras,* Proc. Symp. Pure Math., **58.2** (1995), Amer. Math. Soc., Providence, R.I., 279–303.

[16] M.-A. Knus, A.S. Merkurjev, M. Rost, J.-P. Tignol, *The Book of Involutions,* to appear.

[17] M.-A. Knus, R. Parimala, R. Sridharan, *A classification of rank 6 quadratic spaces via pfaffians,* J. reine angew. Math., **398** (1989), 187–218.

[18] M.-A. Knus, R. Parimala, R. Sridharan, *Pfaffians, central simple algebras and similitudes,* Math. Z., **206** (1991), 589–604.

[19] M.-A. Knus, R. Parimala, R. Sridharan, *On the discriminant of an involution,* Bull. Soc. Math. Belgique, Sér. A, **43** (1991), 89-98.

[20] M.-A. Knus, R. Parimala, R. Sridharan, *Involutions on rank 16 central simple algebras,* J. Ind. Math. Soc., **57** (1991), 143–151.

[21] T.Y. Lam, D.B. Leep, J.-P. Tignol, *Biquaternion algebras and quartic extensions,* Pub. Math. I.H.E.S., **77** (1993), 61–102.

[22] D.W. Lewis, J.-P. Tignol, *On the signature of an involution,* Arch. Math., **60** (1993), 128–135.

[23] D.W. Lewis, *Trace forms and involutions of the first kind,* preprint.

[24] A.S. Merkurjev, *Simple algebras and quadratic forms,* Izv. Akad. Nauk SSSR, Ser. Mat., **55** (1991), 218–224; English translation: Math. USSR Izvestiya, **38** (1992), 215–221.

[25] A.S. Merkurjev, *K-theory of simple algebras,* Proc. Symp. Pure Math., **58.1** (1995), Amer. Math. Soc., Providence, R.I., 65–82.

[26] A.S. Merkurjev, I. Panin, A.R. Wadsworth, *Index reduction formulas for twisted flag varieties, I, II,* to appear.

[27] A. Quéguiner, *Signature des involutions de deuxième espèce,* Arch. Math., **65** (1995), 408–412.

[28] A. Quéguiner, *Cohomological invariants of algebras with involution,* preprint.

[29] M. Racine, *A simple proof of a theorem of Albert,* Proc. Amer. Math. Soc., **43** (1974), 487–488.

[30] D. Saltman, *The Schur index and Moody's theorem, K*-theory, **7** (1993), 309–332.

[31] W. Scharlau, *Quadratic and Hermitian Forms,* Grundlehren math. Wiss. **270**, Springer-Verlag, Berlin, 1985.

[32] A. Schofield, M. Van den Bergh, *The index of a Brauer class on a Brauer-Severi variety,* Trans. Amer. Math. Soc., **333** (1992), 729–739.

[33] J-P. Serre. *Cohomologie Galoisienne,* Lecture Notes in Math., **5** (5e édition), Springer-Verlag, Berlin, 1994.

[34] J-P. Serre, *Cohomologie galoisienne : progrès et problèmes,* Séminaire Bourbaki, 46ème année, 1993–94, n° 783.

[35] D. Tao, *A variety associated to an algebra with involution,* J. Algebra, **168** (1994), 479–520.

[36] J. Tits, *Classification of algebraic semisimple groups,* Proc. Symp. Pure Math, **9** (1966), Amer. Math. Soc., Providence, R.I., 33–62.

[37] J. Tits, *Formes quadratiques, groupes orthogonaux et algèbres de Clifford,* Invent. Math., **5** (1968), 19–41.

[38] J. Tits, *Représentations linéaires irréductibles d'un groupe réductif sur un corps quelconque,* J. reine angew. Math., **247** (1971), 196–220.

[39] B.L. Van der Waerden, *Gruppen von linearen Transformationen,* Ergebnisse Math., **4**, 2, Springer, Berlin, 1935.

[40] A. Weil, *Algebras with involutions and the classical groups,* J. Ind. Math. Soc., **24** (1960), 589-623 (pp. 413–447 in: *Œuvres scientifiques–Collected Papers, Vol. II (1951–1964)*, Springer-Verlag New York, Heidelberg, Berlin 1979).

Progress in Mathematics, Vol. 169, © 1998 Birkhäuser Verlag Basel/Switzerland

Huygens' Principle and Integrability

A.P. VESELOV

Department of Mathematical Sciences
Loughborough University
Loughborough, Leicestershire, LE11 3TU, UK
and
Department of Mathematics and Mechanics
Moscow State University, 119899, Moscow, Russia

1. Introduction

The physical notion of Huygens' Principle goes back to the classical "Traité de la Lumière" by Christian Huygens, published in 1690. Various aspects of this fundamental principle in the theory of wave propagation were later discussed in the works of Kirchhof, Poisson, Beltrami and other scientists. But it was Jacques Hadamard [1], who was the first to propose in 1923 a rigorous mathematical definition of the phenomenon he called *Huygens' Principle in the narrow sense ("minor premise")*. This is the meaning of the term "Huygens' Principle" (or, in short, HP) we use in this paper.

Unlike the original principle of superposition of secondary waves, which holds for all wave propagation phenomena, HP in the narrow sense of Hadamard applies only to a very restricted range of wave processes. Its physical interpretation is that *instantaneous signals remain instantaneous for every observer for each later time*. This property ensures the sharpness of signal transmission and thus is of the greatest importance for our everyday life: it is due to HP that we can hear each other and enjoy music.

In modern mathematical language we say that a linear hyperbolic partial differential equation satisfies HP if its fundamental solution is located on the characteristic conoid, i.e. this solution vanishes in the conoid's complement (see [2],[3]). Such a domain, where the fundamental solution vanishes identically, is called a *lacuna*, thus HP is equivalent to the fact that the interior of the characteristic conoid is a lacuna for the corresponding hyperbolic equation (the exterior is a lacuna for all hyperbolic equations).

The pure wave equation on three-dimensional euclidean space satisfies this property as it follows from the classical Kirchhof formula. The generalization of this formula found by Tedone shows that the wave equation on euclidean space $\mathbf{R}^N$

$$\square_{N+1}\varphi = 0, \tag{1}$$

where $\square_{N+1} = \partial_0^2 - \partial_1^2 - \cdots - \partial_N^2$, $\partial_i = \frac{\partial}{\partial x_i}$, $x_0 = t$, satisfies HP iff N is odd and $N \geq 3$.

The question of how exceptional these examples are, i.e. which second order linear hyperbolic equations satisfy HP, is known as the *Hadamard problem.* It still remains open even in the class of the hyperbolic equations of the form

$$(\Box_{N+1} + u(x))\varphi = 0. \tag{2}$$

For a long time it was thought that pure wave equations are the only equations satisfying HP. This conjecture, usually ascribed to Hadamard, was justified in 1939 by M. Mathisson [4] and L. Asgeirsson [5], who proved this statement for the equations (2) and $N = 3$. But in 1953 K. Stellmacher discovered his famous counterexamples to this conjecture [7]. He showed that the equations (2) with the potentials

$$u(x) = \frac{m(m+1)}{x_1^2} \tag{3}$$

and any odd N, $N \geq 2m + 3$ satisfy HP. Thus already in dimension $N = 5$ one has a nontrivial huygensian equation (23) with $u(x) = 2x_1^{-2}$. In fact, Stellmacher has shown that up to natural equivalence this is the only nontrivial equation of the form (2) with HP in this dimension.

Later in 1967–69 K. Stellmacher and J. Lagnese essentially generalised these examples and described all potentials $u = u(x_1)$, depending only on one of the coordinates, which do not destroy Huygens' Principle [8, 9]. Approximately at the same time the remarkable properties of the Korteweg-de Vries equation and the inverse scattering transform method of integration of nonlinear differential equations were discovered. At that time it seems nobody thought that these two subjects could be related, but they are ! As it was realised later (see [10]), the class of Stellmacher-Lagnese potentials coincides with the class of rational solutions of the KdV equation [11],[12] (see Section 4 below).

Over recent years some new evidence for the close relation between Huygens' Principle and Integrability has been found [13, 14, 15, 16, 17, 18, 19]. We are going to discuss here these new examples of huygensian hyperbolic equations and the related theory of algebraically integrable Schrödinger operators, in particular, quantum Calogero-Moser system and its generalisations. Our approach has its origin in the beautiful theory of "finite-gap" Schrödinger operators in one dimension developed in 1974-1976 in connection with the periodic problem for KdV equation (see the reviews [21, 22]).

2. Coxeter groups, Dunkl operator and Huygens' Principle

The first examples of the huygensian equations (2) with the potentials essentially depending on more than one of the coordinates were found by Yu.Yu. Berest and the author in [13]. They can be considered as generalisations of Stellmacher's examples (3) related to an arbitrary Coxeter system.

By *a Coxeter system* we mean a finite set of vectors ("roots") $\mathfrak{R} = \left\{\alpha\right\} \subset \mathbf{R}^n$, that have pairwise different directions, span $\mathbf{R}^n$ and satisfy the condition that reflections $s_\alpha(\beta) = \beta - 2\frac{(\alpha,\beta)}{(\alpha,\alpha)}\alpha$ map $\mathfrak{R}$ into itself. The reflections s_α generate a finite group G, referred to as the *Coxeter group.*

Define the potential associated with $\mathfrak{R}$ by setting:

$$u(x) = \sum_{\alpha \in \mathfrak{R}_+} \frac{m_\alpha(m_\alpha + 1)(\alpha, \alpha)}{(\alpha, x)^2} , \tag{4}$$

where m_α are integers (referred to as multiplicities of the roots) such that the function $m(\alpha) = m_\alpha$ is G-invariant and $\mathfrak{R}_+ \subset \mathfrak{R}$ denotes the set of roots positive with respect to a suitable linear form on $\mathbf{R}^n$.

Consider the hyperbolic operator

$$\mathcal{L} = \Box_{N+1} + u(x_1, \dots, x_n) ,$$

with the potential (4) and an odd number N, satisfying the following inequality

$$N \geq 3 + 2 \sum_{\alpha \in \mathfrak{R}_+} m_\alpha .$$

Theorem 1 [**13**]. *For any Coxeter group G and integer-valued G-invariant function m on the corresponding root system $\mathfrak{R}$ the equation $\mathcal{L}\varphi = 0$, where the operator $\mathcal{L}$ is described above, satisfies Huygens' principle.*

For the simplest case of the Coxeter system of the type A_1 one has Stellmacher's examples (3).

The Schrödinger operator

$$L = -\Delta + \sum_{\alpha \in \mathfrak{R}_+} \frac{m_\alpha(m_\alpha + 1)(\alpha, \alpha)}{(\alpha, x)^2}, \tag{5}$$

where Δ is the Laplace operator in $\mathbf{R}^n$, without the condition that the m_α are integers, was considered for the first time by M.A. Olshanetsky and A.M. Perelomov [29] (see also [30]) as a generalization of the Calogero-Moser operator [23], [24]

$$L^C = -\Delta + g \sum_{i \neq j} \frac{1}{(x_i - x_j)^2} ,$$

which corresponds to the particular case in which the root system is of $\mathcal{A}_n$ type.

The importance of the integer-valuedness condition for m_α was clarified in [26],[28], where it was shown that this leads to the existence of the additional commuting operators (algebraic integrability, see below) and the remarkable algebraic structure of the eigenfunction of the corresponding operator (5), which has been used in the proof of Theorem 1.

The following construction [26], [28] plays a fundamental role in our approach.

Let $\mathcal{A}$ be a finite set of noncollinear vectors α in $\mathbf{R}^n$ with multiplicities $m_\alpha \in \mathbf{N}$. This set determines the configuration Π of the hyperplanes Π_α given by the equations $(k, \alpha) = 0$ taken with multiplicities m_α.

Let us define the *Baker-Akhiezer function associated with the set* $\mathcal{A}$ as the function $\psi(k, x)$, $k, x \in \mathbf{R}^n$ of the form

$$\psi = P(k, x) e^{(k,x)}, \tag{6}$$

where $P(k,x)$ is a polynomial in k with the highest term $A(k) = \prod_{\alpha\in\mathcal{A}} (k,\alpha)^{m_\alpha}$, which satisfies the following conditions of "quasiinvariance": for any $\alpha \in \mathcal{A}$ the difference $\psi(k,x) - \psi(s_\alpha k, x)$ is divisible by $(k,\alpha)^{2m_\alpha}$. Here s_α, as above, is a reflection with respect to the hyperplane Π_α. The last condition means that all normal derivatives of ψ of any odd order less than $2m_\alpha$ vanish at the hyperplane Π_α.

The following basic result explains the important role of such a function both in the theory of integrable Schrödinger operators in many dimensions and in the theory of Huygens' Principle.

Theorem 2. *If the Baker-Akhiezer function ψ exists then it is unique and satisfies the algebraically integrable Schrödinger equation $L\psi = -k^2\psi$, where*

$$L = -\Delta + \sum_{\alpha\in\mathcal{A}} \frac{m_\alpha(m_\alpha+1)(\alpha,\alpha)}{(\alpha,x)^2} \tag{7}$$

In this case the hyperbolic equation

$$(\Box_{N+1} + \sum_{\alpha\in\mathcal{A}} \frac{m_\alpha(m_\alpha+1)(\alpha,\alpha)}{(\alpha,x)^2})\varphi = 0 \tag{8}$$

with odd N such that

$$N \geq 3 + 2\sum_{\alpha\in\mathcal{A}} m_\alpha.$$

satisfies the Huygens' Principle.

We say that a Schrödinger operator $L = -\Delta + u(x_1,\ldots,x_n)$ in n-dimensional space is *integrable*, if there exist n pairwise commuting differential operators $L_1 = L, L_2, \ldots, L_n$ with algebraically independent constant highest symbols $P_i(k_1,\ldots,k_n)$. L is called *algebraically integrable*, if there is at least one more operator L_{n+1}, which commutes with all L_i and such that its highest symbol $P_{n+1}(k_1,\ldots,k_n)$ takes different values at the solutions of the system of the equations $P_i(k) = c_i (i = 1,\ldots,n)$ (see [22, 26, 28]). In dimension $n = 1$ all algebraically integrable Schrödinger operators are described by the "finite-gap" theory (see for example [21]). In dimension more than 1 it is an open problem.

The first statement of the theorem was proved in [26, 28]. The proof of the second statement essentially repeats the arguments given in [13] for the Coxeter configurations: using the special form of the Baker-Akhiezer function one can write down the fundamental solution of the hyperbolic equation (8) as a finite combination of the delta-function and its derivatives located on the characteristic cone. It is worth noting that if we know the Baker-Akhiezer function we can write down an explicit formula for the fundamental solution of the corresponding hyperbolic equation.

Thus, the main problem is the existence of the Baker-Akhiezer function. At first we thought that all the configurations when ψ does exist are the Coxeter ones. Indeed the following result can be proved.

Theorem 3 [28]. *If the configuration* Π *consists of the hyperplanes of reflections ("mirrors") of any Coxeter group with the multiplicities, which are invariant under the action of this group, then the Baker-Akhiezer function does exist. If all* $m_\alpha = 1$ *then a converse statement is true: if* ψ *exists then* Π *is a configuration of mirrors of some Coxeter group.*

The proof of the direct statement is based on the results of Ch. Dunkl [31], G. Heckman [32] and E. Opdam [33]. The proof of the existence is effective in the sense that an explicit formula for the Baker-Akhiezer function is given in terms of the following *Dunkl operators* [31].

Let $\Re \subset \mathbf{R}^n$ be a Coxeter system of a finite reflection group G. Choosing a hyperplane in $\mathbf{R}^n$ which does not contain any root, we define the set $\mathcal{A} = \Re_+ \subset \Re$ of *positive roots* lying in one of the half-spaces. On the root system $\Re$ consider any G-invariant function $\mu(\alpha) = \mu_\alpha$ (not necessarily integer-valued).

The Dunkl operators are defined as follows

$$\nabla_\xi = \partial_\xi - \sum_{\alpha \in \Re_+} \mu_\alpha \frac{(\alpha, \xi)}{(\alpha, x)} \hat{s}_\alpha \,, \tag{9}$$

where ξ is an arbitrary vector $\mathbf{R}^n$, $\partial_\xi = (\xi, \partial) = \sum_{i=1}^n \xi^i \frac{\partial}{\partial x_i}$ is the derivative with respect to the direction ξ, $\hat{s}_\alpha$ is an operator acting on the functions on $\mathbf{R}^n$ according to the rule

$$\left(\hat{s}_\alpha f\right)(x) = f\left(s_\alpha(x)\right) ,$$

and s_α as above is the reflection corresponding to the root α.

The Dunkl operators have the following important properties:

1) $\left[\nabla_\xi, \nabla_\eta\right] = 0$ for all ξ, η

2) $$Res\left(\sum_{i=1}^n \nabla_i^2\right) = L, L = -\Delta + \sum_{\alpha \in \Re_+} \frac{\mu_\alpha(\mu_\alpha - 1)(\alpha, \alpha)}{(\alpha, x)^2} \,. \tag{10}$$

Here $\nabla_i = \nabla_{e_i}$ for some orthonormal basis $e_1, \dots, e_n$ in $\mathbf{R}^n$, Res denotes taking the purely differential part by restricting the operator to the space $\mathcal{F}^G$ of G-invariant functions (see [31], [32]). This procedure is well-defined in the commutative algebra generated by the Dunkl operators. Indeed, one can place all nonlocal operators on the right-hand side using the relations

$$\begin{aligned} \hat{s}_\alpha \circ \partial_\xi &= \partial_{s_\alpha(\xi)} \circ \hat{s}_\alpha \\ \hat{s}_\alpha \circ \hat{f} &= \hat{s}_\alpha(f) \cdot \hat{s}_\alpha \,, \end{aligned}$$

where $\hat{f}$ is the operator of multiplication by a function f. After this procedure the restriction operation Res becomes evident. Note that a differential operator in $\mathbf{R}^n$ is completely determined by its action on the space $\mathcal{F}^G$.

We introduce another method for taking the differential part of an operator A lying in the algebra generated by Dunkl operators. Let $\varepsilon : \mathrm{G} \to \pm 1$ be an arbitrary character of the Coxeter group. It is completely determined by the set

$\mathcal{E} \subset \Re$ of roots such that $\varepsilon(s_\alpha) = -1$, if $\alpha \in \mathcal{E}$, and $\varepsilon(s_\alpha) = +1$ otherwise. The set $\mathcal{E}$ has to be G-invariant, i.e. $\mathcal{E}$ consists of a certain number of G-orbits. Let $\mathcal{F}^G_\varepsilon$ be the space of functions on $\mathbf{R}^n$ such that

$$\left(\hat{s}_\alpha f(x)\right) = f\left(s_\alpha(x)\right) = \varepsilon(s_\alpha) f(x) \ ,$$

and let $Res^\varepsilon A$ denotes the differential operator that coincides with A on the space $\mathcal{F}^G_\varepsilon$. The Res^ε-procedure is well-defined in the algebra generated by the Dunkl operators (9) for the same reasons as mentioned above.

The following formula [13] can be proved by straightforward calculations

$$Res^\varepsilon\left(\sum_{i=1}^n \nabla_i^2\right) = -L^\varepsilon, \tag{11}$$

where $L^\varepsilon = -\Delta + u^\varepsilon(x)$,

$$u^\varepsilon(x) = \sum_{\alpha\in\Re_+} \frac{\mu^\varepsilon_\alpha(\mu^\varepsilon_\alpha - 1)(\alpha,\alpha)}{(\alpha,x)^2} \ , \tag{12}$$

and $\mu^\varepsilon_\alpha = \mu_\alpha + 1$, if $\alpha \in \mathcal{E}$, and $\mu^\varepsilon_\alpha = \mu_\alpha$ otherwise.

Define the so-called *shift operator* (cf. [33], [32])

$$\mathcal{D}^\varepsilon = Res\left(\prod_{\alpha\in\mathcal{E}_+} \nabla_\alpha\right) , \tag{13}$$

where $\mathcal{E}_+ = \mathcal{E} \bigcap \Re_+$. It satisfies the following relation

$$\mathcal{D}^\varepsilon \circ L = L^\varepsilon \circ \mathcal{D}^\varepsilon \ , \tag{14}$$

where L and L^ε are defined by (10) and (12).

Indeed, since the Dunkl operators commute we have

$$\left(\prod_{\alpha\in\mathcal{E}_+} \nabla_\alpha\right)\left(\sum_{i=1}^n \nabla_i^2\right) = \left(\sum_{i=1}^n \nabla_i^2\right)\left(\prod_{\alpha\in\mathcal{E}_+} \nabla_\alpha\right).$$

Applying the Res-operation to both sides of the latter equation and using the fact that the operator $\prod_{\alpha\in\mathcal{E}_+} \nabla_\alpha$ maps $\mathcal{F}^G$ into $\mathcal{F}^G_\varepsilon$, we obtain

$$Res\left(\prod_{\alpha\in\mathcal{E}_+} \nabla_\alpha\right) Res\left(\sum_{i=1}^n \nabla_i^2\right) = Res^\varepsilon\left(\sum_{i=1}^n \nabla_i^2\right)\left(Res \prod_{\alpha\in\mathcal{E}_+} \nabla_\alpha\right),$$

and therefore the relation (14).

Iterating the shift operators for appropriate sets $\mathcal{E}$ and integer parameters μ_α we come to the following

Theorem 4 [33, 32]. *For the generalised Calogero-Moser operator L with the potential (4) related to any Coxeter system with integer multiplicities m_α, the shift operator $\mathcal{D}$ exists with highest symbol $\prod_{\alpha\in\mathcal{R}_+}(k,\alpha)^{m_\alpha}$ such that*

$$\mathcal{D}\circ L_0 = L\circ\mathcal{D} \tag{15}$$

where $L_0 = -\Delta$.

As a corollary we have that such an operator L has an eigenfunction $\psi(k,x)$ of the form

$$\psi(k,x) = \mathcal{P}(k,x)exp(k,x), \qquad k\ ,x\in\mathbf{R}^n\ , \tag{16}$$

where $\mathcal{P}(k,x)$ is a polynomial in k with leading term $A(k) = \prod_{\alpha\in\Re_+}(k,\alpha)^{m_\alpha}$ and coefficients rational functions of x. Indeed, as follows from the relation (15) the function $\psi = \mathcal{D}exp(k,x)$ is an eigenfunction of the operator L of the form (16).

It can be shown that this function ψ is none other than the Baker-Akhiezer function associated with the corresponding Coxeter system (see [28]).

Coming back to the definition of the Baker-Akhiezer function associated with a general configuration of the hyperplanes Π, it is natural to ask about the following generalisation of the Dunkl operators (9):

$$\nabla_\xi = \partial_\xi - \sum_{\alpha\in\mathcal{A}}\mu_\alpha\frac{(\alpha,\xi)}{(\alpha,x)}\hat{s}_\alpha, \tag{17}$$

where ξ is an arbitrary vector $\mathbf{R}^n$, μ_α are some non-zero coefficients. This problem was investigated in [34, 35] and the result is the following

Theorem 5 [34, 35]. *The operators (17) ∇_ξ and ∇_η commute for arbitrary ξ,η only the in Dunkl case, i.e. if and only if Π is the configuration of mirrors for some Coxeter group G and the function μ is G-invariant.*

Therefore it came as a surprise that there exist Baker-Akhiezer functions associated with non-Coxeter configurations [17]. We will discuss them in the next section.

3. New integrable generalisations of the quantum Calogero-Moser system and Huygens' Principle

The following non-Coxeter configuration related to Huygens' Principle was discovered by O.Chalykh, M.Feigin and the author in [17].

Let $\mathcal{A}_n(m)$ be the set of the following vectors in $\mathbf{R}^{n+1}$: $e_i - e_j$ $(i<j;\ i,j = 1,\dots,n)$ taken with multiplicity m and $e_i - \sqrt{m}e_{n+1}$ $(i = 1,\dots,n)$ with multiplicity 1. In the first nontrivial case we have the configuration $\mathcal{A}_2(m)$, which is shown in Fig. 1.

Theorem 6 [17]. *For the system $\mathcal{A}_n(m)$, the Baker-Akhiezer function ψ* exists. *The corresponding algebraically integrable Schrödinger operator has the form*

$$L = -\Delta + \sum_{i<j}^{n}\frac{2m(m+1)}{(x_i-x_j)^2} + \sum_{i=1}^{n}\frac{2(m+1)}{(x_i-\sqrt{m}x_{n+1})^2}. \tag{18}$$

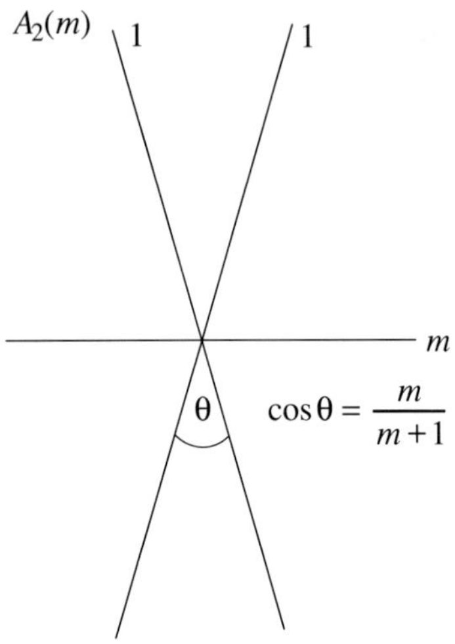

Figure 1.

Notice that in the coordinates $q_i = x_i, i = 1, \dots, n, q_{n+1} = \sqrt{m} x_{n+1}$ the operator L can be written as

$$L = \sum_{i=1}^{n+1} \frac{\hat{p}_i^2}{2m_i} + m(m+1) \sum_{i<j}^{n+1} \frac{m_i m_j}{(q_i - q_j)^2} \tag{19}$$

where $m_i = 1$ if $i = 1, \dots, n$ and $m_{n+1} = 1/m$, $\hat{p}_j = i\frac{\partial}{\partial q_j}$. Thus this new quantum integrable system describes the interaction of the particles on the line of Calogero type but now one of the particles has a "mass" different from the others. One can prove that if the parameter m is not an integer then the Schrödinger operators (18), (19) are still integrable, but not algebraically integrable (see [17]).

The idea of the proof of this theorem is similar to the one we used in [26] when the Dunkl operator technique was not known. We write down the corresponding shift operator as the product of the shift operator related to the Coxeter subsystem $\mathcal{A}_{n-1}$ and the following operator $D_{n+1}^{1,\dots,n}$.

We introduce a set of differential operators $D_N^{i_1,\dots,i_k}$ by the following recursion relations (cf.[26]):

$$\begin{aligned} D_N &= 1, \quad D_N^i = \partial_i - \sqrt{m}\partial_N - (m+1)(x_i - \sqrt{m}x_N)^{-1}, \\ D_N^{i_1,\dots,i_k} &= D_N^{i_1,\dots,i_{k-1}} \circ (\partial_{i_k} - \sqrt{m}\partial_N) - (m+1)(x_{i_k} - \sqrt{m}x_N)^{-1} D_N^{i_1,\dots,i_{k-1}} \\ &+ \sum_{s=1}^{k-1} \frac{m(m+1)}{(x_{i_s} - x_{i_k})^2} D_N^{i_1,\dots,\widehat{i_s},\dots,i_{k-1}}. \end{aligned}$$

Here $\partial_i = \partial/\partial x_i$, $i_1, \dots, i_k$ are different natural indices less than N and the symbol $\widehat{i_s}$ means that the index $\widehat{i_s}$ must be omitted.

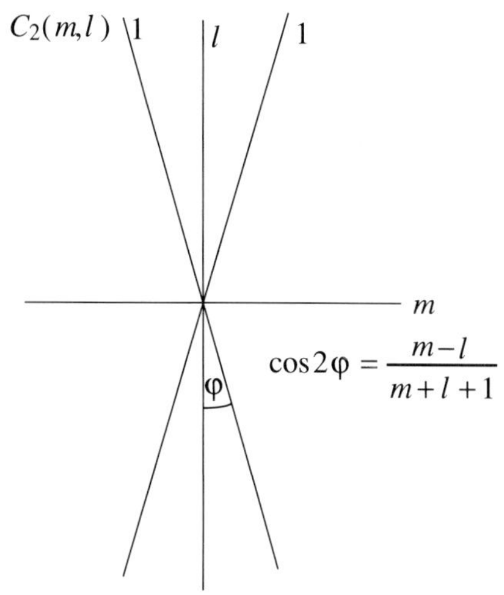

Figure 2.

Theorem 7 [17]. *The Baker-Akhiezer function associated with the system $\mathcal{A}_n(m)$ has the form*

$$\psi(k,x) = D_{n+1}^{1,\dots,n}\psi_0(k,x),$$

where $\psi_0(k,x)$ is the Baker-Akhiezer function associated with the Coxeter subsystem $\mathcal{A}_{n-1}$, which consists of all vectors from $\mathcal{A}_n(m)$ with multiplicity m.

Corollary [17]. *The hyperbolic equation*

$$(\Box_{N+1} + \sum_{i<j}^{n} \frac{2m(m+1)}{(x_i - x_j)^2} + \sum_{i=1}^{n} \frac{2(m+1)}{(x_i - \sqrt{m}x_{n+1})^2})\varphi = 0, \tag{20}$$

satisfies Huygens' Principle if N is odd and $N \geq 3 + 2n + mn(n-1)$.

Another non-Coxeter but huygensian configuration is the following two-dimensional system $\mathcal{C}_2(m,l)$, which consists of the vectors $\alpha_1 = \sqrt{2m+1}\,e_1$ with multiplicity $l, \alpha_2 = \sqrt{2l+1}\,e_2$ with multiplicity m and $\alpha_3 = 1/2(\alpha_1+\alpha_2)$, $\alpha_4 = 1/2(\alpha_1 - \alpha_2)$ with multiplicity 1 (see Fig.2).

It is shown in [17] that the Baker-Akhiezer function associated with this configuration does exist and an explicit formula for ψ is given. The corresponding algebraically integrable Schrödinger operator has the form

$$\begin{aligned} L &= -\Delta + \frac{l(l+1)}{x_1^2} + \frac{m(m+1)}{x_2^2} \\ &\quad + \frac{4(l+m+1)}{(\sqrt{2m+1}\,x_1 + \sqrt{2l+1}\,x_2)^2} + \frac{4(l+m+1)}{(\sqrt{2m+1}\,x_1 - \sqrt{2l+1}\,x_2)^2}. \end{aligned} \tag{21}$$

Just as in the previous case, this operator is integrable for arbitrary values of the parameters m, l, but only for integer values do we have an algebraic integrability and Huygens' Principle for the hyperbolic equation

$$(\Box_{N+1} + u(x_1, x_2))\phi = 0 \tag{22}$$

with the potential (21) and odd $N \geq 7 + 2(m+l)$.

The natural questions arise: what is the classification of all configurations for which the associated Baker-Akhiezer function exists and what kind of geometry is behind this?

At the moment we know the answer only in the case when all multiplicities are equal to 1: it is just the Coxeter classification (see [28] and Theorem 3 above).

For arbitrary multiplicities we have only the following partial result: for the configuration of three lines in the plane the Baker-Akhiezer function exists if and only if the multiplicities are (1,1,m) or (m,m,m). In the first case we have the system $\mathcal{A}_2(m)$ (see Fig.1), in the second – the standard Coxeter system $\mathcal{A}_2$.

4. Restricted Hadamard problems and the Korteweg-de Vries equation

As we have mentioned already in the Introduction, in 1967–69 Lagnese and Stellmacher [8] solved the Hadamard problem within the class of the hyperbolic operators

$$\mathcal{L} = \Box_{N+1} + u(x^1)\ , \tag{23}$$

with real analytic potentials $u\left(x^1\right)$ depending on a single variable only (the restricted Hadamard problem). The corresponding potentials are rational functions given explicitly in terms of some polynomials $\mathcal{P}_k\left(z, \tau_1, \ldots, \tau_{k-1}\right)$ as

$$u(z) = 2\left(\frac{d}{dz}\right)^2 \log \mathcal{P}_k\left(z\right)\ , \quad k = 0, 1, 2, \ldots . \tag{24}$$

These polynomials are defined by the following differential-recurrence relation:

$$\mathcal{P}'_{k+1}\mathcal{P}_{k-1} - \mathcal{P}'_{k-1}\mathcal{P}_{k+1} = (2k+1)\mathcal{P}_k^2\ , \tag{25}$$

with $\mathcal{P}_0 = 1$, $\mathcal{P}_1 = z$ and polynomially depend on arbitrary parameters $\tau_1, \ldots,$ τ_{k-1}. They had been introduced by Burchnall and Chaundy in 1929 (see [36]) but Lagnese and Stellmacher were not aware of this. (In fact, these polynomials are none other than Schur polynomials for the representations of $GL(k)$ with triangular Young tableux, but this only became clear much later due to Sato and his followers.)

The same potentials appeared later in 1977–78 in the papers by H. Airault, H. McKean, J. Moser and M. Adler [11], [12], where the rational solutions of the Korteweg-de Vries equation (KdV)

$$u_t = 6uu_x - u_{xxx} \tag{26}$$

were investigated. The relation of the KdV theory with the Lagnese-Stellmacher results was not realised at that time. It was R. Schimming who was probably the first to make this observation [10].

It turns out that Huygens' Principle is related not only with rational but also with trigonometric solutions of the KdV equation, as it was recently discovered by Yu. Berest and I. Lutsenko [18], [19].

Let $m \in \mathbf{Z}_{\geq 0}$ be a non-negative integer, and let $(k_j)_{j=1}^{l}$ be its monotonic partition of height l, i.e. a finite strictly decreasing sequence of non-negative integers such that

$$m \equiv k_l > k_{l-1} > \ldots > k_2 > k_1 \geq 0\ . \tag{27}$$

We associate with (27) a finite set of elementary 2π-periodic functions on the real line:

$$\chi_j(\varphi) := \cos(k_j\,\varphi + \varphi_j)\ , \tag{28}$$

depending additionally on l arbitrary real parameters ("phases") $\varphi_i \in \mathbf{R}$. Since the partition (k_j) is assumed to be monotonic, the functions $\chi_j(\varphi)$ are linearly independent, and the Wronskian $\mathcal{W}[\chi_1, \chi_2, \ldots, \chi_l]$ of the set $\{\chi_j\}$ does not vanish identically.

Following [18] we consider the hyperbolic equation

$$(\Box_{N+1} + u(x_1, x_2))\varphi = 0 \tag{29}$$

with the following potential u, depending on *two* spatial variables. This potential u is a rational function, which is homogeneous of degree (-2): $u(\alpha x_1, \alpha x_2) = \alpha^{-2}u(x_1, x_2)$, $\alpha > 0$ and in polar coordinates r, ϕ on the plane (x_1, x_2) has the form

$$u = -\frac{2}{r^2}\left(\frac{d}{d\varphi}\right)^2 \log \mathcal{W}[\chi_1(\varphi), \chi_2(\varphi), \ldots, \chi_l(\varphi)]\ , \tag{30}$$

where $\mathcal{W}[\chi_1(\varphi), \chi_2(\varphi), \ldots, \chi_l(\varphi)]$ is a Wronskian of the functions (28).

The angular part of this potential

$$v(\varphi) = -2\left(\frac{d}{d\varphi}\right)^2 \log \mathcal{W}[\chi_1(\varphi), \chi_2(\varphi), \ldots, \chi_l(\varphi)]$$

is given by the same formula as the l-soliton solution of the KdV equation (see e.g.[25]) if we replace the trigonometric functions by their hyperbolic analogues. Notice that the "amplitudes" k_i should be integers to provide the 2π-periodicity of v. Thus we can say that the angular parts of the Berest-Lutsenko potentials (30) are the "trigonometric solitons" of the KdV equation. We should mention that in contrast to real solitons these potentials have pole singularities.

Theorem 8 [18, 19]. *The hyperbolic equation (29) with the potential (30) satisfies the Huygens' Principle if N is odd and $N \geq 3 + 2m$.*

It turns out that this construction solves the Hadamard problem within the class of the operators (29) with homogeneous potentials. The following remarkable result was proved recently by Yu. Berest [19].

Theorem 9. [19] *A hyperbolic equation* (29) *with real locally analytic potential u, depending on two spatial variables and homogeneous of degree* (-2), *satisfies Huygens' Principle only if the potential has a form* (30) *for some partition k_i and a set of real numbers $\{\varphi_1, \varphi_2, \ldots \varphi_l\}$.*

The proof is based on the ideas and constructions developed within the "finite-gap" theory (see [21]).

Notice that both Stellmacher-Lagnese potentials and the angular parts of Berest-Lutsenko potentials are "finite-gap" in the sense that the corresponding one-dimensional Schrödinger operators are algebraically integrable: they are the rational and trigonometric solutions of the stationary higher KdV equations (Novikov's equations [20]).

At the same time we should notice that generic Berest-Lutsenko operators do not satisfy the Baker-Akhiezer axiomatics described in the second section of this paper. This means that the notion of Baker-Akhiezer function in many dimensions is to be extended to cover all the Berest-Lutsenko examples. [1]

5. Integrability and Huygens' Principle on symmetric spaces

The algebraic integrability of the radial part of the Laplace-Beltrami operator on certain symmetric spaces is also the reason for Huygens' Principle being satisfied for the corresponding modified wave equation, as was observed by O. Chalykh and the author in [14, 16].

The *modified wave equation on the symmetric space* **X** is defined as follows:

$$\varphi_{tt} - \mathcal{L}_X \varphi + c\varphi = 0\,, \tag{31}$$

where $\mathcal{L}_X$ is the Laplace-Beltrami operator on the symmetric space **X**, c is a certain constant, for irreducible spaces having the form

$$c = \pm|\rho|^2\,, \quad \rho = \frac{1}{2}\sum_{\alpha\in\mathcal{R}_+} \mu_\alpha\, \alpha \tag{32}$$

Here $\mathcal{R}_+$ is the set of positive restricted roots of the space **X**, ρ is their half-sum with multiplicities μ, and the sign $\pm$ corresponds to the sign of the curvature of **X**: plus for the spaces of compact type, minus for the noncompact type (see e.g.[42]). For the reducible symmetric spaces, c is equal to the sum of the constants for the corresponding irreducible components. The value of c is very important: one can show that the following theorem is valid only when the constant c is chosen as described above (see [16]).

Theorem 10. *Let* **X** *be an odd-dimensional riemannian symmetric space such that all its roots have even multiplicities* m_α. *Then the modified wave equation* (31),(32) *on* **X** *satisfies Huygens' principle.*

In the case when **X** has a noncompact type, this result is due to Solomatina [41], Olafsson and Schlichtkrull [43], Helgason [44], Branson, Olafsson and Schlichtkrull [45]. The case of spheres was considered earlier by Lax and Phillips in [39]. For compact Lie groups and spaces with complex groups of motion this was established by Helgason [40]. The general case follows from the explicit formulas for the fundamental solutions of the corresponding modified wave equations found

[1] Such a generalisation of this notion has been found recently by O. Chalykh, M. Feigin and the author in [54]

by O. Chalykh and the author [14, 16]. Another proof based on duality arguments was found in [46].

Here is a list of the irreducible symmetric spaces of compact type with even multiplicities:

1) odd-dimensional spheres $\mathbf{S}^{2n+1}$,
2) simple compact Lie groups,
3) symmetric spaces of type **AII** in Cartan's notation

$$\mathbf{X} = \mathbf{SU}(2n)/\mathbf{Sp}(n)\,.$$

4) special 26-dimensional symmetric space of type **EIV**: $\mathbf{X} = \mathbf{E}_6/\mathbf{F}_4$.

The noncompact symmetric spaces dual to them are the following:

1) odd-dimensional Lobachevsky spaces $\mathbf{H}^{2n+1}$;
2) symmetric spaces with complex groups of motion $\mathbf{X} = \mathbf{G}/\mathbf{K}$, where $\mathbf{G}$ is a simple complex Lie group, considered as a real one, $\mathbf{K}$ is its maximal compact subgroup;
3) symmetric spaces of type **AII**:

$$\mathbf{X} = \mathbf{SU}^*(2n)/\mathbf{Sp}(n)\,, \quad \dim \mathbf{X} = (n-1)(2n+1).$$

Here the group $\mathbf{SU}^*(2n)$ is the quaternion analogue of $\mathbf{SL}(n)$:

$$\mathbf{SU}^*(2n) = \mathbf{SL}(2n, \mathbf{C}) \cap \mathbf{GL}(n, \mathbf{H})$$

and $\mathbf{Sp}(n)$ is its intersection with $\mathbf{U}(2n)$ (see [42]);

4) special symmetric space of type **EIV**, dual to $\mathbf{X} = \mathbf{E}_6/\mathbf{F}_4\,, \dim \mathbf{X} = 26$, and related to Cayley numbers or octonions.

The "integrability" of the modified wave equations on these spaces is related to the fact that the radial part of the corresponding Laplace-Beltrami operator is conjugated to the following trigonometric (or hyperbolic) version of the generalised Calogero-Moser operators (5), (4) (Sutherland operators) :

$$L = -\Delta + \sum_{\alpha \in \mathcal{R}_+} \frac{m_\alpha(m_\alpha - 1)(\alpha, \alpha)}{\sin^2(\alpha, x)} \tag{33}$$

where the parameters $m_\alpha = 1/2\mu_\alpha$ are equal to half of the root multiplicities (see [30]).

Notice that when all the multiplicities μ_α are even all the parameters m_α are integers. In this case, the corresponding operator (33) is known to be algebraically integrable (see [28]) and one can use the known explicit formulas for the corresponding Baker-Akhiezer function to write down explicitly the fundamental solutions of the modified wave equations on all symmetric spaces listed above (see [16]). From these formulas Huygens' Principle and Theorem 10 follow immediately.

It is worth noting that in this relation that the trigonometric Baker-Akhiezer function for the new non-Coxeter configurations described in Section 2 also exists, and thus the natural trigonometric analogues of the operators (18), (21) are also algebraically integrable (see [17]).

6. Concluding remarks

I would like to mention some connections between Huygens' Principle and Integrability, which have not been discussed in this paper.

1. Self-dual Yang-Mills fields and Huygens' Principle (J. Hadamard [6], L. Asgeirsson [5], M. Atiyah [47], R. Schimming [38]).
2. Huygens' Principle in four-dimensional space-time and the plane-wave exact solutions of Einstein equations (P. Günther [48], R.G. McLenaghan [49]).
3. Hadamard's criterion for Huygens' Principle and the Painlevé test for Integrability of PDE's (in the sense of Weiss-Tabor-Carnevale [37]) (N.H. Ibragimov, A.O. Oganesyan [50], Yu.Yu. Berest, A.P. Veselov [14]).
4. The bispectral problem and Hadamard's conjecture (J.J. Duistermaat, A. Grünbaum [51], Yu.Yu. Berest [52])
5. Huygens' Principle and separation of variables (Yu.Yu. Berest, P. Winternitz [53]).

For other connections as well as the relevant references Günther's fundamental book [3] can be recommended.

I would like to point out that at the moment, in spite of considerable efforts, Hadamard's problem of describing all huygensian equations is still far from a complete solution and remains one of the most intriguing open problems in this area. It is clear that future progress in this direction is connected with the further development of the theory of integrable systems.

Note added in proof.
Some progress in this direction has already been made in the recent papers [55], [56], [57], [58], [59].

References

[1] J. Hadamard *Lectures on Cauchy's Problem in Linear Partial Differential Equations.* (New Haven: Yale Univ. Press, 1923).

[2] R. Courant and D. Hilbert *Methods of Mathematical Physics.*— V.II. (New York, 1964).

[3] P. Günther *Huygens' Principle and Hyperbolic Equations.* (Boston: Acad. Press, 1988).

[4] M. Mathisson, *Le problème de M. Hadamard relatif à la diffusion des ondes*, Acta Math., **71** (1939), 249–282.

[5] L. Asgeirsson, *Some hints on Huygens' principle and Hadamard's conjecture*, Comm. Pure Appl. Math., **9** (1956), 307–326.

[6] J. Hadamard, *The problem of diffusion of waves*, Annals of Math., **43**(3) (1942), 510–523.

[7] K. Stellmacher, *Ein Beispiel einer Huygensschen Differentialgleichung*, Nachr. Akad. Wiss. Göttingen, Math.-Phys. K1, IIa, Bd. **10** (1953), 133–138.

[8] J.E. Lagnese, K.L. Stellmacher, *A method of generating classes of Huygens' operators*, J. Math. & Mech. (1967), V. 17 , N. 5, 461–472.

[9] J.E. Lagnese, *A solution of Hadamard's problem for a restricted class of operators*, Proc. Amer. Math. Soc. (1968), V. 19, 981–988.

[10] R. Schimming *Korteweg-de Vries-Hierarchie und Huygenssches Prinzip*, Sitz. ber. Dresdener Seminar zur Theor. Physik (1986), Nr. 26.

[11] H. Airault, H. McKean, and J. Moser, *Rational and elliptic solutions of the KdV-equation and related many body problems*, Comm. Pure Appl. Math. (1977), V. 30, 95–148.

[12] M. Adler, J. Moser, *On a class of polynomials connected with the Korteweg-de Vries equation*, Comm. Math. Phys.(1978), V. 61, N. 1, 1–30.

[13] Yu.Yu. Berest, A.P. Veselov, *Hadamard's problem and Coxeter groups: new examples of the huygensian equations*, Funct. Anal. Appl. (1994), V. 28, N. 1, 3–15.

[14] Yu.Yu. Berest, A.P. Veselov, *Huygens' principle and Integrability*, Russ. Math. Surveys (1994), V. 49, N. 6, 5–77.

[15] A.P. Veselov, *Huygens' Principle and algebraic Schrödinger operators*, Amer. Math. Soc. Transl. (2) (1995), V. 170, 199–206.

[16] O.A. Chalykh, A.P. Veselov, *Integrability and Huygens' Principle on symmetric spaces*, Comm. Math. Phys. (1996), V. 178, 311–338.

[17] A.P. Veselov, M.V. Feigin, O.A. Chalykh, *New integrable deformations of quantum Calogero-Moser problem*, Uspekhi Mat. Nauk (1996), V. 51, N. 3, 185–186 (Russian).

[18] Yu.Yu. Berest, I.M. Lutsenko, *Huygens' principle in Minkowski spaces and soliton solutions of the Korteweg-de Vries equation.* To appear in Comm. Math. Phys.

[19] Yu.Yu. Berest, *Solution of a Restricted Hadamard's Problem in Minkowski Spaces.* Comm. Pure Appl. Math. (1997), V. 50, N. 10, 1021–1054.

[20] S.P. Novikov, *Periodic problem for KdV equation.I*, Funct. Anal. Appl., (1974), V. 8, N. 3, 54–63.

[21] B.A. Dubrovin, V.B. Matveev, S.P. Novikov, *Nonlinear equations of Korteweg-de Vries type, finite-gap linear operators and abelian varieties*, Russian Math. Surveys. (1976), V. 31, N. 1, 51–125.

[22] I.M. Krichever, *Methods of algebraic geometry in the theory of nonlinear equations*, Russian Math. Surveys. (1977), V. 32, N. 6, 180–208.

[23] F. Calogero, *Solution of the one-dimensional n-body problem with quadratic and/or inversely quadratic pair potential*, J. Math. Phys. (1971), V. 12, 419–436.

[24] J. Moser, *Three integrable Hamiltonian systems connected with isospectral deformations*, Adv. in Math. (1975), V. 16, 1–23.

[25] S.P. Novikov, S.V. Manakov, L.P. Pitaevskii, V.E. Zakharov, *Theory of solitons: The inverse scattering method*, New York: Consultants Bureau, 1984.

[26] O.A. Chalykh, A.P. Veselov, *Commutative rings of partial differential operators and Lie algebras*, Preprint of FIM (ETH, Zürich), 1988. Comm. Math. Phys. (1990), V. 126, 597–611.

[27] O.A. Chalykh, A.P. Veselov, *Integrability in the theory of the Schrödinger operators and Harmonic Analysis*, Comm. Math. Phys. (1993), V. 152, 29–40.

[28] A.P. Veselov, K.L. Styrkas and O.A. Chalykh, *Algebraic integrability for Schrödinger equation and the groups generated by reflections*, Theor. Math. Phys. (1993), V. 94, N. 2, 182–197.

[29] M.A. Olshanetsky, A.M. Perelomov, *Quantum completely integrable systems connected with semisimple Lie algebras*, Lett. Math. Phys. (1977), V. 2, 7–13.

[30] M.A. Olshanetsky, A.M. Perelomov, *Quantum integrable systems related to Lie algebras*, Phys. Rep. (1983) V. 94, 313–404.

[31] C.F. Dunkl, *Differential-difference operators associated to reflection groups*, Trans. AMS. (1989), V. 311, 167–183.

[32] G.J. Heckman, *A remark on the Dunkl differential-difference operators*, Prog. in Math. (1991), V. 101, 181–191.

[33] E. Opdam, *Root systems and hypergeometric functions IV*, Compositio Math. (1988), V. 67, 191–209.
[34] A.P. Veselov, *Calogero quantum problem, KZ equation and Huygens' Principle*, Theor.Math.Phys. (1994), V. 98, N. 3, 524–535.
[35] V.M. Buchstaber, G. Felder, A.P. Veselov, *Elliptic Dunkl operators, root systems and functional equations*, Duke Math. J. (1994), V. 76, N. 3, 885–911.
[36] J.L. Burchnall, and T.W. Chaundy, *A set of differential equations which can be solved by polynomials*, Proc. London Math. Soc. (1929–30), V. 30, 401–414.
[37] J. Weiss, M. Tabor, G. Carnevale, *The Painlevé property for partial differential equations*, J. Math. Phys. (1983), V. 24, 522–526.
[38] R. Schimming, *Laplace-like linear differential operators with logarithm-free elementary solution*, Math. Nachr. (1990), V. 148, 145–174.
[39] P.D. Lax, R.S. Phillips, *An example of Huygens' Principle*, Comm. Pure and Appl. Math. (1978), V. 31, 415–421.
[40] S. Helgason, *Wave equations on homogeneous spaces*, Lect. Notes in Math. Springer-Verlag Berlin (1984), V. 1077, 252–287.
[41] L.E. Solomatina, *Translation representation and Huygens' principle for an invariant wave equation in a Riemannian symmetric spaces*, Soviet Math. Izv. (1986), V. 30, 108–111.
[42] S. Helgason, *Differential geometry, Lie groups and symmetric spaces*, Academic Press, New York, 1978.
[43] G. Olafsson, H. Schlichtkrull, *Wave propagation on Riemannian symmetric spaces*, J. Funct. Anal. (1992), V. 107, 270–278.
[44] S. Helgason, *Huygens' Principle for wave equations on Symmetric spaces*, J. Funct. Anal. (1992), V. 107, 279–288.
[45] T. Branson, G. Olafsson, H. Schlichtkrull, *Huygens' principle in Riemannian symmetric spaces*, Math. Ann. (1995), V. 301, 445–462.
[46] T. Branson, G. Olafsson, *Helmgoltz operator and symmetric spaces duality*, Preprint, June 1995.
[47] M. Atiyah, *Green's functions for self-dual four manifolds*, Math. Analysis and Appl., Part A, Adv. in Math. Suppl. Stud. (1981), V. 7A, 129–158.
[48] P. Günther, *Ein Beispiel einer nichttrivialen Huygensschen Differentialgleichung mit vier unabhängigen Variablen*, Arch. Rat. Mech. Anal. (1965), V. 18(2), 103–106.
[49] R.G. McLenaghan, *An explicit determination of the empty space-times on which the wave equation satisfies Huygens' Principle*, Proc. Cambridge Phil. Soc., (1969), V. 65, 139–155.
[50] N.H. Ibragimov, A.O. Oganesyan, *The hierarchy of Huygens' equations in spaces with a non-trivial conformal group*, Russian Math. Surveys (1991), V. 46, N. 3, 111–146.
[51] J.J. Duistermaat, F.A. Grünbaum, *Differential equations with the spectral parameter*, Comm. Math. Phys. (1986), V. 103, 177–240.
[52] Yu.Yu. Berest, *Nonlinear gauge transformations, Huygens' Principle and Hadamard conjecture revisited*, Preprint CRM-2296 (1995).
[53] Yu.Yu. Berest, P. Winternitz, *Huygens' Principle and Separation of Variables*, Preprint CRM (1996). To appear in Rev. Math. Phys.
[54] O.A. Chalykh, M.V. Feigin, A.P. Veselov, *Multidimensional Baker-Akhiezer functions and Huygens' Principle*, in preparation.
[55] O.A. Chalykh, M.V. Feigin, A.P. Veselov, *New integrable generalizations of the Calogero-Moser quantum problem*, J. Math. Phys. (1998), V. 39, N. 2, 1–9.

[56] A.P. Veselov, *Baker-Akhiezer functions and the bispectral problem in many dimensions*, CRM Proceedings and Lecture Notes, American Mathematical Society, Providence, RI (1998), V. 14, 123–129.

[57] Yu.Yu. Berest, *Huygens' principle and the bispectral problem*, CRM Proceedings and Lecture Notes, American Mathematical Society, Providence, RI (1998), V. 14, 11–30.

[58] Yu.Yu. Berest, A.P. Veselov, *On the singularities of the potentials of exactly solvable Schrödinger operators and Hadamard's problem*, Uspekhi Matematicheskikh Nauk (1998), V. 53, N. 1, 211–212 (Russian).

[59] O.A. Chalikh, *Darboux transformations for multidimensional Schrödinger operators*, Uspekhi Matematicheskikh Nauk (1998), V. 53, N. 2 (Russian).

Progress in Mathematics, Vol. 169, © 1998 Birkhäuser Verlag Basel/Switzerland

Some Problems and Results on the Controllability of Partial Differential Equations

ENRIQUE ZUAZUA*

Departamento de Matemática Aplicada
Universidad Complutense
28040 Madrid. Spain
e-mail: zuazua@sunma4.mat.ucm.es

ABSTRACT. We present some recent results on the controllability of Partial Differential Equations. We discuss the different notions of controllability and comment on how feasible they are depending on the nature of the system under consideration. We discuss both the wave and heat equations as model examples of conservative and irreversible systems respectively and we describe the different tools that have been developed to address these problems. We also present some recent results on the controllability of the linear system of thermoelasticity which is the simplest one coupling both the hyperbolic and the parabolic nature of the wave and heat equation respectively.

1. Introduction

In this paper we present some results obtained recently on the controllability of Partial Differential Equations (PDE). Roughly speaking the problem of controllability can be formulated as follows: consider an evolution equation or system in which some of the data (the control) is at our disposal (for instance, some boundary condition, the right-hand side of the equation, etc.). Given a time interval $(0, T)$ and two values u^0 and v^0 in the space of states where the equation evolves, can one find a control such that the solution u with initial data u^0 and this control takes the value v^0 at $t = T$? The way we have raised the question may be understood as an *exact controllability* problem. The control requirement $u(T) = v^0$ can be relaxed in various ways leading to various notions and degrees of controllability.

Many different problems can fit in this general formulation. On the other hand, the solvability of the problem depends very much on the nature of the precise question under consideration and, in particular, the following features may play a crucial role: linearity or nonlinearity of the system, time reversibility, the structure of the set of admissible controls, etc.

The controllability of PDE has been the object of intensive researches during the last decade. However the subject is older than that. In 1978, D.L. Russell [R1] made a rather complete survey of the most relevant results that were available in

*This work was partially supported by DGICYT (Spain) grant PB93-1203 and grant CHRX-CT94-0471 of the European Union.

the literature at that time. In that paper the author described a number of different tools that were developed to address controllability problems, often inspired and related to other subjects of PDE: multipliers, moment problems, nonharmonic Fourier series, etc.

More recently J.-L. Lions introduced the so-called HUM (Hilbert Uniqueness Method) (see [L1] and [L2] for instance). This was the starting point of a fruitful decade for this subject.

It would be impossible to present in one single paper all the relevant results that have been proved in this area. This is one of the reasons for the decision to focus on the wave and heat equations. The second one is that these two systems are model examples of conservative and strongly irreversible systems respectively. When studying these two simple models one already encounters some of the most relevant intrinsic difficulties of controllability. We will explain how important the role that reversibility plays is when addressing controllability problems. We will also describe the different methods that have been developed to address controllability issues in these two frameworks: wave and heat equations. In both cases we briefly discuss semilinear equations too.

We also study the linear system of three-dimensional thermoelasticity. It is the simplest system combining the hyperbolic and parabolic nature of the wave and heat equations respectively. In a first approach to the problem one might be tempted to think that controllability results for the system of thermoelasticity can be obtained just by adding what is known about the wave and heat equations separately. But this is not the case. Indeed, the techniques available in the literature for the wave and heat equations are of a rather different nature and it is difficult to combine them to address the system of thermoelasticity. We shall see that some decoupling techniques may be a very useful tool to solve this problem.

As we said above, this survey paper is far from being complete. Even in what concerns the three examples we have chosen to develop (wave and heat equations and the system of thermoelasticity) our list of references is rather limited. We hope that other authors having proved relevant results in the area will forgive us for the omission. A more complete list of references can be obtained by looking at the bibliographies of the survey papers and the books in our references.

We have left out some other interesting topics like, for instance, controllability of plates, Schrödinger and KdV equations, the system of three-dimensional elasticity, the Stokes and Navier-Stokes equations, Maxwell's system, etc. On the other hand, we have not considered other important related topics like: numerical computation and simulation in controllability problems, stabilizability, networks of flexible structures, connections with finite dimensional controllability theory, the theory of optimal control, etc. We have included in our bibliography a limited number of references to these issues.

2. The linear wave equation

2.1. Problem formulation Let Ω be a bounded domain of $\mathbb{R}^n$, $n \geq 1$, with boundary Γ of class C^2.

Let ω be an open and non-empty subset of Ω. Let $T > 0$ and consider the linear controlled wave equation in the cylinder $Q = \Omega \times (0, T)$:

$$\begin{cases} u_{tt} - \Delta u = f 1_\omega & \text{in} \quad Q \\ u = 0 & \text{on} \quad \Sigma \\ u(x,0) = u^0(x), u_t(x,0) = u^1(x) & \text{in} \quad \Omega. \end{cases} \tag{2.1}$$

In (2.1) Σ represents the lateral boundary of the cylinder Q, i.e. $\Sigma = \Gamma \times (0,T)$, 1_ω is the characteristic function of the set ω, $u = u(x,t)$ is the state and $f = f(x,t)$ is the control variable. Since f is multiplied by 1_ω the action of the control is concentrated in ω.

In space dimension $n = 2$, $u = u(x,t)$ may represent, for instance, the vertical displacement of a vibrating membrane occupying Ω and fixed on its boundary Γ under the action of the force f concentrated in ω.

When $(u^0, u^1) \in H_0^1(\Omega) \times L^2(\Omega)$ and $f \in L^2(Q)$ system (2.1) has a unique solution $u \in C\left([0,T]; H_0^1(\Omega)\right) \cap C^1\left([0,T]; L^2(\Omega)\right)$.

The problem of controllability consists, roughly, in describing the set of reachable final states $(u(T), u_t(T))$ when the control f varies in a given class, for instance, $L^2(Q)$.

One may distinguish the following degrees of controllability:

(a) *Approximate controllability*: System (2.1) is said to be approximately controllable in time T if the set of reachable states

$$R\left(T; (u^0, u^1)\right) = \left\{(u(T), u_t(T)) : f \in L^2(Q)\right\}$$

is dense in $H_0^1(\Omega) \times L^2(\Omega)$ for every $(u^0, u^1) \in H_0^1(\Omega) \times L^2(\Omega)$.

(b) *Exact controllability*: System (2.1) is said to be exactly controllable at time T if $R\left(T; (u^0, u^1)\right) = H_0^1(\Omega) \times L^2(\Omega)$ for all $(u^0, u^1) \in H_0^1(\Omega) \times L^2(\Omega)$.

(c) *Null controllability*: System (2.1) is said to be null controllable at time T if $(0,0) \in R\left(T; (u^0, u^1)\right)$ for all $(u^0, u^1) \in H_0^1(\Omega) \times L^2(\Omega)$.

Remark 2.1

(a) Since we are dealing with solutions of the wave equation, for any of these properties to hold the time T has to be sufficiently large due to the finite speed of propagation.

(b) Since system (2.1) is linear and reversible in time null and exact controllability are equivalent notions.

(c) Clearly every exactly controllable system is approximately controllable too. However, system (2.1) may be approximately but not exactly controllable. In those cases for every $\left(u^0, u^1\right), \left(v^0, v^1\right) \in H_0^1(\Omega) \times L^2(\Omega)$ and $\varepsilon > 0$ there exists a control $f_\varepsilon \in L^2(Q)$ such that the solution of (2.1) satisfies

$$\| \left(u(T), u_t(T)\right) - \left(v^0, v^1\right) \|_{H_0^1(\Omega) \times L^2(\Omega)} \leq \varepsilon. \tag{2.2}$$

When the system is approximately but not exactly controllable $\| f_\varepsilon \|_{L^2(Q)} \to \infty$ as $\varepsilon \to 0$ for some initial and final data.

(d) Describing the nature of the set of controllable data when exact controllability does not hold is a challenging open problem. We refer to [Le1] for partial but interesting results in this direction.

(e) We have stated controllability problems in $H_0^1(\Omega) \times L^2(\Omega)$ with controls in $L^2(Q)$. Of course this is an arbitrary choice of the functional setting. Many other frameworks make sense.

(f) Null controllability is a physically interesting notion since the state $(0,0)$ is an equilibrium for system (2.1). More precisely, if the solution of (2.1) reaches this null state at time $t = T$ and we do not introduce further control for $t \geq T$, the solution remains in this equilibrium configuration for all $t \geq T$. □

Most of the literature on the controllability of the wave equation has been written on the framework of the *boundary control* problem, i.e. when (2.1) is replaced by

$$\begin{cases} u_{tt} - \Delta u = 0 & \text{in} \quad Q \\ u = v & \text{on} \quad \Sigma \\ u(x,0) = u^0(x), u_t(x,0) = u^1(x) & \text{in} \quad \Omega. \end{cases}$$

In this case the control $v = v(x,t)$ acts on the system through the Dirichlet boundary condition.

The control problems above for system (2.1) are refered to as *internal controllability* problems since in that case the control acts on the subset ω of Ω.

In this paper we will only discuss the internal control problem to avoid additional technical difficulties related to the well-posedness of the boundary value problem.

2.2. Approximate controllability It is easy to see that approximate controllability is equivalent to a unique continuation property of the adjoint system:

$$\begin{cases} \varphi_{tt} - \Delta \varphi = 0 & \text{in} \quad Q \\ \varphi = 0 & \text{on} \quad \Sigma \\ \varphi(x,T) = \varphi^0(x), \varphi_t(x,T) = \varphi^1(x) & \text{in} \quad \Omega. \end{cases} \tag{2.3}$$

Indeed, system (2.1) is approximately controllable if and only if the following holds:

$$\varphi \equiv 0 \text{ in } \omega \times (0,T) \Rightarrow \left(\varphi^0, \varphi^1\right) \equiv (0,0). \tag{2.4}$$

By using Holmgren's Uniqueness Theorem it can be easily seen that (2.4) holds if T is large enough. A careful but simple argument leads to the following estimate of the minimal time for approximate controllability:

$$T > 2\delta(\Omega;\omega) \tag{2.5}$$

where

$$\delta(\Omega;\omega) = \sup_{x\in\Omega} \delta(x;\omega), \tag{2.6}$$

$\delta(x,\omega)$ being the infimum of the lengths of the curves in Ω joining x and ω.

The fact that (2.4) holds is a consequence of the following unique continuation result which is of local nature and a consequence of Holgrem's Uniqueness Theorem: *If φ solves the wave equation and*

$$\varphi = 0 \quad in \quad B_\delta \times (0,T)$$

then

$$\varphi = 0 \quad in \quad \bigcup_{0\le r<T/2} \{B_{\delta+r} \times (r, T-r)\}.$$

By B_δ we denote the ball of radious δ in $\mathbb{R}^n$ with center in $x=0$.

When the geometry of Ω is complicated, this minimal time of controllability may be very large. But, as T. Cazenave [C] proved, this estimate is sharp.

There are at least two ways of checking that (2.4) implies the approximate controllability:

(a) The application of the Hahn-Banach Theorem.

(b) The variational approach developed in [L3].

We prefer to present here the second proof since it is easy to adapt to other situations.

First of all we observe that, since system (2.1) is linear, it is sufficient to consider the case $u^0 \equiv u^1 \equiv 0$. Then, given any $(v^0, v^1) \in H_0^1(\Omega) \times L^2(\Omega)$ and $\varepsilon > 0$ we look for $f \in L^2(Q)$ such that the solution of (2.1) satisfies (2.2).

To do that we introduce the following functional $J_\varepsilon : L^2(\Omega) \times H^{-1}(\Omega) \to \mathbb{R}$:

$$J_\varepsilon\left(\varphi^0, \varphi^1\right) = \frac{1}{2}\int_0^T \int_\omega |\, \varphi \,|^2 \, dxdt + \varepsilon \left\| \left(\varphi^0, \varphi^1\right) \right\|_{L^2(\Omega)\times H^{-1}(\Omega)} \qquad (2.7)$$
$$-\langle \left(\varphi^0, \varphi^1\right), \left(v^1, -v^0\right) \rangle$$

where $\langle \cdot, \cdot \rangle$ denotes the duality pairing between $L^2(\Omega) \times H^{-1}(\Omega)$ and its dual $L^2(\Omega) \times H_0^1(\Omega)$.

The functional J_ε is continuous and convex in the Hilbert space $L^2(\Omega) \times H^{-1}(\Omega)$. On the other hand when the unique continuation property (2.4) holds it is easy to see that it is coercive too. More precisely,

$$\lim_{\|(\varphi^0,\varphi^1)\|_{L^2(\Omega)\times H^{-1}(\Omega)}\to\infty} \frac{J_\varepsilon\left(\varphi^0, \varphi^1\right)}{\|\left(\varphi^0, \varphi^1\right)\|_{L^2(\Omega)\times H^{-1}(\Omega)}} \ge \varepsilon. \qquad (2.8)$$

We refer to [FPZ1] for the details of the proof in the frame of the heat equation.

Under these conditions J_ε has a unique minimizer in $L^2(\Omega) \times H^{-1}(\Omega)$. Let us denote it by $(\widehat{\varphi}^0, \widehat{\varphi}^1)$, i.e.

$$\left(\widehat{\varphi}^0, \widehat{\varphi}^1\right) \in L^2(\Omega) \times H^{-1}(\Omega),\ J_\varepsilon\left(\widehat{\varphi}^0, \widehat{\varphi}^1\right) = \min_{(\varphi^0,\varphi^1)\in L^2(\Omega)\times H^{-1}(\Omega)} J_\varepsilon\left(\varphi^0, \varphi^1\right). \qquad (2.9)$$

Then if $\widehat{\varphi}$ is the solution of (2.3) with data $(\widehat{\varphi}^0, \widehat{\varphi}^1)$, we have

$$\left| \int_0^T \int_\omega \widehat{\varphi}\varphi dxdt - \langle \left(\varphi^0, \varphi^1\right), \left(v^1, -v^0\right) \rangle \right| \le \varepsilon \,\|\, \left(\varphi^0, \varphi^1\right) \|_{L^2(\Omega)\times H^{-1}(\Omega)} \qquad (2.10)$$

for all $(\varphi^0, \varphi^1) \in L^2(\Omega) \times H^{-1}(\Omega)$.

On the other hand, if we set $f = \widehat{\varphi}$ in (2.1) multiplying in (2.3) by u, integrating by parts and taking into account that $u^0 \equiv u^1 \equiv 0$ we get

$$\int_0^T \int_\omega \widehat{\varphi}\varphi dxdt - \left\langle \left(\varphi^0, \varphi^1\right), \left(u_t(T), -u(T)\right)\right\rangle = 0. \tag{2.11}$$

Combining (2.10) and (2.11) it follows that

$$\left|\left\langle \left(\varphi^0, \varphi^1\right), \left(u_t(T) - v^1, v^0 - u(T)\right)\right\rangle\right| \leq \varepsilon \parallel \left(\varphi^0, \varphi^1\right) \parallel_{L^2(\Omega)\times H^{-1}(\Omega)} \tag{2.12}$$

for all $\left(\varphi^0, \varphi^1\right) \in L^2(\Omega) \times H^{-1}(\Omega)$ and this is equivalent to (2.2).

Remark 2.2 The control one obtains when minimizing J_ε is the one of minimal L^2-norm among all the admissible ones. The set of admissible controls is given by.

$$\mathcal{U}_{ad} = \left\{ f \in L^2(Q) : \parallel \left(u(T) - v^0, u_t(T) - v^1\right) \parallel_{H_0^1(\Omega)\times L^2(\Omega)} \leq \varepsilon \right\}.$$

The minimization of J_ε was introduced in [L3] as the dual problem of finding the control of minimal norm in $\mathcal{U}_{ad}$ by Fenchel and Rockafellar's duality principle. □

The unique continuation property (2.4) allows the proof of a stronger result. Indeed, in addition to (2.2), one can guarantee that the solution of (2.1) exactly satisfies a finite number of constraints. More precisely, the following holds:

Theorem 2.1 *Let ω be any open and non-empty subset of Ω. Assume that $T > 2\delta(\Omega;\omega)$. Let E be any finite-dimensional subspace of $H_0^1(\Omega) \times L^2(\Omega)$ and let us denote by π_E the orthogonal projection over E.*

Then, for any $\left(u^0, u^1\right)$, $\left(v^0, v^1\right) \in H_0^1(\Omega) \times L^2(\Omega)$ and $\varepsilon > 0$ there exists $f \in L^2(Q)$ such that the solution of (2.1) satisfies

$$\left\|\left(u(T) - v^0, u_t(T) - v^1\right)\right\|_{H_0^1(\Omega)\times L^2(\Omega)} \leq \varepsilon;\ \pi_E\left(u(T), u_t(T)\right) = \pi_E\left(v^0, v^1\right). \tag{2.13}$$

This result may be proved in at least two different ways:

(a) By a suitable modification of the functional J_ε to be minimized (in Section 3 we give the details in the case of the heat equation).

(b) As a direct consequence of the approximate controllability and the following Theorem of functional Analysis (we refer to [LZ2] for a proof):

Theorem 2.2 *Let V and H be two Hilbert spaces and L a bounded linear operator from V to H with dense range. Let E be a finite-dimensional subspace of H and π_E the corresponding orthogonal projection.*

Then, given any $e_0 \in E$, when v runs over the set of elements of v such that $\pi_E L v = e_0$, Lv describes a dense set in $e_0 + E^\perp$.

The results above hold for wave equations with analytic coefficients too. Indeed, the control problem can be reduced to the unique continuation one and the latter may be solved by means of Holmgren's Uniqueness Theorem.

However, the problem is not completely solved in the frame of the wave equation with lower order potentials $a \in L^\infty(Q)$ of the form

$$u_{tt} - \Delta u + a(x,t)u = f 1_\omega \text{ in } Q.$$

Once again the problem may be reduced to show the unique continuation property (2.4) for the adjoint system:

$$\begin{cases} \varphi_{tt} - \Delta\varphi + a(x,t)\varphi = 0 & \text{in} \quad Q \\ \varphi = 0 & \text{on} \quad \Sigma \\ \varphi(x,T) = \varphi^0(x), \varphi_t(x,T) = \varphi^1(x) & \text{in} \quad \Omega. \end{cases}$$

The following is known:

(a) When $a = a(x)$, D. Tataru [T2] has proved a deep result showing that (2.4) holds with the same estimate in the uniqueness time ($T > 2\delta(\Omega,\omega)$) for bounded, time-independent potentials. Tataru's proof uses Carleman's estimates in an essential way and is of local nature.

(b) When $a = a(t)$, (2.4) can be proved for $T > 2\delta(\Omega;\omega)$. In [H] a simple proof is given by developing solutions on the spherical harmonics. This reduces the problem to a family of one-dimensional wave equations in the time and radial variables. In one space dimension the unique continuation property is easy to prove. It suffices to view the wave equation as an evolution equation in the radial variable while t plays the role of the space variable. We refer to [Z5] for other applications of this argument.

(c) When $a = a(x,t)$, it is known that the unique continuation property (2.4) does not hold in the class of C^∞ potentials. Indeed, the local uniqueness fails (see S. Alinhac [A]).

(d) When ω is a neighborhood of the boundary of Ω, unique continuation holds for any bounded potential $a = a(x,t)$. We refer to A. Ruiz [Ru] for a proof of this fact that uses Carleman's inequalities. Notice that this result is not of local nature.

2.3. Exact controllability As we said above, since the wave equation is reversible in time, exact controllability and null controllability are equivalent notions. Thus, throughout this section, without loss of generality, we will assume that $v^0 \equiv v^1 \equiv 0$.

If instead of the functional J_ε of Section 2.2 we consider

$$\widetilde{J}_\varepsilon\left(\varphi^0,\varphi^1\right) = \frac{1}{2}\int_0^T\int_\omega \varphi^2 dxdt + \varepsilon \parallel \left(\varphi^0,\varphi^1\right) \parallel_{L^2(\Omega)\times H^{-1}(\Omega)} \tag{2.14}$$
$$+\left\langle \left(\varphi(0),\varphi_t(0)\right),\left(u^1,-u^0\right)\right\rangle$$

it is easy to see that the unique continuation property (2.4) allows to prove that $\widetilde{J}_\varepsilon$ is coercive too. More precisely $\widetilde{J}_\varepsilon$ satisfies (2.8). The minimizer of $\widetilde{J}_\varepsilon$ provides a control f_ε such that the solution of (2.1) satisfies

$$\|(u(T),\, u_t(T))\|_{H_0^1(\Omega)\times L^2(\Omega)} \le \varepsilon. \tag{2.15}$$

The problem of exact controllability consists in analyzing whether one can take $\varepsilon = 0$ in (2.15), or more precisely, if the control f_ε remains bounded in $L^2(Q)$ as $\varepsilon \to 0$.

The unique continuation property (2.4) by itself does not allow this question to be answered. To guarantee that the controls f_ε are bounded we need to know that the minimizers of $\widetilde{J}_\varepsilon$ in $L^2(\Omega) \times H^{-1}(\Omega)$ are bounded. To prove the boundedness of the minimizers for all (u^0, u^1) in $H_0^1(\Omega) \times L^2(\Omega)$ we need the following inequality:

$$\|(\varphi(0), \varphi_t(0))\|^2_{L^2(\Omega)\times H^{-1}(\Omega)} \le C \int_0^T \int_\omega \varphi^2 dx dt \tag{2.16}$$

for all solutions of (2.3).

This inequality allows the total energy of the solution of (2.3) to be estimated by means of a measurement in the control region $\omega \times (0, T)$. Thus, it establishes the *continuous observability* of system (2.3). The energy $\|(\varphi(t), \varphi_t(t))\|^2_{L^2(\Omega)\times H^{-1}(\Omega)}$ of solutions of (2.3) is conserved along trajectories. Thus, (2.16) is equivalent to

$$\left\|\left(\varphi^0, \varphi^1\right)\right\|^2_{L^2(\Omega)\times H^{-1}(\Omega)} \le C \int_0^T \int_\omega \varphi^2 dx dt. \tag{2.17}$$

Remark 2.3 When the system to be controlled is irreversible in time, the corresponding adjoint system is irreversible too. In these cases the analogue of (2.16) is of a different nature to that corresponding to (2.17). For instance, as we will see in Section 3, the analogue of (2.16) is true for the adjoint of the heat equation but the analogue of (2.17) is false. □

When (2.16) holds one can directly minimize the functional $\widetilde{J}_\varepsilon$ with $\varepsilon = 0$, i.e.

$$\widetilde{J}\left(\varphi^0, \varphi^1\right) = \frac{1}{2} \int_0^T \int_\omega \varphi^2 dx dt + \left\langle (\varphi(0), \varphi_t(0)), \left(u^1, -u^0\right)\right\rangle. \tag{2.18}$$

The following is easy to prove: *When the observability inequality* (2.16) *holds, the functional* $\widetilde{J}$ *has a unique minimizer* $\left(\widehat{\varphi}^0, \widehat{\varphi}^1\right)$ *in* $L^2(\Omega) \times H^{-1}(\Omega)$ *for all* $\left(u^0, u^1\right) \in H_0^1(\Omega) \times L^2(\Omega)$. *The control* $f = \widehat{\varphi}$ *with* $\widehat{\varphi}$ *solution of* (2.3) *corresponding to* $\left(\widehat{\varphi}^0, \widehat{\varphi}^1\right)$ *is such that the solution of* (2.1) *satisfies*

$$u(T) \equiv u_t(T) \equiv 0. \tag{2.19}$$

Remark 2.4 The control obtained when minimizing $\widetilde{J}$ is the one of minimal L^2-norm among the admissible ones (see [L1,2]). □

Let us now discuss what is known about the observability inequality (2.16):

(a) To our knowledge an inequality of the form (2.16) was proved for the first time by L. F. Ho [Ho] by means of multiplier techniques in the spirit of C. Morawetz [M]. In [Ho] it was shown that if one considers subsets of Γ of the form

$$\Gamma(x^0) = \left\{x \in \Gamma : (x - x^0) \cdot n(x) > 0\right\}$$

for some $x^0 \in \mathbb{R}^n$ (by $n(x)$ we denote the outward unit normal to Ω in $x \in \Gamma$ and by $\cdot$ the scalar product in $\mathbb{R}^n$) and if $T > 0$ is large enough, the following boundary observability inequality holds:

$$\|(\varphi(0), \varphi_t(0))\|^2_{H_0^1(\Omega)\times L^2(\Omega)} \leq C \int_0^T \int_{\Gamma(x^0)} \left|\frac{\partial \varphi}{\partial n}\right|^2 d\Gamma dt \tag{2.20}$$

for all $(\varphi^0, \varphi^1) \in H_0^1(\Omega) \times L^2(\Omega)$.

This is the observability inequality that is required to solve the boundary controllability problem mentioned in Section 2.1.

Later inequality (2.20) was proved in [L1,2], for any $T > T(x^0) = 2 \| x - x^0 \|_{L^\infty(\Omega)}$. This is the optimal observability time that one may derive by means of multipliers.

Proceeding as in [L2], vol. 1, one can easily prove that (2.20) implies (2.16) when ω is a neighborhood of $\Gamma(x^0)$ in Ω, i.e. $\omega = \Omega \cap \Theta$ where Θ is a neighborhood of $\Gamma(x^0)$ in $\mathbb{R}^n$, with $T > T(x^0)$.

(b) C. Bardos, G. Lebeau and J. Rauch [BLR] proved that, in the class of C^∞ domains, the observability inequality (2.16) holds if and only if (ω, T) satisfy the following *geometric control condition* in Ω: *Every ray of geometric optics that propagates in Ω and is reflected on its boundary Γ enters ω in time less than T.*

This result was proved by means of microlocal analysis techniques and has the advantage that:

1. It characterizes all the control sets ω and control times T that guarantee the exact controllability. In this respect observe that the control sets $\Gamma(x^0)$ that one obtains by multipliers are always large enough subsets of Γ. For instance, when Ω is a ball of $\mathbb{R}^n$, $\Gamma(x^0)$ is a connected set that always contains more than half of its boundary. On the other hand the estimate one obtains by multipliers on the time of control $T(x^0)$ is not sharp in general.

2. The methods of [BLR] apply in a large class of wave equations with variable coefficients provided they are smooth and time independent. It is easy to see that the multiplier techniques fail for wave equations with smooth variable coefficients, some particular cases being excepted.

Some of the advantages that the multiplier technique presents are the following:

1. It applies when the domain is of class C^2 or even when the domain has singularities (see P. Grisvard [G]). However, the microlocal approach has recently been greatly simplified by N. Burq [Bu] by using the microlocal defect measures introduced by P. Gerard [Ge] in the context of the homogenization and the kinetic equations. In [Bu] the geometric control condition was shown to be sufficient for exact controllability for domains Ω of class C^3 and equations with C^2 coefficients.

2. The multiplier technique can be systematically applied to other systems and equations to easily obtain suboptimal results: plate and Schrödinger equations, KdV equations, Maxwell's system and the system of elasticity, etc. We refer to [L1,2] and to the recent monograph by V. Komornik [K] for an introduction to these techniques.

   At this point the works by D. Tataru [T1] are also worth mentioning. In [T1] by using Carleman's inequalities the observability was proved for more general equations than those that one may cover by microlocal tools (since less regularity of the coefficients and on the domain is needed) and for a class of control sets that is larger than the very restrictive one covered by multipliers.

(c) An important problem arising, in particular, when dealing with semilinear problems, is the dependence of the constants in the observability inequality on the coefficients.

Let us consider for instance the adjoint equation

$$\begin{cases} \varphi_{tt} - \Delta\varphi + a(x,t)\varphi = 0 & \text{in} \quad Q \\ \varphi = 0 & \text{on} \quad \Sigma \\ \varphi(x,t) = \varphi^0(x),\ \varphi_t(x,T) = \varphi^1(x) & \text{in} \quad \Omega \end{cases} \tag{2.21}$$

with a bounded potential $a \in L^\infty(Q)$.

Assume that the subset ω of Ω and $T > 0$ are such that the observability holds. More precisely, suppose that there exists a constant $C > 0$ such that

$$\|(\varphi(0), \varphi_t(0))\|^2_{L^2(\Omega)\times H^{-1}(\Omega)} \leq C(a) \int_0^T \int_\omega \varphi^2 dxdt \tag{2.22}$$

for all $(\varphi^0, \varphi^1) \in L^2(\Omega) \times H^{-1}(\Omega)$.

This holds, for instance, when ω is a neighborhood of the boundary Γ and $T > \operatorname{diam}(\Omega\backslash\omega)$ (see [Z4]). However the proof of (2.22) uses compactness arguments to absorb lower order terms. Thus, no explicit bounds are obtained. In one space dimension in [Z5] it is proved that (2.22) holds with

$$C(a) = O\left(\exp\left(C \parallel a \parallel_\infty^{1/2}\right)\right) \quad \text{as} \quad \parallel a \parallel_\infty \to \infty. \tag{2.23}$$

This estimate turns to be sharp and it is reasonable to expect it to hold in several space dimensions too. □

We have described here the HUM and some tools to prove observability inequalities. In particular, we have explained how HUM can be combined systematically with multiplier methods. One can also combine HUM with the theory of nonharmonic Fourier series when the coefficients of the system are time independent. We refer to [K] for a complete presentation of this approach. Other relevant works in this direction are also worth mentioning: A. Haraux [Ha2], M. Horvarth [Hr] and I. Joo [J], for instance.

However, other methods have been developed to address controllability problems: Moment problems, nonharmonic Fourier series, fundamental solutions, etc. We will not present them here. We refer to the survey paper by D.L. Russell [R1] for the interested reader.

2.4. The semilinear wave equation Let us consider the semilinear wave equation

$$\begin{cases} u_{tt} - \Delta u + h(u) = f 1_\omega & \text{in} \quad Q \\ u = 0 & \text{on} \quad \Sigma \\ u(x,0) = u^0(x),\, u_t(x,0) = u^1 & \text{in} \quad \Omega \end{cases} \tag{2.24}$$

where $h : \mathbb{R} \to \mathbb{R}$ is a locally Lipschitz function such that

$$| h(s_1) - h(s_2) | \leq C \left(1 + | s_1 |^{p-1} + | s_2 |^{p-1}\right) | s_1 - s_2 |, \forall s_1, s_2 \in \mathbb{R} \tag{2.25}$$

with

$$1 < p \leq \frac{n}{n-2} \text{ if } n \geq 3;\ 1 < p < \infty \text{ if } n = 1, 2. \tag{2.26}$$

Under assumptions (2.25)–(2.26) system (2.24) is locally well-posed: there exists $\delta > 0$ such that for any $\left(u^0, u^1\right) \in H_0^1(\Omega) \times L^2(\Omega), f \in L^2(Q)$ so that

$$\left\| \left(u^0, u^1\right) \right\|_{H_0^1(\Omega) \times L^2(\Omega)} + \| f \|_{L^2(Q)} \leq \delta \tag{2.27}$$

system (2.24) admits a unique solution $u \in C\left([0,T]; H_0^1(\Omega)\right) \cap C^1\left([0,T]; L^2(\Omega)\right)$.

If the non-linearity h also satisfies a "good-sign" condition, for instance

$$H(s) \geq -C \mid s \mid^2 \quad \text{as} \quad \mid s \mid \to \infty \tag{2.28}$$

for some $C > 0$ with $H(s) = \int_0^s h(z)dz$, then system (2.24) is globally well-posed and admits a unique solution u in the class above for any initial data $\left(u^0, u^1\right)$ and right-hand side f.

The approximate and exact controllability problem can be formulated as in the context of the linear wave equation. At this level it is worth mentioning that, in the non-linear context, the equivalence of the exact and null controllability notions is no longer true.

Concerning the exact controllability the following is known: *if ω is a neighborhood of the boundary Γ, $T >$ diam$(\Omega \backslash \omega)$ and h is globally Lipschitz, then system (2.24) is exactly controllable.*

This result may be proved by at least two different methods: a fixed point argument based on Schauder's Theorem (see [Z4]) and a global version of the Inverse Function Theorem (see [LT1]).

Very little is known about the exact controllability of (2.24) when h grows superlinearly at infinity. The following was proved in [Z6]:

Theorem 2.3 *Assume that $n = 1$ and $\Omega = (0, l)$. Let $\omega = (a, b)$ be a subinterval of Ω. Assume that $T > 2\max(a, l - b)$ and*

$$\lim_{|s| \to \infty} \frac{| h(s) |}{| s | \log^2 | s |} = 0. \tag{2.29}$$

Then, system (2.24) is exactly controllable.

Remark 2.5

(a) The control time $T > 2\max(a, l-b)$ is sharp.

(b) The restriction (2.29) on the growth of the non-linearity h seems to be too restrictive. However, it turns out to be sharp. Indeed, if $h(s) = -s\log^p(1+|s|)$ with $p > 2$, then blow-up phenomena occur and due to the finite speed of propagation system (2.24) is not controllable except in the trivial case $\Omega = \omega$ (see [Z6]).

Whether exact controllability holds in the critical case in which

$$|h(s)| \sim C|s|\log^2|s| \quad \text{as} \quad |s| \to \infty$$

with $C > 0$ large is an open problem.

(c) The proof of this result uses the fixed point argument developed in [Z4] and the observability inequality (2.22) with an explicit bound of the form (2.23).

(d) The method of proof of Theorem 2.3 may also be applied in several space dimensions. However we do not know if the observability inequality (2.22) holds with a constant of the order of $\exp(C\|a\|_\infty^{1/2})$ in several space dimensions.

(e) The proof of Theorem 2.3 is not adapted to take into account whether the non-linearity h satisfies the "good-sign" condition (2.29) or not. □

When the non-linearity does not satisfy the conditions of the results above, combining the exact controllability of the linearized system and the Inverse Function Theorem one can prove a controllability result of local nature:

Theorem 2.4 *Let Ω, ω and T be such that the exact controllability of the linear system* (2.1) *holds. Assume that the non-linearity h is of class C^1, $h(0) = 0$, and that the growth conditions* (2.25)–(2.26) *are satisfied. Then, there exists $\delta > 0$ such that for any (u^0, u^1), $(v^0, v^1) \in H_0^1(\Omega) \times L^2(\Omega)$ with*

$$\|(u^0, u^1)\|_{H_0^1(\Omega)\times L^2(\Omega)} + \|(v^0, v^1)\|_{H_0^1(\Omega)\times L^2(\Omega)} \leq \delta \tag{2.30}$$

there exists a control $f \in L^2(Q)$ such that the solution of (2.24) *satisfies*

$$u(x,T) = v^0(x),\ u_t(x,T) = v^1(x) \ \textit{in}\ \Omega. \tag{2.31}$$

Remark 2.6 Observe that the control time of the small data satisfying (2.30) is the same as for the linear wave equation. This is due to the fact that, as we said above, the result is proved by means of the Inverse Function Theorem by linearizing around the equilibrium $u \equiv 0$. □

When the non-linearity does not satisfy a "good-sign" condition of the form (2.28) the result is sharp. Indeed, if one considers non-linearities that grow superlinearly at infinity blow-up may occur for large data and therefore the system may not be exactly controllable due to the finite speed of propagation. However, when the non-linearity satisfies a condition like (2.28) the local controllability result may be completed by means of stabilization results.

Let us explain this in the following model example of a homogeneous non-linearity:

$$\begin{cases} u_{tt} - \Delta u + \mid u \mid^{p-1} u = f 1_\omega & \text{in} \quad Q \\ u = 0 & \text{on} \quad \Sigma \\ u(x,0) = u^0(x), u_t(x,0) = u^1(x) & \text{in} \quad \Omega, \end{cases} \tag{2.32}$$

with $1 < p \leq n/(n-2)$ when $n \geq 3$ and $1 < p < \infty$ when $n = 1, 2$.

The following holds:

Theorem 2.5 *Let ω be a neighborhood of Γ in Ω. Then, if p is as above, for every (u^0, u^1), $(v^0, v^1) \in H_0^1(\Omega) \times L^2(\Omega)$ if T is large enough of the order of*

$$T = O\left(\log\left(\left\|\left(u^0, u^1\right)\right\|_{H_0^1(\Omega)\times L^2(\Omega)} + \left\|\left(v^0, v^1\right)\right\|_{H_0^1(\Omega)\times L^2(\Omega)}\right)\right) \tag{2.33}$$

for large data, there exists $f \in L^2(Q)$ such that the solution of (2.32) satisfies

$$u(x,T) = v^0(x),\ u_t(x,T) = v^1(x) \text{ in } \Omega. \tag{2.34}$$

Whether system (2.32) is exactly controllable in a uniform time or not is an open problem.

The proof of Theorem 2.5 combines the local result of Theorem 2.4 and the fact that solutions of the following locally damped wave equation decay uniformly as $t \to \infty$:

$$\begin{cases} u_{tt} - \Delta u + \mid u \mid^{p-1} u + 1_\omega u_t = 0 & \text{in} \quad \Omega \times \mathbb{R}^+ \\ u = 0 & \text{on} \quad \Gamma \times \mathbb{R}^+ \\ u(x,0) = u^0(x),\ u_t(x,0) = u^1(x) & \text{in} \quad \Omega. \end{cases} \tag{2.35}$$

Let us discuss briefly the decay properties of (2.35).

The energy of (2.35) is given by

$$E(t) = \frac{1}{2}\int_\Omega \left[\mid u_t(x,t) \mid^2 + \mid \nabla u(x,t) \mid^2\right] dx + \frac{1}{p+1}\int_\Omega \mid u(x,t) \mid^{p+1} dx. \tag{2.36}$$

It follows that

$$\frac{dE}{dt}(t) = -\int_\omega \mid u_t(x,t) \mid^2 dx \leq 0. \tag{2.37}$$

Thus, the energy of solutions decreases as time increases. More precisely:

Theorem 2.6 *When ω is a neighborhood of the boundary Γ and $1 < p$, $(n-2)p \leq n$, there exist positive constants $C, \alpha > 0$ such that*

$$E(t) \leq Ce^{-\alpha t} E(0),\ \forall t > 0, \tag{2.38}$$

for every solution of (2.35).

It is easy to see that Theorem 2.5 is a consequence of Theorems 2.4 and 2.6.

The stabilization result of Theorem 2.6 was proved in [Z1] by multiplier techniques. As we said above multipliers are a rather common tool for handling control problems but also in the theory of stabilizability (we refer to [K] and [La] for an introduction to this topic).

The decay property (2.38) is the same as the one that holds for the linear dissipative wave equation. In particular the constants C and α in (2.38) do not depend on the initial data. Obtaining (2.38) requires structural conditions on the nonlinearity and applies in the particular example (2.35) and slight variations of it. Much more general equations and non-linearities can be handled if one relaxes the decay requirement (2.38) to the following: for every $R > 0$ there exists $C(R), \alpha(R) > 0$ such that (2.38) holds for every initial data with energy $E(0) \leq R$.

This type of result can be proved easily as a consequence of the uniform decay of the linearized system and a compactness argument. However, according to this type of result the rate of decay may vanish as $E(0) \to \infty$ and therefore, in this way, one cannot obtain explicit bounds on the controllability time of the form (2.33).

3. The heat equation

3.1. Problem formulation With the same notation as in Section 2 we consider the linear controlled heat equation:

$$\begin{cases} u_t - \Delta u = f 1_\omega & \text{in} \quad Q \\ u = 0 & \text{on} \quad \Sigma \\ u(x,0) = u^0(x) & \text{in} \quad \Omega. \end{cases} \tag{3.1}$$

We assume that $u^0 \in L^2(\Omega)$ and $f \in L^2(Q)$ so that (3.1) admits an unique solution $u \in C\left([0,T]\,; L^2(\Omega)\right) \cap L^2\left(0,T; H_0^1(\Omega)\right)$.

We set $R(T; u^0) = \left\{u(T) : f \in L^2(Q)\right\}$. The controllability problems can be formulated as follows:

(a) System (3.1) is said to be approximately controllable if $R(T; u^0)$ is dense in $L^2(\Omega)$ for all $u^0 \in L^2(\Omega)$.

(b) System (3.1) is exactly controllable if $R(T; u^0) = L^2(\Omega)$ for all $u^0 \in L^2(\Omega)$.

(c) System (3.1) is null controllable if $0 \in R(T; u^0)$ for all $u^0 \in L^2(\Omega)$.

Remark 3.1

(a) As we will see in the next section approximate controllability holds for every open non-empty subset ω of Ω and for every $T > 0$.

(b) It is easy to see that exact controllability may not hold except possibly in the case in which $\omega = \Omega$. Indeed, due to the regularizing effect of the heat equation solutions of (3.1) at time $t = T$ are smooth in $\Omega\backslash\overline{\omega}$. Therefore $R(T; u^0)$ is strictly contained in $L^2(\Omega)$ for all $u^0 \in L^2(\Omega)$.

(c) Null controllability implies that all the range of the semigroup generated by the heat equation is reachable too. Let us make this statement more precise.

Let us denote by $S(t)$ the semigroup generated by the heat equation (3.1) without control, i.e. with $f = 0$. Then, if null controllability holds, it follows that for any $u^0 \in L^2(\Omega)$ and $v^0 \in S(T)[L^2(\Omega)]$, there exists a control $f \in L^2(Q)$ such that the solution of (3.1) satisfies $u(T) = v^0$. In other words, $S(T)\left[L^2(\Omega)\right] \subset R(T; u^0)$ for all $u^0 \in L^2(\Omega)$.

To see this, given $u^0 \in L^2(\Omega)$ and $v^0 \in S(T)\left[L^2(\Omega)\right]$ there exists $w^0 \in L^2(\Omega)$ such that $v^0 = S(T)w^0$. We define $w = S(t)w^0$, the solution of (3.1) with $f = 0$ and initial data w^0. Then, the solution u of (3.1) can be decomposed as $u = w + v$ where

$$\begin{cases} v_t - \Delta v = f 1_\omega & \text{in} \quad Q \\ v = 0 & \text{on} \quad \Sigma \\ v(0) = u^0 - w^0. \end{cases} \tag{3.2}$$

Since system (3.1) is null controllable there exists $f \in L^2(Q)$ such the solution v of (3.2) satisfies $v(T) = 0$ and this is equivalent to $u(T) = w(T) = v^0$.

The set $S(T)[L^2(\Omega)]$ is dense in $L^2(\Omega)$ and therefore null controllability implies approximate controllability. Observe however that the reachable states we obtain by this argument are smooth due to the regularizing effect of the heat equation.

(d) As we have seen above, proving that null controllability implies approximate controllability requires the use of $S(T)[L^2(\Omega)]$ in $L^2(\Omega)$. In the case of the linear heat equation this can be easily seen developing solutions in Fourier series. However, if the equation contains time dependent coefficients this is no longer true. In these cases the density of the range of the semigroup, by duality, may be reduced to a backward uniqueness property in the spirit of J.-L. Lions and B. Malgrange [LM].

(e) The density of the range of the semigroup associated to semilinear heat equations has been studied by C. Fabre, J.P. Puel and E. Zuazua in [FPZ3] where it has been shown that density holds in various functional settings if the nonlinearity is globally Lipschitz. The proof of [FPZ3] is based on the fixed point method we present in Section 3.4 below. □

3.2. Approximate controllability System (3.1) is approximately controllable for any open, non-empty subset ω of Ω and $T > 0$. To see this one can apply Hahn-Banach's Theorem or use the variational approach developed in Section 2.2 for the wave equation. In both cases the approximate controllability is reduced to the unique continuation property of the adjoint system

$$\begin{cases} -\varphi_t - \Delta \varphi = 0 & \text{in} \quad Q \\ \varphi = 0 & \text{on} \quad \Sigma \\ \varphi(x, T) = \varphi^0(x) & \text{in} \quad \Omega. \end{cases} \tag{3.3}$$

More precisely, approximate controllability holds if and only if the following uniqueness property is true: *if* φ *solves* (3.3) *and* $\varphi = 0$ *in* $\omega \times (0, T)$ *then, necessarily,* $\varphi \equiv 0$, i.e. $\varphi^0 \equiv 0$.

This uniqueness property holds for every open non-empty subset ω of Ω and $T > 0$ by Holmgren's Uniqueness Theorem.

Following the variational approach of Section 2.2 the control can be constructed as follows. First of all we observe that it is sufficient to consider the particular case $u^0 \equiv 0$. Then, for any v^0 on $L^2(\Omega), \varepsilon > 0$ and E finite-dimensional subspace of $L^2(\Omega)$ we introduce the functional

$$J_\varepsilon(\varphi^0) = \frac{1}{2}\int_0^T\int_\omega \varphi^2 dxdt + \varepsilon\left\|(I - \pi_E)\varphi^0\right\|_{L^2(\Omega)} - \int_\Omega \varphi^0 v^0 dx \tag{3.4}$$

where π_E denotes the orthogonal projection from $L^2(\Omega)$ over E.

The functional J_ε is continuous and convex in $L^2(\Omega)$. On the other hand, in view of the unique continuation property above, one can prove that

$$\lim_{\|\varphi^0\|_{L^2(\Omega)}\to\infty} \frac{J_\varepsilon(\varphi^0)}{\| \varphi^0 \|_{L^2(\Omega)}} \geq \varepsilon \tag{3.5}$$

(we refer to [Z6] for the details of the proof).

Then, J_ε admits a unique minimizer $\widehat{\varphi}^0$ in $L^2(\Omega)$. The control $f = \widehat{\varphi}$ where $\widehat{\varphi}$ solves (3.3) with $\widehat{\varphi}^0$ as data is such that the solution u of (3.1) with $u^0 = 0$ satisfies

$$\| u(T) - u^0 \|_{L^2(\Omega)} \leq \varepsilon, \; \pi_E\left(u(T)\right) = \pi_E(v^0). \tag{3.6}$$

A slight change on the functional J_ε allows to build *bang-bang* controls. Indeed, we set

$$\widetilde{J}_\varepsilon(\varphi^0) = \frac{1}{2}\left(\int_0^T\int_\omega |\, \varphi \,|\, dxdt\right)^2 + \varepsilon \parallel (I - \pi_E)\varphi^0 \parallel_{L^2(\Omega)} - \int_\Omega v^0\varphi^0 dx. \tag{3.7}$$

The functional $\widetilde{J}_\varepsilon$ is continuous and convex in $L^2(\Omega)$ and satisfies the coercivity property (3.5) too.

Let $\widehat{\varphi}^0$ be the minimizer of $\widetilde{J}_\varepsilon$ in $L^2(\Omega)$ and $\widehat{\varphi}$ the corresponding solution of (3.3). Then, it can be seen that there exists

$$f \in \int_0^T\int_\omega |\, \widehat{\varphi} \,|\, dxdt\, \mathrm{sgn}(\widehat{\varphi}) \tag{3.8}$$

where sgn is the multivalued sign function: $\mathrm{sgn}(s) = 1$ if $s > 0$, $\mathrm{sng}(0) = [-1, 1]$ and $\mathrm{sng}(s) = -1$ when $s < 0$, such that the solution u of (3.1) with null initial data satisfies (3.6).

Due to the regularizing effect of the heat equation the zero set of non-trivial solutions of (3.3) is of zero $(n+1)$-dimensional Lebesgue measure. Thus, the control f of (3.8) is of *bang-bang* form, i.e. $f = \pm\lambda$ a.e. in Q where $\lambda = \int_0^T\int_\omega |\, \widehat{\varphi} \,|\, dxdt$.

We have proved the following:

Theorem 3.1 *Let ω be any open non-empty subset of Ω and $T > 0$ be any positive control time. Then, for any $u^0, v^0 \in L^2(\Omega)$, $\varepsilon > 0$ and E finite-dimensional subspace of $L^2(\Omega)$ there exists a bang-bang control $f \in L^2(Q)$ such that the solution of* (3.1) *satisfies* (3.6).

Remark 3.2 The control (3.8) obtained by minimizing $\widetilde{J}_\varepsilon$ is the one of minimal $L^\infty(Q)$-norm among the admissible ones (we refer to [FPZ2] for the details of the proof in the particular case where $E = \{0\}$). □

3.3. Null controllability The null controllability problem for system (3.1) is equivalent to the following observability inequality for the adjoint system (3.3):

$$\| \varphi(0) \|^2_{L^2(\Omega)} \leq C \int_0^T \int_\omega \varphi^2 dx dt, \quad \forall \varphi^0 \in L^2(\Omega). \tag{3.9}$$

Due to the regularizing effect of the heat equation the norm in the left-hand side of (3.9) is rather weak. However, due to the irreversibility of the system, (3.9) is not easy to prove. For instance, multiplier methods do not apply.

In [R1] the boundary null controllability of the heat equation was proved in one space dimension using moment problems and classical results on the linear independence in $L^2(0,T)$ of families of real exponentials.

Later on in [R2] a deep general result was proved. Roughly speaking it was shown that *if the wave equation is controllable for some $T > 0$ with controls supported in ω, then the heat equation* (3.1) *is null controllable for all $T > 0$ with controls supported in ω*. As a consequence of this, and in view of the controllability result presented in Section 2.3, it follows that the heat equation (3.1) is null controllable for all $T > 0$ provided ω satisfies the geometric control condition.

However, the geometric control condition does not seem to be natural at all in the context of the heat equation and therefore this result is not totally satisfactory.

More recently G. Lebeau and L. Robbiano [LeR] have proved that *the heat equation* (3.1) *is null controllable for every open, non-empty subset ω of Ω and $T > 0$*. This result shows, as expected, that the geometric control condition is unnecessary in the context of the heat equation.

A slightly simplified proof of this result was given in [LeZ] where the linear system of thermoelasticity was addressed. Let us briefly describe this proof. The main ingredient is an observability estimate for the eigenfunctions of the Laplace operator:

$$\begin{cases} -\Delta\psi_j = \lambda_j \psi_j & \text{in} \quad \Omega \\ \psi_j = 0 & \text{on} \quad \partial\Omega. \end{cases} \tag{3.10}$$

Recall that the eigenvalues $\{\lambda_j\}$ form an increasing sequence of positive numbers such that $\lambda_j \to \infty$ as $j \to \infty$ and that the eigenfunctions $\{\psi_j\}$ may be chosen such that they form an orthonormal basis of $L^2(\Omega)$.

The following holds:

Theorem 3.2 *For any open subset ω of Ω there exists positive constants $C_1, C_2 > 0$ such that*

$$\int_\omega \left| \sum_{\lambda_j \le \mu} a_j \psi_j(x) \right|^2 dx \ge C_1 e^{-C_2\sqrt{\mu}} \sum_{\lambda_j \le \mu} | a_j |^2 \tag{3.11}$$

for all $\{a_j\} \in \ell^2$ and for all $\mu > 0$.

This result was implicitly used in [LeR] and is proved in [LeZ] by means of Carleman's inequalities.

As a consequence of (3.11) one can prove that the observability inequality (3.9) holds for solutions of (3.3) with initial data in $E_\mu = \text{span}\, \{\varphi_j\}_{\lambda_j \le \mu}$, the constant being of the order of $\exp\left(C\sqrt{\mu}\right)$. This shows that the projection of solutions over E_μ can be controlled to zero with a control of size $\exp\left(C\sqrt{\mu}\right)$. Thus, when controlling the frequencies $\lambda_j \le \mu$ one increases the $L^2(\Omega)$-norm of the high frequencies $\lambda_j > \mu$ by a multiplicative factor of the order of $\exp\left(C\sqrt{\mu}\right)$. However, as it was observed in [LeR], solutions of the heat equation (3.1) without control ($f = 0$) and such that the projection of the initial data over E_μ vanishes, decay in $L^2(\Omega)$ at a rate of the order of $\exp(-\mu t)$.

Thus, if we divide the time interval $[0, T]$ in two parts $[0, T/2]$ and $[T/2, T]$, we control to zero the frequencies $\lambda_j \le \mu$ in the interval $[0, T/2]$ and then allow the equation to evolve without control in the interval $[T/2, T]$, it follows that at time $t = T$ the projection of the solution u over E_μ vanishes and the norm of the high frequencies does not exceed the norm of the initial data u^0.

This argument allows the projection over E_μ for any $\mu > 0$ to be controlled to zero but not the whole solution. To do that an iterative argument is needed. We decompose the interval $[0, T)$ into disjoint subintervals of the form $[T_j, T_{j+1})$ for $j \in \mathbb{N}$ with a suitable choice of the sequence $\{T_j\}$. In each interval $[T_j, T_{j+1}]$ we control to zero the frequencies $\lambda_k \le 2j$. By letting $j \to \infty$ we obtain a control $f \in L^2(Q)$ such that the solution of (3.1) satisfies $u(T) \equiv 0$.

Remark 3.3

(a) Observe that the null controllability has not been obtained as a consequence of the observability inequality (3.9). Actually, (3.9) is derived *a fortiori* as a consequence of the null controllability property.

(b) Once (3.9) is known to hold one can obtain the control with minimal $L^2(Q)$-norm among the admissible ones. To do that it is sufficient to minimize the functional

$$J(\varphi^0) = \frac{1}{2} \int_0^T \int_\omega \varphi^2 dx dt + \int_\Omega \varphi(0) u^0 dx \tag{3.12}$$

over the Hilbert space $H = \{\varphi^0 : \| \varphi^0 \|_H^2 = \int_0^T \int_\omega \varphi^2 dx dt < \infty\}$.

Observe that J is continuous and convex in H. On the other hand (3.9) guarantees the coercivity of J and the existence of its minimizer.

(c) Once we know that (3.9) holds for any open non-empty subset ω and $T > 0$, by the regularizing effect of the heat equation we deduce that for all $k \in \mathbb{N}$ and $\delta > 0$ there exists $C_k > 0$ such that

$$\| \varphi(0) \|^2_{H^k(\Omega)} \leq C_k \int_\delta^{T-\delta} \int_\omega \varphi^2 dx dt. \tag{3.13}$$

This allows it to be proved that the null controllability may be achieved by means of C^∞-controls.

(d) As a consequence of the internal null controllability of the heat equation one can easily deduce the null boundary controllability with controls in an arbitrarily small open subset of the boundary. Indeed, let Γ_0 be an open and non-empty subset of $\Gamma = \partial\Omega$. We can extend Ω to a larger open connected set $\widetilde{\Omega}$ such that $\widetilde{\Omega} = \Omega \cup \Gamma_0 \cup \omega$ where ω is a non-empty open set of $\mathbb{R}^n \backslash \Omega$. Given $u^0 \in L^2(\Omega)$ we extend it to $\widetilde{\Omega}$ by setting $\widetilde{u}^0 = 0$ in $\widetilde{\Omega} \backslash \Omega$. Then one can control the initial data $\widetilde{u}^0$ for the heat equation in the cylinder $\widetilde{Q} = \widetilde{\Omega} \times (0,T)$, with control f supported in $\omega \times (0,T)$. The restriction of the controlled solution to $Q = \Omega \times (0,T)$ satisfies

$$\begin{cases} u_t - \Delta u = 0 & \text{in} \quad Q \\ u = 0 & \text{on} \quad \Sigma_1 = (\Gamma \backslash \Gamma_0) \times (0,T) \\ u = v & \text{on} \quad \Sigma_0 = \Gamma_0 \times (0,T) \\ u(x,0) = u^0(x) & \text{in} \quad \Omega \\ u(x,T) = 0 & \text{in} \quad \Omega \end{cases} \tag{3.14}$$

with control v, the restriction of $\widetilde{u}$ to $\Sigma_0 = \Gamma_0 \times (0,T)$.

As a consequence of the boundary null controllability we deduce the following boundary observability property:

$$\| \varphi(0) \|^2_{L^2(\Omega)} \leq C \int_0^T \int_{\Gamma_0} \left| \frac{\partial \varphi}{\partial n} \right|^2 d\sigma dt. \tag{3.15}$$

This observability property (3.15) was derived in [LeR] by using Carleman's inequalities and the iterative argument described above to obtain the null controllability.

(e) The method of proof of the null controllability we have described is based on the possibility of developing solutions in Fourier series. Thus it can be applied in a more general class of heat equations with variable time-independent coefficients. The methods of [R2] apply to variable coefficient equations as well but are also limited to time independent coefficients.

(f) The null controllability of the heat equation with lower order time-dependent terms of the form

$$\begin{cases} u_t - \Delta u + a(x,t)u = f 1_\omega & \text{in} \quad Q \\ u = 0 & \text{on} \quad \Sigma \\ u(x,0) = u^0(x) & \text{in} \quad \Omega \end{cases} \tag{3.16}$$

has been studied by A. Fursikov and O.Yu. Imanuvilov (see for instance [CIK], [F], [FI1,2] and [I]). Their approach is different from the one we have presented here. As a consequence of their null controllability results it follows that an observability inequality of the form (3.15) holds for the solution of the adjoint system

$$\begin{cases} -\varphi_t - \Delta\varphi + a(x,t)\varphi = 0 & \text{in} \quad Q \\ \varphi = 0 & \text{on} \quad \Sigma \\ \varphi(x,T) = \varphi^0(x) & \text{in} \quad \Omega \end{cases} \tag{3.17}$$

when Γ_0 is any open, non-empty subset of Γ.

As a consequence of this and proceeding as in [L1], vol. 1, (where it is shown that, in general, boundary observability implies internal observability when ω is a neighborhood of the boundary) one can deduce that (3.9) holds when ω is any open and non-empty subset of Ω.

Thus, system (3.16) is null-controllable for any $a \in L^\infty(Q)$ when ω is any open and non-empty subset of Ω and for any $T > 0$.

(g) More recently, E. Fernández-Cara [FC], following the methods of Fursikov and Imanuvilov, has shown that an observability inequality of the form (3.15) holds for solutions of system (3.17) with a constant C of the order of $\exp(C \parallel a \parallel_\infty)$.

(h) We refer to J.-L. Lions and E. Zuazua [LZ2] for various applications of inequalities of the form (3.11) to the analysis of the cost of controlling parabolic equations in the presence of unstabilities. More precisely in [LZ2] the behavior of the controls for systems of the form

$$u_t - \Delta u - k(-\Delta)^\theta u = 0$$

as $k \to \infty$ is analyzed, θ being a fixed positive constant $0 < \theta < 1$. □

3.4. The semilinear heat equation Let us consider the following semilinear heat equation

$$\begin{cases} u_t - \Delta u + h(u) = f 1_\omega & \text{in} \quad Q \\ u = 0 & \text{on} \quad \Sigma \\ u(x,0) = u^0(x) & \text{in} \quad \Omega \end{cases} \tag{3.18}$$

where $h \in C^1(\mathbb{R};\mathbb{R})$ and satisfies

$$|h(s_1) - h(s_2) - h'(0)\,(s_1 - s_2)| \le C\left(| \, s_1 \,|^{p-1} + | \, s_2 \,|^{p-1}\right) | \, s_1 - s_2 \,|, \; \forall s_1, s_2 \in \mathbb{R} \tag{3.19}$$

for some $C > 0$ with $1 < p \le (n+4)/n$.

Under assumption (3.19) system (3.18) is locally well-posed for $u^0 \in L^2(\Omega)$ and $f \in L^2(Q)$. More precisely there exists $\delta > 0$ such that system (3.18) has an unique solution $u \in C\left([0,T]; L^2(\Omega)\right)$ for any $(u^0, f) \in L^2(\Omega) \times L^2(Q)$ such that

$$\parallel u^0 \parallel_{L^2(\Omega)} + \parallel f \parallel_{L^2(Q)} \le \delta. \tag{3.20}$$

On the other hand, when h is globally Lipschitz, i.e. when (3.19) holds with $p = 1$ system (3.18) is globally well-posed and an unique solution exists for any initial data $u^0 \in L^2(\Omega)$ and control $f \in L^2(Q)$.

The following was proved in [Z6]:

Theorem 3.3 *Assume that h is globally Lipschitz. Let ω be any open non-empty subset Ω and $T > 0$. Then, for any $u^0, v^0 \in L^2(\Omega), \varepsilon > 0$ and E finite-dimensional subspace of $L^2(\Omega)$ there exists $f \in L^2(Q)$ such that the solution u of* (3.18) *satisfies* (3.6).

Remark 3.4

(a) This result was proved in [FPZ1] in the context of approximate controllability, i.e. when $E = \{0\}$. The method of proof is rather similar in both cases. As we will see it is based on the variational approach to approximate controllability described in Section 3.2 and a fixed point argument that was introduced in [Z4] to prove the exact controllability of the semilinear wave equation.

(b) The method of proof of Theorem 3.3 does not apply when the nonlinearity grows at infinity in a superlinear way, i.e. when (3.19) holds with $p > 1$. This is due to the fact that the method of proof does not provide any estimate on the control in terms of the Lipschitz constant of the nonlinearity h. □

Sketch of the proof of Theorem 3.3
Let us fix $u^0, v^0 \in L^2(\Omega), \varepsilon > 0$ and the finite-dimensional subspace E of $L^2(\Omega)$.

For simplicity we assume that $u^0 = 0$ (notice that since the system under consideration is not linear, the problem cannot be reduced to the particular case $u^0 = 0$. However the difficulties are similar in all cases). We also assume that $h(0) = 0$.

We now observe that the heat equation in (3.18) can be written as follows:

$$u_t - \Delta u + g(u)u = f 1_\omega \tag{3.21}$$

where

$$g(s) = h(s)/s \text{ if } s \neq 0;\ g(0) = h'(0). \tag{3.22}$$

Since h is globally Lipschitz the function g is bounded and

$$\| g \|_{L^\infty(\mathbb{R})} \leq \| h' \|_{L^\infty(\mathbb{R})} . \tag{3.23}$$

For any $w \in L^2(Q)$ we consider the linear system

$$\begin{cases} u_t - \Delta u + g(w)u = f 1_\omega & \text{in} \quad Q \\ u = 0 & \text{on} \quad \Sigma \\ u(x,0) = 0 & \text{in} \quad \Omega. \end{cases} \tag{3.24}$$

Notice that the potential $g(w) \in L^\infty(Q)$ and that

$$\| g(w) \|_{L^\infty(Q)} \leq \| g \|_{L^\infty(\mathbb{R})}, \; \forall w \in L^2(Q). \tag{3.25}$$

We look for $f \in L^2(Q)$ such that the solution of (3.24) satisfies (3.6). Of course the control f will depend on w.

To construct this control we introduce the functional

$$J_{\varepsilon,w}(\varphi^0) = \frac{1}{2}\int_0^T \int_\omega \varphi^2 dxdt + \varepsilon \| (I-\pi_E)\varphi^0 \|_{L^2(\Omega)} - \int_\Omega \varphi^0 v^0 dx \tag{3.26}$$

where $\varphi = \varphi(x,t)$ solves the adjoint equation

$$\begin{cases} -\varphi_t - \Delta\varphi + g(w)\varphi = 0 & \text{in} \quad Q \\ \varphi = 0 & \text{on} \quad \Sigma \\ \varphi(x,T) = \varphi^0(x) & \text{in} \quad \Omega. \end{cases} \tag{3.27}$$

The functional $J_{\varepsilon,w}$ is convex and continuous in $L^2(\Omega)$. On the other hand the unique continuation property holds for solutions of (3.27), i.e. *if* φ *solves* (3.27) *and* $\varphi = 0$ *in* $\omega \times (0,T)$, *then* $\varphi \equiv 0$ *and* $\varphi^0 \equiv 0$ (see [SS]). Then $J_{\varepsilon,w}$ is coercive in $L^2(\Omega)$ and satisfies (3.5). Thus, $J_{\varepsilon,w}$ admits a unique minimizer $\widehat{\varphi}^0$ in $L^2(\Omega)$. The control $f = \widehat{\varphi}$ where $\widehat{\varphi}$ solves (3.27) with this data $\widehat{\varphi}^0$, is such that the solution of (3.24) satisfies (3.6).

We have constructed a map $N : L^2(Q) \to L^2(Q)$ such that $N(w) = u$, i.e. to each $w \in L^2(Q)$ we associate the solution u of (3.24) satisfying (3.6) with the control f obtained by minimizing the functional $J_{\varepsilon,w}$. It is easy to see that N is continuous and compact.

On the other hand, it can be proved that the functionals $J_{\varepsilon,w}$ are uniformly coercive and therefore the minimizers $\widehat{\varphi}^0$ are uniformly bounded in $L^2(\Omega)$ for all $w \in L^2(Q)$. This is due to the fact that the potentials $g(w)$ are uniformly bounded in $L^\infty(Q)$. Then, the controls f are uniformly bounded in $L^2(Q)$ and the solutions u of (3.24) too. This shows that the range of N is bounded in $L^2(Q)$.

Thus, we can apply Schauder's fixed point Theorem to N and deduce that it admits a fixed point u. This fixed point solves (3.18) with $u^0 \equiv 0$ and, by construction, satisfies (3.6). □

Remark 3.5 In the proof of Theorem 3.3 we can consider the functional $\widetilde{J}_\varepsilon$ as in (3.7) and deduce that the control f can be constructed to be of the form (3.8) with $\widehat{\varphi}$ solution of (3.27) for some $w \in L^2(Q)$. However to our knowledge it is not known whether the zero set of solutions of (3.27) is of zero $(n+1)$-dimensional Lebesgue measure. □

As we said above the method of proof of Theorem 3.3 does not apply when $p > 1$. However one can apply the Inverse Function Theorem if the control requirement (3.6) is relaxed to

$$\pi_E(u(T)) = \pi_E(v^0), \tag{3.28}$$

i.e. if we only control finite-dimensional projections. Indeed, in this setting one can prove that for any ω non-empty open subset of $\Omega, T > 0$ and finite-dimensional subspace E of $L^2(\Omega)$ there exists $\delta > 0$ such that for any $u^0, v^0 \in L^2(\Omega)$ with

$$\| u^0 \|_{L^2(\Omega)} + \| v^0 \|_{L^2(\Omega)} \leq \delta \tag{3.29}$$

there exists a control $f \in L^2(Q)$ such that the solution of (3.18) satisfies (3.28).

When the non-linearity has the "good-sign", for instance, when dealing with

$$\begin{cases} u_t - \Delta u + \mid u \mid^{p-1} u = f 1_\omega & \text{in} \quad Q \\ u = 0 & \text{on} \quad \Sigma \\ u(x,0) = u^0(x) & \text{in} \quad \Omega \end{cases} \tag{3.30}$$

and taking into account that solutions of (3.30) without control ($f = 0$) decay uniformly in $L^2(\Omega)$ as t increases one deduces that *if* $T > 0$ *is large enough for any* $u^0 \in L^2(\Omega)$ *there exists* $f \in L^2(Q)$ *such that the solution of* (3.30) *satisfies*

$$\pi_E(u(T)) = 0. \tag{3.31}$$

Remark 3.6

(a) This method does not allow us to say anything about the controllability of the projection over $E^\perp$, the orthogonal complement of E.

(b) To our knowledge nothing is known about the controllability of blowing-up solutions. For instance let us consider the pairs $\{u, f\}$ such that

$$u_t - \Delta u - \mid u \mid^{p-1} u = f 1_\omega, \quad \text{in} \quad \Omega \times (0,T)$$

with $p > 1$, $f \in L^2(Q)$, satisfying an initial condition $u(0) = u^0 \in L^2(\Omega)$ and homogeneous Dirichlet boundary conditions.

Observe that for some choices of u^0, the solution of this problem for $f \equiv 0$ may blow-up in time $t \leq T$. Thus, the existence of admissible pairs $\{u, f\}$ is not always obvious.

One of the problems we may consider in this setting is whether the set of $\{u(T)\}$ where $\{u, f\}$ runs over the set of admissible pairs is dense in $L^2(\Omega)$ or not.

As we mentioned in Section 2.4, in the context of the wave equation, due to the finite speed of propagation blowing up solutions may not be avoided, the trivial case in which $\omega = \Omega$ being excepted and therefore, for some choices of the initial data, the set of admissible pairs $\{u, f\}$ is empty. However, the possibility of avoiding the blow up for a suitable choice of f may not be excluded in the frame of the heat equation. □

Recently some local null controllability results have been proved. Lin Guo and Littman in [LiL] by means of a nonlinear version of Cauchy Kovalevski's Theorem have proved that, in one space dimension and when the control acts on one extreme of the boundary, null controllability holds for bounded continuous and sufficiently small initial data provided the nonlinearity is analytic and belongs to Gevrey class 2. On the other hand, Fursikov and Imanuvilov in [FI1] have proved the local null controllability for the semilinear heat equation (3.18) by using Carleman's inequalities. In both cases, when the nonlinearity has the good sign as in (3.30) it can be shown that every initial data u^0 can be driven to zero in a sufficiently large time. More recently E. Fernández-Cara [FC] has shown that the semilinear heat equation (3.18) is null controllable in a uniform time T for any $T > 0$, provided

$$| h(s) | / | s | \log(| s |) \to 0, \qquad \text{as} \quad |s| \to \infty. \tag{3.32}$$

This is the analogue of the exact controllability result for the semilinear wave equation of Theorem 2.3. Observe that (3.32) is an almost sharp condition to avoid blow-up phenomena. Indeed, when $h(s) \sim -s log^p(|s|)$ as $|s| \to \infty$ with $p > 1$, blow-up may occur. The method of proof of [FC] combines a fixed point argument and the explicit observability estimates mentioned in Remark 3.3, (g).

To our knowledge the only controllability result for a non-linear system with superlinear nonlinearity that holds for all initial data in an uniform time is the one due to J.-M. Coron [Co2] in the context of Navier-Stokes equations in two space dimensions. The results in [Co2] state the approximate controllability of all data in an arbitrarily small control time and are obtained through an approximation argument as a consequence of the exact controllability of Euler equations in the class of smooth data proved in [Co1]. The approach by J.-M. Coron is completely different to the classical one that views the nonlinear control problem as a perturbation of the linear one. In [Co2] the Navier-Stokes equations are viewed as a perturbation of Euler equations and the methods employed are genuinely nonlinear.

4. The linear system of thermoelasticity

4.1. Problem formulation Let Ω be a bounded domain of $\mathbb{R}^3$ with boundary Γ and let ω be an open non-empty subset of Ω.

Let us consider the system

$$\begin{cases} u_{tt} - \mu\Delta u - (\lambda+\mu)\nabla \operatorname{div} u + \alpha\nabla\theta = f1_\omega & \text{in} \quad Q \\ \theta_t - \Delta\theta + \beta \operatorname{div} u_t = 0 & \text{in} \quad Q \\ u = 0\,,\, \theta = 0 & \text{on} \quad \Sigma \\ u(x,0) = u^0(x), u_t(x,0) = u^1(x) & \text{in} \quad \Omega \\ \theta(x,0) = \theta^0(x) & \text{in} \quad \Omega. \end{cases} \tag{4.1}$$

In (4.1), $u = (u_1, u_2, u_3)$ describes the displacement and θ the temperature. λ and μ are Lamé's constants and, as well as the coupling constants α and β, are assumed to be positive. The control $f = (f_1, f_2, f_3)$ acts on the system as a volume force.

Notice that we have not introduced any control in the heat equation. Thus, we try to control both displacement and temperature by means of the sole control f. Of course, this requires the system to be strongly coupled.

System (4.1) combines the hyperbolic nature of Lamé's system in linear elasticity and the parabolic character of the heat equation.

System (4.1) is well-posed for $\left(u^0, u^1, \theta^0\right) \in H = \left(H_0^1(\Omega)\right)^3 \times \left(L^2(\Omega)\right)^3 \times L^2(\Omega)$ and $f \in \left(L^2(Q)\right)^3$ and admits a unique solution in the class $(u, u_t, \theta) \in C\left([0,T]; H\right)$.

The different notions of controllability make sense for this system too. However, the exact controllability has to be excluded by the irreversible character of the system. Approximate controllability can easily be deduced when T is large enough as a consequence of Holmgren's Uniqueness Theorem.

Other notions of controllability can also be introduced. For instance, Lagnese and Lions in [LaL] proved *partial controllability* results for various systems of thermoelasticity. System (4.1) is said to be partially controllable if for every $\left(u^0, u^1, \theta^0\right) \in H$ there exists $f \in \left(L^2(Q)\right)^3$ such that the solution of (4.1) satisfies

$$u(x,T) = u_t(x,T) = 0 \text{ in } \Omega.$$

In other words, partial controllability guarantees that the displacement u can be driven to zero but nothing is said about the temperature.

In [LaL] it was shown that when ω and T are such that Lamé's system is exactly controllable then system (4.1) is partially controllable when α and β are sufficiently small.

More recently in [Z2] the following notion of *exact-approximate controllability* was introduced: system (4.1) is said to be exact-approximately controllable if for any $\left(u^0, u^1, \theta^0\right)$, $\left(v^0, v^1, \eta^0\right) \in H$ and $\varepsilon > 0$ there exists $f \in \left(L^2(Q)\right)^3$ such that the solution of (4.1) satisfies

$$\left\{ \begin{array}{l} u(x,T) = v^0(x),\ u_t(x,T) = v^1(x) \quad \text{in} \quad \Omega \\ \| \theta(T) - \eta^0 \|_{L^2(\Omega)} \leq \varepsilon. \end{array} \right. \tag{4.2}$$

Observe that in (4.2) the displacement is required to reach exactly $\left(v^0, v^1\right)$ while the temperature only reaches the target in an approximate way.

Clearly exact-approximate controllability is a stronger notion than partial controllability. On the other hand one expects null controllability to be stronger than exact-approximate controllability. However to prove this fact one needs the backward uniqueness for the system of thermoelasticity (4.1) with $f = 0$ and, to our knowledge, this is an open problem.

In Section 4.2 we will briefly describe the results from [Z2] on exact-approximate controllability. In Section 4.3 we mention the results from [LeZ] on null controllability that apply when an extra control is added in the heat equation or when the Dirichlet boundary conditions are replaced by periodic ones.

One expects the system (4.1) to be null controllable when ω and T are such that the Lamé system is exactly controllable. However this is by now an open problem.

We will not describe here some other relevant results on the controllability of systems in thermoelasticity. For instance the work by S. Hansen [H] shows that the system of thermoelasticity in one space dimension is null boundary controllable. The methods of [H] are different from ours and are based on the reduction of the control problem to a moment problem and the theory of nonharmonic Fourier series.

4.2. Exact-approximate controllability The following result was proved in [Z2]:

Theorem 4.1 *Let Ω be a bounded domain of $\mathbb{R}^3$ with boundary Γ of class C^2. Assume that ω is a neighborhood of Γ in Ω and $T > \operatorname{diam}(\Omega\backslash\omega)/\sqrt{\mu}$.*
Then, system (4.1) is exact-approximately controllable, i.e. *for every $\left(u^0, u^1, \theta^0\right)$, $\left(v^0, v^1, \eta^0\right) \in H$ and $\varepsilon > 0$ there exists $f \in \left(L^2(Q)\right)^3$ such that the solution of (4.1) satisfies (4.2)*

In [Z2] this result was reduced to a suitable observability inequality for the adjoint system:

$$\begin{cases} \varphi_{tt} - \mu\Delta\varphi - (\lambda+\mu)\nabla \operatorname{div}\varphi + \beta\nabla\psi_t = 0 & \text{in} \quad Q \\ -\psi_t - \Delta\psi - \alpha \operatorname{div}\varphi = 0 & \text{in} \quad Q \\ \varphi = 0, \psi = 0 & \text{on} \quad \Sigma \\ \varphi(x,T) = \varphi^0(x), \varphi_t(x,T) = \varphi^1(x) & \text{in} \quad \Omega \\ \psi(x,T) = \psi^0(x) & \text{in} \quad \Omega. \end{cases} \tag{4.3}$$

More precisely it was shown that Theorem 4.1 is a consequence of the following two results:

Theorem 4.2 *Under the assumptions of Theorem 4.1, for every bounded set B of $L^2(\Omega)$ there exists $\delta = \delta(B) > 0$ such that*

$$\delta \leq \int_0^T \int_\omega |\, \varphi \,|^2 \, dxdt \tag{4.4}$$

holds for every solution of (4.3) with initial data such that

$$\|\, (\varphi^0, \varphi^1 + \beta\nabla\psi^0) \,\|_{(L^2(\Omega))^3 \times (H^{-1}(\Omega))^3} \geq 1, \psi^0 \in B. \tag{4.5}$$

Theorem 4.3 *Under the assumptions of Theorem 4.1, if $\varphi = 0$ in $\omega \times (0,T)$ then, necessarily, $\varphi^0 \equiv \varphi^1 \equiv 0$ and $\psi^0 \equiv 0$ in Ω.*

Remark 4.1

(a) The partial controllability result of [LaL] mentioned in Section 4.1 is a consequence of the observability estimate (4.4) in the particular case $B = \{0\}$. In this case, assuming that $\alpha\beta$ is sufficiently small, (4.4) can be derived as a consequence of the observability of the linear system of elasticity.

Indeed, under the assumptions of Theorem 4.1 it follows that

$$\|\, (\varphi^0, \varphi^1) \,\|^2_{(L^2(\Omega))^3 \times (H^{-1}(\Omega))^3} \leq C \int_0^T \int_\omega |\, \varphi \,|^2 \, dxdt \tag{4.6}$$

for every solution of

$$\begin{cases} \varphi_{tt} - \mu\Delta\varphi - (\lambda+\mu)\nabla \operatorname{div}\varphi = 0 & \text{in} \quad Q \\ \varphi = 0 & \text{on} \quad \Sigma \\ \varphi(x,T) = \varphi^0(x), \varphi_t(x,T) = \varphi^1(x) & \text{in} \quad \Omega. \end{cases} \tag{4.7}$$

From (4.6) it follows that (4.4) holds when $B = \{0\}$ and $\alpha\beta$ is small enough.

(b) In Theorem 4.2 B is any bounded set and no restriction on the product of the coupling parameters $\alpha\beta$ is imposed. Therefore, Theorem 4.2 cannot be derived by a perturbation argument from the observability inequality (4.6) in Lamé's system (4.7).

(c) Theorem 4.2 provides an observability estimate that is uniform provided ψ^0 remains a priori bounded. But it does not say anything about the behavior of the constant δ in (4.4) as $\| \psi^0 \|_{L^2(\Omega)} \to \infty$.

Observe however that, due to the irreversibility of system (4.3), an estimate of the form

$$\| \left(\varphi^0, \varphi^1 + \beta \nabla \psi^0, \psi^0\right) \|^2_{(L^2(\Omega))^3 \times (H^{-1}(\Omega))^3 \times L^2(\Omega)} \leq C \int_0^T \int_\omega | \varphi |^2 \, dx dt$$

may not hold for all $\left(\varphi^0, \varphi^1, \psi^0\right)$ with a uniform constant C.

However the following estimate is very probably true

$$\|(\varphi(0), \varphi_t + \beta \nabla \psi(0), \psi(0))\|^2_{(L^2(\Omega))^3 \times (H^{-1}(\Omega))^3 \times L^2(\Omega)} \leq C \int_0^T \int_\omega |\varphi|^2 dx dt \quad (4.8)$$

and is equivalent to the null controllability of the system.

(d) Theorem 4.3 is a consequence of Homgren's Uniqueness Theorem and guarantees the approximate controllability of the system. □

Now, assuming that Theorem 4.2 and 4.3 hold, let us briefly sketch the proof of Theorem 4.1.

First of all we observe that the problem can be reduced to the case where $u^0 \equiv u^1 \equiv 0$ and $\theta^0 \equiv 0$. Then, given $\left(v^0, v^1, \eta^0\right) \in H$ and $\varepsilon > 0$ we introduce the functional

$$J_\varepsilon \left(\varphi^0, \varphi^1, \psi^0\right) = \frac{1}{2} \int_0^T \int_\omega | \varphi |^2 \, dx dt + \varepsilon \| \psi^0 \|_{L^2(\Omega)} -$$
$$- \int_\Omega v^1 \cdot \varphi^0 dx + \langle v^0, \varphi^1 \rangle - \int_\Omega \left(\eta^0 + \beta \operatorname{div} v^0\right) \psi^0 dx \quad (4.9)$$

defined for $\left(\varphi^0, \varphi^1, \psi^0\right) \in H' = \left(L^2(\Omega)\right)^3 \times \left(H^{-1}(\Omega)\right)^3 \times L^2(\Omega)$ where $\cdot$ denotes the scalar product in $\mathbb{R}^3$ and $\langle \cdot, \cdot \rangle$ the duality pairing between $\left(H_0^1(\Omega)\right)^3$ and $\left(H^{-1}(\Omega)\right)^3$.

In view of Theorems 4.2 and 4.3 one can prove the following coercivity property:

$$\lim_{\|(\varphi^0, \varphi^1 + \beta \nabla \psi^0, \psi^0)\|_{H'} \to \infty} \frac{J_\varepsilon \left(\varphi^0, \varphi^1, \psi^0\right)}{\| \left(\varphi^0, \varphi^1 + \beta \nabla \psi^0, \psi^0\right) \|_{H'}} \geq \varepsilon. \quad (4.10)$$

Since J_ε is continuous and convex in H', from (4.10) it follows that J_ε has a unique minimizer in H' that we denote by $\left(\widehat{\varphi}^0, \widehat{\varphi}^1, \widehat{\psi}^0\right)$. The control $f = \widehat{\varphi}$ where

$\left(\widehat{\varphi}, \widehat{\psi}\right)$ solves (4.3) with the minimizer as initial data is such that the solution of (4.1) with null initial data satisfies (4.2).

Remark 4.2 It is easy to see that Theorem 4.1 can be improved to guarantee that the temperature reaches exactly a finite number of constraints as well. In other words, if E is a finite-dimensional subspace of $L^2(\Omega)$ the control f can be chosen such that the solution u of (4.1) satisfies

$$\begin{cases} u(x,T) = u^0(x),\ u_t(x,T) = u^1(x) & \text{in} \quad \Omega \\ \| \theta(T) - \eta^0 \|_{L^2(\Omega)} \leq \varepsilon,\ \pi_E\left(\theta(T)\right) = \pi_E(\eta^0). & \end{cases} \tag{4.11}$$

To see this it is sufficient to replace the functional J_ε above by

$$\widetilde{J}_\varepsilon\left(\varphi^0, \varphi^1, \psi^0\right) = \frac{1}{2}\int_0^T \int_\omega |\, \varphi\,|^2\, dxdt + \varepsilon \,\|\, (I - \pi_E)\psi^0 \,\|_{L^2(\Omega)}$$

$$- \int_\Omega v^1 \cdot \varphi^0 dx + < v^0, \varphi^1 > - \int_\Omega \left(\eta^0 + \beta \operatorname{div} v^0\right) \psi^0 dx. \qquad \square$$

As we said above Theorem 4.3 can be proved by means of Holmgren's Uniqueness Theorem. Thus, let us briefly indicate the main steps in the proof of Theorem 4.2.

Step 1. First of all we introduce the new variable

$$\phi(x,t) = -\int_t^T \varphi(x,s)ds + \phi^0(x)$$

with

$$-\mu\Delta\phi^0 - (\lambda+\mu)\nabla \operatorname{div}\phi^0 = -\varphi^1 - \beta\nabla\psi^0 \ \text{ in } \Omega; \phi^0 \in \left(H_0^1(\Omega)\right)^3$$

and observe that (ϕ, ψ) satisfy

$$\begin{cases} \phi_{tt} - \mu\Delta\phi - (\lambda+\mu)\nabla \operatorname{div}\phi + \beta\nabla\psi = 0 & \text{in} \quad Q \\ -\psi_t - \Delta\psi - \alpha \operatorname{div}\phi_t = 0 & \text{in} \quad Q \\ \phi = 0, \psi = 0 & \text{on} \quad \Sigma \\ \phi(x,T) = \phi^0(x), \phi_t(x,T) = \phi^1(x), \psi(x,T) = \psi^0(x) & \text{in} \quad \Omega. \end{cases} \tag{4.12}$$

The problem is reduced to proving that for every bounded set B of $L^2(\Omega)$ there exists $\delta = \delta(B) > 0$ such that

$$\delta \leq \int_0^T \int_\omega |\, \phi_t \,|^2\, dxdt \tag{4.13}$$

for every solution of (4.12) with data

$$\|\, \left(\phi^0, \phi^1\right) \|_{\left(H_0^1(\Omega)\right)^3 \times (L^2(\Omega))^3} \geq 1,\ \psi^0 \in B. \tag{4.14}$$

Step 2. Following [HLP] we introduce the decoupled system

$$\begin{cases} \widetilde{\phi}_{tt} - \mu\Delta\widetilde{\phi} - (\lambda+\mu)\nabla \operatorname{div} \widetilde{\phi} - \alpha\beta P\widetilde{\phi}_t = 0 & \text{in} \quad Q \\ -\widetilde{\psi}_t - \Delta\widetilde{\psi} - \alpha \operatorname{div} \widetilde{\phi}_t = 0 & \text{in} \quad Q \\ \widetilde{\phi} = 0, \widetilde{\psi} = 0 & \text{on} \quad \Sigma \\ \widetilde{\phi}(x,T) = \phi^0(x), \widetilde{\phi}_t(x,T) = \phi^1(x), \widetilde{\psi}(x,T) = \psi^0(x) & \text{in} \quad \Omega \end{cases} \tag{4.15}$$

where P is the orthogonal projection from $\left(L^2(\Omega)\right)^3$ into the closed subspace

$$\left\{\nabla\varphi : \varphi \in H_0^1(\Omega)\right\}.$$

Both system (4.12) and (4.15) generate a continuous semigroup in H that we denote by $S(t)$ and $\widetilde{S}(t)$ respectively.

From [HLP] we know that $S - \widetilde{S} : H \to C([0,T]; H)$ is a compact map.

Step 3. In view of the compactness of the difference between the two semigroups S and $\widetilde{S}$ it is sufficient to prove the observability for the hyperbolic system that $\widetilde{\phi}$ satisfies in (4.15).

More precisely, let us consider the system

$$\begin{cases} \widetilde{\phi}_{tt} - \mu\Delta\widetilde{\phi} - (\lambda+\mu)\nabla \operatorname{div} \widetilde{\phi} - \alpha\beta P\widetilde{\phi}_t = 0 & \text{in} \quad Q \\ \widetilde{\phi} = 0 & \text{on} \quad \Sigma \\ \widetilde{\phi}(T) = \phi^0, \widetilde{\phi}_t(T) = \phi^1 & \text{in} \quad \Omega, \end{cases} \tag{4.16}$$

and let us assume that the following observability inequality holds:

$$\| \left(\phi^0, \phi^1\right) \|^2_{\left(H_0^1(\Omega)\right)^3 \times (L^2(\Omega))^3} \leq C \int_0^T \int_\omega | \widetilde{\phi}_t |^2 \, dxdt. \tag{4.17}$$

Then, by a classical compactness-uniqueness argument (see for instance [L2], vol. 1, Appendix I) in view of the compactness of $S - \widetilde{S}$ and the uniqueness result of Theorem 4.3 one can deduce that (4.13) holds for solutions of (4.12) with initial data in (4.14).

Step 4. In [Z2] using multiplier techniques the observability inequality (4.17) has been proved when ω is a neighborhood of the boundary and $T > \operatorname{diam}(\Omega \backslash \omega)/\sqrt{\mu}$. □

Remark 4.3 The proof of Theorem 4.1 shows that the exact-approximate controllability of system (4.1) holds whenever one has the observability inequality (4.17) for system (4.16). In [Z2] we proved (4.17) in the simple case in which ω is a neighborhood of the boundary and $T > \operatorname{diam}(\Omega \backslash \omega)/\sqrt{\mu}$. However, the problem of characterizing the pairs (ω, T) such that (4.17) holds in the spirit of the geometric control condition of [BLR] for the controllability of the wave equation is open. □

4.3. Null controllability Let us assume that Ω is the unit cube of $\mathbb{R}^3$, i.e. $\Omega = (0,1)^3$ and consider system (4.1) with periodic boundary conditions instead of Dirichlet ones. This case fits into the framework of [LeZ] where M is assumed to be a compact, connected and C^∞ Riemannian manifold without boundary of dimension three.

Let us consider the linear system of thermoelasticity in the manifold M:

$$\begin{cases} u_{tt} - \mu\Delta u - (\lambda+\mu)\nabla \operatorname{div} u + \alpha\nabla\theta = f 1_\omega & \text{in} \quad Q \\ \theta_t - \Delta\theta + \beta \operatorname{div} u_t = 0 & \text{in} \quad Q \\ u(0) = u^0, u_t(0) = u^1, \theta(0) = \theta^1 & \text{in} \quad M, \end{cases} \tag{4.18}$$

where now $Q = M \times (0,T)$.

Let ω be an open subset of M. We are interested on controlling (4.18) by means of control functions f with support in $\overline{\omega} \times [0,T]$.

In (4.18) θ is again a scalar function and u is a vector-field. We denote by $\widetilde{H^s}(M)$ the space of vector fields with H^s regularity.

The following result was proved in [LeZ]:

Theorem 4.4 *Suppose that any geodesic curve in M of length $\sqrt{\mu}T$ intersects ω. Then for any $(u^0,u^1,\theta^0) \in H = \widetilde{H^0}(M) \times \widetilde{H^{-1}}(M) \times H^{-1}(M)$ such that $\int_M \theta^0 dx = 0$ there exists $f \in L^2\left(0,T;\widetilde{H^{-1}}(M)\right)$ with support in $\overline{\omega} \times [0,T]$ such that the solution of (4.18) satisfies*

$$u(T) = u_t(T) = 0,\ \theta(T) = 0 \ \textit{in}\ M. \tag{4.19}$$

Moreover, there exists $C > 0$ such that

$$\| f \|_{L^2(0,T;\widetilde{H^{-1}}(M))} \leq C \| \left(u^0,u^1,\theta^0\right) \|_H, \tag{4.20}$$

for all $(u^0,u^1,\theta^0) \in H$.

Remark 4.4

(a) The condition we have imposed on (ω, T) is the geometric control condition for the linear system of elasticity in the manifold without boundary M. In this sense the result is sharp.

(b) The condition $\int_M \theta^0 dx = 0$ is necessary to drive the solution of (4.18) to zero by means of the sole control f acting on the hyperbolic equation. Indeed, integrating the heat equation in (4.18) on M one observes that $\int_M \theta(x,t)dx$ remains constant along trajectories. Therefore $\int_M \theta^0(x)dx = 0$ is a necessary condition for the existence of f such that the solution of (4.18) satisfies $\int_M \theta(x,T)dx = 0$.

(c) For simplicity of the proof we have chosen to state the result in the context of weak solutions rather than finite energy solutions as in Theorem 4.1. Consequently the control f is less regular and belongs to $L^2\left(0,T;\widetilde{H^{-1}}(M)\right)$. For this reason we have not multiplied the control by the characteristic function of the set ω. However, the controls under consideration have the support contained in $\overline{\omega}\times[0,T]$ as explicitely stated in the Theorem.

(d) As a consequence of Theorem 4.4 and when considering the Dirichlet problem in a bounded domain, one can obtain the null controllability with Dirichlet boundary controls acting on the whole boundary on both the displacement u and the temperature θ in time $T>\mathrm{diam}(\Omega)/\sqrt{\mu}$. To see this it is sufficient to build a cube $\widetilde{\Omega}$ containing Ω and to construct controls for the system in $\widetilde{\Omega}$ with support in $\widetilde{\Omega}\backslash\Omega$. Then the restriction of the controlled solution to Ω satisfies the required conditions.

The extension of Theorem 4.4 to the case of Dirichlet boundary conditions with controls acting only on the hyperbolic equation and located on a subset ω of Ω satisfying the geometric control condition is an open problem. □

Sketch of the proof of Theorem 4.4

First of all we observe that the null controllability of system (4.18) is equivalent to an observability inequality for the adjoint system

$$\begin{cases} \varphi_{tt}-\mu\Delta\varphi-(\lambda+\mu)\nabla\operatorname{div}\varphi+\beta\nabla\psi_t=0 & \text{in} \quad Q\\ -\psi_t-\Delta\psi-\alpha\operatorname{div}\varphi=0 & \text{in} \quad Q\\ \varphi(T)=\varphi^0,\varphi_t(T)=\varphi^1,\psi(T)=\psi^0 & \text{in} \quad M.\end{cases} \tag{4.21}$$

More precisely the controllability result of Theorem 4.4 is equivalent to the existence of a positive constant $C>0$ such that:

$$\|\varphi(0)\|^2_{\widetilde{H^1}(M)}+\|\varphi_t(0)+\beta\nabla\psi(0)\|^2_{\widetilde{H^0}(M)}+\|\psi(0)\|^2_{H^1(M)}\leq C\int_0^T\|\varphi\|^2_{\widetilde{H^1}(\omega)}dt, \tag{4.22}$$

for every solution of (4.19) with $\int_M\psi^0dx=0$.

To prove (4.20) we proceed in several steps.

Step 1. We consider first the vector field $\sigma=\operatorname{curl}\varphi$ that satisfies

$$\begin{cases} \sigma_{tt}-\mu\Delta\sigma=0 & \text{in} \quad Q\\ \sigma(T)=\operatorname{curl}\varphi^0=\sigma^0,\sigma_t(T)=\operatorname{curl}\varphi^1=\sigma^1 & \text{in} \quad M.\end{cases} \tag{4.23}$$

By [BLR] we know that, under the geometric control condition imposed on ω and T , there exists $C>0$ such that

$$\|\sigma(0)\|^2_{\widetilde{H^0}(M)}+\|\sigma_t(0)\|^2_{\widetilde{H^{-1}}(M)}\leq C\int_0^T\int_\omega|\sigma|^2\,dxdt \tag{4.24}$$

holds for every solution of (4.23).

Step 2. We consider the pair (ρ, ψ) with $\rho = \operatorname{div} \varphi$ that satisfies

$$\left\{ \begin{array}{lll} \rho_{tt} - (\lambda + 2\mu)\Delta\varphi + \beta\Delta\psi_t = 0 & \text{in} & Q \\ -\psi_t - \Delta\psi - \alpha\rho = 0 & \text{in} & Q \\ \rho(T) = \operatorname{div} \varphi^0 = \rho^0, \rho_t(T) = \operatorname{div} \varphi^1 = \rho^1, \psi(T) = \psi^0 & \text{in} & M. \end{array} \right. \tag{4.25}$$

By means of the change of variable $\xi = -\Delta\psi$ this system is reduced to

$$\left\{ \begin{array}{lll} \rho_{tt} - (\lambda + 2\mu + \alpha\beta)\Delta\rho + \beta\Delta\xi = 0 & \text{in} & Q \\ -\xi_t - \Delta\xi + \alpha\Delta\rho = 0 & \text{in} & Q \\ \rho(T) = \rho^0, \rho_t(T) = \rho^1, \xi(T) = -\Delta\psi^0 & \text{in} & M. \end{array} \right. \tag{4.26}$$

In view of (4.24) to deduce (4.20) it is sufficient to prove that

$$\| \rho(0) \|^2_{L^2(M)} + \| \rho_t(0) - \beta\xi(0) \|^2_{H^{-1}(M)} + \| \xi(0) \|^2_{H^{-1}(M)} \leq C \int_0^T \int_\omega \rho^2 dx dt. \tag{4.27}$$

Step 3. The observability inequality (4.27) is equivalent to the null controllability of the system

$$\left\{ \begin{array}{lll} w_{tt} - c^2\Delta w + \alpha\Delta\eta = g 1_\omega & \text{in} & Q \\ \eta_t - \Delta\eta + \beta w_t = 0 & \text{in} & Q \\ w(0) = w^0, w_t(0) = w^1, \eta(0) = \eta^0 & \text{in} & M, \end{array} \right. \tag{4.28}$$

with $c^2 = \lambda + 2\mu$. Observe that in (4.28) both w and η are scalar functions.

More precisely (4.27) holds if and only if for every $(w^0, w^1, \eta^0) \in H^1(M) \times L^2(M) \times H^1(M)$ with $\int_M (\eta^0 + \beta u^0) dx = 0$ there exists $g \in L^2(Q)$ such that the solution of (4.28) satisfies

$$w(T) = w_t(T) = \eta(T) = 0 \text{ in } M$$

and there exists $C > 0$ such that

$$\| g \|_{L^2(Q)} \leq C \| (w^0, w^1, \eta^0) \|_{H^1(M) \times L^2(M) \times H^1(M)}$$

for every (w^0, w^1, η^0) as above.

Step 4. The null controllability of (4.28) was proved in [LeZ] both in the case where $\partial M = \emptyset$ and $\partial M \neq \emptyset$ with Dirichlet boundary conditions.

The proof is based on a spectral decomposition that allows the semigroup generated by (4.28) with $f = 0$ to be split into its hyperbolic and parabolic projection. This can be easily done since the spectrum can be computed almost explicitly because all the differential operators (with respect to the space variables) involved in (4.28) are powers of the laplacian.

Once the semigroup is decomposed into its hyperbolic and parabolic projection one proceeds as follows:

(a) The hyperbolic component is shown to be exactly controllable with controls supported in ω provided ω satisfies the geometric control condition of Theorem 4.4 in time $\sqrt{\lambda + 2\mu} T$ as a consequence of the results in [BLR].

(b) The parabolic component is shown to be null controllable for any open subset ω of M and $T > 0$ by the method described in Section 3.3 to prove null controllability of the heat equation.

(c) Finally the problem of controlling to zero both the hyperbolic and parabolic components simultaneously is formulated and solved by means of Fredholm's alternative. At this level the compactness of the map that associates the $L^2(Q)$-control to every $L^2(M)$-initial data of the parabolic component is essential. In this way the problem is reduced to controlling a finite-dimensional subspace and this is done by means of a compactness-uniqueness argument. □

Remark 4.5

(a) The main point at which the fact that M has no boundary is used is in deriving the equations satisfied by $\operatorname{curl}\varphi = \sigma$ and $(\rho = \operatorname{div}\varphi, \psi)$. Obviously, when $\partial M \neq \emptyset$ systems (4.23) and (4.25) are coupled on the boundary ∂M and therefore they can not be treated separately as above.

(b) Observe that the restriction on the control time is imposed by the observability of (4.23) since system (4.25) is controllable in a shorter interval of time due to the higher speed of propagation.

(c) Similar problems can be formulated when the control acts through the parabolic equation, i.e. when the equations in (4.18) are replaced by

$$\begin{cases} u_{tt} - \mu\Delta u - (\lambda+\mu)\nabla \operatorname{div} u + \alpha\nabla\theta = 0 & \text{in} \quad Q \\ \theta_t - \Delta\theta + \beta \operatorname{div} u_t = g & \text{in} \quad Q. \end{cases}$$

When $\partial M = \emptyset$, $v = \operatorname{curl} u$ satisfies

$$v_{tt} - \mu\Delta u = 0 \text{ in } Q.$$

Thus, v is not controlled at all, and one cannot expect the null controllability to hold. However, in this context and under the assumptions of Theorem 4.4 one can prove that $\operatorname{div} u$ and θ can indeed be driven to zero. □

References

[A] S. Alinhac, Non unicité du problème de Cauchy, Annals of Mathematics, **117** (1983), 77–108.

[BLR] C. Bardos, G. Lebeau and J. Rauch, Sharp sufficient conditions for the observation, control and stabilization of waves from the boundary, SIAM J. Cont. Optim., **30** (1992), 1024–1065.

[Bu] N. Burq, Contrôle de l'équation des ondes dans des ouverts peu réguliers, Asymptotic Analysis **14** (1997), 157–191.

[C] T. Cazenave, On the propagation of confined waves along the geodesics, J. Math. Anal. Appl., **146** (1990), 591–603.

[CIK] D. Chae, O.Yu. Imanuvilov and S.M. Kim, Exact controllability for semilinear parabolic equations with Neumann boundary conditions, RIM GARC Preprint Series 95-76, Seoul National University, Korea, 1995.

[Ch] G. Chen, Energy decay estimates and exact boundary value controllability for the wave equation in a bounded domain, J. Math. Pures Appl., **58** (1979), 249–274.

[Co1] J.-M. Coron, On the controllability of 2-D incompressible perfect fluids, J. Maths Pures Appl., **75**(2)(1996), 155–188.

[Co2] J.-M. Coron, On the controllability of the 2-D incompressible Navier Stokes equations with the Navier slip boundary conditions, ESAIM: COCV, **1** (1996), 35–75.

[FPZ1] C. Fabre, J.P. Puel and E. Zuazua, Approximate controllability of the semilinear heat equation, Proc. Royal Soc. Edinburgh, **125 A** (1995), 31–61.

[FPZ2] C. Fabre, J.P. Puel and E. Zuazua, Contrôlabilité approchée de l'équation de la chaleur linéaire avec des contrôles de norme L^∞ minimale, C. R. Acad. Sci. Paris, **316** (1993), 679–684.

[FPZ3] C. Fabre, J.P. Puel and E. Zuazua, On the density of the range of the semigroup for semilinear heat equations, in *Control and Optimal Design of Distributed Parameter Systems*, IMA Volumes, **70** (1995), 73–92.

[FC] E. Fernández-Cara, Null controllability of the heat equation, ESAIM: COCV, **1** (1997), 87–103.

[F] A.V. Fursikov, Exact boundary zero controllability of three dimensional Navier-Stokes equations, Journal of Dynamical and Control Systems, **1** (3) (1995), 325–350.

[FI1] A.V. Fursikov and O.Yu. Imanuvilov, Local exact controllability of the Boussinesqu equation, RIM-GARC Preprint Series 95-50, Seoul National University, Korea, 1995.

[FI2] A.V. Fursikov and O.Yu. Imanuvilov, On exact boundary zero-controllability of two-dimensional Navier-Stokes equations, Acta Applicandae Mathematicae, **37** (1994), 67–76.

[FI3] A.V. Fursikov and O.Yu. Imanuvilov, *Controllability of evolution equations*, Lecture Notes Series # 34, Research Institute of Mathematics, Global Analysis Research Center, Seoul National University, 1996.

[Ge] P. Gérard, Microlocal defect measures, Comm. P.D.E., **16** (1991), 1761–1794.

[GLiL] R. Glowinski, C.H. Li and J.-L. Lions, A numerical approach to the exact boundary controllability of the wave equation (I). Dirichlet controls: Description of the numerical methods, Japan J. Appl. Math., **7** (1990), 1–76.

[Gr] P. Grisvard, Contrôlabilité exacte des solutions de l'équation des ondes en présence de singularités, J. Math. Pures Appl., **68** (1989), 215–259.

[H] S. Hansen, Boundary control of a one-dimensional, linear, thermoelastic rod, SIAM J. Cont. Optim., **32** (4) (1994), 1052–1074.

[Ha1] A. Haraux, On a completion problem in the theory of distributed control of the wave equation, in *Nonlinear Partial Differential Equations and their Applications*, Collège de France Seminar, vol. X, H. Brezis and J.-L. Lions eds, Pitman Research Notes in Mathematics, **220** (1991), 241–271.

[Ha2] A. Haraux, Quelques propriétés des séries lacunaires utiles dans l'étude des vibrations élastiques, in *Nonlinear Partial Differential Equations and their Applications*, Collège de France Seminar 1987–1988, H. Brezis and J.-L. Lions eds., Pitman Research Notes in Mathematics.

[HLP] D. Henry, O. Lopes and A. Perissimotto, On the essential espectrum of a semigroup of thermoelasticity, Nonlinear Analisys TMA, **21** (1) (1993), 65–75.

[Ho] L.F. Ho, Observabilité frontière de l'équation des ondes, C. R. Acad. Sci. Paris, **302** (1986), 443–446.

[Hr] M. Horvath, Vibrating strings with free ends, Acta Math. Hung., **51** (1988), 171–198.

[I] O.Yu. Imanuvilov, Boundary controllability of parabolic equations, Russian Acad. Sci. Sb. Math., **186** (1995), 109–132 (in Russian).

[J] I. Joo, Contrôlabilité exacte et propriétés d'oscillation de léquation des ondes par analyse nonharmonique, C. R. Acad. Sci. Paris, **312** (1991), 119–122.

[K] V. Komornik, *Exact controllability and stabilization: the multiplier method*, Masson & John Wiley, RAM 36, 1994.

[La] J.E. Lagnese, *Boundary stabilization of thin plates*, SIAM Studies in Appl. Math., Philadelphia, 1989.

[LaLeS] J.E. Lagnese, G. Leugering and E.J.P.G. Schmidt, *Modeling, analysis and control of dynamic elastic multi-link structures*, Birkhäuser, Boston-Basel-Berlin, 1994.

[LaL] J. Lagnese and J.-L. Lions, *Modelling, analysis and control of thin plates*, Masson, Paris, 1988.

[LT1] I. Lasiecka and R. Triggiani, Exact controllability of semilinear abstract systems with applications to waves and plates boundary control, Appl. Math. & Optim., **23** (1991), 109–154.

[LT2] I. Lasiecka and R. Triggiani, Uniform exponential decay in a bounded region with $L^2(0,T;L^2(\Sigma))$-feedback control in the Dirichlet boundary conditions, J. Diff. Equations, **66** (1987), 340–390.

[Le1] G. Lebeau, Contrôle analitique I: Estimations a priori, Duke Math. J., **68** (1) (1992), 1–30.

[Le2] G. Lebeau, Contrôle for distributed systems: The microlocal approach, in *Current trends in applied mathematics*. M. A. Herrero and E. Zuazua eds., Editorial Complutense, Madrid, 1996.

[LeR] G. Lebeau and L. Robbiano, Contrôle exact de l'équation de la chaleur, Comm. P.D.E., **20** (1995), 335–356.

[LeZ] G. Lebeau and E. Zuazua, Null controllability of a system of linear thermoelasticity, Archive Rat. Mech. Anal, to appear.

[LiY] X. Li and J. Yong, *Optimal Control Theory for Infinite Dimensional Systems*, Birkhäuser, Basel, 1994.

[LiL] Y.-J. Lin Guo and W. Littman, Null Boundary Controllability for Semilinear Heat Equations, Appl. Math. Optim., **32** (1995), 281–316.

[L1] J.-L. Lions, Exact controllability, stabilizability and perturbations for distributed systems, SIAM Rev. **30** (1988), 1–68.

[L2] J.-L. Lions, *Contrôlabilité exacte, stabilisation et perturbations de systèmes distribués. Tomes 1 & 2.* Masson, RMA **8** & **9**, Paris, 1988.

[L3] J.-L. Lions, Remarks on approximate controllability, J. Analyse Math., **59** (1992), 103–116.

[LM] J.-L. Lions and B. Malgrange, Sur l'unicité rétrograde dans les problèmes mixtes paraboliques, Math. Scan., **8** (1960), 277–286.

[LZ1] J.-L. Lions and E. Zuazua, A generic uniqueness result for the Stokes system and its control theoretical consequences, in *Partial Differential Equations and Applications*, P. Marcellini, G. Talenti and E. Visentini eds., Marcel Dekker Inc., LNPAS **177**, 1996, 221–235.

[LZ2] J.-L. Lions and E. Zuazua, The cost of controlling unstable systems: Time irreversible systems, Revista Mat. Univ. Complutense de Madrid **10** (2) (1997), 481–523.

[M] C. Morawetz, *Notes on time decay and scatterig for some hyperbolic problems*, Regional conference series in applied mathematics, vol. 19, SIAM, Philadelphia, 1975.

[Ru] A. Ruíz, Unique continuation for weak solutions of the wave equation plus a potential, J. Math. Pures Appl., **71**(5) (1992), 455–467.

[R1] D.L. Russell, Controllability and stabilizability theory for linear partial differential equations. Recent progress and open questions, SIAM Rev. **20** (1978), 639–739.
[R2] D.L. Russell, A unified boundary controllability theory for hyperbolic and parabolic partial differential equations, Studies in Appl. Math., **52** (1973), 189–221.
[SS] J.C. Saut and B. Scheurer, Unique continuation for some evolution equations, J. Diff. Equations, **66** (1) (1987),
[T1] D. Tataru, A priori estimates of Carleman's type in domains with boundary, J. Math. Pures Appl., **73** (1994), 355–387.
[T2] D. Tataru, Unique continuation for solutions to PDE's: between Hörmander's theorem and Holmgren's theorem, Comm. PDE, **20** (6–7)(1996), 855–884.
[TZ] L. de Teresa and E. Zuazua, Controllability of the linear system of thermoelastic plates, Advances Diff. Equations, **1** (3) (1996), 369–402.
[Za] J. Zabczyk, *Mathematical Control Theory: An Introduction*, Birkhäuser, Basel, 1992.
[Z1] E. Zuazua, Exponential decay for semilinear wave equations with localized damping, Comm. in P.D.E., **15** (2) (1990), 205–235.
[Z2] E. Zuazua, Controllability of the linear system of thermoelasticity, J. Math. Pures Appl., **74** (1995), 303–346.
[Z3] E. Zuazua, Exact controllability for the semilinear wave equation, J. Math. Pures appl., **69** (1) (1990), 33–55.
[Z4] E. Zuazua, Exact boundary controllability for the semilinear wave equation, in *Nonlinear Partial Differential Equations and their Applications*, vol. X, H. Brezis and J.-L. Lions eds., Pitman, 1991, 357–391.
[Z5] E. Zuazua, Exact controllability for semilinear wave equations, Ann. Inst. Henri Poincaré, **10** (1) (1993), 109–129.
[Z6] E. Zuazua, Finite dimensional null controllability for the semilinear heat equation, J. Math. Pures Appl., **76** (1997), 570–594.

Round Tables

Progress in Mathematics, Vol. 169, © 1998 Birkhäuser Verlag Basel/Switzerland

Electronic Publishing and Electronic Publications in Mathematics

EDITED BY BERND WEGNER

Participants (in order of their contributions to the Round Table): Peter Michor, Arnoud DeKemp, John Ewing, Jean-Paul Allouche, Hans J. Becker, Bernd Wegner (chairman)

ABSTRACT. The increasing development of electronic devices for the publication of papers and books in mathematics has led to a drastic change in the communication facilities between authors and editors, to new ways of distributing mathematical publications to research mathematicians – like electronic journals, and to an extension of the offers of information on mathematical research to the mathematical community. At this Round Table reports were given on different aspects related to methods of electronic publishing, and an overview of the production and distribution of electronic publications in mathematics to their users has been provided.

Most of the contributions given by the invited speakers of the Round Table covered several of these aspects and the corresponding written version of their contributions will be found in this report. The problem of authors' rights was mainly addressed by Peter Michor (Secretary of the EMS). Two representatives of important mathematical publishers joined the Round Table: Arnoud DeKemp from Springer-Verlag, and John Ewing from the AMS. They gave an overview of the electronic products and the corresponding editorial policies of their publishers. This was complemented by the contribution of Jean-Paul Allouche who reported on a French project dealing with the foundation and installation of some new electronic journals. The most import and classical providers of mathematical literature are the libraries. How they will improve and extend their services by using electronic media and offering access to electronic versions of journals was described in the presentation of Hans J. Becker from the library in Göttingen. Finally, Bernd Wegner (organizer of the European Mathematical Information Service EMIS) described how the system of internet-servers under EMIS is used for the moderated electronic distribution of mathematical literature. This latter service has been developed in 1995 by the EMS as a freely accessible www-tool for information in mathematics and is extended permanently. The written version of this contribution has been combined with the article of Peter Michor.

There was a long discussion during the second part of the Round Table. But no joint conclusions or activities emerged from this discussion. Hence the character of the discussion is represented by two written contributions taken from the many comments made by the audience of the Round Table.

On the Future of Mathematical Publications – Not a Charter*

PETER W. MICHOR, BERND WEGNER

There is an ongoing discussion on the future of Mathematical Publications in the age of electronic communication, and many things are happening: electronic journals appear which are no longer printed, and preprint servers take over the role of providing timely information on new developments in science. Much scientific news is spread via e-mail, as in the heroic days of the enlightenment, when personal letters were the main vehicles of communicating scientific discoveries.

For mathematics a lot is at stake, since it depends very much on the compiled literature, its reliability and accessibility. Mathematics is very vulnerable to any downgrading of the reliability of its literature which could be a companion of the more easy means of publishing today. But mathematics is also happy to have a *de facto* standard of publication: TEX. We think that TEX will be stable in the future. In certain respects the invention of TEX is like the invention of the alphabet. There is high value in its stability, so it will be stable. Even the development of the TEX dialects (LaTEX, AMSTEX) is slowing down considerably, not many users are migrating. Of course the raw TEX is not the means to distribute journals. Here, the standard seems to be PostScript now, and might be Adobe PDF later – so one must keep the chance to transform to changed standards later. We do not think that standards will change as rapidly in the future as they have in the past, since here stability is also of high value to the community. We might even have PostScript many decades from now, if not forever.

The main vehicle for the publication of mathematical research are the scientific journals; some of them have a centuries-old tradition. There is an established method of refereeing, deciding, and editing, and some fear that this is in danger. The real opponents of journals are the preprint servers. The hep-tp server is already starting to implement refereeing. The unbundled articles alone seem to be better units of scientific information transmission than issues of journals.

Mathematics is different. The literature is of high value, each paper is unique, results are rarely published several times. Journals with boards of editors and with established refereeing and publishing policies exist, are recognized, and will stay if they adapt. Old traditions almost never die, they just become feeble if they do not serve any purpose. We think that journals could also be the means of scientific publication in the electronic age, at least in mathematics. By adapting, we mean that journals must be easily accessible and quicker in publishing, and they should be printed and distributed at lower cost.

This is possible: have a look at the 'Electronic Library of Mathematics' of the 'European Mathematical Information Service' (EMIS) at `http://www.emis.de` and its 19 mirrors (as of October 1996): Berlin, master server at the editorial office of

---

*) Rewritten and enlarged version of "On the Future of Mathematical Publications – Not a Charter" by Peter W. Michor, European Mathematical Society Newsletter 20 (June 1996), 10–12

Zentralblatt für Mathematik (run by FIZ Karlsruhe).
University of Adelaide (AU).
Masaryk University Brno (CZ).
Budapest, Magyar Tudomnyos Akademia (HU).
Staats- und Universitätsbibliothek Göttingen (DE).
University of Helsinki (FI).
Københavns Universitet (DK).
Krakow, Uniwersytet Jagiellonski (PL).
Lawton, OK, Cameron University, (USA).
Universidade de Lisboa (PT).
Marseille, CIRM (FR).
Universidad Nacional Autónoma de México (MX).
Moskva, Russian Academy of Sciences (RU).
Ottawa, Canadian Mathematical Society (CA).
Rehovot, Weizmann Institute of Science (IL).
University of Southampton (UK).
Torino, CSI-Piemonte (IT).
Universität Wien, Rechenzentrum (AT).
ETH Zürich (CH).

The main content of EMIS is the Electronic Library of Mathematics with the following subsections: Electronic Journals, Electronic Conference Proceedings, Electronic Newsletters and also Electronic Monographs. At present there are two conference proceedings in it and 22 journals:

Acta Mathematica Universitatis Comenianae (Bratislava).
Annales Academiae Scientiarum Fennicae Series A. I. Mathematica.
Archivum Mathematicum (Brno).
Beiträge zur Algebra und Geometrie (Heldermann Verlag).
The Electronic Journal of Combinatorics.
The Electronic Journal of Differential Equations.
Electronic Journal of Probability and
  Electronic Communications in Probability.
Electronic Research Announcements of the American Mathematical Society.
Electronic Transactions on Numerical Analysis.
Geometry and Topology.
Journal of Convex Analysis (Heldermann Verlag).
Journal of Lie Theory (Heldermann Verlag).
Matematicki Vesnik (Belgrade).
Mathematica Pannonica.
Revista Colombiana de Matemáticas.
Séminaire Lotharingien de Combinatoire.
Southwest Journal of Pure and Applied Mathematics.
Theory and Applications of Categories.

Soon some more journals will be included:

Journal of Graph Algorithms and Applications.
Commentationes Mathematicae Universitatis Carolinae.
Webs and Quasigroups.
Journal de Théorie des Nombres de Bordeaux.

We think that for reasonably priced printed versions of electronically accessible journals there will always be a market: libraries, even individual subscriptions to special journals in one's own field. The available experience shows that the number of subscriptions to print journals having an electronic copy in EMIS went up in some cases and remained stable for others, while on the average the subscription rates to mathematical journals went down due to budget restrictions for most of the libraries. The additional offer of electronic versions through the system of mirrors of EMIS is used frequently: all servers together report some 200,000 hits per week.

About archiving: The "Electronic Library of Mathematics" is a sort of model for this. Mirror sites have to agree to keep the whole content of the server, they will be spread out over all continents, so even major breakdowns cannot destroy all of them. Hence, if we succeed in extending the distribution of mirrors world-wide, then we will have solved the problem of archiving in the best possible way.

Those who conceived EMIS see it as a self-organizing electronic distribution and archiving system for refereed mathematical literature. To be a little more precise about what we have in mind, we include below as an appendix the guidelines for journals and conference proceedings to be included in EMIS. It gives some hints how EMIS is working and why.

There is always an argument about copyright, and about the interests of the commercial publishers which should be protected. On the other hand practically no one speaks for the interests of the main actors involved: authors and readers. Authors of scientific articles want their work widely distributed, easily accessible, and safely kept in the future. They do not earn money with most of their articles, only scientific reputation, and this is not provided by commercial publishers who severely restrict the circulation of journals by very high prices, it is only provided by the reputation of the editorial board and by the readership of a journal. Readers of scientific articles want quick information and easy access. One has the feeling that today more is written than read, so readers are very valuable: it should be easy for them to get at their reading. Who is speaking on behalf of authors and readers? It should be the learned societies. But the largest of them are indeed like publishing houses and they seem more interested in the well-being of their revenue-generating enterprises than in the interests of their members: the authors and readers of mathematical articles.

The following charter would take care of the interests of authors and readers.

Draft of a charter for free electronic access to publications

The mathematical literature and access to it is the single most important asset of the world of mathematics. But it is under pressure: prices of journals are rising higher and higher, whereas the budgets of libraries are increasing somewhat below the rate of inflation or even decreasing. There are more and more new journals. Even in big libraries one cannot find all serious journals today. The role of a printed copy of a journal nowadays mainly consists in sitting in a library waiting for one of its articles to be photocopied. A conservative estimate is that on the average an article is photocopied 50 times, all over the world, and read 25 times. To typeset, print, and distribute it is a waste of human and natural resources: the same service could be done via computer quicker and with greater reach.

This charter is not meant to formulate the future of all the mathematical communication; its aim is only to pave a way for the transition of traditional journals to freely accessible electronic ones. The result might be cheaper and better accessible mathematical literature in the future.

The rationale behind this charta is the following: the aim is to try to lower the overall cost of mathematical literature. One way could be the following: we should distinguish carefully between the final electronic version of a paper into which, in addition to the work of writing and refereeing, only marginal costs have been invested, and the printed and distributed version, which makes use of a rather industrial process. The costs of the former, since they are small, should be carried locally, by academic institutions or libraries. Important libraries could adopt existing journals and take over their costs up to the electronic product. Free personal access to the electronic product should be ensured. The costs of the latter industrial process should be borne by the consumer (library) who prefers to have at hand a nicely printed version. There will be not so few of them, perhaps.

Draft of the charter

(1) The right of access to the electronic file of a paper which was prepared electronically by the author or his institution lies with the author(s).

(2) If the author submits this paper to a preprint server and/or if the paper has gone through a refereeing process positively, the author automatically acknowledges the right of free electronic access to this paper to the (mathematical) public.

(3) If a journal prints a refereed article, the right to distribute and sell this printed paper lies completely with the publisher of the journal. But the right of free electronic access to the file remains with the public.

(4) Everybody who uses the right of electronic access to a paper is entitled to print freely some copies for personal or restricted (educational) use. He is not entitled to sell copies of this paper for profit.

(5) Libraries are entitled and asked to keep permanently accessible files of electronically accessible journals, in order to ensure their perpetuity, besides or without printed copies.

Status of the discussion on the charter

The draft above received mixed responses. Among them:

A short charter like this has its charm, but if we start making all the sentences really precise it will get as ugly as the declarations of the European Community.

Some say that it is too early to fix any rules, or that it is really dangerous to do so: it might wake sleeping dogs.

The idea of having a charter such as this one did not find enough support from influential people thinking about the future of mathematical publications, especially in Europe. It seemed better to drop the idea. On the other hand, now is still the time when things may be formed and influenced; later, habits are fixed, and the chance of lowering the cost of mathematical literature as a whole may be over. Thus the formulation of the charter is given to the public in the form of this article. The reader should be aware that no assembly of any legitimation voted on it – it is just the personal opinion and guideline of the authors, but it led to the creation of the Electronic Library of the European Mathematical Society.

A question to all publishers

Let us ask here a concrete question to the representatives of commercial and non-commercial publishing houses, which are present in this room: Are you willing to put the files of your mathematical journals on EMIS, with a delay of some years, which you can determine and which should make you feel safe against possible loss of subscriptions? No immediate answer is requested. But in this way you put the labor of archiving into the hands of those really interested in it, and you would insure, finally, the existence of the mathematical literature as a whole.

Another proposal: if a mathematical book is available on the Internet, then it will probably sell better: any interested person can see whether it contains what he wants, and if so he will order it, if the price is not beyond a certain threshold, which we guess is at around 10 cents per page. Why do we believe this? It is not nice to read from the screen. And a printout is just a heap of paper, very awkward to read, and not free either. A nicely bound book is valuable in itself and is a pleasure to keep on a bookshelf.

Which publishing house will be running the first experiment on this?

Appendix

ELibM Guidelines for Journals and Proceedings Volumes. The following are guidelines for mathematical journals and proceedings volumes to be included in the Electronic Library of Mathematics (ELibM). This library is a broadly based electronic distribution and archiving network for mathematics. It is provided by the European Mathematical Information Service (EMIS) under the auspices of the European Mathematical Society (EMS).

1. Scientific standards. An editorial board must ensure that articles accepted for publication are of good scientific quality.

Every article must be refereed and recommended for publication by one or more experts.

The choice of articles must involve no discrimination on the basis of politics, nationality, race, sex, or religion.

2. Editorial and typographic standards. Articles should appear in a polished form similar to that of printed mathematical journals, to the extent this is practicable. In particular, they should have an approximately uniform typography, and each should have a unique identification that is suitable for use in bibliographies of other articles. It is strongly recommended that an abstract be included for each article. The dates of receipt of articles and the dates of publication of articles, issues, or volumes should be visible.

3. Logistics. In case a journal or proceedings volume is being offered for sale (in any form), the release of the electronic version to ELibM may be delayed a specified period so as to favor sales. This delay period will be at the discretion of the editors or publisher.

The editors of an electronic journal are responsible for delivery of their journal's articles to ELibM. Appropriate technical facilities for automatic document delivery will be proposed by the manager of the ELibM-Server.

While standards may change in the future, currently a complete PostScript file containing all the font information should be made available for every article in a journal participating in ELibM.

Provision of additional file formats (.TEX, .DVI, etc.) will be at the discretion of the editors of the corresponding journal or proceedings. The contents of all versions should essentially agree, although there is no requirement that they all share any particular style.

The proliferation of file formats has not posed a space problem thus far. However, ELibM reserves the right to limit bulk in case total file size becomes burdensome. They are beginning to impose a burden now.

To relieve editors of awkward physical responsibilities, ELibM is willing to hold the master posting (including all versions in all provided formats). This offer will be of particular interest for proceedings volumes; on the other hand, many journals are so well equipped that they prefer to retain their master posting, which ELibM then mirrors to all servers of its network.

4. Costs. Reader access to the electronic journal and proceedings volumes of ELibM will be free of charge. Likewise, to be included in ELibM, these publications must be provided to ELibM free of charge.

As a service to the mathematical community, the costs and maintenance of ELibM are covered by the EMS in cooperation with Fachinformationszentrum Karlsruhe / Zentralblatt für Mathematik.

5. Copyright. Distribution through ELibM servers is authorized by the original copyright holders. It does not imply transfer of any form of copyright to EMIS or any related agency, nor to any person or agency acquiring the data through ELibM. For a journal with a printed version, a person obtaining an article from EMIS has no greater legal right to reproduce it than if he had borrowed the journal from a library.

6. Permanence. Science has important personal and historical dimensions requiring that future generations be quite sure what was published and when. The provisions of the next paragraph help to assure this.

One of the first published versions of each article or volume should be designated as original and archival. (This is a must for journals not also published on paper.) The archival original must be maintained unchanged forever, while alternative versions in other formats may be introduced or deleted. The constancy of the archival version will be continually monitored by ELibM as the library's basic guarantee of document integrity to the mathematical community.

The editors and publishers are free to designate the archival version; failing that, the choice is left to EMIS at the time of posting in ELibM.

7. Distribution agreement. The above guidelines are an amalgam of information, offers of assistance, assured freedoms, advice, and obligations. The obligations are clearly signaled by the word "must". EMIS invites editors and publishers to enjoy the freedoms and to request assistance at any time. On the other hand EMIS requests prompt mention of points where the above advice may not be followed. EMIS insists that unaccepted obligations be renegotiated in detail before finalization of a ELibM distribution agreement.

Each ELibM distribution agreement (formal or informal) may be terminated on short notice, either by the copyright holder or by EMIS.

* * *

The Impact of Information Technology and Networks: New Perspectives for Scientific, Technical and Medical (STM) Publishing

Arnoud de Kemp

Abstract. This text is a strongly abbreviated version of a much longer presentation at the Round Table on Electronic Literature at the European Congress of Mathematicians (ECM2 – 1996) in Budapest. The process of publishing will undergo dramatic changes due to the influence of information technology and networks. The publishing business as a whole will shift from traditional print and paper based organisations to a fully digital workflow from author to end user. Electronic publishing has moved from pre-print activities to digital content on a variety of servers, but still most scientific documentation is printed and not only for archival purposes. In this Round Table a plea will be made for new rules in scientific communication, that authors, editors, publishers, societies, libraries and users can recognize. In addition, in the electronic age we need more protection for original work, transactions over networks and against misuse in general.

A Short History

In 1665 the first journals appeared: "Le Journal des Scavans" in Paris and "Philosophical Transactions" in London. They were the first publications with quality

---

*) (This is a modified version of the author's article in "Vistas in Astronomy" Vol. 40/3 (1996))

control, introducing "Approbation" and "Imprimatur". Today we see approximately 70,000 regular primary publications. Catalogues from subscription agencies like Lange&Springer, Swets, Ebsco and Blackwell list well over 300,000 serial titles. Little has been done to improve accessibility and retrievability. Book catalogues, whether printed or electronic as OPACs (Online Public Access Catalogues), are still based on rather simple cataloguing rules. The journal literature was abstracted and indexed by so-called secondary or A&I services. One of the most practical tools for accessing new articles in primary journals is Current Contents. There is no library any more that can acquire all relevant publications, not even the Library of Congress (USA), the British Library (UK) or the National Diet Library (Japan). The number of publications seems to double every 15 years and it would be interesting to correlate the number of active members of learned societies like the AMS and EMS and the number of publications offered and accepted over the years. The only effective answer to this increasing problem seems to be the introduction of electronic publishing and communication over networks.

What publishers do

The publishing process, until very recently, was traditional and reclusive. There are very many smaller publishing organizations. Publishers can be companies, societies like the AMS, universities, cities, departments, institutions, states or individuals on a part-time basis. They are as varied as the aims and scope can vary and many are so small, that the publisher is an all-rounder taking care of editorial work, production, marketing, sales and distribution him- or herself ... Publishing as a whole is big business with several mega-companies that seem to dominate the landscape. The many mergers and acquisitions during the last fifty years have worried many scientists and their societies.

In STM publishing we define the following categories: scholarly publishing, professional publishing and textbook publishing. Not all the works published are unsolicited texts that scientists offer for publication. Instead, we see many publications, that are written at the request of publishers or by editorial staff in their offices. I would like to mention reference works, dictionaries, multi-volume compendia, of which some have a very long publishing tradition. Publishers are not homogeneous, but they share the belief of freedom to publish.

The basic or core activities in publishing are selection, editing (also copy and language editing), design and illustrations, translating and indexing, production, promotion, marketing, sales and distribution. The computer has become a powerful tool and the large reference works, along with the abstracting and indexing sources, were the first to be converted to electronic databases.

Still, it has taken a long time for fulltext to become totally digital. The main reason may be the fact that computer screens were not designed for reading text. Although the acceptance of reading lines on a monitor has increased, most information is printed on the fly and then read. The next generation will certainly be more receptive. Publishing always concentrated on the printing or multiplication of documentation, recorded information, carefully checked and peer reviewed, original, integer and authentic. In the electronic age, it seems that the barriers between informal communication and formal publications are disappearing. I would like to make a plea for a careful definition of:

- communications, papers offered for publication,
- papers accepted for publication in a specific journal or serial title and
- publications with page numbers, a publication date etc., that can be cited in bibliographies.

Also, in the middle area, before the publishing organisation can announce the fact of a final version, it is recommended that a date and a version be specified.

Last but not least, it should be mentioned that the publishing process in the USA is partly financed out of research grants, and several publications levy page charges, which is rather uncommon in Europe.

New Technologies

Quoting Marshall MacLuhan, who in the fifties, said that automation conferences looked like a gathering of coachmen discussing the re-education of horses in times of more and more horseless carriages, it is difficult to see what the newest technologies will bring and what they will mean. We know that repography first has challenged and then threatened the principle of intellectual property. In times of increasing immaterialization, it is difficult to define what an original is. Do we still need originals? Major developments have been word processing, text and pixel retrieval, storage, local and other networks, the personal computer, the workstation, the laser printer, the network computer, Internet, Intranets, Java, digital broadcasting, etc. The future might look like InterCast, a combination of cable or digital broadcast television with a telephone line for interactive communication. Who knows?

The fact is that in STM publishing we recognize and work with over 120 different text editors and word processors. STM, the International Association of Scientific, Technical and Medical Publishing, has carried out some interesting investigations. We see increased use of SGML, but there are already different standards, interpretations and adaptations. Well-known and commonly used formats are PostScript, PDF, TeX, LaTeX, TIFF, GIFF, JPEG, MPEG, MHEG, Quicktime, etc. Over 500 different search engines are used in CD-ROM production. I expect that SGML (ISO 12083) and TeX will become more important and will be used more and more as the source format for electronic publishing and dissemination.

New Products

New products emerge, based on hypertext and other navigation systems, with click-through facilities. Alerting, based on personal profiles, will come as a service. The information industry is working on meta indexes and intelligent agents. At Springer we are involved in a project called MEDOC, with Ariadne as an intelligent agent for computer scientists. We see the first distributed information systems. Information, thanks to SGML, can be searched and presented in different layers, which makes rich indexing possible. Multimedia opens a whole new world for 3-D, simulation, video- and sound-embedded documents. We already see first living books and documents. Pixels become voxels and bits will be supplemented by trits.

Virtual reality allows for a new vision and can be used extremely well in design, production, education, telelearning and teleteaching.

Publishing will get several new dimensions and turn more and more into media production, which can become very expensive. Creation and production involve new skills and bring in added value. In addition to intellectual property, industrial property will become very important.

Conclusion

There will be changes everywhere and all the time. Old rules and values may disappear or be upgraded. The laws that protect intellectual property need to become global. The legal protection of information, transactions and users on networks needs to be enforced and supported by encryption techniques. First attempts are being made with electronic watermarking, tattooing and fingerprinting.

In addition, we need document or information identifiers. Hopefully we will manage to keep and identify quality, and respect originality, integrity and authenticity against fraud and misuse.

My personal vision is that mathematics will become so media and platform independent, that I can click on a mathematical object, cut and paste it and then use in my personal environment, before I ship my article to somebody in cyberspace. That somebody is in my vision most likely Mr. Springer (or Dr. Heinze, our maths publishing director).

An extensive bibliography will be mailed on request. For discussion please contact the author at `dekemp@adkathome.de`.

* * *

Publishing at the AMS

John Ewing

We publish journals, books, and a database at the American Mathematical Society. And all of them will be affected by changing technology.

Journals and books you likely understand, but the database? That's *Math Reviews*, of course, and like *Zentralblatt* it's different from a journal in almost every respect. *Math Reviews* is produced by a staff of over 75, processing over 100,000 items each year, dealing with thousands of reviewers, librarians, and publishers, to produce nearly 50,000 reviews. There is a fanatical demand for accuracy, precision, and completeness. Delivering that database, either in paper or on disk or over the Internet, requires even more people, and of course the Internet version of *Math Reviews* is a sophisticated application, developed by many people with varied talents. It is not merely a collection of hyperlinked files residing on a server.

Because Math Reviews is a database, rather than a journal, it moves into the electronic age most easily. The great strengths of electronic delivery – searching and linking – are ideally suited for a database. Of course, moving "easily" into the electronic age doesn't mean moving cheaply; we invested heavily in developing MathSciNet, and we continue to do so.

Books? Books will take advantage of new technology more slowly and more subtly. It is still hard to curl up with a computer to read in bed, and it's not a lot of fun to pore over a computer screen chapter by chapter. Some books are natural: those that resemble databases are made more valuable by searching and linking. But books in general will move into the electronic age more slowly, over a longer period of time, taking advantage of yet unrealized technology.

That leaves journals, the most difficult. Why are they difficult? Because people have too many expectations, and too little vision. I had better explain those remarks.

Why aren't journals cheaper?

People expect journals to be cheaper and, simultaneously, better. Why cheaper? Because, increasingly, journal articles are author-prepared, and because, potentially, journals can be delivered electronically. At the AMS, we now have lots of experience with author-prepared manuscripts, since slightly over half arrive in electronic form. And we have some experience with electronic journals – all our paper journals are now also in electronic form and we are adding electronic-only journals as well.

Preparation of manuscripts is only a fraction of the cost of producing a journal. Even so, for traditional paper journals, if all material were really author-prepared costs would surely decrease. Alas, along with most publishers, we find that roughly half of author-prepared manuscripts require substantial TEX work. Because that intervention is by highly skilled people who might have to track down a stray macro, for example, it can cost as much as keyboarding the paper from scratch. The savings are therefore smaller than you might think.

Will electronic journals be cheaper? That depends on what you mean by the term electronic journal. If you mean merely a mechanism for mathematicians to retrieve copies of papers, say a DVI or a PostScript file, then Yes, electronic journals should be cheaper. But this is document delivery, not an electronic journal.

This brings up the other expectation: that electronic journals should be a better way to deliver information, with more function. Links, for example: click on a reference number and see the detail, click again and reach the review or even the reference itself. The author claims this is true by Lemma 5? Click to see what that lemma is. Or click to see a derivation of some stated fact, which the author tosses out as folklore. Adding links creatively makes electronic publication a new and better way to communicate mathematics. And journals, or collections of journals, should be searchable: Just type in the phrase "group action" and you can find all papers in all AMS journals this year with that phrase in them. (There are six, by the way.) Electronic journals – real electronic journals – should substantially improve the way we do mathematics.

Doing all this has costs. Putting in the links, providing search engines, developing the environment, costs large amounts of money.

Worse, author preparation makes electronic delivery more difficult, not less. To add links efficiently, for example, one needs uniformity in manuscript preparation. Mathematicians are especially creative when it comes to using TEX. And while what appears on the printed page may look the same from paper to paper, the TEX code that's inside the paper is certainly not uniform.

We face a dilemma: The more author preparation we promote, the more costs we incur to prepare the papers for true electronic publication.

What about the other costs of electronic publishing? We have a top-of-line computer on which our server and search software runs. Most likely, it will be replaced in 2 or 3 years. The computer is supported by (expensive) technical personnel. We provide customer support, so that bewildered users can phone or e-mail to get help. We provide access control, so that users with the proper IP addresses or passwords can gain access. We must support mirror sites. We make CD-Rom archives, so that libraries have a hard-copy of the journal at the end of the year. And we are engaged in ongoing development in order to keep up with the next wave of browsers and Internet software.

What about the savings in printing and mailing? Well, eventually, we may realize some savings. Right now, while we publish both paper and electronic versions, we save only the marginal cost of printing slightly fewer copies of the journal. It's not much: we are investing in the future.

Vision

Where are we headed in the future? No one knows, not with any certainty. The only "experts" in electronic publication are the self-proclaimed kind, and we ought to be skeptical. At the moment, we are setting forth on a journey, and we don't have a roadmap.

But even if we don't know precisely what lies ahead, we can still anticipate the dangers. We can still ask what future consequences are likely to follow from our present actions. We can – we must – have a vision for scholarly literature.

Will everyone be publishing his or her own papers? This isn't a new phenomenon; preprints have been around for many years as an informal system to exchange information. In the past, finding preprints was complicated, and now electronic dissemination has the potential to make that much easier. But part of the function of journals is to organize the literature – journals are supposed to ensure not just accessibility, but quality and stability as well; it's part of the reason they cost money. Replacing our present system of journals with a freeform system of self-published preprints leads to academic anarchy.

Could we give structure to that chaos? Could we create a system to assign groups of preprint numbers to anyone running a server, and then provide a service to register them. That costs money too. Who pays that money? Publishers, from the revenues of journals?

Can we be certain that papers will survive after 100 or 50 or even 10 years? I don't know; no one does. But I do know that hardware and operating systems will be wildly different in 20 years, and we'd better be prepared. I also know that TEX, in its present form, will not be around in 20 years. If we must convert the one million electronic papers that have accumulated, we had better insist that they have some uniform structure, so that we can recognize titles and theorems and references when we do that conversion. We had better be prepared for the future. Being prepared also costs money. At the AMS we are now setting aside 1% of all subscription revenue each year to create a fund so that we are prepared. But who will take care of all the "free" electronic journals in 20 years?

Journals are more than a mechanism to advertise our recent theorems to our colleagues around the world. They must reach across time as well, to future mathematicians. As we replace our present system of communicating mathematics, we need to take a broad view of the roles and responsibilities of publishers in the past, and we need vision about how those roles will be carried out in the future.

Our legacy for the future

Can we control the costs of journals? Absolutely, we must. Can we make electronic journals less expensive than paper? I believe we can – I think we must.

But electronic journals are not free, and we cannot avoid costs by either pretending those costs are not there, or by passing them on to our university or to a government grant or to an enthusiastic junior colleague who maintains a web server. That's not sustainable. In publishing, as in life, you don't solve problems by avoiding them.

Can we find ways to create, to organize, and to maintain electronic publications so that articles published today can be found (and read) in the future? We ought to make that our highest priority; we should not – we can not – relinquish our responsibility as caretakers of a 2,500 year-old discipline. We need to leave a scholarly legacy for future generations in the same way that past generations left one for us.

* * *

Electronic Mathematics Journals in France

JEAN-PAUL ALLOUCHE

1. The general project

A World-Wide-Web server for French electronic mathematics journals opened in 1995. This is a joint project directed by the SMAI and the SMF (Société de Mathématiques Appliquées et Industrielles and Société Mathématique de France), and supported by the French government (Direction de l'Information Scientifique et Technique et des Bibliothèques du Ministère en charge de l'Enseignement Supérieur et de la Recherche). Technical support is provided by INRIA (Institut National de Recherche en Informatique et Automatique, Sophia).

This server presently hosts

- two purely electronic journals
  journals from ESAIM, European Series in Applied and Industrial Mathematics, *Contrôle, Optimisation et Calcul des Variations* and *Probabilités et Statistiques*;
- a purely electronic proceedings series
  ESAIM Proceedings;

- the electronic version of a paper journal
  Journal de Théorie des Nombres de Bordeaux;
- the abstracts of the articles of a paper journal
  Annales de l'Institut Fourier.

At the moment, the access to this server is free:

http://www.emath.fr

The author of this contribution is a former editor of the paper version of the Journal de Théorie des Nombres de Bordeaux, and now responsible for the electronic version of this journal.

2. Submission and refereeing

All these journals contain only papers that have been accepted after being read by one or more referees. This was, of course, already the case for the paper versions of the *Journal de Théorie des Nombres de Bordeaux* and the *Annales de l'Institut Fourier.*

The submission is essentially electronic although, in particular for the journals having a paper version, other submissions are accepted, if not encouraged. Almost all of the correspondence between editors and referees, and between authors and editors is electronic.

3. Overview of problems and questions

Here are but a few questions or guidelines as a first step for discussion. They should be considered only as rather vague starting points.

- *Financial and political aspects*
  Although of relatively low cost, an electronic journal is not free (one cannot bypass the expense of secretaries, buying and maintaining machines, disks, ...). Should it be accessible for free, or with a subscription fee, or with a "pay per view" system? It seems reasonable, if a journal is not free, to charge only sites, and not individual machines on a given site. Note that statistical data on connections are easily available. Note also that it is easy to lock or unlock access for a given machine or site.

- *Scientific aspects*
  Since the papers are submitted through a refereeing system, peer review is exactly the same as for a paper journal. But the evaluation instances have to be aware of the quality of the electronic journals and to admit that electronic journals can have standards as high as paper journals.

The electronic journals are probably the only answer to the huge increase of scientific literature ... provided that the "highway syndrome" does not strike. Indeed, each time a highway is constructed to decrease traffic jams, the long term effect is ultimately the attraction of many more vehicles. Nevertheless, the easy access to electronic journals and the possibility of computerized key-word search are unbeatable features. Note in passing the necessity of asking librarians to catalogue electronic journals.

- *Technical aspects*
  Although I am not an expert in technical aspects, I would like to point out the following facts. TeX and its avatars are certainly a very important tool. Clearly, only DVI- or PostScript-files should be available, for security reasons. Note that inserting pictures is not really machine-independent. Note also that unconditional fans of MacIntosh machines might have trouble in reading DVI- or even PostScript-files. A tool translating LaTeX to HTML is available, but not extremely efficient. The future versions of HTML might contain mathematical features that might even (?) make TeX obsolete (see http://www.lfw.org/math/). ACROBAT and JAVA are probably becoming new standards. The *Journal de Bordeaux* intends to make it possible for the e-reader to run algorithms while viewing a paper: this can be done by using JAVA.

4. Future developments and conclusion

The French server is ready to host other French journals and to cooperate on international actions (experience sharing, exchange of free access, mirror sites, ...). The back issues of the paper versions of the journals that are already on the server might become progressively available.

Since many e-journals already exist and many e-journals are being created, the question is evidently no longer to be for or against electronic publishing. The question is: how to keep the process under control from the political, financial, technical, and, *last but not least*, scientific point of view. The next question will be: how to take advantage scientifically of the electronic revolution? For example, an important question that only begins to be addressed (see Re-engineering peer review, *The Economist*, June 22nd–28th 1996) is certainly the opportunity of making the peer review process more precise, more efficient, and may be less "consanguine".

* * *

Electronic Publishing and the Role of Libraries*

HANS J. BECKER

First of all I want to thank Prof. Wegner for the opportunity of giving a short description of some aspects of the role of libraries in the development of electronic publishing using the projects of the Lower Saxony State- and University Library of Göttingen.

In the Library at Göttingen is one of the important collections of mathematical literature in Europe. Since the eighteenth century the mathematics literature has been one of the important subject fields of the collection, supported by some of the most eminent German mathematicians of the last two centuries:

---

*) e.g. the Lower Saxony State- and University Library of Göttingen

- the famous scientific library of Gauss is a part of the collection; and
- Klein and Hilbert supported the library by demanding extra grants from the Prussian ministry in Berlin to enable the library to collect materials especially from foreign countries, particularly from Great Britain and the United States.

After World War II the library was supported by the "Deutsche Forschungsgemeinschaft", establishing the so-called "Sondersammelgebiet Mathematik" and after 1975: "Sondersammelgebiet Reine Mathematik" in Göttingen.

You are surely wondering why I'm reporting on the development of the collection during the last two centuries, as you are anxious to hear about the newest developments in electronic publishing ... But in my view as a librarian and as a mathematician, especially interested in the history of mathematics, there should be a very definite connection between the current electronic form of mathematical literature and the printed literature before the electronic revolution. Mathematics as a special form of science is building up on the results of former mathematicians reported in literature which is not as accessible as the newer one. But in order to understand the whole structure of our science we have to make all efforts not to forget the achievements of our predecessors. It is true, as mentioned, that a huge part of mathematical literature was written during the last forty years, but a great part of the works of our great mathematicians was written before these years and we should not forget them simply because they were not written in electronic formats.

Therefore there is a great challenge for librarians and libraries and, I think, for the mathematical associations and publishers, too, to connect the past with the present to get a richer future.

Therefore we wish to ask: How can we define projects to realize this connection and what projects are the most effective ones to start with? I do not think that it will be such an easy task to put our mathematical heritage in electronic form.

This is the starting point for some of our efforts at the library in Göttingen. But before I give a detailed description I have to mention three other components: first, the role of the "Deutsche Forschungsgemeinschaft", especially of the department responsible for libraries, the "Bibliotheksreferat". For the last five years this department, together with librarians and scientists from different fields, has tried to initiate the spread of the electronic evolution to German libraries by formulating important plans.

On the other hand, the "Deutsche Mathematiker Vereinigung" (DMV) is looking for a change in publishing especially journals and pre-prints. There has been a project by the BMFT (Ministry of Education, Research and Technology) together with the DMV, resulting in:

- established World Wide Web servers at nearly every German mathematical institute,
- many experienced people in the German mathematical community, and
- an electronic journal, the *Documenta mathematica.*

The third component is the development of the *Zentralblatt für Mathematik* in different electronic forms, now easily accessible to the mathematical community through the Internet.

The projects in the electronic area of the "Sondersammelgebiet Reine Mathematik" at the Göttingen library are the following:

1. We are trying to establish a World-Wide-Web server especially for the mathematical community in Germany with gateways to the printed literature with delivery/copy function and lending function to the electronic forms in the Internet; we started with a special grant from the "Deutsche Forschungsgemeinschaft" to evaluate the Internet in order to give access not to all addresses but to the important and significant addresses; you are familiar with these efforts under the keywords: subject guides, Clearinghouses, Value-added packages etc.

We hope that we can open our homepage with these special features in January 1997. We will try to cooperate with other World-Wide-Web-suppliers in the field of mathematical literature.

Especially in the field of electronic and printed journals we will work together with mathematical associations and with publishers such as Springer, Academic Press and Kluwer.

2. In association with Professor Wegner and the Deutsche Forschungsgemeinschaft and in cooperation with Professor Dennis we intend to start the second project (guided by the European Mathematical Society and the American Mathematical Society): "The Electronic Jahrbuch über die Fortschritte der Mathematik 1868–1942, an enhanced and commented version"

The project will have three stages:

1. The first goal is to produce a searchable database as a digital version of the text of the "Jahrbuch".
2. After this there will be an intellectual review by a committee of mathematicians, who will formulate digital input in form of:
   - classification of the articles according to the MSC (Mathematics Subject Classification), which is used by both *Mathematical Reviews* and *Zentralblatt f. Mathematik*;
   - free keywords in English;
   - additional English titles;
   - comments and evaluation.

   The role of the library in Göttingen will be centred on the bibliographic work. There will be groups of librarians from the Göttingen library and the AMS library to look at the bibliographical aspects:

   a) There should be a standardization of the abbreviations of journal titles, like that in modern databases, and in case the title of a journal has changed the various versions of the title should be identified. In some cases the publishers have renamed their journals, and in others, in the course of the many years that *Jahrbuch über die Fortschritte der Mathematik* appeared, the abbreviation of source names took place according to different systems.

A modern user often knows only the current names and abbreviations. Such identification is important if one wishes to consider the source as a reliable search datum.

b) Some of the sources are journals that have continued publication to the present day, but some are series of publications that are now known only from references. In these latter cases getting hold of the literature itself is usually a difficult problem. For this reason the journal sources data which can be derived from the *Jahrbuch* should be complemented in an appropriate way with pointers to obtaining the originals.

c) Author identification should be undertaken as carefully as possible. Over the years there may be changes in orthography and transliteration of an author's name. The identification of the various forms of an individual's name is of great service to someone searching a database.

The third part of the project consists of:

a) the selection of important journals;

b) the selection of important collected works; and

c) the selection of some other important categories of publications, e.g. dissertations, Schulschriften, etc. by a committee of mathematicians and librarians in order to scan these printed works as images in cooperation with authors, publishers and mathematical associations.

When the project is completed the mathematical researcher will be able to search for an author, a title or some classification. If he is interested in the original work after looking through the details in the database, he should have some links to the electronic image, if the work has been selected for scanning, or he should have some links to obtain the work through conventional lending methods or copying from certain libraries. There will be some more problems such as copyrights and charging problems.

To conclude I want to mention another project of the library. We are planning to install an electronic archive as described by Frank Quinn from Virginia Tech in his paper: "A Digital Archive for Mathematics". We think that this will be a "natural" extension of storing handwritten and printed material in our library and that it will be useful for the German mathematical community.

To close my speech I would like to cite James Michael: "It was the best of times. It was the worst of times". Sounds familiar? This is the opening sentence of Dickens' "A Tale of Two Cities" – a story of revolution.

This is what we are facing today as information users, information providers, information seekers, and information creators. We are facing a network information revolution. With revolution comes chaos, uncertainty about the future, and myriad other problems. Carefully planned change is required. Strategic change is required.

* * *

Summary of the Plenary Discussion of the Preceding Contributions

Bernd Wegner

The discussion in the previous reports reflected two controversial positions: that of the authors and readers of mathematical publications who mainly wanted to see some reduction of publication costs and subscription prices resulting from well-prepared electronic files of their work, application of desktop publishing technologies and particular involvement of the authors in the publication process, and that of the mathematical publishers who wanted to point out that costs for the publication of mathematical texts even will increase due to offers of the texts through several media, enhanced facilities for electronic offers, archiving and standardization problems for electronic publications and increasing costs for maintenance of the technical background for electronic offers.

Obviously this discussion could not lead to any joint conclusion or coordinated joint activity involving both parties. The reason is that most of the electronic publication activities are in a preliminary state or just a vision for the near future. The few electronic journals in mathematics which have gained some experience in this part of the publication business are run on an almost voluntary basis, supported by a mathematics department or some consortium of mathematicians. Their economic structure cannot be taken as a model which could serve as the basis of an estimate of the costs of electronic publishing at a commercial publisher. On the contrary, the figures of commercial publishers presented in this context can be considered as a very rough estimate only, because they are pure speculation. The period of commercial electronic offers of mathematical journals is too short to provide more precise figures. Most of these journals are electronic versions of well-established print journals. Pure electronic journals offered on a commercial level can be considered as experimental only.

Two of the longer contributions to the discussion are represented in full text below. They give some impression of the opinions of those who did not attend the Round Table and may roughly be considered as representative of the comments from the audience.

Comments from

David A. Brannan

Open University, UK, e-mail: d.a.brannan@open.ac.uk

There are real costs involved in producing a quality mathematics research journal, whether paper or electronic.

There are a number of myths around that support the flawed view that journals should be free or priced at minimal cost:

1 All authors write clearly in good grammatical English with proper spelling and punctuation so that even readers new to the field or readers whose native language is not English can read and understand the mathematics without difficulty or ambiguity arising.

⇒ Not true! High quality journals have to return something like 50% of papers to authors for them to improve the exposition and language. Even then many authors cannot achieve a satisfactory standard. A high quality journal thus has to PAY for a Technical Editor to go through every paper to "debug" the exposition.

2 Authors will do the TEX typesetting work for a journal, so the journal typesetting costs will be nil or negligible.

⇒ Not true! Even with TEX there are so many dialects that journals often have to spend large amounts of time modifying formats to standard formats. Too many authors insist on embedding their own personal macros or using special fonts [often proprietary fonts, that cannot be used in an electronically-available version of the journal without a licence]. Quite often authors use TEX badly – for example, their files cannot be read. Many good authors do not use TEX anyway, and many journals do not wish to exclude those authors. Thus a high quality journal has to pay for a Technical Typesetter to go through every paper to TEX it, to debug an existing TEX file, or to re-TEX it if the author's file is unusable. Such typesetting is an expensive skill, that journals have to PAY for.

3 Individual journals will be available on the Editor's server or his University server and can be accessed at any time.

⇒ Not true! Editors move from university to university; servers get replaced every few years; universities have financial crises; technology moves on and files can no longer be read or printed easily; networks suffer e-congestion; working mathematicians get baffled as to where to find journals if there are too many possible locations, etc. Also it is unrealistic to imagine that individuals [or perhaps even universities] servers can act as the "archive of record" for mathematical material. Mathematicians are sufficiently vain to imagine that their published work will be of interest to someone in 50–1000 years; and indeed some of what is published at present will be – the problem is that we cannot easily identify which papers these are! Mathematics is a subject that develops like a pyramid (recall that Newton said that he "stood on the shoulders of giants"); therefore it is our duty for posterity to ensure that published work continues to be available indefinitely. This is something that publishers accept responsibility for, and so PAY for.

4 Editors do their work at zero cost, as their universities absorb the cost.

⇒ Not true! Editors run up quite large bills for postage (especially air mail postage), photocopying, and secretarial assistance; especially if they insist on papers being modified to be clearly written. Many Editors proofread, as they believe that their reputations require this guarantee of accuracy. Increasingly universities are unwilling to allow academics to run up real costs (e.g. postage, secretarial time) without recompense; some ask for partial replacement for teaching too. The journal has to PAY for these costs.

Therefore quality costs money:

1 To prepare the electronic files, whether for paper, server, or both;

2 To create accessibility of the material – that is, its distribution in the short term;

3 To maintain its accessibility in the long term.

If publishers have no income, or little income, the necessary consequence is that:

1 Papers will be less well prepared and less easy to understand;

2 In the long term, papers will simply "disappear". This is not satisfactory; therefore I believe that publishers (whether commercial, learned societies or universities) will continue to have to charge for mathematics journals.

Finally, it is my personal opinion that there will be a market for both paper and electronic versions of journals for years to come, as each format has its own role; and that probably paper will continue to be the principal archiving format.

* * *

Summary of Remarks at the Round Table on "Electronic Literature in Math"

AVIEZRI S. FRAENKEL

First set of comments

1. Peter Michor said that the author should retain the right to his own paper. I'd support this position strongly. This view has recently been realized by Bill Thurston, Director of MSRI, who signed a contract with CUP (Cambridge University Press) for publishing the MSRI Proceedings, after lengthy negotiations with some of the leading Math publishers. The contract seems to protect the rights of the authors for republishing, etc.
2. Two kinds of electronic journals were mentioned by some of the panel members; those published by commercial publishers or professional societies such as the AMS, and those published by scientist-volunteers. As to the former, it would be useful to hear their projected cost; in fact, I was surprised that no panel member addressed this question so far. The latter were referred to by John Ewing as "unsustainable". I think it is too early to grieve over them. Some of them, such as the ElJC (*Electronic J. of Combinatorics*) are highly successful, having now over 1,000 subscribers. The "battle" between the commercial publishers and the volunteers is yet to be decided.

Second set of comments

1. John Ewing said that he is afraid some of the volunteer-electronic journals might fold, and Arnoud de Kemp from Springer said that a journal published both electronically and on paper, as many of the Springer journals are now produced, costs 25% more than the paper version only. It might thus be

of interest to note that there exist today electronic journals, produced by volunteers, such as ElJC, which are completely charge-free to the end-user, strictly refereed and archived. I believe that ElJC, whose Editor-in-Chief is Herb Wilf, and of which I am one of the editors, is archived by the AMS. So even if it folds, the archives are there.

2. It was mentioned that at AMS, 50% of TEX-submitted papers need extensive editing; and that at Springer:
   (i) much money is spent in unifying/standardizing the electronic articles,
   (ii) a paper accepted for electronic publication still has to wait for publication until the entire issue is formatted. Again it is of interest to contrast this to what happens at volunteer-electronic journals. At ElJC, a paper can be submitted in any flavor of TEX or as a PostScript file. After acceptance, no further unifying/standardizing is performed. The paper is published the next day, in the current volume, as there are no issues, which seem to belong to the pre-electronic era. The same day also all the subscribers are informed automatically of the new paper.
3. Bernd Wegner mentioned that most papers in EMIS appear in .PS or .DVI form. It seems to me worthwhile to encourage submission of the .TEX version as well, even if it cannot be TEX-ed by the end-user, as it lets one make searches in a simple way.
4. The extra value added by the electronic media is exciting and promising. This includes
   (i) refereed addenda that are added to the papers (errors, further developments, etc), which transform the paper from a static to a dynamic entity.
   (ii) Having the references, or a subset thereof "clickable" (hypertext).
   (iii) Dynamic Surveys of entire subfields. Such surveys used to be static and published only at long time intervals. They can now be updated periodically, at relatively frequent intervals. All of these things have been implemented at ElJC, which currently features four dynamic surveys.

Progress in Mathematics, Vol. 169, © 1998 Birkhäuser Verlag Basel/Switzerland

Report of the Round Table on Mathematical Games

The following took part in the Round Table. David Singmaster was the Organizer.

Aviezri Fraenkel, Applied Mathematics and Computer Science, Weizmann Institute, Rehovot 76100, Israel. Email: fraenkel@wisdom.weizmann.ac.il. Combinatorialist and a leading researcher in the theory of combinatorial games.

Mogens Esrom Larsen, Kobenhavns Univ. Matem. Inst., Universitetsparken 5, DK-2100 København 0, Denmark. Email: mel@math.ku.dk. One of Scandinavia's leading writers on recreational mathematics and has been setting puzzle problems in a multilingual science magazine for ten years.

Ernő Rubik, Studio Rubik, Varosmajor ut. 74, H-1122 Budapest, Hungary. Inventor of Rubik's Cube and numerous other mathematical puzzles.

David Singmaster, SCISM, South Bank University, London, SE1 0AA, England. Email: zingmast@sbu.ac.uk. Writer on recreational mathematics, particularly on Rubik's Cube and the history of recreational mathematics. Has provided problems for several newspapers and magazines.

Tibor Szentivanyi, Varosmajor ut. 49, H-1122 Budapest, Hungary. Retired computer scientist, leading Hungarian writer on recreational mathematics.

Claude Berge, Richard Guy, Miguel de Guzman, Andy Liu, and Rüdiger Thiele were unable to participate.

Initial thoughts, circulated by the Organizer

Let me emphasize that I consider "games" as including puzzles and other mathematical recreations. We do not have a specific direction or viewpoint on the topic, but perhaps I could suggest that we think about something like "The significance of mathematical games for mathematics." This leads to a number of directions that you might like to consider.

1. The role of recreations as a source of serious mathematics – e.g. probability, combinatorics, number theory, topology, etc.
2. The role of recreations in teaching.
3. The significance of recreations as showing the historical and multicultural aspects of mathematics.
4. The role of recreations in the popularization of mathematics and in communicating with the general public.

Statements from participants

Singmaster's statement was abbreviated for the Conference Program. Larsen's statement has been expanded. Szentivanyi's statement was not available for the Conference Program.

David Singmaster. I view "mathematical games" as covering the entire field of recreational mathematics.

The general field of recreational mathematics is of immense significance for mathematics for several reasons. Indeed, one of my talks at the Paris ECM was on "The Unreasonable Utility of Recreational Mathematics" – a deliberate variation on the famous paper of the Hungarian Nobel Prize winning physicist Eugene Wigner: "The unreasonable effectiveness of mathematics in the physical sciences".

Recreational mathematics is important as the basis of several major fields of mathematics. Most significantly, problems of gambling led Pascal and Fermat to create probability theory and, within a century, this was the basis of the insurance industry. Graph theory was long considered a recreation, but has emerged as a major discipline and has been a source for topology. The earliest combinatorial problems appear to be the river crossing problems and the problem of getting food across a desert using camels which must be fed – these are first known to occur in the *Propositiones ad acuendos juvenes* attributed to Alcuin, c800. Most number theory has begun as recreation – attempts to solve Fermat's Last Theorem were a major source of algebraic number theory. Some more specific topics arising from recreational mathematics are Latin squares, conic sections, regular and not-so-regular polyhedra, presentations of a group, knot theory, the Möbius strip, the Gray code, Penrose pieces and quasicrystals.

Recreational mathematics has great pedagogic utility. It provides a treasury of problems which make mathematics fun and motivate students. "A good problem is worth a thousand exercises." (Ancient proverb due to David Singmaster.)

Recreational mathematics has great historical utility - specific problems can be distinctly traced, showing the development and transmission of mathematics (and culture) in time and place. Hence it is ideal for showing the historical and multicultural aspects of mathematics to the student.

Finally, and perhaps most important at the moment, recreational mathematics provides a way to communicate the nature of mathematics to the public at large. We know that the typical response of the public is that mathematics is difficult and mathematicians are weird. Yet there is a large audience interested in mathematical puzzles and games, perhaps most dramatically illustrated by the fact that something like 200 million Rubik Cubes were sold in three years! The sales in Hungary amounted to several times the population. Martin Gardner's "Mathematical Games" column was a major factor in the popularity of *Scientific American* and probably inspired more students to study mathematics than any other influence. I have heard that circulation dropped significantly when he retired. There really is considerable interest in mathematics out there and if we enjoy our subject, it should be our duty and our pleasure to try to encourage and feed this interest. Indeed, it may be necessary for our self-preservation – the number of closing mathematics departments is larger than it ought to be.

Aviezri Fraenkel. Applications of Game Theory to Other Branches of Mathematics.

Modern combinatorial game theory has largely been a parasite: it drew tools and results from fields such as logic, computational complexity, graph theory, matroid theory, combinatorics, algebra and number theory to generate results for itself. More recently, it has also begun to contribute back to some of its benefactors, such as to surreal numbers, a subject created by John Conway [Conway 1976], and to linear error-correcting codes (which is linear algebra) [Conway &

Sloane 1986], [Conway 1990], [Brualdi & Pless 1993], [Fraenkel 1996]. We may also include in this list areas in combinatorial game-theory itself, such as Mathematical Go [Berlekamp 1991], which have benefited from other areas of combinatorial game theory. We plan to review some of these developments, adding recent new results on digraph kernels and error-correcting codes as they evolve from games.

Mogens Esrom Larsen. Mathematics is like a disease, you may get infected with it – but it is also possible to be vaccinated against it. It is essential for mathematics that it is possible to understand everything – there is no need to quote any authority but the logic of your own arguments. Nevertheless mathematics is sometimes taught in a way which gives the impression of overwhelming authority, and this is what often acts as a vaccination. When we popularize mathematics by telling the great stories, then we may fall into the same trap. In order to reach the summit of a theory we make jumps to conclusions leaving the reader to trust us. If we want to infect our readers with the spirit of mathematics, then we must restrain ourselves to those challenging problems, where we can explain the solution in complete detail. This is a major role of recreational mathematics.

One problem is worth a thousand exercises. (Ancient proverb – see above.)

The recipe for popularizing science is simple enough: omit the mathematics. A similar approach to the popularization of mathematics is possible – but not advisable. If we try to tell the story of mathematics by emphasizing the highlights and omitting the boring details, we commit two mortal sins. The first is that we lose the spirit of mathematics; the second is that the effort runs the risk of a kind of vaccination against serious learning of the subject.

Although the development of mathematics exhibits recognizable patterns, it is not "the royal road" to understanding to learn these abstract patterns and then fill in the details later. Rather it is necessary to learn and master a great many details to be able to appreciate the patterns. [The architect Mies van der Rohe is credited with saying "God is in the details", an observation as appropriate in mathematics as in architecture – DS.] The ideas of Jean Piaget that the first subjects to learn must be the very simplest of them all, like topology and categories, are not only futile, but also prevent students from learning anything at all. This is mainly because of the lack of the spirit!

The essence of mathematics, its spirit so to speak, is the ability to understand everything. That's why the problems of H.E. Dudeney and Sam Loyd are so popular and essential. A good problem is self-contained and needs no previous knowledge or tools to speak of. The solution is reached by the use of logic and imagination only – no quotations of previous scholars are called for. That is what defines the problem in contrast to an exercise. In the exercise, the solution is made by the tools learned in the course. In the problem, you are free to use any tool you have! The only demand is that the logical explanation be convincing.

Tibor Szentivanyi

The impact of recreational mathematics on society

(How people are influenced by it and react to it.)

This contribution tackles a practical question, which arises automatically when we observe the present day's extreme interest in Recreational Mathematics. The enormous demand for rebuses, brain teasers, enigmas and puzzles (including mechanical puzzles) shows how Recreational Mathematics penetrates into our daily life. People are eager to grapple with knotty new problems to test their wits and their tempers. Why is Recreational Mathematics so popular nowadays?

There are different reasons and premises for this phenomenon. In the first place, people are looking for more and quicker relaxation, i.e. getting away from it all, than ever before. But the answer is not as simple as it seems to be at first glance. For detailed discussion of one side of the question, psychology is called in.

Taking account of the reasons, in the very beginning we find the principal usage of Recreational Mathematics in education, helping the development of mental ability, improving the flexibility of thinking, both of the young and the old. Recreational Mathematics helps to achieve so-called nonconventional thinking. It also teaches the method of searching for alternative settings of problems and not to be satisfied with those solutions which are immediately at hand. As a consequence, the student attains the ability to get a rapid review and recognition of alternative solutions of problems.

Recreational Mathematics is frequently used by the media and the advertising industry, far more than we realise. Recreational Mathematics gives a means for the management of companies to select the best people for a given job. The problem solving methods applied in the process reveal useful information about the qualities or features of the person whom we deal with. Even in short games or problems, we can learn more about our fellows (partners, colleagues, etc.) than in many hours of conversation.

Recreational Mathematics and related subjects are not only valuable pastimes but enable one to achieve experience of success. But Recreational Mathematics may have drawbacks too – these consequences can't be neglected. Think of the harmful passions induced by the excessive usage of games or puzzles based on recreational mathematics. There are cases where people become dependent on normally amusing passions like gambling, to the point where medical treatment is needed.

The author analyses the basic types of problem solvers, how they start and develop. All these results influence the design of toys, especially puzzles, which are developed on the basis of the findings, thus having an important effect on them.

Some notes from the presentations

David Singmaster. The following OHP slides were not shown due to the OHP not working at first. Though not all points were covered in the talk, these summarize most of the talk, which basically followed the statement.

The nature of recreational mathematics?

1. Fun and popular
2. Fun, used pedagogically
   a. Diversion and recreation
   b. Making complicated situations fun

- Games of chance and of strategy
- Mechanical puzzles

- Art, beauty, symmetry

Utility of recreational mathematics

1. Basis of serious mathematics: Probability, Graph Theory, Number Theory, Topology, Geometry, Algebra, even Calculus
2. Source of useful ideas
3. Pedagogic utility
   a. A treasury of time-tested problems
   b. "A good problem is worth a thousand exercises"
   c. Gives historical and multicultural perspective
4. Historical utility: Traces ideas through time and space

Some useful recreational mathematics

A. Probability
B. Geometry: Conic sections; C_{60}
C. Non-Euclidean geometry
D. Graph theory: Seven bridges of Königsberg; Icosian game
E. Number theory: Fermat's Last Theorem, Ideals, etc.; Factoring, Public key cryptography
F. Algebra: Solving equations, negatives; Galois fields
G. Analysis: curves
H. Topology: Knots; Möbius strip
I. Combinatorics: Gray code; Memory wheels
J. Penrose pieces: Quasicrystals

Mogens Esrom Larsen. Some examples of simple but non-obvious problems which can be completely understood by a student.

A. Consider a triangle. We wish to divide the area into two equal parts by drawing a line, possibly curved, from one side to another. We call this a fair curve. Which fair curve has the shortest length?
B. Consider two intersecting circles. Draw a line through a point of intersection. This meets both circles in two other points and the line between these is a "double chord". Which double chord has greatest length?

Aviezri Fraenkel. The speaker described several games, showing the connection with error correcting codes. A principal objective of research in the field is finding games intermediate in complexity between Nim (for which a complete and simple theory is known) and chess (which seems like a theory is generations away).

Tibor Szentivanyi. Winning the lottery is less likely than a car accident, but people do not view them in that order!

Puzzle solvers fall into a number of categories: Intuitive; Trial and error; Experienced (ones who transfer from past experience); Theoretical (non-tactile); Speculative.

This analysis leads to a number of rules for the design of successful puzzles. For example, eight pieces of the following shape make a square.

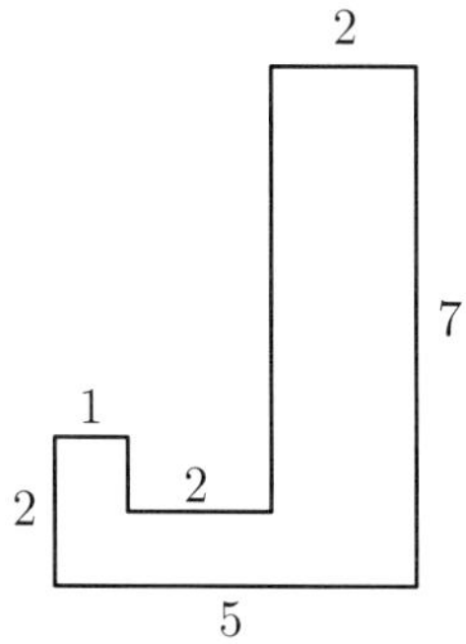

The thickness of the lower section is 1.

Ernő Rubik. Recreational Mathematics appeals to all ages – problem solving already appeals to infants. Concentrated thought frees you from all other problems. Understanding a problem is close to solving.

How does one create a good puzzle? It should be basically simple – complexity can arise from the number of pieces or the number of possibilities.

Play, problem solving, recreational mathematics, all have educational value.

Interestingly, Recreational Mathematics is not being used in video games as much as it should be, with the result that such games are often pretty boring!

Some notes from the discussion

As usual the discussion took a while to get started and unfortunately had to stop when it began to get interesting because the building was closing!

Richard Pinch (Cambridge, UK) asked how recreational problems can be used in teaching as most students find them unsolvable and hence frustrating, whereas ordinary exercises are solvable. And many recreational problems have tricks which students never find on their own.

Panel members felt that this is a matter of careful selection and fitting problems to the context. Some recreational problems are simply elaborations of straightforward problems by means of setting them in some kind of story situation. But many recreational problems are not straightforward exercises and one cannot use them as a substitute for ordinary exercises. This type of problem is intended to stretch the student's mind into new areas. Some of these do involve tricks, but they are intended to make the student realise that problem solving sometimes does involve non-straightforward approaches. With other problems, a solution may require extended work, particularly if one generalises from the specific problem. These make the student realise that an apparently fairly straightforward problem may require a lot of investigation and thought. The teacher or setter needs to judge the ability of the student and the difficulty of the problem before setting such a problem. Perhaps it is best to set a variety of problems so that students can pick an appropriate one. It is pleasantly surprising to find that students attack problems which one would not have expected them to do.

Mogens Esrom Larsen added that the non-educational setting of a magazine puzzle column is often more appropriate for mathematical recreations than a classroom setting with all its pressures.

There was a question to Mogens Esrom Larsen: how did he manage to produce new problems every month.

Larsen replied that he didn't really know, he had only been doing it for ten years! But seriously, problems are rarely completely new. There is a huge literature in the field and the problemist is constantly reading problem books for inspiration, which can range from simple piracy to complete innovation. He had perhaps 400 books on mathematical recreations. Singmaster agreed and said that he has perhaps a 1000 books of mathematical puzzles and several thousand related books. Several dozen books of mathematical puzzles appear every year.

An American lady asked Ernő Rubik to explain how he had invented the Rubik Cube.

Rubik gave a brief description, which was amplified through further discussion. [The most detailed description appears in: Ernő Rubik et al.; *Rubik's Cubic Compendium*; Oxford Univ. Press, 1987, chap. 1. This book also exists in several other languages. I have drawn on this to make the description a bit more complete.] His background is in art, design and geometry and his father is an engineer so he also had engineering and workshop experience, e.g. he was used to making models which were mobile and transformable. When he was about 30 years old, in early 1974, he was teaching three dimensional design at the Academy of Applied Arts. A natural object to study is the cube and he was contemplating the subdivision of a cube into eight smaller cubes. He thought of making this movable in some way. He made a model of this and realised that it was a puzzle. Rubik has said that his first model was a set of eight cubes with holes drilled diagonally through them and then joined together with rubber bands. Though the rubber bands soon broke, they lasted long enough to show that restoring the cube from a few random moves was an interesting puzzle. Over the next six weeks or so, he developed the mechanism for the $3 \times 3 \times 3$ cube [now universally acknowledged as a masterpiece of ingenuity], and he worked out how to restore it from any position. The latter was necessary as he was worried that a solver might have to precisely undo all the moves which had been made and that would be an impossible and uninteresting puzzle.

The development of the mathematics of Rubik's Cube is an excellent example of the processes of problem solving and I will give a brief synopsis of it. But the historical development did not proceed in anything like the logical development!

The zero-th rule of puzzle solving is not to mess up the puzzle until you know what you are doing – however, hardly anyone obeys this rule!

In order to know what you are doing, you need a good notation. The value of a good notation is obvious once you see it, but it can take some time to realise that it is needed. In the meantime, one uses ad-hoc notations or diagrams. In late 1978 I came up with the notation which soon became standard. I had tried a different system earlier and found it too awkward and I now wonder why I ever tried it.

My notation represented both positions or states of the Cube and processes or sequences of movements.

Once one has a notation, one can explore the problem. But experience should teach one not to try the whole problem at once, but instead try subproblems. On the Cube, these correspond to subgroups of the whole group. The group generated by one move is pretty simple, but the group generated by two moves is already pretty complicated. However, one can analyse the group generated by the squares of two moves and this turns out to have just 12 elements, one of which is a pair of 2-cycles of edges. Since every move is a pair of 4-cycles, hence even, a pair of 2-cycles is as simple a move as one can find.

At this point, one asks if one can obtain any pair of 2-cycles of edges. After some work, one realises one can move any four edge pieces into the positions which the above pair of 2-cycles acts upon. So one moves them, then applies the pair of 2-cycles and then one moves the edges back where they came from to obtain the desired pair of 2-cycles. This is simply conjugation in the group of the Cube. But it is conjugation which you do with your hands and so you really do understand what is happening. And if you do understand this, you have understood a basic problem-solving principle. For example, the exact same idea is used in the mathematical techniques of logarithms, similarity of matrices, the use of Laplace or Fourier or other integral transforms, the use of generating functions, the use of characteristic functions of probability distributions, even the use of analytic geometry. In all cases, we have a problem and we transform it from an unfamiliar setting to a familiar setting where we can apply standard methods to obtain a solution in the familiar setting. Then we have to translate it back to the original setting. Indeed this is the basic idea behind all mathematical modelling. One can even make an analogy with an assembly line or a workshop. I have a piece of wood on my bench which has to be sawn, so I take the wood to the saw, saw it and then bring it back to my bench! Little of this was ever pointed out in my mathematics courses.

Now one seeks for more simple moves and group theory suggests commutators. Again, the Cube allows one to see how a commutator tends to produce a relatively simple move. The powers of the commutator of two adjacent faces turn out to give simple moves, but it took about a year before we realised that one could commute this commutator with a face in which it only moved one piece to obtain an easy 3-cycle of corners and an easy way of twisting corners. With these simple moves, one can restore the cube in numerous ways (I gave up counting after about a hundred), each of which breaks the whole problem down into a number of simple stages.

Another question asked whether problems can stimulate abstraction in contrast to more direct logical thinking.

It was admitted that recreational problems tend not to demand abstraction on the part of the solver, but it is also true that the accumulation of examples is the usual background to an abstraction. That is, an abstraction is usually not developed until there is a need for it, i.e. until there are several examples of it. "God is in the details" applies here also.

A questioner wondered whether puzzles are too solitary an activity. Shouldn't students be learning to work in groups?

First, puzzles are rarely the only thing that a student does and a student should be learning both individual thought and group work. The student who doesn't participate in class or group work probably has psychological problems, though he may just be too bright or too dumb for the class. Second, recreational mathematics also includes games for two or more players. Third, a puzzle is not entirely a solitary activity – it's a contest against the poser and it can also be part of a competitive process. So the panel did not feel this was a serious problem.

A questioner asked about setting prize problems.

Larsen said this is difficult. One has to carefully judge the level of the problem and the ability of the audience so there are not too many solutions sent in. Alternatively, one has to allow a lot of time for examining the solutions sent and then one has to select prize winners at random. Magazines rarely allow enough time or money for this.

Progress in Mathematics, Vol. 169, © 1998 Birkhäuser Verlag Basel/Switzerland

Women and Mathematics

EDITED BY KARI HAG

Norwegian University of Science and Technology, Trondheim, Norway

The purpose of this Round Table was to address the underrepresentation of women in mathematics. We wanted in particular to focus on how differences in culture and tradition affect this situation since the proportion of women in mathematics varies considerably from country to country.

Some of the issues proposed for the Round Table were:

- Importance of a friendly, supportive atmosphere, mentoring.
- Attitudes towards equal opportunities and affirmative action.
- Different structures in academia – early permanent jobs versus late tenure.
- Rigid versus flexible paths for career development.
- Different conceptions about the role of women in society and mathematics.
- The status of positions versus the number of women holding them.
- Network for women in mathematics.

Close to 100 people took part in the Round Table. The presentations and general discussion were lively and a number of issues and suggestions were put forward. These will be followed up by the EMS-Committee on Women and Mathematics. To supplement the Round Table on interesting developments in Sweden and the United States two new articles have subsequently been added.

1 Program of the Round Table

1.1 Opening. Kari Hag opened the Round Table by presenting the EMS-Committee on Women and Mathematics which had undertaken to arrange the Round Table:

Eva Bayer, France, (past Chair)
Christine Bessenrodt, Germany (Chair),
Kari Hag, Norway,
Ina Kersten, Germany
Stewart Robertson, United Kingdom,
Laura Tedeschini Lalli, Italy, and
Vera Trnkova, Czechoslovakia.

Kari Hag had taken on the responsibility of chairing the Round Table. In preparing the program she received assistance from members of the Committee but worked most closely with **Bodil Branner**, Denmark and **Marjatta Näätänen**, Finland.

Kari Hag referred to the important and comprehensive report by Eva Bayer of the previous Round Table on Women and Mathematics. (*First European Congress of Mathematics, III, Round Tables*, Birkhäuser Verlag 1994, 37–74.)

This document underlines the importance of collecting data on women in mathematics at regular intervals and Kari Hag reported that the Committee had requested up-dated statistics from the mathematical societies in Europe. She also presented the main points from the following report by **Christine Bessenrodt**.

Women Mathematicians in Switzerland

In the spring 1994, the EMS-Committee on Women and Mathematics sent out letters to all Swiss universities and polytechnic schools asking them to comment on the low number of women mathematicians in Switzerland and on programs to improve the situation. Answers were received from all institutions except the University of Neuchâtel.

With one exception, all shared our concerns about the underrepresentation of women among the mathematicians in Switzerland and expressed the desire to improve the situation. But only a few institutions actually described programs for achieving this.

The University of Geneva at least has a law saying that persons of the underrepresented sex are to be given preference if they have equivalent scientific and teaching qualifications. The Swiss Federal Institute of Technology (EPF) at Lausanne has established a special program to advance equality between the sexes. For women scientists, this program includes:

- financial help for graduate students with children;
- the creation of doctoral fellowships reserved for women;
- the creation of a few assistant professor positions reserved for women; and
- making it easier for women to obtain part–time employment, even as scientists on jobs with large responsibilities.

These propositions seem very promising, and it is to be hoped that they will serve as a model for other institutions.

A few specific institutions claimed that their small pool of women students in general accounted for this underrepresentation. Other academic institutes reported that 30% of their mathematics students, through to the Ph.D. level, were women. Also, the low number of female applicants for positions was mentioned; it was deplored, moreover, that there were very few Swiss candidates. The reason for this may be a structural one: young Swiss mathematicians may feel that preference may be given either to local applicants or to internationally renowned scientists from abroad, and hence they may be discouraged from applying. It is well known that women are especially apt to react in this way.

As all suggested explanations seemed to be based on rather incomplete or institute-specific data, we planned a more detailed follow-up study on the statistics of women Ph.D.'s in mathematics in Switzerland. As a first step, we obtained quite extensive data concerning the University of Geneva. We then wrote again to the same Swiss institutions reporting the results of the first inquiry and asking explicitly for the number of Ph.D.'s in mathematics obtained at these institutions between 1975 and 1990 and, in particular, how many were granted to women. The statistics for the University of Geneva were enclosed as an example.

In reply, only four institutions sent us further information concerning Ph.D. statistics, including also some information on women holding faculty positions.

During the period 1975–1990, 33 mathematicians obtained their Ph.D. at the University of Bern. There was only one woman among them and there is currently only one female Ph.D. candidate. On the positive side, of Bern's 5 full professors of mathematics, one is a woman.

The complete information collected for the University of Geneva describes a very different situation: there 8 out of 39 Ph.D.'s in the period 1975–1990 were women.

We received fairly complete statistics from the EPF Lausanne for this period: 3 out of 63 Ph.D.'s in mathematics were women. Moreover, females constitute 33% of the students in mathematics and about 25% of 90 assistants are female. However, there are no women among the 18 professors.

The figures coming from the ETH Zürich seemed to indicate some improvement in recent years: during the period 1975–1990, 14 women among 179 candidates obtained their Ph.D. while the corresponding ratio for 1991–1994 was 6 out of 46. Unfortunately, there were only 4 women among the 78 assistants (but 8 among the 59 "Hilfsassistenten"). Among the 32 professors there was one woman assistant professor (for a 2-year term).

There also seems to be some change in recent years at the University of Zürich-Irchel: 6 women among the 75 Ph.D.'s were reported for 1975–1990 and 4 out of 22 for 1991–1994. Here, this figure corresponds to the number of 4 women among the 24 assistants. However, there are no women in higher ranks.

Summarizing what information we have received, we see that the picture for all parts of Switzerland is not completely homogeneous with respect to the percentage of women Ph.D.'s. One can say, however, that throughout the country, the percentage is still much lower than the percentage of female students in mathematics. Some positive examples and improvements in recent years indicate clearly that there is great potential for the departments change, and should be supported and encouraged by further actions. The situation, however, is nearly the same everywhere when one looks for women mathematicians holding tenured positions – at best there is a single woman professor in a department. We hope very much that in the near future the number of women faculty members will increase at Swiss universities and polytechnic schools.

1.2 **Video.** Women and Mathematics across Cultures, EWM – European Women in Mathematics

The video was made by **Marjatta Näätänen**, University of Helsinki, Finland. It consists of a short presentation of the organization European Women in Mathematics, EWM, followed by four interviews. (Information on EWM is given in the Appendix of this report.) The videotaping was done at the Madrid meeting of EWM September 1995 where Marjatta Näätänen interviewed the following women mathematicians, all of whom have studied and worked in several countries:

Marjorie Batchelor, United Kingdom,
Laura Fainsilber, France,
Isabel Labouriau, Portugal, and
Mara Neusel, Germany.

The list of issues for the Round Table given in the preliminary remarks is an extract of points put forward by these four women. We expand the points with the understanding that the video, as a collection of personal interviews, should be **seen** and not just summarized.

- **Importance of a friendly, supportive atmosphere, mentoring.** The isolation is tremendous, for example, in the mathematics departments in Germany and Switzerland. There are places where women students simply feel ignored. Attention and contact with professors mean a lot to women students trying to build a career in a masculine field.
- **Attitudes towards equal opportunities and affirmative action.** This was said to be quite different in England from what it is in the United States. In England, there seems to be an absence of concern that equal opportunities and affirmative action will be implemented while in the United States they are "frightened by" the law. Therefore they actively seek and interview women for jobs.
- **Different structures in academia – early permanent jobs versus late tenure.** In many countries junior mathematicians go from postdoc to postdoc at a time which is particularly crucial to women. In France the situation is different. And there you have many impressive women mathematicians although some of them may have had years when they were not able to do much research.
- **Rigid versus flexible paths for career development.** How should one compare the CV of a man working full time on mathematics to the CV of a woman raising children and working half time on mathematics? According to one woman's experience this question is never raised in England.
- **Different conceptions about the role of women in society and mathematics.** The difference between England and Latin countries was said to be striking. In a Latin country, being a woman and a mathematician is not an issue. Maybe more women have careers there than in England. To look into this question should be an interesting topic for a social historian.
- **The status of positions versus the number of women holding them.** The following empirical "law" seems to hold in many protestant countries: the status of a position times the number of women holding such a position is constant. (The status of a professor is particularly high in Germany.)
- **Networks for women in mathematics.** One needs a strong women's network to combat the old boys' club in countries like Germany. EWM is very valuable. It is important for women from northern Europe to come and see women working together on mathematics.

1.3 Short presentations. Ragni Piene of the University of Oslo, Norway, was the moderator for the panel which had three invited speakers in addition to Kari Hag:

> **Dusa McDuff**, State University of New York at Stony Brook, USA,
> **Roswitha März**, Humboldt-University, Berlin, Germany, and
> **Sylvie Paycha**, University of Clermont-Ferrand, France.

Report by Dusa McDuff

The question of how to be both a mathematician and a woman has certainly been (and still is) a live issue in my life. However, I don't think that that is peculiar to me but that it reflects social conditions in both Britain and the United States (and no doubt in other countries, too). Since I think that self-doubts can have a huge negative impact on creativity, I think that this is one of the most important issues for women to talk about. Therefore, my remarks will be largely personal.

In some ways, I grew up with all the advantages: my family was academic, my mother was a trained architect with a fulltime job, and I was always expected to get an education and have a career. However my family was definitely different from those of my schoolfriends in Edinburgh, and I grew up in relative isolation, always thinking of myself as exceptional, not part of any group or community or used to working together with other people.

I was an undergraduate in Edinburgh and a graduate student in Cambridge, where I solved a well known problem in the theory of von Neumann algebras. I then followed my husband to Moscow, he being a specialist in Russian literature. There I met Gelfand who was a great influence for many years. He encouraged me to become a topologist. However, on returning to Cambridge in 1970 after 6 months in Moscow, I was again very isolated and took several years to find my way back into mathematical research. My first child was born during this time, but this was not what was holding me back: the time after childbirth can be very productive (as I found after my second child was born in 1984). The real trouble was that at that point I approached mathematics in far too passive a way, as something beautiful but distant, and that I was not talking to anyone. Looking back, I think that my isolation was quite largely self-imposed but that it may well have been necessary for my survival as a mathematician, since it made it possible to for me to distance myself from the restrictive cultural messages about women that existed all around me. Eventually I started working with Graeme Segal, with whom I essentially did a second Ph.D.

Quite out of the blue, I was invited to MIT for the year 1974–5, to take up a visiting position especially reserved for women. This was a wonderful opportunity which made a big change in my life. I began to have a few mathematical ideas of my own again, and became aware that one has to build one's own career (e.g. by applying for fellowships) and that this was my chance to do so. Meeting Ragni Piene there (the first woman mathematician with whom I felt I had anything in common) was also important.

After that, I probably would have been able to survive mathematically whatever happened. However, after moving to the States in 1978, I appreciated the positive encouragement that I felt the math community extended to me. The foreignness of the culture also meant that I was not so aware of general expectations about what I should be (with my accent I could never fit in unnoticed, anyway), although it took me a long time to develop any sense of connection with the community I live in.

Recently I went back to Cambridge and was dismayed to find that things there had hardly changed since my time. I took the initiative to encourage British Women in Mathematics to hold yearly meetings, but have not had time to do much more than attend those meetings. I think it very important to make women visible and to form networks to support women in order to counteract the isolation that women can still suffer.

Report by Roswitha März

First of all, I do not think that there is a special problem for women in *mathematics*. I am sure that we are still far from having equal opportunities for women and men *in science and in scientific institutions in general.*

As far as I know, a similar situation can be observed in all European countries. The presence of women decreases exponentially the higher the position, the higher the income, the greater the power and opportunities of influence. Mary Osborn, (1994, SCIENCE 263, 1389–1391) has studied the share of women among full professors in Europe and the United States. Germany, with only 2.5%, is considerably behind the United States with about 14% or Turkey with about 20%. Only the Netherlands comes off worse than Germany.

Let me quote an interesting observation of Mary Osborn: "Surprisingly, the percentage of women at all levels seems smaller in the more economically developed northern European countries, with the exception of France, than in the southern European countries. In addition, in northern European countries such as England, Germany, and the Netherlands, the numbers have remained static for a decade or more, whereas in the United States the percentage of full professors who are female has increased by 60% in the last 10 years."

Next, I would like to present some further interesting numbers. At the Technical University Dresden, about 300 full professorships were available in the last few years. These positions were advertised. After looking through and evaluating the applications, the University made up lists of candidates, each of whom was graded 1 to 3. These lists were handed over to the state secretary, who decided on each appointment. Here are the related figures (1994, KONSENS 5, 4–5). Among all applicants, 6.3% were women. Among the applicants who were nominated for the list (places 1 to 3), 3.8% were women. Among those who were finally appointed, only 1.3% were women. While 76% of the men heading the lists were appointed, only 33% of the women heading the lists were appointed. 10 men were appointed although they had only "back" places (2–3). However, only one woman in a "back" position was conceded this privilege.

Does this indicate discrimination against female scientists? Obviously, it does! These structural discriminations are realized not only via the given academic structure, but also by politics.

My second point concerns the issue of "early permanent jobs versus late tenure". I have had personal experience with two extreme cases. Formerly in the GDR, we had early permanent jobs almost exclusively. Today, we have hardly any permanent jobs, except for the professorships. Both extremes seem to be bad.

Surely, too many early appointments to permanent jobs yields negative consequences. Inflexible, no-longer creative, complacent people "clog" the university system. On the other hand, the consequences of only granting short-time jobs are also bad. Everybody **has to be** successful in order to secure the next job in pressured time. One is to get ahead in life rather than to consider serious, complicated problems the utility of which may not be immediately known. This is bad for science! Moreover, teaching seems to be less important. This is, in particular, bad for women in science because their careerism is less distinctive.

In my opinion, a compromise would be the best solution. A greater number of permanent jobs should be available for scientists. In particular, all women and men who have completed the habilitation should get, if they wish, an appointment at their university.

My final point concerns the issue "networks". Yes, I think that women should unite in networks – either official or unofficial ones. Women should act together, keep each other well-informed and help each other to gain more self-confidence.

Timely information about vacancies, possible applicants and advice for evaluations as well as tips regarding possible grants – I consider it to be very important that women should encourage each other to become better known and more visible in the scientific community. Women should invite each other to give papers at colloquia, to chair sessions, to make known their results. In this sense, women should deliberately search out women and help them to improve their professional opportunities.

A first success: In 1995, an Oberwolfach conference was held on differential-algebraic equations, where 24 of the 43 participants were female. Also the program of lectures was slightly dominated by female mathematicians.

Women's networks should reveal discrimination against women, discuss it and make society more sensitive to it.

Report by Sylvie Paycha

I represent here both EWM, as its present convenor, and the French organization "femmes et mathématiques", as former chairwoman. The latter was founded in 1987, and now includes 100 members or so.

Before I tell you how this French organization was founded, let me try to describe some specific features of the French tuition system and comment on the situation of women mathematicians in France.

In France permanent positions start when you are relatively young, approx. 30 years old for maître de conference positions. This is an advantage for women when compared to other systems in Europe in which permanent positions start later, on the whole. This is one of the reasons which contribute to the fact that the proportion of women mathematicians among mathematicians in French universities, although still very low (and it is getting lower still as we shall see later on) is higher than in other northern European countries.

However, it is not the only reason and one should take into account the specific organization of the French higher educational system. There are two parallel systems in France: Universities on the one hand and "grandes écoles" on the other hand, with very competitive entrance exams. Earlier, the entrance exams for men and women were separate and most women mathematicians trained in France at that time came from a "grande école". The fusion of the competitive exams for men and women that took place about ten years ago was followed by a rapid decrease in the number of women starting a career as a mathematician. A group of women mathematicians, concerned about this and about the fact that women are a (small!) minority among mathematicians, founded the French organization "femmes et mathématiques".

Let me now say a few words about my own path as a French woman mathematician. It starts rather atypically since I do not come from any of these "grande écoles". I graduated at a university in Paris, after which I went to Germany to prepare a Ph.D. in mathematical physics, working there as an assistant for four years. This turned out to be an interesting experience for me as a young woman mathematician. There were very few women in the mathematics department where I was working, which could have made me feel quite uncomfortable, especially as I was only an assistant, at the very bottom of the German university hierarchy! However, I just made my way along, and not being German, I was quite unaware that I was probably transgressing some (if not most!) of the implicit rules, such as the ones which underlie the hierarchical structure of a German university department. This "unawareness" gave me a freedom as a young woman mathematician to try out new things such as organizing an interdisciplinary seminar and inviting full professors to join in, something that I would probably not have dared try out, had I known the unspoken rules!

After this, coming back to France was not an easy job since I did not belong to any of the French schools or research groups. It was very near the border between France and Germany, in Strasbourg, that I eventually got a position and spent 6 years as a maître de conférence. After that I was appointed as professor (a year ago)... right in the center of France!

I was quite lucky to get a professorship at a time when the proportion of women among mathematicians in France is decreasing with the decrease of the number of positions on the whole and more strikingly so for professorships.

Let us first glance at the "maître de conference" positions. All the figures given here are available in the "Rapport Basdevant (DSPT1), 1995". In 1991, 23 out of a total of 113 "maître de conference" appointed in France that year in France were women (20.5%). In 1992, 28 out of 119 were women (23%), 10 in pure mathematics and 18 in applied mathematics. In 1993, 39 out of 145 were women (26.8%), 16 in pure and 23 in applied mathematics. In 1994, 25 out of 147 were women (17%), 7 in pure and 18 in applied mathematics. As you go up in the hierarchy, the figures sink drastically and tend to zero. Let us look at the number of women mathematicians recently appointed as "professors". In 1991, 14 out of a total of 78 professors appointed that year in France were women (18%). In 1992, 8 out of 82 were women (9.7%), 2 in pure and 6 in applied mathematics. In 1993, 13 out of 88 were women (14.5%), 4 in pure and 9 in applied mathematics. In 1994, 4 out of 51 were women (7%), 4 in applied and none in pure mathematics.

When the jobs get rare, women are the first ones to feel the consequences!

Report by Kari Hag

In Norway more than 40% of the members in parliament and government are women and we have had a woman Prime Minister for many years. She enjoys telling how small boys now ask her, "Can a man become a Prime Minister in Norway?"

But in academia we still have some way to go. According to recent figures, among the full professors, 10% are women; among the tenured faculty, 20%; and among those who hold fellowships or scholarships, 40%. For mathematics the figures are lower as you can see from our statistics.

On the positive side, the number of Ph.D.'s has grown in recent years both in total and in the number held by women. For example, 59 students including 13 women have finished a Ph.D. in mathematics during the last 4 years. However, very few women find employment at the universities. Only 1 woman has a post doctoral fellowship and she has gone to Italy and would like to stay there. More than half the female Ph.D.'s are in applied mathematics and statistics. Very few go into pure mathematics. In Norway, the youngest woman with a position in pure mathematics is about 50.

In Italy the situation is quite different! They have few women in politics and many women in mathematics. The percentage of women mathematics students who earn the first university degree in mathematics is 70. For most other subjects the difference between Norway and Italy is not large. In physics the percentage is slightly higher in Italy (30 compared to 23) but in chemistry it is approximately 50 in both countries. In the biological sciences, medicine and law the percentages are the same as well (65, 50 and 55 respectively). Does this mean that there are fewer women in engineering in Italy than in Norway? The statistics say that in both countries 23% of these students are women. (Looking more closely at the numbers one finds that engineering includes architecture, an area very popular among female students, and architecture constitutes almost a third of the engineering student body in Italy.)

Why then so many women in mathematics in Italy? Let me paraphrase from e-mail discussions with Laura Tedeschini Lalli of Italy:

About women's choices in Italy... Of course, here we only have guesses, based on personal conversations with many friends and students and so on... Keep in mind that in Italy you apply to the same school of mathematics and get the same first two years' training, whether you want to go into research, industry, or teaching at any level.

In those first two years, students have to take 8 full-year courses, covering a wide range of math, which are on average tougher than the same level of courses at most other European universities, as we are discovering via the European exchange programs. While this might be wrong for a number of reasons and purposes, it is worth stressing that it ultimately works for this aspect: By the end of the second year, those who have survived and had enrolled in mathematics intending to teach, realize instead that they can go on with theoretical courses which they have mastered and enjoyed. This really happens a lot. The typical woman mathematician in Italy has enrolled in mathematics thinking it is a good subject for a teaching job, there are more jobs teaching mathematics than other subjects. This is also why her family was happy and didn't question her choice. In two years, she might become bold, and go on. The family only finds out at the end, when she gets the degree, and starts looking for fellowships instead of jobs in schools. By then, everybody is proud. I really think this is a pattern, except when coming from a very educated family that might trust their children more...

It sounds as though in other countries mathematics is considered "unfeminine". Not only do girls expect not to be able to understand it, they even seem to think it would make them less feminine. This seems very scary, especially in adolescence. This one thing is REALLY not true in Italy...

Maybe it would be interesting to develop a questionnaire on how girls see themselves when studying different subjects in different countries. One has to be extra careful in these things, but it would be nice to know.

I agree with Laura, and would like to start with Italy and the Scandinavian countries.

1.4 General discussion. The report below is based on notes taken at the Round Table. In some cases, participants later submitted further comments which have been subsequently incorporated.

On the situation in Eastern Europe

Nadija Zinchenko, Ukraine, Kiev University
In Eastern Europe and the former Soviet Union, many mathematicians are losing their jobs and women are the first to do so. There is a general crisis in science since the governments do not believe that science is important for society. Non-academic institutions are dying. There are about 50% female students, perhaps more, who finish a candidate degree, similar to a Ph.D. degree, in mathematics. Not so many women finish the higher doctoral degree and, since it is a requirement for obtaining a full professorship, there are not many female full professors. The number of female mathematicians will decrease now. Women must work in our society, but they need not create a career. The society used to be closed. A positive change is the better contact with people outside. It is our own problem to solve economic and educational problems. Help with making contact and exchanging information from outside is needed.

Polina Agranovich, Ukraine, Kharkov
We have some scientific grants and funds, for example the INTAS-grant and the INCO-COPERNICUS grant. I think this is very important and a good way to help our scientific work. I suppose this form is interesting for foreign scientists as well. In the frame of grants we can take part in conferences, seminars. It is a real possibility for youth to study mathematics in this very difficult time for us. It is, in fact, easier for us to be in e-mail contact with people from the west than with those from Kiev!

Eva Bayer, France
There are now many international cooperation programs. It would be good to encourage women to take a very active part in these. This could lead, for instance, to the creation of networks that contain many women – of course, not exclusively women. For example, there exists a program called "reseaux formation-recherche" with which one can form networks between one or more French regions and some regions of Eastern Europe or the former Soviet Union countries. The networks provide money for short visits and also doctoral scholarships.

On affirmative action

Donald Collins, United Kingdom
I would like comments on three concrete suggestions for affirmative action which had been discussed in his department and which had not been supported.

- Special Ph.D. fellowships for women.
- Grants for women graduate students to participate in the meeting of BWM.
- A prize for the best female first year student.

Dusa McDuff
I believe one has to think very carefully when creating anything which is "just" for women, though sometimes it is justified. As for the question of the prize, I understand the reluctance of the department to institute a prize for women only (especially when there is no analogous prize for men). It would be better if such a prize were donated by a private person. I once received a special prize: the best mathematical achievement in the last five years by a woman, and I really appreciated it. Such a prize has real value in helping make the work of women visible. A department can try to preserve the spirit of the award while not restricting the prize to women by, for example, having several prizes, one year two women and one man will get prizes, next year it may be opposite, etc. Or, it could keep track of who gets the prize and try to make a generous proportion of the winners women.

Sylvie Paycha
Affirmative action in Germany has had a "backfire" effect. Women applicants for jobs are more likely to be invited to lecture because of affirmative action; afterwards a negative report may be written in order to explain why this woman is not considered for the list. A reason has to be given.

Annette Huber, Germany
Basically, the only permanent positions in Germany are professorships. There is, in general, no formal application process for the remaining junior positions up to the age of 35 to 40. Some departments have to state that they could not find a qualified woman to fill such a position when appointing a man. From what I see in my department this rule does not have any effect. Still, I think that affirmative action on the junior level could be effective. And it would be a good way of making women more visible in the teaching of mathematics.

Eva Bayer
Some universities include in the announcements of openings a text like "women are encouraged to apply". Very often, this does not mean anything, although it gives the impression that they make an effort to hire women. The Swiss Federal Institute for Technology in Lausanne announced very interesting programs to encourage women. It would be good to inquire whether anything really happened.

Dusa Mcduff
Affirmative action is a subtle thing. There is a very good NSF-supported program for female visiting professors. They serve as role models at the top universities as well as having the opportunity to further their own research. It is, in particular, difficult to have affirmative action for permanent jobs. It is much better to do

something at the temporary level. And such temporary positions do help in career-building which is extremely important.

On CV's

Eva Bayer
At the council meeting of the EMS there was a discussion following the report from the committee on Women and Mathematics. It was mentioned that CV's for men and women often look very different. Many women do their best work later in life. Females are generally less competitive and take more time to gain full confidence. It was suggested that the committee should collect a large number of CV's to illustrate this phenomenon. Without such a study it is hard to treat men and women differently for an appointment. A position should not just be an award for past work, prediction of future work is important.

Dusa McDuff
It takes more time for a woman to feel she is a mathematician.

On the North-South situation

Sylvie Paycha
I can confirm what Dusa said. But I would like to ask the audience the following question. In Portugal there are many more women in mathematics. The job is not well paid and does not have a high status. Does that explain the difference between North and South?

Francisco Craveiro de Carvalho, Portugal
To say that the job is not well paid may give the wrong impression that women are underpaid in comparison to men in Portugal. That is not so. Men and women have the same salary and equal opportunities. The problems that women mathematicians face in the Portuguese society have to do with their being women and not with their being mathematicians.

Maria Celeste Gouveia, Portugal
Portuguese women mathematicians also have children and family as do other women in Europe, but they show through hard work that they can do the same as men in their jobs. So they have the same opportunities and, in the universities, jobs are given taking into account the scientific qualifications and not the sex. That is the reason why Portugal has so many women in universities. I think that the situation in Spain is like that in Portugal. So the crisis in Europe is not only caused by men but also by women who don't fight for their rights.

Irene Sciriha, Malta
Malta is Italy's neighbor. Having been under British rule for a number of years, the Maltese have inherited the British system of education. Recently aspects of the Italian system have been introduced as well. The Latin mentality among the Maltese is significant and is very similar to the situation as explained by Laura Tedeschini Lalli of Italy. Unfortunately, in Malta female mathematicians face the drawbacks experienced in England as well. I am the only female mathematician in my department. I was appointed because I had good references as a tutor and I have to fight to be given time to do research.

Rudolf Rentschler, France
(Back to the situation in Italy and Scandinavia.) Why are there so many women politicians in Scandinavia, but so few mathematicians?

Ragni Piene
There is no single reason. In the Scandinavian countries women choose other fields than mathematics. Hardly anybody wants to become a teacher.

Aderemi Kuku, Nigeria
All over the world it is a problem that very few young people go into mathematics.

Closing remarks

Sylvie Paycha
It is not enough to make geographic comparisons. Time comparisons inside each country are also necessary. In France the number of women having a position in mathematics is decreasing. The more difficult it gets the more difficult it gets for women.

Dusa McDuff
I would like to stress the importance of networking and of encouragement within all kinds of mathematics.

Roswitha März
I agree with Dusa. Also do a lot of things "just" for women.

Kari Hag
Make women more visible. Show female students that you expect something from them!

2 Supplementary articles

To supplement the Round Table on the situation in Sweden, where the government has taken some new initiatives, and in the United States, where the Association for Women in Mathematics has done pioneering work, a few women from these countries were invited to submit written contributions.

Women and mathematics – the Swedish situation by Gerd Brandell and Barbro Grevholm

Luleå University of Technology, Luleå (Sweden)
Kristianstad University College, Kristianstad (Sweden)

Introduction. The gender imbalance is still great in general at the Swedish universities. Women are in a majority at the undergraduate level and succeed better on average in their studies, as measured by results in exams. Nevertheless only 8% of the Swedish professors are female (Statistiska Centralbyrån, 1995, 1996). Positions as professors are very rare in Sweden. The great majority of academic teachers with a research degree and research possibilities have positions as senior lecturers. Among those only 20% are women. (The situation in Denmark is perhaps even worse with 5% female professors and 18% female senior lecturers (Drotner,

1996)). The Swedish government has stressed the importance of changing this situation and making it more gender balanced.

Mathematics is regarded as a highly prestigious subject in Sweden and academic positions are relatively few. In 1993 there were only 30 professors in mathematics at all Swedish universities. The number of senior lecturers is about 350. Research assistants (postdocs) are very few (14 in 1993). When positions are rare the competition to get one is of course harder and this seems to effect women very negatively in Sweden. The participation of women in mathematics at a professional level is extremely low. There has not been a female professor of mathematics since Sofia Kovalevskaia died in 1891. There is no female research assistant and there are only about 20 female senior lecturers.

The mathematical community is in general anonymous and isolated. In a research study (Grevholm, 1994b), among the questions put to teachers was whether they knew the name of any professor (alive, not historical) in mathematics. Hardly any could name a professor and many answered that they had never met a professor in mathematics. These teachers had spent three or four semesters as students at a mathematics department. Professors in general devote themselves to research and to teaching and supervising graduate students. This same fact was also pointed out in an evaluation of the undergraduate education in mathematics (Universitetskanslern, 1995, p. 15). The isolation of the mathematical community was criticised in the same report (U, 1995, p. 18).

Professors of mathematics seldom take part in the official debate in society or in the public discussion of school mathematics. This issue is addressed in an international review of Swedish research in mathematics (Swedish Natural Science Research Council, 1995):

> The communication to the public of the nature and value of mathematics is much more difficult than for other subjects. The Swedish mathematics community should attempt to do better whenever possible. It is important that the public, from politicians to school children, understand the central role of mathematics in modern society. The ground for a solid recruitment in mathematics is laid already in primary school. The lack of women among Swedish mathematicians is an explicit example of the lack of communication with the public. (p. 49)

There seems to be a view commonly held by many Swedish mathematicians that there is no labour market outside the universities. The same report noted that:

> Similarly, Swedish mathematics should benefit from closer interactions with other sciences as well as industry and other relevant fields. The isolation of mathematics often prevents Swedish mathematicians to get into contact with interesting problems. (p. 53)

The isolation and lack of communication with the rest of society probably effects the recruitment of women into mathematics negatively. There are a number of other factors that prevent women from choosing mathematics. Work as a mathematician is regarded as an isolated job. Pure mathematics is glorified compared to applied mathematics by many mathematicians. This hinders students from seeing the value of research. Often there is a lack of outspoken support for female students and women are made invisible in their departments. Many mathematicians and others claim that there is no great need for more mathematicians in society. Some women might feel awkward being the only woman in a group. All these factors certainly play an important role when women reject mathematics as their professional field.

Official recognition of the gender issue

The two reports mentioned above agree that the lack of women in Swedish mathematics is a major problem.

In the evaluation of undergraduate education the lack of women is analysed and it is pointed out that the relative number of women is decreasing as we move upwards in the system, from students to Ph.D. students to academic teachers (U, 1995, p. 17). Affirmative action is discussed as one way of stimulating more women to participate. The following is one of 14 recommendations given in the conclusion of the report:

> The departments should in various ways take measures that stimulate increased recruitment of female students at all levels. (p. 21)

In the research evaluation (SNSCR, 1995) one noted:

> There is a remarkable lack of women among Swedish mathematicians [reference to figures and tables]. Female students therefore constitute a large and poorly utilised pool for recruitment. They should be encouraged at all levels to engage in mathematical studies. (p. 51)

And one of the recommendations in the summary urged:

> There should be much stronger emphasis upon the recruitment of women into the mathematical sciences. (p. 14)

The awareness of the gender question as a problem in connection to mathematics research and academic education is clearly stated in both these evaluations, for the first time at an official level. Although evaluations as such do not bring about anything unless they are followed up by actions, this recognition of the gender issue is of great importance and can be used in the future to encourage and legitimate different actions. They also help to raise the awareness about women's situations among academic staff at mathematics departments.

Initiatives for more women to enter mathematics

The situation is still clearly unsatisfactory in our country and the progress is slow. Nevertheless several initiatives have been taken during the last few years with the purpose of improving women's participation in mathematics education and research.

The network Women and Mathematics

In 1990 the network Women and Mathematics was created to support women in mathematics at all levels of the educational system. This network now has over 700 members and a regular Newsletter is sent out (Grevholm, 1996b). The network has arranged three conferences and the proceedings have been documented

(Grevholm, 1992; Brandell et al., 1994a; Lindberg, in press). It cooperates with several important organisations and authorities. It has shown organisers of conferences that there are many competent women who can be engaged as speakers in mathematics and mathematics education. The network has been given the opportunity to report on its activities in every edition of the Swedish mathematics teacher journal (Grevholm, 1991).

Information is collected on women in mathematics today and in history and the network strives to make these women visible (Grevholm, 1994a). The first woman to be awarded a doctor's degree in mathematics in Sweden was Louise Petrén at Lund University in 1911. At that time women could not have a position at a university and consequently she became a school teacher in Lund. It was 36 years before the next woman did research in mathematics. That was Ingrid Lindström, also at Lund University. She became headmistress at an upper secondary school in Örebro. The third woman to receive her degree was Sonja Lyttkens in 1956. She succeeded in obtaining a permanent position as senior lecturer at Uppsala University in 1963. Until 1995 there were in total less than 30 women in Sweden with a research degree in mathematics. This historical picture is conveyed to the young women who are research students today.

The network and the conferences have resulted in the fact that the (very few) women at different mathematics departments have come into contact with each other and developed cooperation. This knowledge has clearly been of great importance both at an individual level and as a basis for understanding the extreme situation for women in mathematics in Sweden (Grevholm, 1995).

Special positions for women

The government and the research councils are now funding special positions for women. Among these are professorships, postdocs and positions for doctoral students. Some of these are in mathematics. Such a professorship has been created at the mathematics department at Uppsala University. Fourteen women of different nationalities, none of them Swedish, have applied for this position. When the appointment is made it will be the first time since Sofia Kovalevskaia that a woman will hold a (tenured) professorship in mathematics in Sweden.

A programme for a female guest professor in mathematics directly funded by the government has been created at Luleå University. Two female professors from the United States and Spain respectively held this position for six months each in 1996. The research councils in science and technology are funding posts for guest professors also during 1997 and 1998.

Each year a famous mathematician is invited by the Swedish National Committee for Mathematics to give a series of lectures at chosen mathematics departments. In 1996 for the first time a woman was invited. Karen Uhlenbeck visited four universities and gave highly appreciated lectures for both faculty and students.

Summer school for graduate students

A Nordic summer school for female graduate students was arranged this year by Umeå and Luleå Universities in cooperation. The idea came up at the Women and Mathematics conference in 1993 during a discussion among women from different universities (Brandell, 1994b). The summer school was funded by the government and was attended by 17 female students from Sweden and Denmark. Among the seven lecturers, all but one female, were Ragni Piene and Bodil Branner. Besides their lectures, the participants gave seminars. Proceedings from the summer school will soon be published.

The students were enthusiastic and hoped that there would be new summer schools in the future. Most important of all seemed to be the fact that these young women got to know several female mathematicians and mathematics research students.

Undergraduate education
Strong recommendations have been made by the Swedish parliament that academic teaching should be carried out in such a way that female students are not excluded by a male academic culture in which they might feel uncomfortable. This is difficult to realise in practice, since the culture and tradition of academic teaching is deeply rooted and the process of changing it therefore must be slow. There is also a certain opposition to the strategy of reforming the educational system from those who are quite satisfied with the formal equal rights that have existed for many decades and do not find it appropriate to go further to meet the needs of the female students.

A project called "Women in Mathematics", the KIM-project was carried out at Stockholm University during 1992 and 1993. The goal was to change teaching practices and organisation of the teaching and learning situation, so as to suit female students better (Jacobsson & Elvin-Nowak, 1994). It involved participation of all teachers in development programmes including those dealing with gender issues. Other mathematics departments – though far from all – have adopted plans for increasing equity and are now implementing these. It is too early to evaluate the outcome of this in actual figures.

Too few students go into science and technology, both at the secondary level in school and at university level. Programmes for recruiting more good students into these areas often give special attention to female students. Since mathematics is a basis for these areas, this has implications for mathematics as well. Five such programmes received substantial funds from the Swedish council for renewal of undergraduate education from 1994 for several years. The intention is to create programmes in engineering and science that will attract more female students (Wistedt, 1996). The funds are used for teacher development in gender issues as well as for development of pedagogical methods that are aimed to attract more women. The methods that are used are often problem-based learning and projects parallel to ordinary courses.

The first students were admitted to these new programmes last year, 1995. It is too early yet to evaluate the programmes, but they are apparently struggling with some problems. Although female participation has increased, the dropout rate is sometimes high and it seems to be mostly women who choose to leave. Nevertheless these programmes represent a great challenge to all those involved and they certainly will give the university system a lot of new experience when it comes to implementing new and radical pedagogical ideas in a gender perspective. One of the teachers comments on a principle that is used, Noah's principle (50% women among the teachers involved), in an evaluation report (Wistedt, 1996, p. 47):

> We are, of course, aware of the fact that there exist deeper problems of equality, but such a simple, almost banal measure, has brought to the fore a range of hidden realities, for example, how many competent women there are... this is a bit shameful to admit... but how many exceedingly competent women there are, that I have never noticed. Isn't it shocking?

Besides these programmes with special funding, several universities have started other new programmes in engineering and science with the aim of attracting more female students (Brandell, 1996). Applied mathematicians mostly receive their training in engineering schools, mainly through programmes in engineering physics and related areas. Therefore more women entering these programmes means more women candidates for jobs as professional mathematicians in the future.

The situation in school

Teachers

There is a shortage of trained teachers of mathematics at all levels of the school system. Many mathematics teachers do not look upon themselves primarily as teachers of mathematics but as teachers of physics, chemistry, science, technology or economics. In upper secondary school (age 16 to 19) one position for teachers is that of lecturer, a position that requires a Ph.D. in the subject. There are hardly any school teachers with this background in mathematics nowadays. This is partly a consequence of the fact that there are so few graduates with a research degree. The new Ph.D.'s are all swallowed by the university system. Besides, the compulsory teacher-education studies in mathematics for teachers of grades 4 to 9 (students aged 10 to 16) has been cut. A serious repercussion is that the mathematical background of the mathematics teachers is insufficient in many cases. As a consequence few teachers are genuinely interested in improving the teaching and learning of mathematics.

The school system

All pupils receive a nine-year compulsory education with mathematics every year. Girls have better marks than boys (Grevholm & Nilsson, 1994). The students think the subject is important but they do not find it interesting. In the final year students have to choose one of 16 programmes in upper secondary school for their continued education. Here a marked gender difference can be noticed. The most advanced course in mathematics is taken in the natural science programme and this is chosen by twice as many boys as girls. This programme is a prerequisite for university studies in mathematics and science. Here we have a serious barrier for girls. When they avoid this programme they have limited their choices for careers in the labour market.

The upper secondary school is not gender neutral. Actually most of the programmes are dominated by one sex. We could in fact partially talk of a girl's school and a boy's school (SC, 1995). The results of the girls are good and they have better marks than boys in programmes preparing for university studies and more girls choose these programmes. Girls more often than boys go on to tertiary studies in general. But with twice as many boys in the natural science programme, the number of women in tertiary mathematics is also only one third of the total. The balance between the sexes in different subject areas seems to level out in time in most subjects, but not in mathematics. In mathematics there are no changes or perhaps a lower percentage of women can be noticed if we compare data from the mid 1980s with the first years of the 1990s (Grevholm, 1996a).

Conclusions

In Sweden there are no legal barriers for women in education and research. More women than men graduate from university (59% women in 1994/95), but in research education men dominate examinations (71% in 1993/94). The higher up we look in the academic hierarchy the fewer women we find (8% professors).

The picture in mathematics is worse, because of the lack of women even at the upper secondary level and still more so at higher levels, where there are no professors. If we make a comparison, for example, with Italy the first impression of that country is much better (Spitaleri, 1996a,b). This fact ought to be investigated further.

A lot of actions have been taken lately as described in the examples above, but we need to follow the results of these actions carefully and if necessary find other ways to improve women's participation in mathematics. We are certainly convinced that both mathematics and society will gain from the participation of women in the subject at all levels.

References

[1] Brandell, G. et al. (Eds.). (1994). *Kvinnor och matematik. Konferensrapport.* Luleå, Luleå University.

[2] Brandell, G. (1994). Finns det uteslutningsmekanismer riktade mot kvinnor vid de matematiska institutionerna? Finns det hinder av annat slag? In G. Brandell et al. (Ed.) *Kvinnor och matematik. Konferensrapport.* (pp. 177–182). Luleå, Luleå University.

[3] Brandell, G. (1996). *Gender in engineering education.* Centre for Women's studies, Luleå, Luleå University.

[4] Drotner, K. (1996). Køn og kvalifikationer. *Politiken,* 11 Nov, 1996.

[5] Grevholm, B. (1991). Kvinnor och matematik – nätverk bildat. *Nämnaren,* nr 2, 43–45.

[6] Grevholm, B. (Ed.). (1992). *Kvinnor och matematik. Rapporter om utbildning, nr 1, 1992.* Malmö , Lärarhögskolan i Malmö.

[7] Grevholm, B. & Nilsson, M. (1994). Sweden. In L. Burton (Ed.), *Who counts? Assessing mathematics in Europe.* (pp. 241–262). Birmingham, Trentham Books.

[8] Grevholm, B. (1994a). Svenska kvinnor i matematiken. In G. Brandell et al. (Ed.) *Kvinnor och matematik. Konferensrapport.* (pp. 62–73). Luleå, Luleå University.

[9] Grevholm, B. (1994b). Gender and mathematics education. Theory into practice. In E. Pekhonen (Ed.),*Proceedings of the Nordic conference of mathematics teaching in Lahti 1994. Research Report 141.* (pp. 78–84). Helsinki, University of Helsinki.

[10] Grevholm, B. (1995). A national network of women: Why, how and for what? In P. Rogers & G. Kaiser (Eds.), *Equity in mathematics education.* (pp. 59–65). London, Falmer Press.

[11] Grevholm, B. (1996a). Women's participation in mathematics education in Sweden. In G. Hanna (Ed.), *Towards gender equity in mathematics education,* (pp. 111–124). Dordrecht, Kluwer Academic Publishers.

[12] Grevholm, B. (Ed.). (1996b). *Kvinnor och matematik. Rundbrev 1990–1995.* Malmö, Lärarhögskolan.

[13] Jacobsson, C. & Elwin-Novak, Y. (1994). *Kvinnor i matematiken. Ett trevligt inslag eller pålika villkor?* Stockholm, Council for the Renewal of Undergraduate Education.

[14] Lindberg, L. (in press). *Kvinnor och matematik. Konferensrapport.* Göteborg, Göteborgs universitet.

[15] Spitaleri, R. M. (1996a). Women's Know-How and authority: Italian women and mathematics. In G. Hanna (Ed.), *Towards gender equity in mathematics education,* (pp. 111–124). Dordrecht, Kluwer Academic Publishers.

[16] Spitaleri, R. M. (1996b). Women's Authority and Career: A Few Meanings. In *EWM's Newsletter* no 3, January 1966 (pp. 10–12).

[17] Statistiska Centralbyrån (SC.) (1995). *Women and men in Sweden. Facts and figures 1995.* Stockholm, Statistics Sweden.

[18] Statistiska Centralbyrån (SC.) (1996). *Women and men in Sweden. Facts and figures 1996.* Stockholm, Statistics Sweden.

[19] Swedish Natural Science Research Council (SNSCR.) (1995). *International Review of Swedish research in mathematical sciences.* Stockholm, Swedish Natural Science Research Council.

[20] Universitetskanslern (U.) (1995). *Nationell utvärdering av grundutbildningen i matematik.* Stockholm, Universitetkanslersämbetet.

[21] Wistedt, I. (1996). *Gender-inclusive higher education in mathematics, physics and technology.* Stockholm, Högskoleverket

Women in Mathematics in the United States by Marie A. Vitulli and Sylvia Wiegand

University of Oregon, Eugene, Oregon (USA)
University of Nebraska, Lincoln, Nebraska (USA)

In this article we describe the current situation for academic women mathematicians in the United States and the impact and activities of the Association for Women in Mathematics (AWM).

Many of the problems facing European women in mathematics also face women mathematicians in the U.S. There has, however, been a remarkable increase in the participation by women in mathematics in the last twenty-five years and women have enjoyed considerable success in their mathematical careers. At mathematics meetings everywhere more women are visible; for example at a regional meeting of the American Mathematical Society (AMS) in Lawrenceville, New Jersey, in October 1996, two of the four invited hour addresses were given by women and half the speakers at the commutative algebra session were women. On the other hand there appears to be a perception by some that opportunities are now slanted unfairly in favor of women. We examine this perception, which is a threat to continued progress.

The Association for Women in Mathematics

One of the contributing factors to the rise in the number of women in mathematics worldwide has been the support and encouragement of the U.S.-based Association for Women in Mathematics, which began in 1971 with the purpose: "to serve and encourage women to study and have active careers in the mathematical sciences". Now, twenty-five years later, AWM is a substantial organization (4500 members) that sponsors outstanding programs for women in mathematics.

When the AWM began its Noether Lecture Series in 1980, named for Emmy Noether, women rarely gave the prestigious invited hour addresses at meetings of the AMS and the Mathematical Association of America (MAA). Since then the situation has changed dramatically (the 1997 meeting in San Diego will include three additional hour addresses by women). The lectures have become a showcase for the most distinguished women mathematicians. Although the speaker is generally over the age of fifty, she need not be from the U.S. (If outstanding foreign women are known to be visiting the U.S., they are considered for the award. It has proven complicated to arrange and finance travel otherwise.)

The AWM regularly schedules a wide variety of other events for the annual Joint Mathematics Meetings of the AMS and the MAA; for example, there are students as well as a business meeting, dinners, and a party. Each year the AWM presents awards to an outstanding woman educator (for contributions to mathematics education), and to an outstanding woman undergraduate (for excellence in mathematics). As a result of all this activity women are now extremely visible at meetings – and feel much more comfortable and welcome than in the old days.

AWM frequently organizes special workshops, such as one held in Berkeley in July 1996 in honor of Julia Robinson. The events at that workshop included mathematics talks given by prominent women mathematicians, a special talk about Julia by her sister, panel discussions on popular topics, and opportunities for recent Ph.D.'s and graduate students to communicate their work. These workshop and other outreach activites sponsored by AWM have been inspirational for young women and have helped them to establish contacts that will be useful later in their careers. (Older women mathematicians also gain inspiration and excitement from these events; the presence of so many impressive women mathematicians of all ages inspires every woman to continue to work hard and realize her potential, and the women are terrifically supportive of each other.)

Other ongoing programs and projects of AWM include: a monthly newsletter; mathematics education programs; career publications; an AWM membership directory; the Directory of Women Mathematicians; the Sofia Kovalevskaia mathematics days (especially for high school girls in mathematics); and many other special events. All these AWM programs are produced by energetic and dedicated volunteer woman mathematicians and the tiny devoted AWM staff. Like most women and women's organizations, they are over-extended and under-funded!

A Presidential Year

During the year 1996 the presidents of nine major mathematics and related organizations in North America were women.[1)] The abundance of women simultaneously in leadership positions of all these organizations was mentioned at a meeting of the Conference Board of the Mathematical Sciences in December 1996; it was the first such meeting to include so many women, and many of these women were present. Some said that there was a "ripple effect"; when some organizations started nominating women, the other organizations followed suit. Others said that feminine traits such as attention to details are evidently becoming more valued for presidents of these organizations.

Affirmative Action Curtailment and Backlash

The increased representation of women in mathematics is also a result of a wide range of local and national affirmative action programs. In the last few years these programs have been attacked from diverse fronts. The National Science Foundation proposed ending the Visiting Professorship for Women Program, a productive and supportive program which existed for about ten years. (It has been reinstated.) Scholarships, fellowship programs and special admissions policies for undergraduate, graduate, and postgraduate programs intended to increase the participation of women and minorities have also come under attack. Many states, such as California, have eliminated affirmative action programs for higher education. Articles and letters to the editor in professional journals have protested affirmative action programs and preferential treatment for women and minorities; other articles document the existence of an affirmative action backlash.[2)] Department heads sometimes tell unsuccessful male job applicants that the department was pressured into hiring a woman; obviously this increases the backlash and does a disservice to the profession.

Employment Study

Mary E. Flahive (Oregon State University) and Marie Vitulli analyzed data from the 1991–1995 AMS-Institute of Mathematical Statistics (IAS)-MAA Annual Surveys for gender differences in initial employment of new Ph.D.'s in mathematics from U.S. institutions between 1990 and 1995.[3)] This data analysis was undertaken partly in response to affirmative action backlash. The questions studied and their findings are summarized in the tables and discussion below.

---

1) The organizations and their presidents are 1. The American Mathmatics Association for Two Year Colleges: Wanda Garner; 2. The AMS: Cathleen Morawetz; 3. The American Statistical Association: Lynn Billard; 4. The Association of State Supervisors of Mathematics: Mari Muri; 5. The Canadian Math Society: Katherine Heinrich; 6. The National Council of Teachers of Mathematics (NCTM): Gail Burrill; 7. The Society of Industrial and Applied Mathematics: Margaret Wright; 8. The Sociedad Matematica Mexicana: Patricia Saavedra; and 9. The AWM; Chuu-Lian Terng. The Executive Directors of the MAA (Marcia Sward) and the NCTM (Linda Rosen) for 1996 are also women.

2) A. Jackson, Notices of the AMS Vol. 40, (Jul.–Aug. 1993) pp. 588–590 and Vol. 41, (April 1994) pp. 286–287.

3) The full report will appear in the February 1997 issue of the Notices of the AMS.

Question 1: Do men and women have the same employment rates? To answer this the rate of joblessness was calculated; that is, the percentage of individuals who were either still seeking or were unemployed and not seeking employment out of all those new Ph.D.'s who reported their initial employment status. The jobless rate for females was 10.2% and for males 12.0% – these rates are not substantially different.

Question 2: For the Ph.D.'s who obtained jobs, what types of jobs were men and women getting? Table 1 below summarizes the frequencies and percentages of first jobs in various categories. "Group I" refers to the top 48 Ph.D.-granting mathematics departments, "Group II" to the next 56 departments, and "Group III" to the remaining 72 Ph.D.-granting mathematics departments. The designations "Bachelor's" and "Master's" refer to departments in which the highest degree offered is the bachelor's (undergraduate degree) and master's degree, respectively. (That is, Ph.D. degrees are not offered.)

Type of Employer	# Females	%	# Males	%	Totals
Group Ia*	108	(15.4%)	491	(19.6%)	597
Group II	34	(4.9%)	149	(5.9%)	183
Group III	65	(9.4%)	183	(7.3%)	248
Master's	82	(11.9%)	235	(9.4%)	317
Bachelor's	185	(26.9%)	422	(16.8%)	607
Other Academic (incl. foreign)	155	(22.5%)	652	(26.0%)	807
Government, Industry	62	(9.0%)	379	(15.1%)	441
Totals	689		2511		3200

Note: *Group Ia=Group I + Research Institutes.

Table 1. Observed frequencies of first jobs for new Ph.D.'s 1990–1995

Apparently U.S. women take first jobs at bachelor's-granting mathematics departments at a much higher rate than men. Whether this noteworthy difference is due to employer-bias or preference on the part of the new Ph.D.'s is unclear; but certainly it is related to the high percentage (62.4% in 1995–96[4)]) of tenure-track jobs at bachelor's-granting departments. (The term *tenure* refers to the promise of lifetime employment at the institution. Positions that could lead to tenure are called *tenure track positions*. Non-tenure track positions are usually terminal after a fixed number of years.) Women's career patterns generally differ from men's. Many women are most creative and productive later in their careers. The early years are important for their careers, to be sure, but they are also important for their families, because they are the prime child-bearing years. Temporary jobs seem to disproportionately and negatively affect women for these reasons. A possible explanation for the large proportion of women in bachelor's departments is the fact that diverse forms of scholarship and service to the institution are valued and rewarded in these departments. Today many research universities are under pressure to increase the quality of undergraduate education but few of these universities reward faculty for good teaching.

Question 3: Have women been equally successful in obtaining academic positions at a department of at least comparable ranking to that of the degree-granting department? After earning the Ph.D. many non-U.S. citizens left the U.S. and the type of foreign employment they have found is unknown. Consequently, this analysis includes only the U.S. citizen cohort, which consists of 412 females and 1303 males. Table 2 below summarizes these comparable employment rates.

	Type of Ph.D.-Granting Department					
	Group I		Group II		Group III	
Employer Type	Female	Male	Female	Male	Female	Male
Group Ia	25.9%	26.7%	na	na	na	na
Group Ia-II	na	na	9.3%	14.9	na	na
Group Ia-V	na	na	na	na	19.8%	16.1%

"na" means that this data entry is not applicable

Table 2. Comparable Employment Rates for U.S. Citizen new Ph.D.'s 1990–1995

Since the data does not distinguish between either one-year and multi-year positions or between tenure track and non-tenure track positions, the analysis for Question 3 is not completely satisfactory. For instance, a tenure track appointment in a Group I department is counted the same as a one-year "hold-over" position for a former student. These observations are particularly important because a very high percentage (84.2% in 1995–96[4)]) of the first jobs in U.S. doctorate-granting mathematics departments are not tenure track. Nevertheless the information in Table 2 suggests that women are about as successful as men at obtaining comparable employment; thus the perception that women have an advantage appears to be false.

Temporary Positions

An increasing number of U.S. mathematics departments are offering recent Ph.D.'s temporary postdoctoral positions (called *postdocs*; this refers both to the positions and to the recipients). Some departments offer postdocs with reduced teaching loads and a chance to work with senior mathematicians in the postdoc's research field. Other departments offer postdocs with nearly the same teaching loads as tenure track positions; they use the postdocs as a way of bringing young people, or sometimes women, to a department. Departments sometimes hire women as postdocs to counter the charge that women are severely underrepresented on their faculty. These departments artificially increase the number of women on their faculty and thereby "look better" in affirmative action studies, which are frequently

4) Notices of the AMS, Vol. 43, (December 1996), pp. 1493–1511. After a 6 year probationary period, a tenure track professor is reviewed for tenure. If tenure is awarded, with few exceptions, the professor has a promise of lifetime employment from the institution. If tenure is denied the professor must leave the institution.

required for good standing with grant-giving agencies, including the federal government. Postdocs with no real benefit to the recipients are surely harmful to young mathematicians in general, and women in particular. If women are getting a large share of the postdocs but are not obtaining tenure track positions as subsequent jobs, then the apparent near-equality of the comparable employment rates for men and women is questionable.

Increases in Numbers of Women in Mathematics

The number of women who receive Ph.D.'s in mathematics and the number who obtain entry-level positions at doctorate-granting mathematics departments have both increased over the past twenty-five years. Table 1 shows that 21.5% of the new Ph.D.'s in mathematics from U.S. institutions between 1990 and 1995 were women. Looking at U.S. citizens only, in 1972 only 6% of the U.S. citizen Ph.D.'s were women; in 1976 women were about 12%. The percentage of U.S. women grew to 23.9% of all U.S. citizen Ph.D.'s between 1990 and 1995 and the percentage for 1996 was over 25%. Women received 20.1% of the jobs for new Ph.D.'s at doctorate-granting U.S. mathematics departments or research institutes between 1990 and 1995. There haven't been comparable increases in the number of women in senior-level positions at doctoral-granting mathematics departments, despite the fact that since 1982–83 at least 20% of the new U.S. citizen Ph.D.'s. in mathematics have been women. In 1993 only 5.8% of the tenured faculty at Ph.D.-granting mathematics departments were women; at the same time 4.9% of the tenured faculty in Group I departments were women[5)]. The AMS-IMS-MAA does not conduct longitudinal studies to follow the careers of new Ph.D.'s, both men and women, five and ten years after the Ph.D. The combined societies are reluctant to take on the project until a funding source can be found. The AWM and the Joint Committee on Women in the Mathematical Sciences have advocated for longitudinal studies for the past several years.

In summary, the prospects for women Ph.D. mathematicians are better in the U.S. than they were 25 years ago, but there is still room for improvement. We hope that the momentum and progress will continue.

3 Statistical data in Europe

The statistics on women in mathematics in Europe is visualized in a striking way in the video "Women and Mathematics across Cultures, EWM – European Women in Mathematics".

Before the second European Congress of Mathematics a letter was sent by the EMS-Committee on Women and Mathematics to all mathematical societies of Europe. We tried to make it clear that it is important to follow changes in the role of women in mathematics within each country. Therefore we asked for the same information in 1995 as was asked for in 1990 (as to Ph.D.'s awarded in 1995 we added that for small numbers we would like the number to be an average, preferably for the last 4 years):

---

5) M. B. Ruskai, "Time for Advancement" MAA Focus 14(6), 25–28 (Dec. 1994)

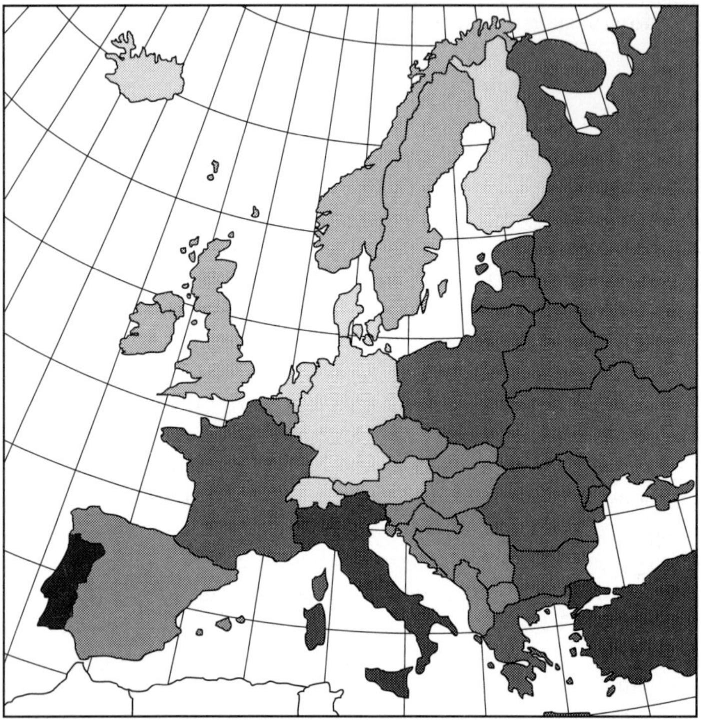

Table 3.[6] Percentage of women among tenured mathematicians at university level

I. The number of mathematicians holding a permanent position in a mathematics department of a university or a research institution in the country in question.
II. The number of full professors ("full professor" being the highest academic position).
III. The number of Ph.D.'s awarded in 1995.
IV. The percentage of women among the students in mathematics.

We requested the numbers of women, as well as the totals, in the questions I, II, III.

Since the 1995 survey was a repetition of that in 1990 and since it is a primary aim of the EMS-Committee on Women and Mathematics to collect such data every four years, we hoped that many countries would have efficient means for doing this. As of October 20 the societies in Denmark, Finland, Israel, Macedonia, Norway, Slovenia, Slovakia, Switzerland and the United Kingdom have furnished us with complete data. We have also received some data from France and Sweden. The

6) This map was made by Marjatta Näätänen and Jarkko Lavinen, University of Helsinki, and taken from the video *Women and Mathematics across Cultures* (see Section 1.2). The video can be purchased by contacting: `ulmanen@sophie.helsinki.fi`

Country	Students	Ph.D.'s	Mathematicians	Full Prof.
Austria '89	39%	3/21=14%	54/762=7%	0/73=0%
Belgium	54%	12/32=38%	30/219=14%	8/134=6%
Bulgaria	40%	4/23=17%	254/855=30%	1/42=2%
C.S.F.R	42%	1/10=10%	60/500=12%	2/65=3%
Denmark	30%	1/5=20%	4/121=3%	1/19=5%
Denmark '95	30%	0.75/4[1]=19%	3/121=2%	1/21=5%
Estonia	60%	0/4=0%	32/109=29%	0/8=0%
Finland	41%	8%	3/127=2%	1/34=3%
Finland '95	37%	1/6=17%	9/146=6%	1/34=3%
France	30%	20%	20-25%	8%
France '95				10%
W. Germany '87	33%	19/208=9%	40/1500=3%	4/490=1%
Greece			67/283=24%	3/56=5%
Georgia	70%	1/28=4%	486/1217=40%	7/72=10%
Hungary	15%	5/38=13%	32/281=11%	4/103=4%
Ireland	30%	2/3=67%	7/135=5%	0/9=0%
Israel '95	45%	10/37=27%	15/295=5%	5/165=3%
Italy	70%		609/1727=35%	84/646=13%
Iceland	33%	0/0	0/10=0%	0/4=0%
Luxemburg	24%		1/7=14%	1/5=20%
Macedonia '95	73%	7/11=64%	25/56=45%	1/12=8%
Netherlands	22%	2/41=5%	19/437=4%	1/88=1%
Norway	20%	0/5=0%	9/103=9%	3/45=7%
Norway '95	29%	3.25/14.75[1]=22%	8/117=7%	3/67=4%
Poland (5 univ.)	60%	3/16=19%	107/359=30%	7/44=16%
Portugal	50-60%		40-50%	5%
Slovakia '95	29%	4/11=36%	149/482=31%	1/27=4%
Slovenia '95	52%	0/2=0%	4/51=8%	0/14=0%
Spain			168/1075=16%	12/279=4%
Sweden	30%	1/10=10%	7/150=5%	0/21=0%
Sweden '95[2]		?/14[1]	22/310=7%	0/41=0%
Switzerland	25%	4/42=10%	3/141=2%	0/91=0%
Switzerland '95	20–30%	10/50=20%	4/150=3%	2/90=2%
U. Kingdom	31%	46/266=17%	97/1379=7%	3/267=1%
U.Kingdom '95[3]	37%	36/198=18%	250/2143=12%	10/325=3%
Ukraine	50%	1/3=33%	31/145=21%	1/33=3%
Yugoslavia	60%	4/24=17%	91/530=17%	12/132=9%

[1] Average figures
[2] Broader definition of mathematicians
[3] Polytechnics are now universities, the number of universities has increased from 50 to 150

Table 4. Percentage of women in mathematics in Europe

German Mathematical Society will include our questionnaire when it updates its Address Book of mathematics departments in the fall.

The United Kingdom now has a higher education statistical agency which was able to provide the necessary figures for them. For many other countries, however, the following letter from Luc Lemaire of Belgium, unfortunately, reflects some of the difficulties encountered in obtaining such data.

As for the questions in your letter, I fear there is no way I could provide the answers within your deadline (or even deadline + large $\varepsilon > 0$).

The way I obtained answers the first time was to write to all mathematics departments, wait a few weeks (months), write again etc. Indeed, I don't believe global statistics are available in Belgium.

If I were to do the same now, I wouldn't get complete answers before at least 6 months. Moreover, the shortcoming of the approach is the following: With regard to the figures of four years ago there was an element of choice by the various chairs of departments who gave them. Is X a mathematician or a physicist, is Y in a mathematics department etc.? If I asked the same questions now to different people, the difference in those choices might well exceed the evolution due to the small number of retirements or appointments in the last four years – so the evolution of the situation will not be faithfully represented...

If we combine the new figures with the figures from the survey in Table 1 on page 43 of *First European Congress of Mathematics, III, Round Tables*, Birkhäuser Verlag 1994, we obtain the following summary. Data is from 1990 unless otherwise indicated.

Acknowledgements

This Round Table could never have taken place without the help of so many dear friends and colleagues. I wish to thank them all and, in particular, Bodil Branner, Lois Gehring, Marjatta Näätänen and Ragni Piene for their invaluable support and counsel in different phases of this undertaking.

Kari Hag

4 Appendix

Description of EWM

European Women in Mathematics (EWM) is an affiliation of women bound by a common interest in the position of women in mathematics. Our purposes are:

- To encourage women to take up and continue their studies in mathematics.
- To support women with or desiring careers in research in mathematics or mathematics related-fields.
- To provide a meeting place for these women.
- To foster international scientific communication among women and men in the mathematical community.
- To cooperate with groups and organizations, in Europe and elsewhere, with similar goals.

Our organization was conceived at the International Congress of Mathematicians in Berkeley, August 1986, as a result of a panel discussion organized by the Association for Women in Mathematics, in which several European women mathematicians took part. There have since been six European meetings: in Paris (1986), in Copenhagen (1987), in Warwick (England) (1988), in Lisbon (1990), in Marseilles (1991), in Warsaw (1993), and in Madrid (1995). The next meeting will be in Trieste, Italy, December 13–17, 1997.

At the time of writing, there are participating members in the following countries:

Belgium, Czech Republic, Denmark, Estonia, Finland, France, Germany, Greece, Italy, Latvia, Lithuania, Malta, the Netherlands, Norway, Poland, Portugal, Romania, Russia, Spain, Sweden, Switzerland, Turkey, Ukraine, and the United Kingdom; and there are contacts in Albania as well as with several non-European countries. Activities and publicity within each country are organized by regional co-ordinators. Each country or region is free to form its own regional or national organization, taking whatever organizational or legal form is appropriate to the local circumstances. Such an organization, Femmes et Mathematiques, already exists in France. Other members are encouraged to consider the possibility of forming such local, regional or national groups themselves. There is also an e-mail network, and EWM has its own Web page. The address is: `http://www.math.helsinki.fi/EWM`.

For further information contact:
The secretary of EWM: Riitta Ulmanen,
Department of Mathematics,
P.O.Box 4 (Yliopistonkatu 5),
FIN-00014, University of Helsinki, Finland;
e-mail: ulmanen@@sophie.helsinki.fi,
Tel 358 9 191 22853, Fax 358 9 191 23213

Progress in Mathematics, Vol. 169, © 1998 Birkhäuser Verlag Basel/Switzerland

Public Image of Mathematics

ROLAND BULIRSCH

A round table discussion:
(Roland Bulirsch, Madame Mireille Chaleyat-Maurel, Gyula Bencze)
Budapest, July, 24, 1996, 14:30–16:00

Introductory remarks by R. Bulirsch

The public image of mathematics is not of the best. The work of mathematicians and to a certain extent the work of computer scientists is not considered as an important contribution to our civilisation. Even in countries like France where mathematics enjoyed a higher reputation than elsewhere, things have changed. Mathematicians are dismayed to find "public" doors closed and there are increasing problems in obtaining financial supports for their research projects.

Part of the difficulty is with mathematics itself: to present rather abstract things to a public inundated by a deluge of pictures makes it a tough job to put mathematics in the right sight. But part of the difficulty is also with the mathematicians themselves: very many of them are unwilling to demonstrate the importance of their work fearing that popularization of mathematics could do damage to the, as they say, flawless image of mathematics, the "purest of the pure sciences".

At the Munich University of Technology some films have been produced to show the non-mathematical public, the educated layman, what kind of useful things mathematics can do in our life.

- Flow and currents in tiny transistors, the coloured images show the solution of a complicated system of nonlinear partial differential equations of elliptic type.
- The optimal design of a spacecraft and its optimal flight trajectory to a planet like Venus. A visualisation of a highly complicated nonlinear multipoint boundary value problem arising from optimal control theory.
- Evolution of the Sun in its 12 billion years life-cycle. Another visualization of the solutions of a highly complicated system of partial differential equation of parabolic type.

After the presentation comments were made by panelists and the audience:

Mr. Gyula Bencze

It is generally believed by the public that mathematics as a science is extremely uninteresting, "dry" (meaning difficult to digest), and devoid of exciting new developments. It is also held that those – especially young students – interested in various mathematical problems are in some sense "strange" and differ from people with "normal" interests. While this is more or less an accurate description of the situation in Hungary, my personal experience in several European countries as well as in the US led me to assume that it is a problem shared by almost every country.

In addition the general public has in fact very little idea of what mathematicians are actually doing and how important mathematics is in forming our way of thinking. This is reflected by the curious phenomenon that in Hungary – as a rule – the most popular TV personalities boast about their bad marks in high school math. Had they known that the inability to learn basic mathematics is strongly correlated with stupidity and low IQ they most certainly wouldn't be so proud of their "achievement". In Hungary, according to the Central European tradition of "classical" education, one can be an "intellectual" without a morsel of knowledge in both mathematics and (natural) sciences. In fact recently in a national TV broadcast our country's "intellectual elite" was mentioned to consist of some politicians(!), journalists, writers as well as economists. Not a single scientist was even considered as a candidate for inclusion in this prestigious ensemble!

While most scientists – including myself as a theoretical physicist – are either practitioners of mathematical methods or at least are fully aware of the importance of mathematics as a tool for both theory and practical applications, the society's decision makers as well as the general public are almost totally ignorant of the relevant facts.

In a democratic society the funding of scientific research is supported indirectly by taxpayers' money on the decision of politicians. It is imperative that both politicians and the public should be fully informed about the importance of mathematics and science in general. This most important task is best done by the scientists themselves. However, it is not very easy to formulate in simple terms the basic goals of mathematics research. In any case, it is the scientists' responsibility that, instead of the sensationalism of the mass media, accurate and easily understandable information is provided on the most important results of mathematics with due emphasis on their practical application. This is the only way that an increased support for basic research can be secured.

In order to emphasise that the education of not only the public but the politicians is equally important, let me just mention that Hungary's minister of defense, according to his own admission, is a firm believer in UFOs! Other high level politicians are also avid fans of the worst kind of pseudoscience while the grave financial problems of "official science" obtain no such (moral) support.

Mathematicians sometimes argue that physics and other branches of natural science can more easily get funds for research due to the more attractively worded research proposals, e. g. "quark-gluon plasma phase transition" or "search for traces of life on other planets". I can assure you that in Hungary mathematics is not a subject of negative discrimination; in fact all branches of (natural) science have the same (insufficient) level of funding.

There is, however, a single advantage that mathematics has over all branches of science: the almost complete absence of pseudoscience and charlatanism flooding almost all the new democracies! The explanation seems to be simple; even the solution of the simplest mathematical problems requires some mental effort, moreover it is easy to distinguish between good mathematics and quackery. On the other hand, as an example of the other extreme, wherever you look, you find people who want to extract the infinite energy of the physical vacuum, or the army of "alternative healers" who claim to be able to treat all the incurable diseases including AIDS!

To conclude this short comment, I agree with the opinion expressed by a number of people here that the public image of mathematics can be changed to its advantage only through concentrated effort of the scientists themselves with the help of properly educated – and sympathetic – representatives of the mass media.

Mr. Carlo Sempi

I wish to argue that part of the responsibility for the poor image of mathematics lies with us. Other scientists, notably physicists and biologists, are willing to present their researches in a form accessible to the general public. Some of the results presented by the physicists are no less abstract than what mathematicians do. The cultivated layman is aware of some of the problems of modern physics – the search for fundamental particles or the exploration of the universe – or of biology, for example the study of genes and of viruses. The general view of mathematics seems to be that of a perfect building, which has always been in the present shape; some parts of the building are still covered and, from time to time, a veil is dropped to reveal a tiny piece. Mathematics appears to be without both life and history. As a consequence, nobody, even among the cultivated public, knows what the problems of modern mathematics are. We are also hindered by our belief in rigour; this makes everything we write or say for the non-specialist liable to be attacked by colleagues ready to point out the lack of precision or of complete assumptions.

Mr. Kuku Aderemi

It cannot be over-emphasised that there is a crying need to repair the dented image of mathematics all over the world through intensive popularisation efforts at national, subregional , regional and global levels. For one thing, the number of young people choosing careers in mathematics is dwindling in quality and quantity because of the generally unfavourable image of the subject resulting in a generally un-complementary attitude, bordering on hostility from government functionaries, funding agencies, parents and the general public.

One major reason for this hostility has to do with the historical development of the subject that makes people rather ignorant of what mathematicians do. Indeed, those who study mathematics up to high school know little more than the mathematics discovered in the 17th century. Moreover, most of the basic and applicable mathematics discovered in the last two hundred years are yet to find their places in the mathematics education curricula at the appropriate levels and/or published for the information of the general public.

It is well known that most research mathematicians, who are best equipped to propagate mathematics through simplification of the ideas, are not interested in doing so. So, it is essential to re-orientate the training of mathematics educators – those working for masters/Ph.D. degrees in mathematics education – to include the ability to simplify contemporary mathematical ideas. This involves exposing them to a lot more mathematics at graduate level.

To popularise mathematics, all modes and media should be exploited: electronic communication, including e-mail, satellite communication, print and electronic media, e.g., newspapers, popular magazines, radio, TV, Videos, Audio Visuals. Mathematical organisations at national and regional levels should flood the whole world with popular writings on what mathematics is good for, etc, etc.

Mr. Jean-Pierre Bourguignon

Two points seem to be worth mentioning:

1. *Opposing fundamental mathematics to applications of mathematics when speaking of mathematics to the public seems inappropriate* to me since the very nature of mathematics is the mixing of the two. Throughout history, mathematicians have been solving problems raised by concrete situations, and many times the solution involved objects that had been introduced earlier (sometimes by other mathematicians) for completely different purposes (think of the ellipse as a geometric object studied as such that later turns out to be fundamental as a dynamical object in the solution of the two-body problem in Celestial Mechanics). This is the real power of abstraction. In this connection it may be worth recalling a definition of mathematics proposed by Henri Poincaré: "to do mathematics is to give the same name to two different objects".

2. One should *not overlook the possibility of creating public events centered around mathematics.* Quite often professional people from the media consider this very risky. My experience based on a number of different situations in France is that any time such an event is created, some official organizers are reluctant, but more people turn up than expected, sometimes even in record breaking numbers: this happened for the day devoted to young people at the Ministry of Research on the occasion of the awarding of the 1994 Fields medals; this happened again for the whole week of mathematical events that were organized in Bures-sur-Yvette "Les Mathématiques dans la Ville", and gathered more people (a great majority of them non-scientists) than any other cultural week in this small town. I also experienced the same thing in smaller towns in France, such as Biarritz or Pau, where local groups put together a series of conferences on mathematics for a couple of weeks with impressive audiences.

Progress in Mathematics, Vol. 169, © 1998 Birkhäuser Verlag Basel/Switzerland

Education: Challenges for Mathematics

EDITED BY VAGN LUNDSGAARD HANSEN

Technical University of Denmark, Lyngby (Denmark)

Mathematics is one of the oldest scientific fields and it is certainly the most widely taught subject in schools all over the world. As mathematicians we have great responsibilities towards the education in mathematics of future generations. What we teach in the classrooms or present in lectures, is not only important for the immediate mathematical needs of the students but has, in the longer perspective, a decisive impact on the image of mathematics in society. Actually mathematics meets a real need of modern societies and individuals.

It is not enough that researchers in mathematical education or didactics of mathematics think about these matters. It is important that all teachers of mathematics – also at the university level – think about them and, above all, are ready to take initiatives to improve the teaching of mathematics at all levels, where necessary. The purpose of the Round Table on Education was to stimulate discussion and awareness among mathematicians of the fundamental importance of the teaching of mathematics.

Issues for the Round Table on Education

1. The contribution of the didactics of mathematics to our understanding of mathematics as a science and a subject.
2. Basic issues of mathematical education.
3. Mathematics in primary and secondary schools, and recruitment to university.
4. Mathematics for talented children.
5. European summer schools for secondary school teachers.
6. Current problems in the teaching of mathematics at the university level.
7. Computerized tests for mathematics qualifications.
8. What are the challenges for the teaching of mathematics in schools and at universities in the 21st century?

Structure of the Round Table on Education

The oral contributions to the round table included two invited presentations on the above issues and questions, and a few other invited contributions in order to open the discussion on issues not covered by the presentations. Following each invited contribution, participants were free to respond to the issues raised. After the round table some of the participants submitted further comments, which have been included in this report together with the written accounts of the invited contributions and extracts from the discussions.

The two invited presentations were:

Lajos Posa: *Convictions and doubts concerning mathematical education.*

Jean-Pierre Boudine and Christian Mauduit: *A European Knowledge Certification Network System.*

The other invited contributions were:

Vagn Lundsgaard Hansen: *Mathematics in primary and secondary schools, and recruitment to university.*

Miklos Laczkovich: *Current problems in the teaching of mathematics at the university level.*

Vagn Lundsgaard Hansen: *New challenges for the teaching of mathematics.*

To supplement the report on points not sufficiently covered by the invited contributions and discussions, a few persons were subsequently invited to present their views on some of the issues for the round table. These contributions are placed at the end of the report and include the following articles:

Vinicio Villani: *Feedback from mathematics education to mathematical research.*

Mogens Niss: *The contribution of the didactics of mathematics to our understanding of mathematics as a science and a subject.*

Jean-Pierre Kahane: *Mathematics education towards mathematics.*

* * *

Convictions and doubts concerning mathematical education

LAJOS POSA

Eötvös University, Budapest (Hungary)

The world over, mathematics is taught as an *important* subject. Why is it so important? I would like to believe (as many of us do) that mathematics is important because it trains children to think. Indeed, much sooner than other branches of learning, and requiring very little knowledge, mathematics provides the possibility of creative thinking at a high level, unrestricted by authoritarian principles, using common sense and the flight of the imagination.

But have these hopes been realized? I am afraid that the freedom and pleasure of thinking have not been imparted to children. This has remained an empty slogan, a mere declaration of our importance. Meanwhile, in everyday life, and in some schools, mathematics is of a low standard, boring and empty; elsewhere it is too difficult and remains the source of constant frustration. Mathematics is seldom what it should be. Why is this so? Here are some possible causes:

1. We have not gone beyond the traditional schooling based on the exposition given by the teacher. This old-fashioned attitude emanates from textbooks as well as from the teachers themselves. No wonder: they were taught in the same way.

2. The usual subject matter of mathematics is not suited to the purpose of original thinking and learning through exploration and discovery. I am afraid that the 'usual' fundamentals that 'should be' taught according to adult society are hopelessly difficult for a significant portion of children. To take just one example: for many children aged twelve, computations with complicated fractions are nothing but phantasms lacking any real meaning. Perhaps some unusual topics – like games of strategy – would be more suitable for our purposes?

3. Is mathematics really for *everyone*? Is the chance of joyful and successful thinking really granted to every child? Or should mathematics be the amusement of children fascinated by this topic, as is the case in music, drawing and sports? (For today's school children it is much more frustrating not being good at mathematics than, say, being unable to sing at the same standard.)

And, finally, another fundamental question. Supposing that a child is a good and effective mathematical thinker, does this *really* improve their thinking in some deeper sense of the word? Does this faculty transcend itself? For example, does it *transfer* to a better understanding of other subjects? Does it really have benefits of any kind?

I would like to believe in this *transfer*, and also that it is practicable for every child. Or, at least, that it will be practicable, if we do it better.

I have also been asked to say a few words about my work with mathematics for talented children.

Unlike the mass of children frustrated by failure in mathematics, there are children who easily learn the mathematical curriculum of normal schools. They are bored by the curriculum and they could easily discover much more. It is those children that I here, for short, call "talented". There are many such children, according to my personal experience.

The problem of the mathematical education of talented children is unsolved. As for special classes, I think on the one hand that we need them, but on the other hand that they do not solve the problem, since: (i) not every talented child is admitted to such a class; (ii) this form of education can be harmful for some children (when they are not among the best in such a class).

It is frequently asked why one should care about the mathematical development of young children aged 12–13. After all, we do not need that many mathematicians! The answer is that for children aged 12–13, talent, as defined above, usually is NOT exclusively mathematical. However, at this age the well-developed intellect can manifest itself only in mathematics; it is mainly in this field that the mind can produce arguments at a high level. When we teach these children, we therefore educate the intelligentsia (or, more generally, thinking people) of the future. This makes the responsibilities in connection with this work so great.

From the discussion

Helge Elbrønd Jensen, Denmark: It is well known that Hungary, like other East European countries, has been successful in training talented children to obtain good and even extraordinarily good results in international competitions in mathematics. But now I ask: What actually happens to these children afterwards? Do they become creative mathematicians either as researchers or teachers? Is it from these special classes that society recruits the persons who take the leadership in research and teaching as adults?

Lajos Posa, Hungary: Only very few children from the special classes go on to study mathematics. Typically, they study engineering or economics or something like that. One should not advise children to come to the special classes unless they are really good at mathematics, say among the top 10% in their class.

Per Hag, Norway: It should be emphazised that today, more than ever, mathematics (and statistics) is an important, and in many cases indispensable, tool in different fields: engineering, science, economics, ecology, medicine and biology. As is stated in the well-known Davis report [Notices Amer. Math. Soc., October 1984] on the situation for mathematics in the US: "Too few people recognize that the high technology, so celebrated today, is essentially a *mathematical* technology."

To say that mathematics is important because it trains children to think is, therefore, misleading. One should rather stress the fact that teaching mathematics (and statistics) is important, foremost, because mathematics is the basic tool in many professions today. Collaboration between teachers and experts in the various fields of applications of mathematics is necessary in this connection!

* * *

A European Knowledge Certification Network System

Jean-Pierre Boudine and Christian Mauduit

Université de la Méditerranée, Marseille (France)

The system we want to present is a response to an initiative of the European Community Commission. On November 26, 1995, the commission published a book entitled *Teaching and learning, toward the Cognitive Society*, signed by Edith Cresson, Padraig Flynn and Martin Bangemann. One of the proposed goals described in this book is "the support of the acquisition of new skills", and related to this goal, "to create new kinds of skills certification". Professor Jean Louis Reiffers, Université de la Méditerranée, France, who was one of the editors of the White Book providing background material for the initiative of the ECC, asked us to work on the concrete realization of this project in the case of mathematical knowledge. Together with a few other French and Italian colleagues, we therefore began to work out computerized tests in mathematics, respecting the highest possible safety regulations and living up to recognized evaluation criterions. The tests are based on automatization in order to make them easily accessible, to secure objectivity, and to keep the costs of the whole system low.

Anybody should be able to take those tests at almost any time and at a designated place within easy reach. The use of a computer should guarantee equal chances to all candidates, since there will be no more or less demanding, well-meaning, or exhausted examiners.

The target group

The target group for this testing system is really the whole population of the European Community, with no age limit. Access will be granted to everyone who wants his knowledge in a particular field to be tested and certified. The results obtained will be stored on a magnetic card. The recognition of the results of these tests by institutions and companies will depend on the existence of a consensus on the objective value of the procedure. In a sense, the system can be thought of as a broad generalization of tests like the TOEFL.

In our opinion, such a system will not represent any threat to recognized academic diplomas. In fact, the system of knowledge certification and academic diplomas should not be considered as rival systems, but as complementary systems. Moreover, it could be of some help on the issue of foreign diploma equivalence, but this is really a minor point in relation to the main purpose.

It is generally believed that the users of computerized testing systems will most often be people who need to acquire and validate new competences after their entry into professional life. It seems reasonable to believe that making the access to certification easier will strongly motivate people to undertake training, and thus increase the whole level of knowledge in Europe.

Certification centers

Once the system has been established (in mathematics as well as in other fields), a network of companies will be in charge of its management. The first task of this consortium will be to advertise the system in every European country with the authorities who will arrange the certification centers: Universities, Chambers of Commerce, city councils and other local administrative boards. We plan several thousands of certification centers, and several millions of examinations per year.

Description of the project for mathematics

The level of the tests must range from the most elementary level (adults with no diplomas or no qualifications) to a qualified technician level (that is to say the first two years of University).

1. One has to structure the field of mathematical knowledge into a limited number of *units*. At the moment we have about thirty units in mind, encompasing such things as: Elementary operations with numbers, ordering, ratios, elementary logic, polygons and basic geometry, basic trigonometry, functions and polynomials, calculus, finite probabilities, Fourier analysis, etc.

This phase is of tremendous importance and difficulty, since we do not want the construction to be relevant to one particular educational system only, BUT for the employment market as a whole, i.e. for use in society. It is clear that we should also recognize mathematical skills other than those specifically, or implicitly, tested or rewarded in schools. For example, the ability to work with a technical drawing, to perform data analysis, or to use computer software for symbolic computations.

2. The second phase is the elaboration of the set of *knowledge units*. They will include open questions (with numerical or algebraic answers) as well as multiple choice questions (with a non-constant number of answers, including sometimes several exact answers). We will also take into account whether the answers (even false) are consistent with each other throughout the whole test. Moreover, it will be necessary to create sufficiently many variants, for each kind of question in each test unit, and to renew this set of questions very often. This will be important, in order to avoid the possibility of cheating, and to make sure that candidates will not just be trained to answer some stereotyped form of questions, but will obtain a real knowledge of the field tested by the unit.

3. The certification process will be the same for mathematics and every other field involved in the whole program. The certification process could be as follows. The candidate, after identification by a local authority, will sit in front of a terminal, insert his card and ask for a particular unit. The computer will be connected to a remote server by the internet and will receive the questions in an interactive form. At the end of the exam, the form will be sent back to the server. The server will mark the answers and send back the results, updating the record card.

The prototype

Up to now, we have started working on those three parts in order to build a prototype. Angelo Lissoni and the Italian team has developed the third part (the computer system). In collaboration with several colleagues, we have sketched a set of thirty units and gathered a few test collections. However, our work should be considered only as a first draft. It is clear that defining the content of each unit in the first place and then test-editing it, is a huge task that requires the participation of specialists from many countries already involved in this kind of work.

The Leonardo Project

Obviously, we need to have a larger framework for this initiative. In particular, we need to involve more people in it. This is the reason that we have applied for support from the European Community through a Leonardo Project. The promoter of the project will be Université de la Méditerranée (Aix Marseille 2), which has accepted to be in charge of it. The international consortium counts among its members:

- the University of Cambridge Local Examinations Syndicate (United Kingdom),
- the Institut für Techno- und Wirtschaftsmathematik e.V. from Kaiserslautern (Germany),
- the Department of Computer Science of Trinity College from the University of Dublin (Ireland),
- the Department of Mathematics and Computer Science from Uniwersytet Mikolaja Kopernika from Torun (Poland),
- the Centro Eleusi of the Universita Commerciale Luigi Bocconi from Milan (Italy),

- several publishers, among which Springer Verlag from Heidelberg (Germany), but also Quadrature (France) and Sigma Editura (Rumania),
- several associations, among which QCM Pro from Ecole Supérieure d'Ingénieurs en Electronique et Electrotechnique and Math pour Tous.

From the discussion

K. Császár, Hungary: Examinations of students by computerized tests can be used only to identify those who are not able to reach even the lowest level of knowledge and so have to fail. By a simple test – after my experience – you can check only whether the student is able to apply a theorem, whether he has clearly understood its content, but it is impossible in a simple way to evaluate logical thinking, i.e. the ability to go through a chain of thoughts. A test may be good as an entry for a written exam but has to be followed by an oral exam where you can check the intellectual power of students who would not become specialists in mathematics but creative specialists in some other branch of science and technology.

Exams organized in this way need a bigger personnel teaching mathematics, you have to work much more at the exams, but it is worth the trouble.

Azad Tagi-zade, Azerbaijan: Experience gained from mathematical olympiads indicates that there is a good correlation between 'test' results and traditional methods for assessment of mathematical knowledge.

Gerhard König, Germany: Computerized tests can be used for testing knowledge, but as everybody knows, there is a probability that the correct answer is given by guesswork. This can, of course, be taken care of by demanding x answers right from n distractors. How do we know, however, that the person being tested understands the problem, or whether he or she only marks an answer? What information will truly indicate how the person being tested has done the problem? Are only those test items chosen which can be easily formulated, but are not important for the subject? Have test experts (from psychology) been engaged in the formulation of the problems, in the actual testing process and in the evaluation of results?

Miklos Simonovits, Hungary: I think that computers may have an important role in many parts of our educational system, e.g. in teaching data connected with history and in the teaching of foreign languages.

I think that teaching mathematics is very complicated and that computers may be useful in developing notions in calculus, geometry, etc. But I do not believe that computers are good in checking knowledge to replace oral exams and traditional written exams.

A Hungarian student in our educational system has often to write half a page to answer one single question in a test. Often we give 4 – 5 questions for a 1 hour test. I do not believe that in the near future computers will be good at checking these answers. Allowing computers in the examination proceedures will therefore force us to change the whole educational system from better to worse.

The question of testing unemployed people is a marginal problem in this context.

John Erdos, UK: One must always keep in mind that the tests and the mode of testing always influence teaching. There is a danger that teaching and especially learning, will be aimed at the material and the skills required by the tests. The more successful and widely used the testing system is, the greater the danger.

The argument that the results of some new test correlate well with the results of traditional tests is totally specious; it is virtually self-evident that it will be so, since the rank order of students on any reasonable test will reflect their talent and education. But if future students aim at tests that only verify that they have learnt simple skills, then the good education that is required to do well in traditional tests might go out of the window.

I do feel that there is a place for computer-based testing and uniform testing over a large population. Apart from the obvious cost advantages, it would be useful for things like comparison of standards. With other tests, such comparisons are more difficult. However, to avoid the dangers I outlined, computer tests – or other types of simplified or "objective" tests – should be combined with other forms of assessment. These forms should, as far as possible, be direct tests of the skills and abilities we value and not some quantity that we hope might correlate with them.

Miklos Laczkovich, Hungary: If we enlarge the discussion on computerized tests to the broader subject of computer-assisted education (and programs teaching calculus, etc.), I should like to remark that (notwithstanding the obvious usefulness and importance of these programs), my experience is that at the very first and very basic level of university mathematics education, there are problems where they cannot really help. What I mean is that the basics of university level mathematics education are: understanding the logical operations (and, or, not, if, if and only if), understanding exactly what we mean by a mathematical statement, by an implication, by a proof, and by mathematics in general. This is very difficult for most students, since they have seen nothing similar at secondary school. To make them understand these basic notions and methods, they need many examples, problems, personal attention, and I doubt if computer programs can be of much help in these basic questions.

Boudine and Mauduit, France: We shall try to give answers to some of the critical remarks and comments put forward in the discussion.

First of all, it is not very surprising for such a new subject, that certain misunderstandings tend to put off the true centre of the discussion. The system of tests by computer which we are building ought to provide validation, not for basic university education, but for professional training. Nor should it provide validation for basic school education, but for further training, not only in mathematical fields, but also in mathematical skills and knowledge fundamental for professionals. In fact, it concerns validation of knowledge of anybody who has to change region, country or profession.

Our colleagues are right in thinking that this system of tests by computer has some inherent limitations. By definition, the system is not allowed to know the capacity of a person for oral and written expression. While this limitation is evident, the other limitations are less obvious and the best present-day tests avoid the ones which have been mentioned. For example, we can in practice easily prevent the strategy of random answers and we can also know if correct reasoning was applied. Moreover, it is probable – in fact, it is most often the case – that before a vacancy is filled, the employer will complete the validation of professional qualifications by an interview and/or a written examination.

By way of conclusion, we want to emphasize the deep social significance of this project, which could be of prime importance for the development of competencies in Europe.

As a follow up to the discussion, Boudine and Mauduit invited mathematicians to join a group with the purpose of working out sufficiently varied and suitable testing problems in mathematics for the computerized system.

* * *

Mathematics in primary and secondary schools, and recruitment to university

Vagn Lundsgaard Hansen

Technical University of Denmark, Lyngby (Denmark)

In the last decade many countries have experienced a decrease in the number of students enrolling for university studies in the natural and engineering sciences, with biology as the only exception. Many explanations have been given. Some people think that teaching in primary and secondary schools does not stimulate interest in these subjects and, furthermore, that it does not prepare students properly for university studies. Other people think that changes in society and attitudes to scholarship, caused among other things by the influence of television, make it difficult to motivate students to study subjects which need hard and dedicated work without immediate rewards except for intellectual insight. A third factor, which is regularly mentioned as a reason, is that it is no longer prestigious to study these subjects; the 'heroes' in society are not scientists but people in business and directors: money before intellect.

There is probably not one single cause of the problem. And as a matter of fact, what is the problem? There are more students than before studying these subjects, except for a short period in the 1960s and 1970s. The point is rather that the total number of students going to university has grown enormously and that the growth has primarily taken place in other subjects. Since most governments link the budgets to the number of students at the various faculties, the natural and engineering sciences certainly have a financial problem. Aside from this financial problem, the real question for me therefore is not the relative decrease in the number of enrollments but rather whether we are getting the bright and motivated students. The absurd thing is that the subjects where mathematics and physics are really fundamental are often the subjects where the entrance requirements are lowest.

The above considerations are valid in my own country, Denmark. I suspect they are valid more generally. I hope the discussion will clarify this. In any case, I would find it useful if we could get short reports from various European countries.

My own experience is that it is very difficult to have a discussion on the role of teaching mathematics in primary and secondary school in relation to the recruitment of students to university studies in which mathematics plays an essential role. The tendency is that the discussion rather quickly ends up with mutual accusations between the various levels of the educational system. It would indeed be helpful if mechanisms could be found for having a balanced discussion on this vital point.

Comments and short reports from various countries

Miklos Laczkovich, Hungary: Sometimes the different levels of mathematical education blame each other: universities and colleges claim that their education is not effective because the level of high school students is low, so they have to deal with masses of mediocre students. On the other hand, high schools complain that the level of mathematics teacher education is low, so they have no good teachers. This vicious circle should be stopped by increasing the level of education at universities and colleges. The education of good teachers should come first, and governments should realize that high priority must be given to good university and college education.

Christian Mauduit, France: I want to mention some activities aimed at a large public that have been developed during the last years, mainly in France, but not only in France. For example, the Kangaroo competitions, based on an idea coming from Australia and Canada, since 1991 have annually involved from 100,000 up to 700,000 children in France, and with several thousands of participants in many other European countries. The most important characteristic of these kinds of activities is that they are not exclusively devoted to the best students, or to the most motivated children. Rather, the goal is to show that mathematics can be attractive to everybody, even to students unable to get good results at school.

One of the most interesting projects in this direction has been developed by Pierre Audin and Pierre Duchet since 1990. A description of the project, called MATh.en.JEANS, can easily be found in the many publications of Audin and Duchet. Let me here just say that its goal is to give young people the possibility of working together with teachers from high schools and professional mathematicians on open questions in mathematics.

Even though I strongly believe in the importance of the development of mathematical activities for everybody, I want to call the attention of the mathematical community to the risk of a sharp drift of these programs. As soon as such an activity attracts significantly many people, a lot of money can be involved (e.g. in the case of the Kangaroo competitions, between 1 and 2 million USD each year). If we are not careful, this situation can attract people or commercial enterprises which, on the pretext of developing popularization, could have as their first goal the making of huge profits (sometimes from the work of benevolent or underpaid colleagues). It is time now for us to decide what rules must govern the connections between money and popularization in mathematics, and I strongly hope that the EMS will soon open a discussion on this very important problem.

Klaus D. Bierstedt, Germany: Two 'caveats' at the beginning: Some of my statements below tend to be exaggerated, due to lack of space. Also, in Germany education and universities are a matter for the Federal States, not for the Federal

Republic; thus, the situation has always been different at universities in different states of the Federal Republic.

Quite generally, however, in recent years the number of students who would like to get a diploma degree in mathematics seems to be too small. More students want to obtain the so-called '(Erstes) Staatsexamen (für die Sekundarstufe II)', to become a teacher of mathematics and some other topic in a Gymnasium, but the amount of mathematics they learn is much smaller, a tendency which has become stronger and stronger during the last decades.

Let me explain this point in more detail, concentrating on the situation in the biggest German state (Nordrhein-Westfalen) and on the curriculum at my own university (Paderborn), which is trying to make the best out of the existing state law. Students who would like to get the 'Staatsexamen' take the courses 'Analysis I, II' and 'Linear Algebra I, II' during the first two semesters. Then they have one course in applied mathematics (i.e., numerical analysis or stochastics or differential equations) and a shorter course on 'mathematics with a computer', before passing the intermediate exam ('Zwischenpruefung'). In the rest of their time at the university (in the so-called 'Hauptstudium'), they take one course on each of the following topics: applied mathematics, complex function theory or functional analysis, topology or geometry, algebra or computer algebra or number theory, and they have to pass one seminar in mathematics. In didactics of mathematics, they would also have a shorter course (didactics of analysis or linear algebra or stochastics) and a seminar.

It is quite clear that these students do not know very much mathematics when they become teachers. They do not get a good picture of what is happening in modern mathematics, or of its many applications. E.g., since they do not usually take the second semester of a course in, say, functional analysis, they will not see the deeper theory, nor most of the more interesting applications. But this is only part of the problem: The students for a 'Staatsexamen' take courses from three different departments; they do not even think of themselves as being 'math students'. They try to judge each course in mathematics mainly by its possible value for what (they believe) they might have to teach in school later on. Thus, for both students and professors, the situation is all but desparate, and there does not seem to be a simple 'remedy'.

Luc Lemaire, Belgium: In the 1980s, Belgian universities experienced a sharp decline in the enrollment of students in mathematics, both in total number and in the number of very bright students. Surprisingly, it was possible to reverse that trend in almost a single year (1990) by various information campaigns.

Secondary schools in Belgium have undergone major financial cuts, and the situation of teachers has deteriorated. Not many young people want to become teachers, and those who do, fear they will not get jobs. In fact, most secondary school students used to think that studies in mathematics only led to a career in teaching, and this was a major reason for the weak recruitment.

Almost none of the very bright students knew that mathematics was a living science, and hence they turned to physics or biology excited by the idea of a research career. The action we have taken has been to find out the careers of (almost) all former students over the last 10 years. At the time (1989) all had jobs, evenly spread between banking, teaching and research. When this was made known in a very precise article in a major newspaper, the effect was immediate. At my university we moved from 35 students in the first year to 70 and this is an essential progress for us. We have also been successful in getting into contact with the very bright students at the finals of the olympiads, for example, giving informal lectures on contemporary mathematical research.

My advice for improving recruitment is therefore: 1) Provide information on the variety of careers mathematics can lead to, including the actual job situation; 2) Be as precise as possible, do not stick to general statements, and always tell the truth about the job market (the future of young people should come first); 3) Explain whenever possible to bright students that mathematical research exists, by talking about some present day problems.

I think this kind of information campaign must be done for mathematics by itself, and not for scientific studies in general. Indeed, it is an important fact that mathematics is used in insurance and banking, whereas other sciences are not. Hence it does not make sense to use the same arguments for all sciences.

Concerning the level of secondary school teaching, the financial cuts in Belgium are so drastic that many teachers will get sacked in 1996. Against all odds, highly motivated teachers fight desperately to maintain standards. I suspect (hope) that the situation is worse in Belgium than in the rest of Europe.

Kari Hag, Norway: Only 13% of the full age group takes the highest level mathematics in Norwegian secondary school. This percentage has remained relatively constant over the ten year period 1986–96. The change in demography in Norway means, however, that the number of 18 year olds has dropped almost 25% over the same period. Moreover, as in other countries, the percentage of students studying physics decreases, while the number of those studying biological sciences increases, despite the drop in the total number of secondary school students.

Though recruitment in science and engineering at the Norwegian universities has remained good, colleges are struggling to fill their quotas and at the engineering colleges one out of four places is vacant.

In Norway students do not enter a university or college to study mathematics (or any other subject) as an undergraduate major. In comparison with the situation in other countries, very few students go on to graduate studies in mathematics, which means entering a Master's Program. Popular areas of study at this level are applied and industrial mathematics including statistics followed by pure mathematics. Unfortunately, practically none of the Master's Degree students want to become teachers. Indeed, of the last 70 candidates at the University of Oslo only 2 are now secondary school teachers.

Although relatively few students finish the Master's Program, which is a requirement to enter a PhD Program, the number of students who receive a PhD in Mathematics is comparable to those in other countries, for example, the United States.

How can we increase the number of students who specialize in mathematics? More information for possible careers is needed. For example, the results of the Belgian information campaign are encouraging. In addition, a long term project is to raise the quality of the mathematics teachers in primary as well as secondary schools. As in most other countries, the government is introducing reforms at all school levels. However, no reform will succeed unless good teachers are available.

David Salinger, UK: In England and Wales, the number of undergraduate students studying mathematics has increased by about a third since 1985–86, the first year for which the data is available. Though the last figures I have are for 1993–94, I am fairly confident that there are now about ten thousand UK students studying mathematics as a first degree in the "old" universities. Over the same period, general recruitment to universities has risen more quickly, particularly among women. If it can be done, we should encourage more women to continue into university mathematics: they form a third of mathematics students, but more than half of university students as a whole.

As far as teacher education is concerned, the picture is one of chronic under-recruitment. For example, fewer than half the mathematics teachers in secondary schools in England have a university degree with mathematics as a main subject.

If the figures for the number of students entering university to study mathematics are mildly encouraging, it is another matter when we consider the mathematical level of starting students. A recent report [*Tackling the Mathematics Problem*, London Mathematical Society, October 1995] stated that mathematicians, scientists and engineers in Higher Education perceived that those students had: (i) a serious lack of essential technical facility – the ability to undertake numerical and algebraic calculation with fluency and accuracy; (ii) a marked decline in analytical powers when faced with problems requiring more than one step; (iii) a changed perception of what mathematics is – in particular of the essential place within it of precision and proof. For a passionately-argued view, the interested reader may consult *Wrong way. Go back!* and *Back to the future* and *"Problem solving" or Problem Solving?* by A. Gardiner in [Mathematical Gazette, Nos. 485–487, 1995–1996].

Surprisingly similar concerns are shared by Marc Rogalski [*Le nouveau public étudiant en sciences: quels choix stratégiques?* Gazette des Mathématiciens, 69, Juillet 1996, 39–44] and others in the French mathematical community. Moreover, this round table debate shows that the issue is not confined to a couple of European countries.

* * *

Current problems in the teaching of mathematics at the university level

Miklos Laczkovich

Eötvös University, Budapest (Hungary)

I shall restrict myself to talk about the undergraduate level only. At this level, the general problems in teaching are mainly related to two circumstances: (i) the lack of time; (ii) the sizes of groups.

As for (i), the pressure of limited time is always in conflict with the desire to cover as much material as possible. Some teachers express the opinion that every university course should contain some new results of the subject, and should give a review of the modern developments. Unfortunately, sometimes we have to give up these hopes. Usually we have to restrict ourselves to the basics. Besides, a deeper knowledge of the basics is much more important than a superficial glance at some sophisticated results which will be forgotten anyway.

With respect to (ii) the sizes of classes should be as small as possible. This is of course a purely financial problem, but it can be crucial, if we want to maintain a certain level of the education.

Remarks on the teaching of mathematics at the university level

K. Császár, Hungary: It is certain that mathematicians at the Technical University Budapest, as well as at Eötvös University in Budapest, would like to teach a few branches of mathematics in depth with greater possibility to continue developing the students' creativity, and the ability to apply knowledge learned to the solution of new problems. I think, wherever mathematics is taught as a background subject, and not as a subject in its own right, our task is first of all to teach basic mathematics needed by the experts teaching branches applying mathematics. If we leave it to these experts in other fields to teach the basic mathematics, students will often not be able to apply a mathematical theorem or method to any other problem than that in which he or she first met it. Talking briefly about important theorems without proving them is hard work for mathematicians as well. However, as they are not bound to one single application and they have a broader knowledge of mathematics, they will be able to teach more successfully even theorems not presented in every detail. Therefore, at the Technical University Budapest we are teaching in mathematics a lot of branches of our science, the very basic ones in all details, the other ones on a large scale but trying to show the logical relations among them and mentioning the most important fields of applications.

Helge Elbrønd Jensen, Denmark: The major problem in teaching at the universities in Denmark is caused by the fact that university studies have gradually changed from an education for the elite, towards mass education. Furthermore, in science and engineering the number of interested students has decreased considerably over the last 5–10 years. Altogether, the *average* student entering university studies requiring mathematics has less knowledge, maturity and academic skills than is optimal for such studies.

The first reaction from the mathematicians is typically to ask for more teaching hours. At a technical university this is in general very difficult to obtain, because there are so many (new) disciplines the students should learn. Another natural reaction is to ask for higher entrance requirements to university. But this may also be difficult because both the society and the university itself wants a certain number of engineers to graduate each year.

At my university we have decided to face facts and teach the compulsory courses in a way which is suitable for the actual average student, i.e. by reducing the curriculum (cutting away some sub-disciplines) and abandoning the detailed axiomatic, proof-oriented approach to mathematics. Instead we focus on basic concepts and results in calculus and linear algebra, and try to teach the students the meaning and content of the concepts and their significance in solving a variety of problems. We also offer a number of courses in mathematics which are more or less voluntary for the students. In these courses we teach in a different way and put emphasis on precision, proofs, axiomatics and the abstract approach. In this way the students can learn mathematics from different points of view and in my opinion this works pretty well.

The mathematical needs differ for various types of engineering students, and it is an advantage both for the teaching and the mathematics that students following a specific course have a serious wish to learn the subject of that course. Particularly, I would like to emphasize that we have many very good students, and for those students it is possible to reach a level comparable to a normal university degree in mathematics.

The second great problem in the teaching of mathematics is to find out how to incorporate the computer. The possibilities of the computer make it less important for an engineer to master detailed calculations of the traditional kind, and consequently we need not focus so much as before on these matters. But then we have a pedagogical problem. The working of many exercises, calculations and examples is a decisive part of learning abstract ideas, and therefore, if we pay less attention to doing things with paper and pencil (and the head!), there is a great risk that more general understanding will subsequently decrease. The right balance must be found.

What really matters in the applications of mathematics at university level is not the ability to carry out large calculations by hand, because the computer can do this, and it is not the ability to carry out complicated long proofs, because this often, for the engineering students, adds very little to understanding. What really matters, in my opinion, is the ability to understand the mathematical concepts and the meaning of the mathematical results, and the ability to carry out mathematical thinking at an abstract level. The computer should be used as a tool to train this ability and not as a tool to avoid it.

* * *

New challenges for the teaching of mathematics

Vagn Lundsgaard Hansen

Technical University of Denmark, Lyngby (Denmark)

In the not too distant future, if present trends continue, a large proportion of the population of many countries will go to university, which will become the common place for a tertiary education. Moreover, many adults with university studies behind them will return to university, from time to time, to be retrained in work connected with their profession. In such countries, the university will become a place for continuing education.

In these circumstances the distinction between *mathematics for all* and *mathematics for those who will go to university* might somehow lose its meaning. But a destinction still has to be made between those who will attend science and technology faculties and those who will attend all other faculties. When selecting the mathematical content in the curriculum at the secondary school level, it will therefore become increasingly important to choose those units of mathematics which foster the right skills, abilities and attitudes for a meaningful and useful tertiary education.

In Europe most countries have a high proportion of the population attending secondary school. In Denmark, for example, about 60% of teenagers currently complete a secondary school education. In countries where education for the majority of the population currently ends in early grades, it is to be hoped that, in the near future, it will be possible to extend mass education, at least until middle schools. In such a mass oriented education, I think one should avoid too much abstraction and too much formalism; and teaching methods and curricula should be linked to local culture and tradition, and not just adopt curricula from some leading country.

One big problem will probably continue to be a point of discussion, namely from which grade in school should there be different curricula reflecting the future choices of the students? Probably it should take place somewhere between grade 8 and grade 10 in school. For those students who have more interest in mathematics, the conceptual and structural aspects of mathematics should be stressed. For those who are less mathematically motivated, it is advisable to put more stress on the applications of mathematics.

In the 21st century computers will be part of daily life and teaching at all levels might to a large extent be based on interactive computer programs. This will probably be beneficial more for those who are already motivated and successful than for the students who have difficulties with the traditional ways of teaching. It will be an important task to rethink the role of teachers and the importance of social contacts in such an educational environment. In these didactical considerations one should not exclude the possibility that the changes might not go very much further than making available a very powerful teaching aid, like a 'dynamic overhead'. In any event, the understanding of concepts and the ability to do creative thinking with these concepts using all available tools will be increasingly important. Since computer programs make strong use of visualization and visual thinking must be learned, I predict that interest in basic geometry will grow.

Comments

Gerhard König, Germany: One of the most important tasks in mathematics education today is the revision of curricula and teaching strategies to take advantage of the expanding electronic technology currently available to both teachers and students.

There are many reasons for those involved in mathematics education to reconsider mathematics teaching and its perspectives for the 21st century:

(a) The challenges of new technologies and their use in teaching;
(b) Developments and trends in the mathematical sciences;
(c) The penetration of informatics/computer science into the curriculum;
(d) Mathematics as a technology and its contribution to general education;
(e) New results in the theory of learning.

The most important of the above reasons is the impact of new technologies (computer algebra systems, supercalculators such as the TI-92 or other graphics calculators), which will undoubtedly result in curricular changes in the near future. The prevailing opinion is that not only a change in curriculum, but also a reform of teaching methods, is necessary. Others argue that with the help of new technology and the use of better teaching methods the current curriculum is quite satisfactory. The question of whether curriculum or pedagogy is the major issue of education remains debatable.

The new technologies open a window on creative learning, and computer algebra systems and graphic calculators may help students to achieve a higher level of learning. But many questions remain open. When mathematics courses are built around the use of computers they tend in general to be more demanding. The downgrading of routine skills will make mathematics still more difficult for everyone in the future. Is this actually desirable? Can more interesting mathematics produce more motivation as compensation? Furthermore, curricular and pedagogical changes cannot occur without accompanying changes in student assessment. Turning to the teachers, the continual development of new technologies requires teachers to constantly enhance their technological skills. How many of them will do this? Where do teachers find the time to learn to use these new tools and the time to train students to use them (if necessary)?

And finally, the problem on the role of basic skills is quite open. Which basic skills are necessary in the future is a matter of discussion in several countries. In this connection it is important to note that it takes time to develop and evaluate major curricular reforms. I think it will take a lot of experiment, research, discussion, and conferences before we have a clear picture of how we can integrate the new tools in the teaching and learning processes.

David Salinger, UK: What follows is a personal view of the situation in England and Wales. (Scotland has a separate educational system.)

With the expansion of knowledge in the twentieth century has come a growth in uncertainty as to what constitutes a cultural core. Mathematics has to jostle in the school curriculum with a host of other subjects. Even the notion of what constitutes mathematics has changed. So mathematics occupies a smaller slot in the school timetable and the subjects 'arithmetic, algebra, and geometry' are diluted to allow place for 'data processing, statistics, project-work'. Geometry has been de-natured to such an extent that proof is effectively absent from it. This is not an accident: teachers found that only the best students gained anything from the study of rigorous Euclidean geometry and campaigned successfully for its removal.

As a result, university mathematicians are querying whether the subject called "mathematics" is, any longer, mathematics. The absence of proof and, hence, of coherent development of topics, leads pupils to view the subject as an unstructured collection of useful results.

However, "mathematics" seems to have become a more attractive subject at school and, in recent years, more students have opted to continue "mathematics" beyond the age of 16. It would not serve a useful purpose if a reform of the school curriculum, inspired by mathematicians' concerns, discouraged students from pursuing mathematics beyond the school level.

Meanwhile, university mathematicians, and other users of mathematics, have been active in several ways. They have, at last, sought to have an official voice in the processes which determine the school curriculum. They have sought to engage in a public dialogue with teachers of mathematics and the theorists of mathematics education. They have sought to stimulate interest in mathematics through masterclasses and competitions. And they have changed university first degree course to give students more time on the early stages, while preserving the level of degree by allowing the better students to take four, rather than three, years to complete it.

Though much remains to be done, I am optimistic about the outcome.

* * *

Feedback from mathematics education to mathematical research

VINICIO VILLANI

Università di Pisa, Pisa (Italy)

It is all too obvious that mathematical research exerts a strong influence on mathematics education. It is maybe less obvious that conversely there exists an equally strong feedback from mathematics education to mathematical research. In this short contribution, I will try to illustrate at least one of the many aspects of this feedback: the effect of mathematics teaching at school on the enrollment of students at university in mathematics classes, and on the attitudes of mathematics researchers towards the most innovative parts of our discipline.

At the end of secondary school, pupils who wish to continue their studies at college or university must choose a faculty and a curriculum. This choice is often determined only by a superficial knowledge and by a purely emotional perception of the characteristic features of the various disciplines learnt at school, as conveyed to them by their teachers from early grades.

If at primary and secondary levels pupils have experienced mathematics as a mere technical tool for utilitarian needs, or as a formal, unchanging and old-fashioned discipline, it is likely that even those pupils who might be potentially interested in studying 'true' mathematics will prefer to embrace different fields of study, in the expectation of subsequently taking up a job or a profession which seems in their eyes more challenging than that of a mathematician, e.g. engineer, physicist or computer scientist. Unfortunately we must admit that we mathematicians are less able than engineers, physicists and computer scientists in conveying to the larger public a correct perception of what a mathematicians' work really consists of!

Conversely, if mathematics has been taught to pupils from the beginning as a living subject, highlighting its ongoing dynamical evolution and its capability of contributing to the solution of a huge variety of challenging problems, both practical and theoretical, it is more likely that at least some of the brighter pupils will be attracted to enter into further mathematical studies.

But what about those (rather few) students who have already chosen to become mathematicians? Whatever their initial motivation for this decision may have been, at some point in their university career they must make further choices concerning their future specific mathematical research fields. These choices, as well as their personal research style are going to be heavily influenced, whether they are aware of it or not, by the "imprinting" they have received during their former studies. If mathematics has been shown to them (as is common even at university!) mainly as a polished end product, where axioms, definitions, theorems and formulae follow each other in a neat order, it is no wonder that they will in turn be most attracted to research in well-established mathematical fields, whose structures have already been wholly shaped and codified. And they will be reluctant to apply themselves boldly and inventively to topics still in gestation, where large parts of a global theoretical framework have yet to be formalized, in spite of the fact that in new fields many more challenging problems remain to be investigated. And if it happens that these same researchers will teach in their turn future primary and secondary teachers, they will be inclined to perpetuate a formal and static conception of mathematics.

At all levels, including universities, a more 'problem oriented' approach to the teaching, learning and assessing process in mathematics could be more suitable to give to new researchers more confidence in their creative attitudes and to encourage them to enter into paths not yet completely marked out. Thus the importance of making guesses and conjectures before passing to formal proofs would be stressed, and even errors would not be seen as an absolute evil, but rather as an unavoidable component of progress in knowledge.

Is it not true that the value of a new discovery is due to its unpredictability and to its capacity to link together aspects that seemed until then unrelated, more than to the mere subtlety of the techniques involved?

Of course formalization and rigour are important, but they are not the only goals of doing mathematics. Formalization without meaning is worthless. Rigour is a goal to which to aspire, but absolute rigour is not achievable and moreover, even if it were, the result would be endlessly boring, just like a computer program where too many technical details end up in hiding, rather than clarifying, the main underlying ideas.

For a wide range of opinions about the nature of mathematical understanding and about problems in the communication of mathematical results, I suggest looking at the interesting debate originated by the article of Jaffe and Quinn: *Theoretical Mathematics: Toward a Cultural Synthesis of Mathematics and Theoretical Physics* [Bull. A.M.S. (New Series), **29**, July 1993, 1–13] and specifically at Thurston's and Thom's contributions [Bull. A.M.S. (New Series), **30**, April 1994, 161–207].

* * *

The contribution of the didactics of mathematics to our understanding of mathematics as a science and a subject

MOGENS NISS

Roskilde University, Roskilde (Denmark)

The teaching and the development of mathematics

In the history of mathematics, it is widely recognised that major conceptual developments in mathematics as a science have been stimulated, and in several cases even generated, by needs arising in the teaching of mathematics. The fact that mathematics has had to be taught, at all levels, to interested, enthusiastic, critical, demanding, and reflective students has required teachers, lecturers, and textbook authors to structure, organise and clarify their thinking and exposition of mathematics in order for their students to grasp, understand and learn (acquire) what is taught. This, in turn, further leads to the necessity, in the teaching of mathematics, to lay a foundation for the exposition through the definition of concepts and objects, the investigation of their properties, logical relationships and hierarchies, the establishment of methods and criteria of justification, as well as the determination of terminology, notation, and symbols.

Thus, the teaching of mathematics at various levels provides instrumental contributions to the conceptual formation and development of mathematics. This was the case, e.g., in ancient Mesopotamia (demonstrated in the work of Jens Høyrup), in classic Greece (Euclid's Elements was probably written primarily as an advanced textbook), in the 19th century in Cauchy's and Weierstrass' university lectures, and in the first several stages of the Bourbaki project of our century in relation to mathematics taught at the university level. Also, throughout the 20th century, the production of authoritative, advanced textbooks introducing new topics or theories for the first time has not only served the purpose of making new developments accessible to a wide readership of newcomers to the area, but quite as much of laying the conceptual, logical and terminological foundation of the new development in a way convincing to the mathematical research community. In a way, such textbooks, rather than the initial underlying research papers, can be perceived as 'certificates of citizenship' of innovations in mathematics as a science.

Of course, these deliberations are not meant to imply that all major developments in mathematics are driven by the needs of teaching. That is simply not the case. What is the case, however, is that during the last hundred years, at least, the emergence of any major new development in mathematics has been accompanied by the publication of such almost 'canonical', first comprehensive introductions to the field, normally written by the 'field creators' themselves. In other words, mathematical exposition, by serving the dual purpose of reclaiming new mathematical land to the discipline and of introducing mathematical topics or theories to new recipients (whether students or colleagues), provides an essential and intimate link between the production and teaching of new mathematical knowledge.

Mathematics elucidated by the didactics of mathematics

So far, we have been dealing with the teaching of mathematics only, not with the didactics of mathematics as the scholarly and scientific field of research and development that studies the teaching and learning of mathematics. The fact that the teaching of mathematics has (had) an important impact on the development and constitution of mathematics as a science does not automatically imply that the same is true with the didactics of mathematics. In what follows I shall, by giving three general examples, argue that the didactics of mathematics does, in fact, contribute to our understanding of mathematics as a science and a subject. This is not equivalent to saying that this field generates innovations in mathematics as a science. However, to the extent that the didactics of mathematics has valuable contributions to offer to our understanding of mathematics, it would not be unreasonable to expect it to give rise to an impact on the mathematics community and its professional practice.

The first example is closely related to the way in which mathematical teaching and exposition has influenced the conceptual constitution of mathematics. In the first stage of the didactics of mathematics, the main focus was on the study and construction of curricula. The investigation of conceptual networks and hierarchies, from both logico-systematic and cognitive perspectives, used to be a predominant activity – in German-speaking countries usually named "Stoff-Didaktik". It was often meant to serve as a foundation for the writing of textbooks and for the presentation of subject matter. Quite an amount of the work done in this area offers penetrating insights into the conceptual foundation and architecture of mathematical topics and sub-disciplines, such as arithmetic and algebraic operations, geometry, and functions, to name just a few. There still is a solid didactical interest in the study of curricula, but as numerous new areas have attracted researchers' attention in the last couple of decades, curriculum studies no longer occupy a predominant position in the field.

As an offspring of the endeavour to analyse or construct topics of mathematics curricula, a substantial interest in mathematical cognition has emerged and developed since the 1970s. Rather than focusing on the logico-systematic aspects of larger networks of concepts, studies in cognition investigate the ways in which students perceive, understand and acquire (or the opposite!) specific mathematical notions or concepts, or clusters of interrelated notions/concepts, e.g. the notions of limit, continuity or differentiability, the concept of fraction, the notion of proof. Several such studies are stimulated by the empirical observation that students' errors and misconceptions more often than not display characteristic patterns within and across individual students, patterns which may provide access to the ways students actually (may) think about the mathematics at issue. On closer analysis, there is, not infrequently, found to be a substantial amount of rationality in students' notions and concepts, even if the rationality happens to rely or act upon insufficient or wrong conceptions that call for remedy. Such studies not only help us understand how students think with respect to mathematics and how they acquire mathematical knowledge. They also serve to uncover, elucidate and clarify the nature and characteristics of mathematical notions, concepts, and thinking per se, and thus to enrich our understanding of mathematics as a science and a subject. Intimately linked to this is the understanding of the nature of mathematical problem solving, of proofs and proving, features of the mathematical discourse (language and symbolism) and so forth, all of which are intensive areas of study in the didactics of mathematics.

While my first two examples address matters that are, primarily, interior to the edifice of mathematics, my third and final example involves exterior matters as well. As mathematics education is being given to human beings who live and (are to) work in society, the didactics of mathematics has to address matters pertinent to the position, role, and function of mathematics in the world (i.e. nature, society, and culture). While it is, of course, impossible to come to grips with mathematics in its relations to the world without considering mathematics as a science, it is equally necessary to consider mathematics from outside. In other words, meta-studies of mathematics are also required. The position of mathematics in the world is determined by its five-fold nature as a pure science, an applied science, a system of instruments for social practice, a field of aesthetics, and last but not least, a subject of teaching and learning. The latter aspect is a reflection of the four former aspects, of which the first three are the most important ones. It is an essential task for the didactics of mathematics to conduct scholarly and scientific investigations into the position, role, and functions of mathematics in the world (which is not to say that the didactics of mathematics should be alone in assuming responsibility for this task). Issues such as 'where is mathematics located in the practice of society?', 'where and how is mathematics applied in other subject and practice areas?', 'what categories of mathematical competence are needed in society, and who should possess which competences?', 'what are the effects of mathematical applications and modelling for the organisation and government of society?', 'what sorts of public appreciation of and insight into mathematics are desirable, and how can they be achieved?' are all crucial in the didactics of mathematics. Answers to them are not only relevant to mathematics education but also to our understanding of mathematics as a science and a subject. The significance of such understanding deserves to be appreciated by the mathematical community at large.

* * *

Mathematics education towards mathematics

Jean-Pierre Kahane

Université Paris-Sud, Paris (France)

The number of human beings involved in mathematical research is of the order of 10^5; in mathematics teaching, 10^7; and in mathematics learning, 10^9. In no other science is the relative importance of education so large. Why is this so? What should be taught, learned, and how? These basic questions for mathematics educators and didacticians are addressed to research mathematicians as well. My personal experience is that educational studies, such as the series of studies performed by ICMI since 1985, are very stimulating in order to investigate and grasp some essential features of mathematics as a science. Before giving examples, let me say that educational concern is part of the mathematical tradition, from Euclid to Bourbaki's *Eléments de mathématique* through Euler's *Lettres à une princesse allemande*, Laplace's *Leçons à l'Ecole Normale*, Cauchy's *Cours de l'Ecole Polytechnique* and Felix Klein's *Elementarmathematik vom höheren Standpunkte aus*. To a large extent, the history of mathematical education is a very good introduction to the history of mathematics, not only because some of the best mathematicians of all times and countries contributed to mathematical education, but also, and mainly, because social effects of and impulses to mathematics can be deciphered from school programs and university courses more easily than from research articles.

Let me come back to the present time. Among the first ICMI studies I would consider *The influence of computers and informatics on mathematics and its teaching*, *Mathematics as a service subject*, and *The popularization of mathematics.* The three of them dealt with key issues in or about mathematics education. At the same time they provided an interesting new view of mathematics itself.

The idea that computers and informatics should have an effect on mathematics was not new in 1985; I heard prophetic statements thereabouts by J.S. Haldane when I was a student, in the late 1940s. However, the first international study on this topic – including the effect on concepts, trends of research, style, methods and proofs – was the ICMI study. It is still a good reference now.

Mathematics as a service subject does not mean second rank mathematics, but mathematics taught for a specific purpose. The choice of subject matter and the method of teaching depend on the main interests and future needs of the students. This forces us to consider mathematics not as a tree rooted somewhere in set theory, but as a graph with many entrance doors. Moreover, discussion showed that almost every mathematical notion can be introduced and developed at different levels with different meanings. Some oppositions between mathematics and physics may prove irrelevant; for example, a training in geometry is needed by physicists because of the deep and intimate link between physics and geometry.

The study on the popularization of mathematics was an opportunity to collect information and to suggest future action – actually the situation is better now than some years ago, and it has been claimed that popularization of mathematics has become popular. As a by-product we pointed out a specific feature of mathematics as a science. For there is one more reason to popularize mathematics more than any other science: not being related to any specific domain of the natural or social world, mathematics comes from everywhere and may go in unpredictable directions, as soon as people are aware of its power. If mathematics is not widespread enough, part of its creative power is lost.

These are very general examples. It would be possible to start from any theme of the ICMI studies, mathematics teaching in the 1990s, mathematics and cognition, assessment in mathematics, gender in mathematics education, research in mathematics education, and show how it involves some fundamental aspects of mathematics as a science.

As a personal conclusion, I should like more mathematicians to be involved in the field of didactics. I have stressed the possible benefit for mathematicians. I am sure that it could help didacticians as well.

Progress in Mathematics

Edited by:

Progress in Mathematics *is a series of books intended for professional mathematicians and scientists, encompassing all areas of pure mathematics. This distinguished series, which began in 1979, includes authored monographs, and edited collections of papers on important research developments as well as expositions of particular subject areas.*

We encourage preparation of manuscripts in such form of TeX for delivery in camera-ready copy which leads to rapid publication, or in electronic form for interfacing with laser printers or typesetters.

Proposals should be sent directly to the editors or to:
Birkhäuser Boston, 675 Massachusetts Avenue, Cambridge, MA 02139, U.S.A.

92 CONNES/DUFLO/JOSEPH/RENTSCHLER. Operator Algebras, Unitary Representa tions, Enveloping Algebras, and Invariant Theory. A Collection of Articles in Honor of the 65th Birthday of Jacques Dixmier
93 AUDIN. The Topology of Torus Actions on Symplectic Manifolds
94 MORA/TRAVERSO (Eds). Effective Methods in Algebraic Geometry
95 MICHLER/RINGEL (Eds). Representation Theory of Finite Groups and Finite–Dimensional Algebras
96 MALGRANGE. Equations Différentielles à Coefficients Polynomiaux
97 MUMFORD/NORMAN/NORI. Tata Lectures on Theta III
98 GODBILLON. Feuilletages, Etudes géométriques
99 DONATO/DUVAL/ELHADAD/ TUYNMAN. Symplectic Geometry and Mathematical Physics. A Collection of Articles in Honor of J.-M. Souriau
100 TAYLOR. Pseudodifferential Operators and Nonlinear PDE
101 BARKER/SALLY. Harmonic Analysis on Reductive Groups
102 DAVID. Séminaire de Théorie des Nombres, Paris 1989-90
103 ANGER/PORTENIER. Radon Integrals
104 ADAMS/BARBASCH/VOGAN. The Langlands Classification and Irreducible Characters for Real Reductive Groups
105 TIRAO/WALLACH. New Developments in Lie Theory and Their Applications
106 BUSER. Geometry and Spectra of Compact Riemann Surfaces
107 BRYLINSKI. Loop Spaces,Characteristic Classes and Geometric Quantization
108 DAVID. Séminaire de Théorie des Nombres, Paris 1990-91
109 EYSSETTE/GALLIGO. Computational Algebraic Geometry
110 LUSZTIG. Introduction to Quantum Groups
111 SCHWARZ. Morse Homology
112 DONG/LEPOWSKY. Generalized Vertex-Algebras and Relative Vertex Operators
113 MOEGLIN/WALDSPURGER. Décomposition Spectrale et Series d'Eisenstein
114 BERENSTEIN/GAY/VIDRAS/YGER. Residue Currents and Bezout Identities
115 BABELON/CARTIER/KOSMANN-SCHWARZBACH (Eds). Integrable Systems. The Verdier Memorial Conference
116 DAVID (Ed.). Séminaire de Théorie des Nombres, Paris, 1991-1992
117 AUDIN/LAFONTAINE (Eds). Holomorphic Curves in Symplectic Geometry
118 VAISMAN. Lectures on the Geometry of Poisson Manifolds
119 JOSEPH/MIGNOT/MURAT/PRUM/RENTSCHLER (Eds) First European Congress of Mathematics (Paris, July 6-10, 1992). Volume I: Invited Lectures (Part 1)
120 JOSEPH/MIGNOT/MURAT/PRUM/RENTSCHLER (Eds). First European Congress of Mathematics (Paris, July 6-10, 1992). Volume II: Invited Lectures (Part 2)

121 JOSEPH/MIGNOT/MURAT/PRUM/RENTSCHLER (Eds). First European Congress of Mathematics (Paris, July 6-10, 1992). Volume III: Round Tables
122 GUILLEMIN. Moment Maps and Combina torial Invariants of Hamiltonian T^n-spaces
123 BRYLINSKI/BRYLINSKI/GUILLEMIN/KAC (Eds). Lie Theory and Geometry: In Honor of Bertram Kostant
124 AEBISCHER/BORER/KÄLIN/LEUENBERGER/REIMANN. Symplectic Geometry. An Introduction based on the Seminar in Bern, 1992
125 LUBOTZKY. Discrete Groups, Expanding Graphs and Invariant Measures
126 RIESEL. Prime Numbers and Computer Methods for Factorization
127 HÖRMANDER. Notions of Convexity
128 SCHMIDT. Dynamical Systems of Algebraic Origin
129 DIJKGRAAF/FABER/VAN DER GEER (Eds). The Moduli Space of Curves
130 DUISTERMAAT. Fourier Integral Operators
131 GINDIKIN/LEPOWSKY/WILSON. Functional Analysis on the Eve of the 21st Century in Honor of the 80th Birthday of I.M. Gelfand. Vol. I
132 GINDIKIN/LEPOWSKY/WILSON. Functional Analysis on the Eve of the 21st Century in Honor of the 80th Birthday of I.M. Gelfand. Vol. II
131/132 Set Vols I+II
133 HOFER/TAUBES/WEINSTEIN/ZEHNDER (Eds). The Floer Memorial Volume
134 CAMPILLO LÓPEZ/ NARVÁEZ MACARRO (Eds). Algebraic Geometry and Singularities
135 Amrein/Boutet de Monvel/Georgescu. C_0-Groups, Commutator Methods and Spectral Theory of N-Body Hamiltonians
136 BROTO/CASACUBERTA/MISLIN (Eds). Algebraic Topology. New Trends in Localization and Periodicity
137 VIGNÉRAS. Représentations l-modulaires d'un groupe réductif p-adique avec $l \neq p$
140 KNAPP. Lie Groups Beyond an Introduction
141 CABANES (Ed.). Finite Reductive Groups: Related Structures and Representations
142 MONK. Cardinal Invariants on Boolean Algebras
143 GONZÁLEZ-VEGA/RECIO (Eds). Algorithms in Algebraic Geometry and Applications
144 BELLAÏCHE / RISLER (Eds). Sub-Riemannian Geometry
145 ALBERT / BROUZET / DUFOUR (Eds). Integrable Systems and Foliations Feuilletages et Systemes Integrables
146 JARDINE. Generalized Etale Cohomology Theories
147 DI BIASE. Fatou Type Theorems Maximal Functions and Approach Regions
148 HUANG. Two-Dimensional Conformal Geometry and Vertex Operator Algebras
149 SOURIAU. Structure of Dynamical Systems. A Symplectic View of Physics
150 SHIOTA. Geometry of Subanalytic and Semialgebraic Sets
151 HUMMEL. Gromov's Compactness Theorem for Pseudo-holomorphic Curves
153 BUESCU. Exotic Attractors. From Liapunov Stability to Riddled Basins
154 BÖTTCHER/KARLOVICH. Carleson Curves, Muckenhoupt Weights, and Toeplitz Operators
155 DRAGOMIR/ORNEA. Locally Conformal Kähler Geometry
156 GUIVARC'H/JI/TAYLOR. Compactifications of Symmetric Spaces
157 MURTY/MURTY. Non-vanishing of L-functions and Applications
158 TIRAO/VOGAN/WOLF (Eds). Geometry and Representation Theory of Real and p-Adic Groups
159 THANGAVELU. Harmonic Analysis on the Heisenberg Group
161 SAGAN. Mathematical Essays in Honor of Gian-Carlo Rota
162 ARNOLD/GREUEL/STEENBRINK. Singularities. The Brieskorn Anniversary Volume
163 BERNDT/SCHMIDT. Elements of the Representation Theory of the Jacobi Group
164 ROUSSARIE.. Bifurcations of Planar Vector Fields and Hilbert's Sixteenth Problem
165 MIGLIORE. Introduction to Liaison Theory and Deficiency Modules
166 ELIAS/GIRAL/MIRÓ-ROIG/ZARZUELA (Eds). Six Lectures on Commutative Algebra
167 FACCHINI.Module Theory
168 BALOG / KATONA / SZA'SZ / RECSKI (Eds). European Congress of Mathematics, Budapest, July 22-26, 1996. Vol. I
169 BALOG / KATONA / SZA'SZ / RECSKI (Eds). European Congress of Mathematics, Budapest, July 22-26, 1996. Vol. II
168/169 Set Vols I+II